THE COD FISHERIES

THE HISTORY OF AN INTERNATIONAL ECONOMY

THE COD FISHERIES

THE HISTORY OF AN INTERNATIONAL ECONOMY

BY

HAROLD A. INNIS

REVISED EDITION

UNIVERSITY OF TORONTO PRESS
TORONTO BUFFALO LONDON

Reprinted in paperback 1978
University of Toronto Press
Toronto Buffalo London

Printed in Canada

ISBN 0-8020-6344-6

This volume was first published in the Series "Canada and the United States," a series of studies prepared under the direction of the Carnegie Endowment for International Peace, Division of Economics and History, of which James T. Shotwell was Director.

TO

ALEXANDER JOHNSTON, LL.D., C.M.G.

MEMBER OF LEGISLATIVE ASSEMBLY OF
NOVA SCOTIA, 1896–1900
MEMBER OF PARLIAMENT FOR CAPE BRETON, 1900–1908
DEPUTY MINISTER OF MARINE AND FISHERIES, 1910–1932
MEMBER OF THE ROYAL COMMISSION,
PROVINCIAL ECONOMIC ENQUIRY, NOVA SCOTIA, 1934
MEMBER OF THE FEDERAL DISTRICT COMMISSION, 1938

FOREWORD

This book is a challenge to the imagination and insight of the reader. It is not too much to say that for most of us it extends the frontiers of North America over a vast area that we have never thought of before as constituting a part—and a fundamental part—of the continent. A thousand miles of misty sea, yielding a harvest which for a long time rivalled in importance the produce of our opening wildernesses, it evoked the courage and daring of competing nations as did the conquest of the mainland. Its history reflects, as Professor Innis points out, the changing economies of Europe during the rise and growth of capitalism. Even today, although its importance has been relatively lessened by the rise of industry and the vast extension of the exploration of the landed surface of the globe, it remains a world of its own in international economy, one in which the interplay of American, Canadian and European interests has created a situation unique in history.

In exploring this great field, Professor Innis has opened up many a rich chapter of half forgotten history. Portuguese, Basque, Gascon, Breton, Norman and English sailors of the "West Country" sold their shiploads of cod in the markets of Europe for money reckoned in all the varied coin that was then beginning to transform the economy of Europe because of the gold and silver brought from America in Spanish galleons. The cod fisheries thus shared with the fur trade and that of Asia in widening the circle of European imports. This half obliterated history has been traced in the logbooks of sea captains and the account books of the merchants. It is possible that some impatient readers will not linger over these crabbed and confusing documents even to the limited extent that this volume brings them again to life. But for some there will be still that full suggestion of romance in trade which must have been felt by those who stood on the wharfs of St. Jean de Luz or St. Malo to watch the home fleet return from the Grand Banks.

The narrative begins with this chapter of European history; but almost from the beginning the theme lay along the fringe of that New World story which deals with the relations of neighboring colonies. The heart of Professor Innis' study, however, is the description of more recent times, when the cod fisheries passed, for a great part, into the hands of the seamen of New England and Nova Scotia, and became a matter of international concern between the two countries. The fishery disputes which arose in this American setting have been treated elsewhere in detail in their political and juridical aspects; here they are viewed in the more real sense of economic interests, of changing methods of fishing, changing markets, and, above all, of the economic prosperity of those most directly concerned. Like the fishermen in whose schooners he sailed and with whom he consorted, the author keeps his eye on the fish.

It is for this reason that the volume occupies a place of its own in this series. International relations are too frequently depicted as but the political relationship between governments, the register of policies already arrived at; here we see the springs of thought and action. It is the only method of approach to the problems of today which reaches behind the façade to an understanding of disagreements or an appreciation of concessions in compromises. Moreover, the setting of this aspect of Canadian-American relations in the wider perspectives of history offers the basis of a sounder judgment of local issues.

In this way it is a contribution, and a real one, to fulfilment of the purposes of the present series. Research is not here denatured by practical exigencies; it is kept at its own task of discovering the hidden or little understood ways of men and nations. The results are a challenge—and a guide—to intelligence.

James T. Shotwell

AUTHOR'S PREFACE

In a region with the extensive waterways which characterize the northern part of North America economic development is powerfully directed toward concentration on staples for export to more highly industrialized regions. It is not too much to say that European civilization left its impress on North America through its demand for staple products and that these in turn affected the success of empires projected from Europe. The author's study of the fur trade is therefore followed by a study of the fishing industry in the hope that it will throw light on the significance of that industry for the economic, political, and social organization of North America and Europe.

The task of indicating the significance of the fishing industry has been difficult, particularly for an Upper Canadian familiar with the centralizing tendencies of the St. Lawrence, which are so strikingly apparent to a student of the fur trade and the Canadian Pacific Railway. The psychological barrier incidental to a sustained interest in the St. Lawrence proved a handicap to an interest in the fishing industry and its regions. I have attempted to overcome it during the past decade by visiting a substantial portion of the regions devoted to the industry. As a member of the Royal Commission of Economic Enquiry in 1934, I have had exceptional opportunities to study Nova Scotia. Former students, especially Mrs. R. F. Grant and Dr. S. A. Saunders, have contributed valuable studies on the Canadian Maritimes. I must beg the indulgence of the residents of the Maritimes if in spite of these efforts I have not succeeded in overcoming the handicaps and in contributing to an appreciation of the necessity of tolerance in approaching the complex difficulties which accompany the divergent points of view in the Dominion.

The student of the fishing industry becomes aware of its peculiarities from the character of his source material. Its position in the mercantile system ensured it a prominent place in the documents of governments of Europe and North America. These documents are essentially instruments of offense and defense and their limitations are inevitable. For example, the profits and production figures of the industry of any nation if presented by an opponent tend to be exaggerated, and if presented by the nation itself tend to be minimized. Similarly, within a nation, interests present exaggerated accounts of the effects of policies which they favor or oppose. The "bearishness" of traders which was peculiar to the fishery increased the difficulty of securing a balanced picture. I have not attempted nor is it possible to arrive at an accurate appraisal, and the reader should in each case note the source of the information in order to allow a discount or premium. I have made generous use of citations in the hope that they may be of assistance. The enormous wealth of diplomatic material is in sharp contrast with the paucity of accurate detailed information upon technique and personnel. The illiteracy of the fisherman is the reverse side of the literacy of the diplomat. The fisherman is intensely individualistic and suspicious. The names of fishing localities are revealing in their variations and have defied the efforts of geographers and philologists.

The conflict of diplomatic documents exemplifies the intensity of the struggle between mercantile systems and empires, and the struggle within mercantile systems. The documents were concerned with the fishery as contributing to naval defense and show what encouragement was given to the industry by the state. Such encouragement brought increase in ships, in food, in trade, and especially in specie. It was an industry which strengthened the state directly and indirectly. In the fishery, the business of national defense was inherent in its development, in contrast with the fur trade where national defense could find a place only at the expense of the development of fur trading. The encouragement of the fishery as one part of navigation policy gave the Atlantic maritime regime of the New World a crucial position in the struggle between the mercantile systems of Europe. The effectiveness of the English mercantile system was a measure of the encouragement given the fishery. France was gradually pushed into poorer areas and forced to rely to an increasing extent on government support. The expansion of the fishery in Newfoundland and New England involved a growth of shipping, exports of manufactured products from England, exports of dried fish to tropical regions, and imports of salt, wine, or commodities of relatively higher bulk and lower value than fish, and consequently specie. The lack of balance between the bulk and value of exported cargoes and the bulk and value of the cargoes imported but added force to the demands of mercantile states for specie. The industrial arts gave their support to commercialism and mercantile policy. Valuable products such as tobacco and sugar were placed in a monopoly position within the system in order that a cheap, bulky commodity such as salt, and, then specie, might be purchased from without. The concentration on monopoly and the help given to the fishing industry, which was essentially competitive, intensified contradictions which broke the English mercantile system. The storm of the American Revolution was followed by rumblings which marked the collapse of the colonial system. This study attempts therefore to add to the significant studies of mercantilism by Professors Heckscher and Viner.

The problem of markets is a phase of the collapse of empires. The breakup of the Spanish and the Portuguese empires, so closely identified with feudalism and land, which followed the breakup of empires so identified with trade, such as the earlier empires of Holland and England, and, to a lesser degree, that of France, has been succeeded by a slow adjustment of the resulting independent states to the demands of an international economy. The financial weakness of Brazil has been accompanied by the corporatism of Portugal, Spain, and Italy. The corporatism of these regions has made necessary corporatism in Newfoundland and government intervention in Canada.

The Maritimes were divided into separate economies centering around the fishing industry and based on subordinate developments such as agriculture. Wheat was the basis for independent growth as became evident in the difficulties of the French Empire, in the success of the American colonies and eventually of Nova Scotia in Confederation. The extreme sensitiveness of the industry shown in the mobility of personnel and capital and the wide range of markets involved a quickening of independent growth evident in tariffs and bounties, which went hand in hand with improved technique. Tariffs and bounties in New England and France accompanied restrictions and the exclusion of American trade in Newfoundland and the West Indies. The emergence of customs regulations in Newfoundland, the checking of the migration of labor and commercial capital, and the growth of population were followed by problems of debt and the rise of private property, courts, and institutions of government, and the disappearance of the English fishing ships. Nova Scotia, engaged in strenuous competition with New England and Newfoundland, attempted to check American trade in the West Indies, to secure revisions of the colonial system, and to drive American ships from British waters. While the fishing industry of New England and France was recovering after the American Revolution and the Napoleonic Wars, settlements expanded in Newfoundland and found more extensive markets and the support of the British Atlantic Maritimes. The position of the sugar interests in the British West Indies was weakened by the demands made by Newfoundland on Nova Scotia, and by the insistence of Nova Scotia on the exclusion of the United States. The admission of the United States to the British West Indian market after 1830 made it easier for the British Atlantic Maritimes to be of help to Newfoundland.

The complexity of the problems of competition has been indicated rather than adequately explained. The same complexity made the old colonial system impossible and hastened the coming of responsible government as an escape from it. The steamship and the ironclad war vessel struck at the heart of the assistance provided by the navigation policy. Under responsible government and without the protection of empires each region pursued, more aggressively, policies of expansion and protection and engaged in a battle of tariffs, bounties, embargoes, exclusion, transportation improvements, and new techniques. With the mobility of population and ships, the ultimate success of these devices was registered in migration, and the failures in the accumulation of debt.

The cycles of disturbance which went with the effective penetration of commercialism supported by the mercantile policy of empires in the first instance and, later, of small isolated regions under responsible government have ended with the spread of industrialism and, in Newfoundland, the loss of responsible government. The strength of commercial interests in New England led to the early development of commercialism in Nova Scotia, and their strength in England to the late development in Newfoundland. The delay that was made longer by limited resources and the late development of shipbuilding was followed by a more rapid development of responsible government and the vigorous measures which accompanied it. These measures promoted the introduction of large-scale methods of production among rivals such as France. A way out opened with the development of new resources and new markets. For example, restrictions on the export to France of herring for bait were accompanied by increased trade with Canada and the United States. But large-scale methods spread to Newfoundland and increased the complexity of her problems. Her soft-cure product competed with the hard-cure produced by her skilled labor. The dominance of commercialism weakened the power of the state to moderate the severities arising from the spread of industrialism. It was unable to meet the problems of competition which followed industrialism in the shape of the steamship, the railway, and the gasoline engine. The last days of responsible government were supported by loans from an oil company. Commercialism was too active to permit responsible government to intrench itself deeply enough to save itself from destruction. The struggle of short-term credit against long-term ended with the spread of industrialism. Capitalism represented by the intervention of governments has softened the acerbities of commercialism. But the absence of machinery to effect an adjustment of commercialism to capitalism can never serve as an excuse for destroying the contribution by commercialism to responsible government.

Flexibility has decreased, and the migration of labor has been checked by quotas and restrictions. The importance of the flow of capital funds has been accentuated in the improved capital equipment of the fishing regions. Trawlers have been introduced on a large scale. American capital has contributed to the development of the fresh-fish industry in the Canadian Maritimes. It will possibly contribute, with British capital, to its development in Newfoundland as the latter has assisted Iceland. The rate of adjustment has varied widely. Adaptability of capital equipment and capitalistic organization to the technique of the industry in relation to products, markets, and geographic background has been least conspicuous in Newfoundland. Government support given to an intensive commercialism exposed the region to the full impact of capitalism. The decline of bankruptcy as a method of adjustment brought the collapse of responsible government. It was significant that a banker occupied a prominent position on the Royal Commission which recommended its abolition.

When Newfoundland's dictatorship as a whole refused to accept the highly questionable policy of one of its members, Mr. Thomas Lodge, he argued that "to have assumed responsibility for the good government of Newfoundland from altruistic motives and to have achieved economic rehabilitation might have cost the British taxpayer a few millions. It would [however] have added to the prestige of the British Empire." But that will not do. For those who believe in democracy the prestige of the British Empire must have suffered a blow with the destruction of its fundamental basis in the oldest colony. We cannot base our argument on the importance of the British Empire to the maintenance of democracy when we calmly allow the light to go out in Newfoundland. It is the hope of the author that this volume may contribute in its own way to a solution of the problems of the region by fostering that coöperation among its component parts which has emerged with industrialism, and by tempering the bitterness which has marred the history of the industry.

Economic history is complementary to political history; but its sources are more discursive and demand a knowledge of a wide range of information not found in the documents of state and church. I am therefore indebted to many people for assistance at numerous points. In Nova Scotia, I am indebted to the Honorable Angus L. Macdonald, the late G. Fred Pearson, Mr. W. C. Ackers, Messrs. Ralph and Winthrop Bell, Mr. H. V. D. Laing, Mr. O. F. Mackenzie, the Reverend Dr. J. J. Tompkins, the Reverend Dr. M. M. Coady, Mr. A. B. Macdonald, Mr. A. H. Whitman, to Professor D. C. Harvey of the Nova Scotia Archives and to his staff; in Newfoundland to Mr. W. A. Munn, Mr. J. T. Cheeseman, Mr. C. Crosbie, Mr. Henry King, Mr. George Hawes, Mr. C. Carter, Mr. Arthur Edgecombe, Mr. John Butler, and Dr. Nigel Rusted; and throughout the whole region devoted to the industry, to many others. In the University of Toronto, Mr. W. S. Wallace as librarian, Dr. A. G. Huntsman, Professor G. A. Cornish, Professor D. G. Creighton, and Mr. R. A. Preston, and in Queen's University, Professor G. S. Graham have generously met demands made upon them. In Ottawa I have had more than kindness from Mr. Gustave Lanctot, Dominion Archivist, and his staff, and particularly Mr. James Kenney; from Mr. W. A. Found of the Department of Fisheries and his staff; from Mr. R. H. Coats, Dominion Bureau of Statistics, and Mr. F. A. Harvey of the

Parliamentary Library. I am grateful to Professor I. M. Biss, Professor J. B. Brebner, and Mr. R. H. Fleming, who read the entire manuscript, and to Professor C. R. Fay, who read the earlier part, all alike with much advantage to me. Mr. R. H. Fleming at much personal sacrifice accompanied me on trips along the North Shore of the St. Lawrence, through the Gaspé Peninsula, and from St. John's to the Labrador. In the extremely tedious task of preparing the manuscript for the press I have been greatly assisted by Miss I. C. Hill. Without the exact and extensive work of Mr. A. E. McFarlane, who edited the text for the Carnegie Endowment, it could not have been printed. I am indebted to him for the Appendix to Chapter II. And, as editor of the economic volumes in this series and particularly as author of this book, I have been under obligations deeper than words can express to Professor James T. Shotwell.

H. A. I.

Peter's Finger, Foote's Bay,
January, 1940.

PREFACE TO THE REVISED EDITION

The Cod Fisheries appeared originally as one of a series of studies entitled "The Relations of Canada and the United States" and prepared under the direction of the Division of Economics and History of the Carnegie Endowment for International Peace. The decision to republish the book both as a memorial to Dr. Innis and as a service to students who find copies of the book all but impossible to obtain has been backed by a generous grant of the Canadian Social Science Research Council. The editors of the revised edition wish to thank both the Carnegie Endowment for International Peace and the Canadian Social Science Research Council for their co-operation and support in this undertaking.

It was Dr. Innis's custom to write in the margins of his published books new facts, references, and ideas bearing on the subject concerned. His copies of the *The Fur Trade in Canada* and *The Cod Fisheries* contain many such notes which enrich their subjects with newly-found material and pick out fresh points of view. In the present volume these notes have been printed exactly as they appear in his text. Occasionally a word has been added in parentheses where it seemed necessary and a few repetitions have been removed. These new notes are indicated in the text by asterisks instead of by numbers to distinguish them from the footnotes of the original edition and they are printed by themselves at the back of the book.

MARY Q. INNIS
S. D. CLARK
W. T. EASTERBROOK

CONTENTS

MAPS

A brave desseigne it is, as Royall as Reall; as honourable as profitable. It promiseth renowne to the king, revenue to the Crowne, treasure to the kingdom, a purchase for the land, a prize for the sea, ships for navigation, navigation for ships, mariners for both, entertainment of the rich, employment for the poore, advantage for adventurers, and encrease of trade to all the subjects.

EDWARD MISSELDEN, MERCHANT

The Circle of Commerce, London, 1623.

It is a certain maxim that all states are powerful at sea as they flourish in the fishing trade.

WILLIAM WOOD

A Survey of Trade, London, 1722.

THE COD FISHERIES

THE HISTORY OF AN INTERNATIONAL ECONOMY

CHAPTER I

THE COD

That leave might be given to hang up the representation of a cod-fish in the room where the House sit, as a memorial to the importance of the cod-fishery to the welfare of this Commonwealth, as had been usual formerly.

MOTION PASSED BY THE HOUSE OF REPRESENTATIVES
BOSTON, MARCH 17, 1784

THE history of the northeastern maritime region of North America has been dominated by the fishing industry, but it is significant that the cod (*Gadus callarias Linnaeus*), the staple fish, has secured recognition only grudgingly as the basis of economic development. Massachusetts has paid tribute to its key position; but Newfoundland and Nova Scotia have shown neglect. Whereas in Canada the beaver was fittingly chosen as a symbol of unity, in Newfoundland the cod was largely responsible for disunity, and its lack of recognition is a result.

An interpretation of its significance in the economic history of the area depends on an understanding of its geographical background and habits. The cod-fishing industry is primarily concerned with coastal waters and with the submerged continental shelf[1] which adjoins the northeastern portion of the North American continent and corresponds to the coastal plain south of New York. The increasing submergence of the continent toward the northeast is registered in the increasing width of the shelf and "the progressive deepening of its outer margin in the same direction." Submergence within recent geological times to a depth of probably 1,200 feet in the Gulf of Maine has left outlying "cuestas" as relatively shallow areas which are the Banks on which the cod fishery has been prosecuted. Remaining above the sea are the peninsula of Nova Scotia, various islands of the Gulf, and the Island of Newfoundland.[2] The weaker rocks of the Triassic period were responsible for the long inlet in the Bay of Fundy, and those of the Carboniferous period for

1. Douglas Johnson, *The New England–Acadian Shoreline* (New York, 1925), *passim*, but especially chap. viii; also Robert Perret, *La Géographie de Terre-Neuve* (Paris, 1913), chap. iii; and G. B. Goode, *The Fisheries and Fishing Industries of the United States* (Washington, 1887), sec. 3.

2. R. G. Lounsbury, *The British Fishery at Newfoundland 1634–1763* (New Haven, 1934), pp. 1–18; J. W. Goldthwait, *Physiography of Nova Scotia* (Ottawa, 1924).

the phenomenon which has necessitated the mining of Cape Breton coal under the sea. The more resistant rocks of the older formations, their location, and their relatively small area have restricted the economic activities of the land. These limitations have combined with a cooler climate and the increasing width of the submerged plain toward the northeast to give growing importance to the fishing industry. While New England, with a more favorable climate and an extensive hinterland, became less dependent on the fishing industry, Newfoundland grew increasingly dependent on it. But throughout the whole area the exploitation of the fisheries has been of dominant importance. No other industry has engaged the activities of any people in North America over such a long period of time and in such restricted areas. In contrast with the fur trade, with its dependence on natural resources subject to constant depletion and on the cultural adjustments of the native population, the fishing industry has been carried on in selected areas, with continuous supply,[3] and by the labor of Europeans. Trade provided a contrast to industry.

The great wealth and complex interdependence of animal life along the seaboard of the Maritimes have as yet baffled the scientist, and only small areas which have yielded to economic exploitation have come under the range of intensive investigation. The Banks are subjected primarily to ocean phenomena, and are not influenced by rivers from Newfoundland or by fresh water from the Gulf of St. Lawrence. The Gulf Stream and the Labrador Current,[4] a variety of conditions of temperature and climate, and a food supply varying from plankton to the larger fish in the vicinity of the Grand Banks are responsible for the abundance and diversity of the animal life which supports the extensive but fluctuating cod fishery. "In Newfoundland as nowhere else can one be made to feel the contrast between a land that is infinitely silent, mo-

3. "The cod appears to be one of the most prolifick kind of fish. Of this there needs no other proof than the great number of ships which annually load with it from this island; and it is only known in these seas; for tho' the British Channel and the German ocean are not without this fish, their numbers are so inconsiderable comparatively to those of Newfoundland, that they may rather be looked upon as stragglers." Antonio de Ulloa, *A Voyage to South America* (London, 1758), II, 410. "It is the land of the codfish! Your eyes and nose, your tongue and throat and ears as well, soon make you realize that in the peninsula of Gaspé the codfish forms the basis alike of food and amusements, of business and general talk, of regrets, hope, good luck, everyday life—I would almost be ready to say of existence itself." *Trans.* from L'Abbé Ferland, *Journal d'un voyage sur les côtes de la Gaspésie* (1871). Cited by Antoine Bernard, *La Gaspésie au soleil* (Montreal, 1925), p. 220.

4. R. R. du Baty and J. Habert, *Terre-Neuve et Islande, mémoires* (série spéciale), No. 7, Office scientifique et technique des pêches maritimes (Paris); and *North American Council on Fishery Investigations Proceedings 1921–1930* (Ottawa, 1932), No. 1, pp. 35 ff.

tionless, poor in vegetation, above all poor in its variety of living creatures, and a sea which harbors every form of life."[5]

The cod prefers a salinity of 34 per thousand and a temperature of 40° to 50°, but its range is far beyond these limits. It frequents chiefly rocky, pebbly, sandy, or gravelly grounds[6] in general from 20 to 70 fathoms in depth, although it has been taken at 250 fathoms and thrives in temperatures as low as 34°. The cod usually spawns in water less than 30 fathoms deep and apparently in fairly restricted areas. A female 40 inches long will produce 3,000,000 eggs, and it has been estimated that a 52-inch fish weighing 51 pounds would produce nearly 9,000,000. The eggs float in the upper layers of water, where they are fertilized and hatched. They have a specific gravity of 1.024 with the result that the Gulf of St. Lawrence, with its larger proportion of fresh water, has proved less satisfactory as a spawning ground, and Conception Bay less so than Trinity Bay, Placentia, and the Western Banks. Experiments have shown that a temperature of 47° will lead to hatching in 10 or 11 days, of 43° in 14 or 15 days, of 38° to 39° in 20 to 23 days, and of 32° in 40 days or more. Warmer water, say from 41° to 47°, tends to favor a more rapid development of the fish and more successful reproduction. The cod spawns usually in winter in the Gulf of Maine and at a later date toward the north. On the Grand Banks and in the Gulf of St. Lawrence it spawns chiefly in summer.[7] The temperature of the water at the surface in Labrador in summer varies from 40° to 45°, but drops rapidly to 32° at five to seven fathoms.

The larvae of the cod feed chiefly on plankton found near the surface of the water. Probably at the age of two months the young fry take to the bottom where they feed on various small crustaceans, and as they increase in size become ground feeders and consume "invertebrates in great variety and enormous amount." The adult fish is distinguished by

5. Perret, *op. cit.*, pp. 172 ff., 183, 185–186.

6. M. Bronkhorst, *La Pêche à la morue* (Paris, 1927), pp. 9 ff. "In some parts it is found in greater numbers than in others. . . . This proceeds from the quality of the bottom; for those parts where the bottom is sandy are fuller of fish than where it is rocky; but if the bottom be muddy, fish are very scarce; likewise in a great depth of water the fish are not caught in that plenty as when it does not exceed thirty or forty fathom." De Ulloa, *op. cit.*, II, 407.

7. *Canadian Fisheries' Expedition 1914–1915* (Ottawa, 1919), especially A. N. Dannevig, *Canadian Fish Eggs and Larvae.* Normal seasons for spawning are: Cape Cod and New Jersey coast, Southern Banks, and Georges Bank, November to December; Gulf of Maine, December to April; Sable Island Banks and Middle Ground, February to May; Banquereau, May to June; the escarpments or shelves south of Whale Deep to the Grand Bank, May to July; and the Grand Bank, July, August, and September. R. A. MacKenzie, "Cod Movements on the Atlantic Coast," *Contributions to Canadian Biology and Fisheries* (Ottawa), New Series, Vol. VIII, No. 31.

its three dorsal and two anal fins, its lack of spines, the location of the ventral fins well forward of the pectorals, the protrusion of the upper jaw beyond the lower, the nearly square tail, and a pale lateral line.[8] It has a heavy body and a head about one fourth the length of the fish, and both head and body are covered with small scales. In color, the cod falls into two main groups, gray and red, both ranging through a wide scale. The upper part of the trunk, the sides of the head, and the fins and tail are generally thickly speckled with small, round, vaguely edged spots.

In the Gulf of Maine[9] shore cod weighing 5 to 20 pounds have been estimated to be from three to six years old. It is claimed that the bank cod are large and mature at four years, while the shore cod of Newfoundland are small but mature at two or three. The fish varies greatly in size. One caught off the Massachusetts coast in May, 1895, on a line trawl was over six feet in length and weighed 211 pounds. A fish which dressed 138 pounds, and probably weighed 180 when alive, was caught on Georges Bank in 1838; and various weights from 100 to 160 pounds have been recorded. In the Gulf of Maine the average "large" fish caught near shore weighs about 35 pounds, and those on the Bank about 25. "Shore fish" in Newfoundland waters average 10 to 12 pounds.[10] In the Gulf of St. Lawrence and on the east coast of Labrador, the fish are of smaller average size, while those to the north of Labrador are shorter and thinner than those taken at the Straits of Belle Isle. Grenfell gives as the largest fish caught on the Labrador coast one that weighed 102 pounds and was five feet six inches long. According to him, "The average Labrador cod taken in the trap-net is about twenty inches long and weighs between three and four pounds. Those caught with hook and line in the autumn are much larger and heavier."[11] Throughout the general range along the Atlantic coast

8. H. B. Bigelow and W. W. Welsh, *Fishes of the Gulf of Maine,* Bulletin of the United States Bureau of Fisheries, XL (1924), Part I, pp. 409 ff.; also Joseph Hatton and Moses Harvey, *Newfoundland* (Boston, 1883), pp. 235 ff. See Bartholomew's *Atlas of Zoögeography* (1911), Plate XXIV, Map I, on the distribution of *Gadidae,* especially in the North Atlantic; and Goode, *op. cit.,* sec. 1, pp. 200 ff.

9. Cod off southern Massachusetts are 7 to 8 inches long at one year, 14 to 17 at two years, 19 to 22 at three, 23 to 26 at four, 27 to 29 at five. W. C. Schroeder, *Migrations and Other Phases in the Life History of the Cod off Southern New England,* U.S. Department of Commerce (Washington, 1930), Bureau of Fisheries, Fisheries Document No. 1081, p. 120.

10. Hatton and Harvey, *op. cit.,* p. 240.

11. W. T. Grenfell and others, *Labrador, the Country and the People* (New York, 1909), p. 286. See also Adolphe Bellet, *La Grande pêche de la morue à Terre-Neuve* (Paris, 1902), pp. 5 ff. The average weight of cod caught on the Banks is given as from two to three kilograms, and is said to be only half the weight of cod caught a century ago. "Tho' their number is still immense, they are evidently diminished. . . .

from New York and New Jersey, but particularly from the Nantucket shoals to northern Labrador, the weight tends to grow less as one goes north.

Mollusks are regarded as the largest item in the diet of the cod in the Gulf of Maine.[12] To the north, in the Gulf of St. Lawrence, and Newfoundland and the Labrador, herring, capelin, and squid are important food items. The herring is said to be extremely sensitive to temperature changes. It winters at the bottom of bays free of ice, and appears on the southwest coast in March. The capelin[13] is a small subarctic species of fish which spawns in multitudes along the shore and is pursued by large numbers of cod. It appears on the southern coasts in June and on more northerly coasts later in the season. Its movements are uncertain but it serves as one of the important links between the smaller species of ocean life and the larger species including the cod. The squid appear late in the summer season.

On the fringes of its optimum range the cod follows pronounced seasonal migrations. During winter and spring[14] sea life withdraws to deep waters, but with the retreat of the ice the cod follows the capelin and comes to spawn from May to September in the gulfs and fiords and in

Much fewer are now caught in the same space of time, than there were twenty-five or thirty years ago." De Ulloa, *op. cit.*, II, 410.

12. Du Baty and Habert, *op. cit.*, chap. iii; also M. G. Massenet, *Technique et pratique des grandes pêches maritimes* (Paris, 1913), pp. 38 ff. Their voracity is a matter of common notoriety, and they are reported as swallowing scissors, oilcans, old boots, books, and keys, as well as live ducks, guillemot, and other birds. Cartwright wrote on July 19, 1776: "Observing many cod-fish to come close in to the shore, where the water was deep, I laid myself flat upon the rock, took a caplin by the tail, and held it in the water, in expectation that a cod would take it out of my fingers; nor was I disappointed, for almost instantly a fish struck at, and seized it; and no sooner had one snatched away the caplin, than another sprang out of the water, at my hand, which I had not withdrawn, and actually caught a slight hold of my finger and thumb. Had I dipped my hand in the water, I am convinced they would soon have made me repent of my folly, for they are a very greedy, bold fish." *Captain Cartwright and his Labrador Journal,* ed. C. W. Townsend (Boston, 1911), p. 204.

13. G. W. Jeffers, *The Life History of the Capelin,* doctorate thesis, University of Toronto; also G. F. Sleggs, *Reports of the Newfoundland Fishery Research Commission,* Vol. I, No. 3 (Plymouth, 1933). "In June capline in such aboundance dryve on shoare as to lade carts . . . and cods [are] so thicke by the shoare that we heardlie have been able to row a boate through them." John Mason, *A Briefe Discourse on New-found-land,* Prince Society Publications (Boston, 1887). The cod "which draw to the coast to spawn and take their fills of smaller fishes which they follow there are commonly above thirty dayes together before they hale off from the shore again, and in such manner there come there severall shoales of the cod-fishes in the summer time. The one of them followes on the herrings, the others follow the capling . . . and the third follows the squid." Richard Whitbourne, *A Discourse and Discovery of Newfoundland* (London, 1622). As will already be evident, the spelling of "capelin" has many variations.

14. See W. H. Greene, *The Wooden Walls among the Ice Floes* (London, 1933), chaps. iii, iv, for a description of the character and limits of the polar ice pack.

the shallower layer of water which covers the Banks. The migrations take place, in a way, "both vertically and horizontally." At Cape Bauld in northern Newfoundland the fishing[15] lasts from about June 20 to October 20, or more than 100 days. From Chateau Bay north to Batteau it begins between June 20 and July 12 and lasts until October 1, or from 80 to 102 days. From Batteau to Okkak it begins between July 12 and July 28 and lasts until October 1, or from 65 to 80 days; and at Hebron it lasts from about August 15 to September 15, or some 32 days. The season varies from year to year. The smaller fish are said to leave first and to be followed by the larger ones in November. Comparatively little is known as to the character of their migrations; but cod with French hooks from the Grand Banks have been caught on the Labrador and in Ipswich Bay, Massachusetts. They apparently follow the temperature lines between warm and cold currents; and the extensive distribution of the polar current in 1927 has been held responsible for the bad fishery of that year.

The long and continuous exploitation of the cod fishery has been in part a result of the cod's wide range, and in part because it is "different from other fishes like the salmon and herring; its flesh is rich and gelatinous without being fatty, and readily lends itself to a simple and efficient cure by salting and being dried in the sun."[16] As a protein commodity it has been called "the beef of the sea." The fishery was dependent, therefore, not only on the location of the fishing areas but also on factors which affected the technique of handling and of marketing the product. Ice conditions and wind direction became important in the routes of sailing vessels, and hours of sunshine and proximity to land determined the various methods of drying and the necessity of "salting down green." The Gulf Stream and the Labrador Current which pro-

15. Over an area north of Conception Bay for 700 miles "the cod approach the shore about one week later for every degree of latitude we advance to the north." They appeared at Notre Dame Bay about June 20 and in northern Labrador from July 20 to July 28. "The fishing season lasts 143 days on the Newfoundland coast, 87 days in southern Labrador and 52 days in northern Labrador." Hatton and Harvey, *op. cit.*, p. 241. It has been suggested that the fairly reliable fishery of the Cape Bauld and Blanc Sablon regions is a result of comparatively steady temperatures, i.e., from 40° to 50°. G. W. Jeffers, *Observations on the Cod-fishery in the Strait of Belle Isle* (Toronto, 1931).

16. *Report of the Commission of Enquiry Investigating the Sea-fisheries of Newfoundland and Labrador, 1937* (St. John's, 1937), pp. 44 ff. "The ideal preservation would be that method which while ensuring durability of the product by removal of sufficient water for that purpose, gave a product capable of reverting to the original, or, in other words, capable of taking up the water which has been removed. This ideal reversibility is never attained, but light-salted, shore cure fish have a greater degree of reversibility than the heavier-salted fish." See also *Reports of the Newfoundland Fishery Research Commission,* Vol. II, No. 1, p. ii; also H. F. Taylor, *Principles Involved in the Preservation of Fish by Salt* (Washington, 1922).

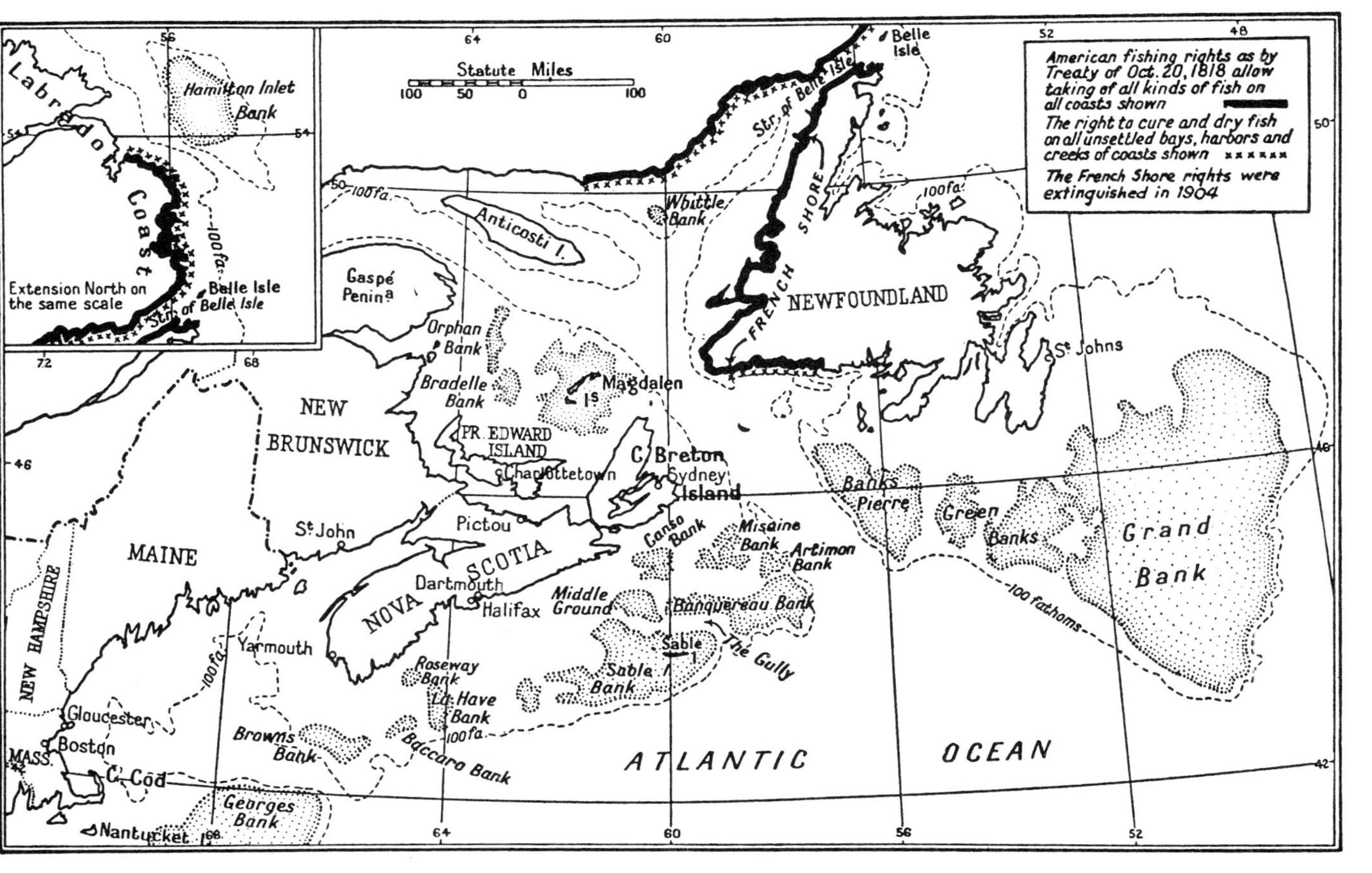

The Fishing Banks from Cape Cod to Labrador

duce diverse temperatures near Newfoundland are held responsible for the continual diversity[17] of the climate. The Gulf Stream, as *le père des vents* (the father of winds), although moderating the climate of Europe, produces off Newfoundland contrasts like those of the Sahara or Thibet. But storms are rarer and less violent from May to October than from November to April. Newfoundland has no dominant winds and there are in all seasons "differentiating factors which, in the entire coast line of the North Atlantic, are unique in their power to create moisture and to change temperatures." The climate is both maritime and continental. The air currents are almost never rectilinear but constantly cyclonic or anticyclonic. Temperature fluctuates continually as a result of changes in the wind and other local factors. Fogs are formed by the condensation of humid air from the Gulf Stream when passing over the cold water of the Labrador coast, and of warm, humid winds blowing overland, with the result that the Straits of Belle Isle have even more fog than the south of Newfoundland. Fogs blowing toward the middle of the island tend to lose moisture; and the majority of fogs have relatively low humidity and are easily dispersed by evaporating winds. "Newfoundland is, above all, the country where fog does not wet one."[18] The contrast between the Gulf Stream and the Labrador Current contributes to the greater frequency of fogs in the region between the State of Maine and Nova Scotia, and in the neighborhood of the Grand Banks. Fog apparently covers greater areas on the Grand Banks between April and August than between September and March, but the changing positions of the currents render the boundaries of such fog areas uncertain. The Avalon Peninsula, especially in the neighborhood of Cape Race, and Petit Nord in the neighborhood of Belle Isle are shown in the meteorological reports to be the regions most susceptible to fogs, and the middle of the island, especially about Notre Dame Bay, to be the least susceptible. Indian summer, from about September 15 to November 15, is characterized by a minimum of fog. Rainfall tends to have little relation to fog and may be heaviest in January, April, or July at St. John's.

As a result of these factors the fishery varies in time and technique.

17. Perret, *op. cit.*, pp. 115–116, 123. See especially his objections to the theory that the summer is dominated by southwest winds and the winter by northwest winds, pp. 120–121. From the standpoint of the fishery, Talon wrote that vessels for Newfoundland planned to leave France in April at the latest to take advantage of north and northeast winds. H. A. Innis, *Select Documents in Canadian Economic History 1497–1783* (Toronto, 1929), p. 318; see also L. R. Jones and P. W. Bryan, *North America* (London, 1924), chap. i.

18. Perret, *op. cit.*, pp. 131, 134. The Cape Race fog average was (1894–98) from September to March, 38 days; from April to August, 86 days. *Idem*, p. 138.

In the southern area the fishing grounds of the New England shore are found at about twelve or fifteen miles from the coast.[19] In the Gulf of Maine there are several small banks at a great distance from it. Beyond these banks are the large banks[20] beginning with Georges and extending northeast to the Grand Banks. Fish are caught on parts of Georges Bank in large numbers during late winter, or from February to April. Northeast and south of Cape Sable is Brown's Bank, and near it are the Seal Island grounds, Roseway Bank, and La Have Bank. Still farther northeast are Sable Island Bank, Banquereau, Misaine Bank, and Canso Bank. These are tributary to New England and Nova Scotia, and are separated by a deep channel from the Bank of St. Pierre, Green Bank, and the Grand Banks. The last have an area of 37,000 square miles, and cod are found especially on the southern half. Fishing begins in the south in April and, throughout the season, or to October, keeps moving northward. Near the shore, in the Bay of Fundy and along the Nova Scotia coast, cod are found only in small numbers. They are plentiful on the Cape North ground, at the northern end of Cape Breton, especially in May and June. Near the Magdalen Islands, at Bradelle Bank and Orphan Bank, near the New Brunswick coast at Miscou, and at Gaspé, cod fishing is important in the summer months. On the east and west coasts of Newfoundland and "on the Labrador," fishing is mostly confined to the areas near shore. On the east coast the fishing is carried on from the middle of June to about the first of November.

The fishing industry developed around the cod.[21] As one of the most prolific and most adaptable of the *Gadus* family, it extends over the widest range, from northern Labrador to New England and even farther south. The most favorable conditions for its development are apparently found in the area from Cape Race to Cape Cod which includes the winter fishery in the Gulf of Maine, the bank fishery, and a large coastal fishery. To the north, seasonal variations have more importance and spawning is more largely confined to the summer months. In the Gulf of St. Lawrence and on the coasts of Newfoundland and Labrador the fish are smaller in size, the seasons are more closely restricted, the cod feeds largely on the capelin, and the shore fishery becomes more im-

19. Raymond McFarland, *A History of the New England Fisheries* (New York, 1911), chap. i.

20. R. F. Grant, *The Canadian Atlantic Fishery* (Toronto, 1934), chap. i; Albert Close, *The Fishermen's and Yachtsmen's Chart, Cape Cod to Newfoundland* (London, 1931); also M. A. Hérubel, *Pêches maritimes* (Paris, [1911]), pp. 32–36; and especially Du Baty and Habert, *op. cit.*, chaps. i, ii; G. C. Curtis, "The Fishing Banks of Our Atlantic Coast," *Bulletin of the American Geographical Society* (Washington, 1913), pp. 413–422.

21. O. E. Sette, *Statistics of the Catch of Cod off the East Coast of North America to 1926* (Ottawa, 1927).

portant. The curing of dry fish[22] has been more directly related to the northern part, particularly the coastal areas of central Newfoundland and small areas such as Gaspé,[23] and the green fishery has been confined for the most part to the more southern portions of the whole area, especially the Banks. The cool drying season of the winter months facilitated the development of the winter fishery in New England, but ice conditions in Newfoundland and in the Gulf of St. Lawrence restricted activity to the summer. The relatively smaller size of fish along the eastern shore of Newfoundland limits the demands for salt essential to a summer cure.

In the areas such as the Banks, with the cod larger and more distant from land, European countries like France, which possessed cheap supplies of solar salt, had striking advantages; in areas with relatively abundant sunshine, and cod smaller and nearer land, countries like England, even without cheap supplies of solar salt, were at a greater advantage. But efforts to reach such adjustments marked the long and checkered history of the industry. In the first half of the sixteenth century the advantages possessed by France gave her a rapid and extensive occupation of the region.

22. "We dry them more than they are dried in the United States but not so hard as in Newfoundland." *Documents and Proceedings of the Halifax Commission* (Washington, 1878).

23. "The Bay of Chaleur cod are more prized in the markets of the Mediterranean and will, at all times, sell there more readily and at higher prices than any other. They are beautifully white, and being very dry can better withstand the effects of a hot climate and long voyages than a more moist fish. The peculiarity of their being smaller than cod caught elsewhere, is also of great importance as regards the South American market." M. H. Perley, *Report on the Sea and River Fisheries of New Brunswick* (Fredericton, 1852), p. 7.

CHAPTER II

THE EXPANSION OF THE FRENCH, 1497–1550

The French, being desirous to try experiments as well as the Biscaines, found a fishing land fifty leagues off to sea from Newfoundland and called it the Bank; where commonly they make two voyages yearly without going ashore to dry their fish, and therefore it is called wet-fish.

Sir William Monson's *Naval Tracts*

The effect of the discovery of North America by John Cabot in 1497 was brought out most immediately and most strikingly by the development of the fishing industry. Of the scattered documents relating to his expedition, the letter of Raimondo di Soncino to the Duke of Milan, written in London and dated December 18, 1497, may be regarded as the most valuable. Assuming that Cabot had returned on August 6 of that year,[1] sufficient time had elapsed for the available evidence to have been sifted and the more important results confirmed. Cabot and his companions, "practically all English, and from Bristol . . . affirmed," writes Soncino, "that the sea there is swarming with fish, which can be taken not only with the net but in baskets let down with a stone, so that it sinks in the water. I have heard this Messer Zoane state so much. These same English, his companions, say that they could bring so many fish that this kingdom would have no further need of Iceland, from which place there comes a very great quantity of the fish called stockfish."[2] Whatever the writer may have believed as to other details of information he had received from Cabot and his companions, the fishery was regarded as certain and worthy of emphasis and report.

Bristol merchants with cargoes of "slight and gross merchandises"

1. J. A. Williamson, *The Voyages of the Cabots and the English Discovery of North America under Henry VII and Henry VIII* (London, 1929), pp. 30–31, 161.

2. H. P. Biggar, *Precursors of Jacques Cartier 1497–1534* (Ottawa, 1911), p. 20. Cod taken in winter in Iceland and dried in the frost were known as "stockfish"; when taken in summer and dried they became the "poor john" of Newfoundland, or "haberdine," a corruption of "Labourd," a Basque region. According to Bristol records, as late as 1471 stockfish was valued at £5 per last (12 barrels) or about one half the value of salt fish. Williamson, *op. cit.*, p. 9. Stockfish are "cured without salt by splitting open and drying hard in the air." The fish are "tied two together at the tails and thrown across a very smoothly planed, quite heavy 'longer' (pole) which rests upon stout beams. Those fish are not removed until thoroughly cured. It is considered useless to hang fish for stock cure later than early April, and the process occupies three or four months according to weather conditions. . . . When cured the article is a veritable club being so hard and dry and tough." Report of Hon. W. F. Penney re his trip to Norway. *Journals of the Assembly, Newfoundland*, 1921, pp. 418 ff.

went with Cabot on his second voyage in 1498. Possibly in recognition of the rights of Portugal acquired in the Treaty of Tordesillas of June 7, 1494,[3] in which Portugal shared the New World with Spain, and of early trade relations with Bristol,[4] a patent dated March 19, 1501, was granted to John and Francis Fernandez and John Gonsalvez, and to Richard Warde, Thomas Asshehurst, and John Thomas;[5] and apparently an expedition was sent in that year and another in 1502. A new agreement dated December 9, 1502, provided for an amalgamation of the Cabot interests and the interests concerned in the charter of 1501 in an association which became "The Company Adventurers to the New Found Lands." Expeditions were sent annually from 1503 to 1505 by the company, but existing documents point to dissension and friction among its members.

Later ambitious attempts at trading and colonization, recorded in official documents, proved unattractive to capital and unsuccessful. The expedition of John Rastell in 1517 was a failure.[6] Although Bristol was enthusiastic over a projected adventure in 1521,[7] only the pressure of Wolsey and the King succeeded in getting a favorable reply to pleas for support from the Drapers and other companies of London.[8] Another expedition consisting of two vessels, the *Mary of Guilford* and the *Samson*, left Plymouth on July 10, 1527, but of this little knowledge has come to light.[9] A letter written on August 3 of that year from St.

3. P. E. Renaud, *Les origines économiques du Canada* (Mamers, 1928), pp. 195 ff.

4. Bristol ships sailed in the spring to Portugal for salt, thence to Iceland for fish, back to Portugal where the fish were sold, and whence cargoes of wine, oil, and salt were brought back to England. See C. B. Judah, *The North American Fisheries and British Policy to 1713* (Urbana, 1933), p. 13.

5. Williamson, *op. cit.*, chap. viii.

6. G. B. Parks, *Richard Hakluyt and the English Voyages* (New York, 1928), pp. 8–9; also A. W. Reed, "John Rastell's Voyage in the Year 1517," *Mariner's Mirror*, IX (1923), 137–147. See also A. W. Reed, "John Rastell, Printer, Lawyer, Venturer, Dramatist, and Controversialist," *Transactions of the Bibliographical Society*, XV, 65–68. John Rastell and Richard Spicer were partners in outfitting the *Barbara*, and William Howting financed the *Mary Barking*. The vessels turned back after reaching Ireland, seemingly as a result of mutiny and of the collusion of the naval authorities. The outfitters sent cargoes of salt, flour, and other supplies, and were apparently interested in the fishery. The mutineers promised as much profit—considering the success of an Irish vessel in capturing a Portuguese ship—from piracy as from fishing: "that hit shuld be as profitable for hym as his fysshyng in the new lands." The ships returned to Falmouth, and thence to La Rochelle and Bordeaux to dispose of the outfit. The hostility of the naval authorities may have been accompanied by the hostility of fishermen from the West Country opposed to any attempt at forming settlements. This may be the beginnings of the struggle which dominated Newfoundland history. See Henry Harrisse, *Découverte et évolution cartographique de Terre-Neuve* (Paris, 1900), pp. iv–v.

7. Biggar, *op. cit.*, pp. 134 ff.

8. Williamson, *op. cit.*, chap. x.

9. *Idem*, chap. xi, pp. 104–105; also H. P. Biggar, *An English Expedition to America in 1527* (Paris, 1913).

John's Harbor, Newfoundland, by John Rut of the *Mary of Guilford* tells us that they "found eleven saile of Normans, and one Brittaine, and two Portugall Barkes, and all a-fishing." In 1536, Richard Hore of London chartered the *William* to go to Newfoundland. She left in April and returned in September or October and "caught fish or caused them to be caught."[10]

With these records of unsuccessful expeditions there is evidence of a successful fishery. In 1522,[11] at the outbreak of war with France, a plea was made for the dispatch of a convoy before the "commyng home of the New found Isle landes flete." An act passed in 1548 (2 and 3 Edw. VI, c. 6) stated that

> forasmuch as within these few yeeres now last past there have bene levied, perceived and taken by certaine of the officers of the Admiraltie, of [from] such Marchants and fishermen as have used and practised the adventures and journeys into Iseland, Newfoundland, Ireland and other places commodious for fishing and the getting of fish, in and upon the seas or otherwise, by way of Marchants in those partes, divers great exactions, as summes of money, doles or shares of fish and such other like things, to the great discouragement and hinderance of the same marchants and fishermen and to no little dammage of the whole common wealth, and thereof also great complaints have bene made, and informations also yerely to the kings Majesties Most honourable councell: for reformation whereof, and to the intent also that the sayd marchants and fishermen may have occasion the rather to practise and use the same trade of marchandizing and fishing freely without any such charges and exactions, as are before limited, whereby it is to be thought that more plentie of fish shall come into this Realme, and thereby to have the same at more reasonable prices.[12]

Such exactions were made subject to severe penalties. This statute would imply that the fishery had become important enough to offer organized resistance to the demands of the navy in its levy of food supplies. But in spite of a market,[13] and encouragement from the government,[14] the growth of the Newfoundland fishery in the first half century

10. Williamson, *op. cit.,* chap. xii; also E. G. R. Taylor, "Master Hore's Voyage of 1536," *Geographical Journal,* May, 1931, pp. 469–470.

11. Two letters, dated August 21, 1522, were sent by Vice-Admiral Fitzwilliam to Cardinal Wolsey and to King Henry VIII advising the immediate dispatch of the *Mary James* westward. Biggar, *Precursors of Jacques Cartier,* pp. 142–143.

12. *Voyages of the English Nation to America before the Year 1600, from Hakluyt's Collection of Voyages (1598–1600),* ed. Edmund Goldsmid (Edinburgh, 1889), I, 298.

13. In 1545, Newfoundland cod was exported from France to England at 20 sols the hundred for the large sort, 10 for the average size, and 4 for the small. Antoine de Montchrétien, *Traicté de l'oeconomie politique* (Paris, 1889), p. 379; also Théophile Malvezin, *Histoire du commerce de Bordeaux* (Bordeaux, 1892), II, 205.

14. An act passed in 1541 (33 Hen. VIII, c. 2), provided that "whosoever shall buy

appears to have been limited. England continued to depend on Iceland. In 1528, 149 vessels, all from east-coast ports—8 being from London—were engaged in the Iceland fishery,[15] and in 1533, 85 vessels returned from that area. As late as 1578 Parkhurst wrote: "The trade that our nation hath to Island [Iceland] maketh that the English are not there [in Newfoundland] in such numbers as other nations."

The scant evidence of English fisheries in Newfoundland in the first half of the sixteenth century contrasts strikingly with the abundant evidence of French and Portuguese fisheries. The Portuguese were directly concerned in the possibilities of trade with the Orient. On October 28, 1499, letters patent were issued by King Manuel of Portugal to John Fernandez[16] and on May 12, 1500, a new patent was issued to Gaspar Corte-Real. Fernandez apparently became concerned in the Bristol venture of 1501. Corte-Real went on an expedition to the New Lands in 1500, and again in 1501, from which second expedition he failed to return. Fresh letters patent were issued in January, 1502, to his brother Miguel, but in that year he also was lost. A fourth expedition sailed in 1503 but without tangible results. The first evidence of interest in the fisheries is found in a document dated October 15, 1506, referring to "a grant of possession . . . made of the tithes on the fish that come from the fisheries of Newfoundland" to the seaports near Vianna.[17] "Some verdicts given by the judges of the Royal Customs" in connection with the tithes probably refer to preceding years. The fishery was sufficiently important by 1506 to warrant an attempt to secure a decision as to the disposition of the tithes that would be favorable to the King. "And since this is a matter of great importance to our service and must not pass without control we order you, as soon as this reaches you, to find out the towns in which the said possession has been thus granted, and not to allow those who hold it to make use thereof."[18] Fagundez made a voyage to the "new lands" in 1520 and letters patent

any fresh fish . . . of any stranger . . . to put to sale to any person within this realm shall forfeit for every time ten pounds, but this shall not extend to any persons which shall buy fish in any part of Iseland, Scotland, Orkney, Shetland, Ireland or Newland." These acts probably reflected the growing importance of the West Country.

15. M. Oppenheim, *A History of the Administration of the Royal Navy and of Merchant Shipping in Relation to the Navy* (London, 1896), p. 89; also D. W. Prowse, *A History of Newfoundland* (London, 1896), pp. 24–29; and Eileen Power and M. M. Postan, *Studies in English Trade in the Fifteenth Century* (London, 1933), pp. 155–182.

16. Williamson, *op. cit.*, chap. vii; also H. P. Biggar, *Voyages of the Cabots and of the Corte-Reals to North America and Greenland* (Paris, 1903).

17. Biggar, *Precursors of Jacques Cartier*, pp. 97–98.

18. *Idem.*

were dated at Vianna on May 22, 1521. The ports about Vianna probably sent annual expeditions. A map accompanying the account of a voyage in 1539[19] locates the Island of Baccalaos (Portuguese) between Cape Race and Bonavista; and the early existence and persistence of Portuguese names on the maps of Newfoundland serve as evidence of Portugal's early and sustained interest. On the other hand, John Rut found only "2 Portugal barkes" as against "eleven saile of Normans and one Brittaine" in St. John's Harbor on August 3, 1527. Roberval found "seventeene shippes of fishers" when he arrived at St. John's on June 8, 1542, and spent a good deal of time "in composing and taking up of a quarrell betweene some of our countreymen and certaine Portugals," from which it might be inferred that the Portuguese were on the increase. But they were probably fewer in numbers than the French.[20]

It would be dangerous to suggest that the number of documents extant is an index of the importance of the fishery, but such an index would point to the development of an important French fishery. It is probable that the French arrived on the fishing grounds at about the same time as the Portuguese or possibly as early as 1504.[21] In 1509 Norman and Breton vessels were going to Newfoundland, and in 1510 "Newland" fish were being sold in Rouen.[22] A document[23] providing for a voyage by John de Agramonte from Spain, dated October 29, 1511, states "that two of the pilots may be Bretons or belong to some other nation which has been there," that is, in Newfoundland. Not only were pilots to be obtained from Brittany but also wines, flour, and meat for the voyage. A document of December 14–17, 1514, refers to an agreement between the inhabitants of the Isle of Bréhat, not far from Ple-

19. Giovanni Battista Ramusio, *Delle navigationi et viaggi* (Venetia, 1565), III, 423–424.

20. See George Patterson, "The Portuguese on the North-east Coast of America and the First European Attempt at Colonization There, A Lost Chapter in American History," *Transactions of the Royal Society of Canada*, 1890, sec. 2, pp. 127–173. One Portuguese port was stated to have sent 150 ships to Newfoundland in 1550. Judah, *op. cit.*, p. 18. Sir Humphrey Gilbert refers to an attempt to establish a settlement on Sable Island about 1550.

21. Bellet, *op. cit.*, chap. i; also Bronkhorst, *op. cit.*, pp. 25 ff., especially the reference to R. P. Fournier, *Traité de hydrographie* (Paris, 1667).

22. A document dated January, 1513, referring to an expedition to Newfoundland in 1510, stated that a ship called the *Jacquette* from Dahoüet, a small harbor near Pleneuf, had gone to Newfoundland and returned with a cargo of fish which was sold at Rouen. The voyage was apparently not a success from the standpoint of the men, and a dispute arose between the mate and one of the crew over wages, with the result that the mate pursued the man who became so frightened that he fell overboard and was drowned.

23. Biggar, *op. cit.*, pp. 107–111.

neuf, and the monks of the Abbey of Beaufort providing that tithes might be exacted on cod and other fish taken from Newfoundland.[24] The voyage of Rastell in 1517 gave rise to his writing the poem "A New Interlude and a Mery," according to which,

> Nowe frenchemen and other have founden the trade,
> That yerely of fyshé there they lade,
> Above an C [hundred] sayle.

An English document refers to a ship from Rouen loaded with 9,000 fish and manned by eleven men, which was captured, probably in 1523; and a later document dated December 2, 1531, refers to the plundering of a vessel from Brittany, carrying a cargo of salted fish from the "new founde iland," which had taken refuge at Ramsgate during a storm. In the decade from 1520 to 1530 Ango and other merchants of Dieppe and Rouen were said to have sent from 60 to 90 ships to Newfoundland. John Rut, as we have seen, found one Breton and 11 Norman fishing ships at St. John's in 1527. It is probable that the fishery was a factor leading to the union of France and Brittany in 1532. Cartier found it necessary to secure an order forbidding fishermen to leave St. Malo before he had a crew in 1534; and he met with opposition for similar reasons in the following year. In 1547 the English captured 25 French ships with cargoes of fish. The evidence points to an important interest of the Channel ports in the fisheries of Newfoundland, probably because of the profitable market at Rouen and Paris.

The ports of Brittany were supported not only by Channel markets but also by La Rochelle in the Bay of Biscay. According to a document[25] of October 15, 1523, Jehan le Moyne of La Rochelle was joined by Jehan Boisseau in a charter party dated March 26 to send the *Margarite* of Blavet to Newfoundland. The agreement entitled Boisseau to 2,750 cod. Following an agreement of the last day of March, he could also purchase 500. On June 17, Le Moyne had provided an outfit for the *Marie* from the port of Croisic for which he was to receive in return a share of the cod, oil, and the profits generally. In the same year Pierre Jourdain, *le jeune*, "marchant et bourgeoys" (outfitter) of La Rochelle and André Morisson entered into a charter party on April 14 to send the *Catherine* of Binic on a voyage to Newfoundland entitling them to a share of the "pesche, huilles, gaings et prouffiets"; that is, of the "fish, oil, gains and profits." The same men had apparently outfitted the *Marguerite* of Pornic and other ships under the same conditions. A

24. The curé of La Hougue made similar demands in 1520. Charles de la Roncière, *Histoire de la marine française* (Paris, 1906), III, 140.
25. Biggar, *op. cit.*, pp. 161–162.

French Fishing Ports

document of October 22, 1523, refers to the disposal of the "third part of the fish, oil, gains and profits made on their voyage to Newfoundland according to the custom of the sea"[26] to three men and others absent in the *Marguerite* of St. Brieuc. La Rochelle was a financial center and ships from Croisic, Binic, Pornic, Blavet, St. Brieuc, and elsewhere were given an outfit in return for a definite number of cod or for a share of the returns. The capital was supplied by a relatively small group of men generally working in partnerships.

Breton experience was combined with La Rochelle capital. Guillaume Legatz of Paimpol in Brittany, master and pilot, was engaged for the sum of 30 livres tournois[27] by a La Rochelle merchant, Nycholas Mailhard, to take command of the *Marguerite Antoinette* on February 19, 1535, the ship and crew to be ready by March 25[28] to sail and "make a first voyage, God willing, to the Newfoundland fishery."[29] He took with him nineteen "compaignons," and it was agreed that "Mailhard is to be held responsible for the payment of their wages and expenses in like manner and to the same amount as, each for each, other merchants and outfitters of La Rochelle would pay; and he has also, as a guaranty that they will be brought back to the said La Rochelle, paid over in advance to the said Legatz 8 livres tournois, which will be deducted from the above account."[30] The arrangements were placed in the hands of the captain and pilot, who were expected to carry them out with the support of the merchants. On April 14 of the same year an agreement was made between Durant Buschet and Jehan Bernyer, "marchans et bourgeois" of La Rochelle, with Gluille Le Cludir of Pel Proux in Brittany to go as captain of the *Xpristofle*, 70 tons, of La Rochelle with twenty-two "compaignons pescheurs et mariniers," the merchants supplying the boats, guns, cables, anchors, and provisions. The ship was required to leave with the first favorable wind and to return as early as possible. Of the returns, the merchants were entitled to two thirds and the captain and *compaignons* to one third, "thus giving each his rightful part

26. "la tierce partie de la pesche, huilles, gaings et prouffictz qu'ils ont faict en leur voiaige de la Terre Neufve, selon le cours de la mer."

27. For a discussion of names and values of monies, see Appendix to this chapter, p. 27; also Adam Shortt, *Documents relating to Canadian Currency, Exchange and Finance during the French Period* (Ottawa, 1925), I, Introduction, and p. 3.

28. This vessel apparently returned before September 3.

29. "Pour aller et fayre le voyage du premier temps au plaisir Dieu, à la pescherie de Terre Neufve."

30. "Mailhard sera tenu payer leurs sallaires et voyage à mesme et semblable pris que les autres marchans et bourgeois de la Rochelle en payeroient ung pour ung et pour la dépense qui ce pourroyt fayre pour les amener a ladite Rochelle, ledict Mailhard a remis par avance audict Legatz 8 livres tournois qu'il a pris et recu en deduction susdite." Biggar, *A Collection of Documents Relating to Jacques Cartier and the Sieur de Roberval* (Ottawa, 1930), pp. 47–48.

of such catch as they may take in the said fishing voyage."[31] Of such provisions as remained the merchants were to have first choice of purchase, "at such a price and for such money as other merchants would pay, honestly and without double-dealing."[32] The men were lent 68 livres, 5 sols tournois for the voyage, to be paid back on their return to La Rochelle.[33] The arrangement was more distinctly a share arrangement, but was financed entirely by the merchants. In the autumn of 1533 Yvon Raymond, "marchant et maistre de la navyre nommée *Xpristofle* de Ploumanac" (merchant and master of the ship called the *Xpristofle* of Ploumanac), borrowed 30 livres tournois from Julien Giraud, "marchant et bourgeois" (merchant and outfitter) of La Rochelle, "for the first voyage that he hoped to make from La Rochelle to Newfoundland or other countries." In return he promised to deliver 2,000 "moullues parées . . . bonnes et marchands (dressed codfish . . . prime and "merchantable"). Masters[34] were apparently emerging as relatively independent merchants.

Musset has suggested some conclusions based on references to 71 vessels from La Rochelle prior to 1550.[35] The majority ranged from 50 to

31. "En payant chacun sa quotité du pillote pour telle part qu'ils prendront en ladite pesche."

32. "Pour tel pris et somme de deniers qu'ils en trouveront d'aultres marchans, bien et loyaulment, sans flaulde."

33. Biggar, *op. cit.*, pp. 56–58.

34. It was not merely that a captain, i.e., a *maitre* (or *maistre*), could also be his own *bourgeois*, or *bourgeoys*, here meaning outfitter. He could also be in part the *armateur*, or supplier of equipment, and in part the *victuailleur*, or provider of food and drink. And the shares or individual investments or rights in the profits were likewise highly complicated. By traditional custom or special arrangement, as indicated in the text, every member of the crew commonly had his allotted share, while other and generally larger shares could be purchased or offered in exchange for services. Most of the technical terms used are fairly plain. One might be *bourgeois*, or *victuailleur*, *pour la totalité*, that is owner outright. Or one's ownership might be only to the extent of a half interest, *pour la moitié*, or a quarter interest, *pour le quart*, or a third interest, *pour un tiers*, or a two-thirds interest, *pour deux tiers*, or three quarters, or half a quarter, *pour un demi-quart*, or a sixteenth, *pour un seizième*, or even a thirty-second.

35. Georges Musset, *Les Rochelais à Terre-Neuve 1500–1550* (Paris, 1893). He suggests that there is little evidence as to the methods of conducting the dry fishery prior to 1600; also Musset, *Les Rochelais à Terre-Neuve* (La Rochelle, 1927).

Vessels Sailing to Newfoundland Prior to 1550

English (only)	11
English-Portuguese	3
Portuguese	12
Spanish	9
French, from La Rochelle	71
From other French ports	22
	128

(*continued on next page*)

35 (*continued*).

VESSELS SAILING TO NEWFOUNDLAND FROM FRENCH PORTS

Year of Sailing	Port of Departure	Number of Vessels
1523	Croisic	1
	Bény (Binic)	1
	Pornic	1
	St. Brieuc	1
	Blavet	1
1533	Ploemeur	1
1534	La Rochelle	1
1535	La Rochelle	2
1536	La Rochelle	1
1537	La Rochelle	4*
	St. Jean de Luz	2
	Ascaing	1
	Bayonne	1
1538	La Rochelle	2
	St. Jean de Luz	1
	Bayonne	1
1539	La Rochelle	2
	St. Jean de Luz	1
1540	La Rochelle	1
1541	La Rochelle	2
	Ré	1
	Barfleur	1
	Normandie (?)	1
	St. Brieuc	3
	St. Jean de Luz	3
	Bayonne	2
1542	La Rochelle	1
	Ré	2
	St. Jean de Luz	1
1543	La Rochelle	1
	D'Olonne	2
	De Jard	2
	Brittany (?)	1
	Rouen	1
1544	La Rochelle	1
	St. Jean de Luz	1
1546	St. Pol de Léon	1
	La Rochelle	2
1547	La Rochelle	2
	De la Flotte (?)	1
	Arvert	1
	St. Just en Marennes	1
	Erquy	1
	St. Jean de Luz	1
1548	La Rochelle	2
	St. Jean de Luz	1
1549	St. Just	1
	Talmont sur Jard	1
1556	La Rochelle	1
	St. Brieuc	1
	Cap Breton	1

* One vessel made two voyages.

80 tons and occasionally to 100 and even 200. A relatively large number of men were required, averaging from 18 to 25 for ships of 70 or 80 tons. The master was at times given a salary or he was "an owner of one fourth." If he was a merchant he was generally accompanied by a pilot. Other possible officers included a second mate, carpenter, surgeon, and a representative of the supplier, to check waste. Ships recruited their men from Brittany as they were preferred to all others. Payment by supply merchants[36] varied from payment in money to payment in kind or by shares. The men were required to give their supply merchants the preference in the sale of their fish at current prices. After 1550 the custom of giving one third to the men, one third to the boat owner, and one third to the supply merchant was apparently still kept up in the case of the "dry" fishery,[37] but only a quarter was given to those engaged in the "green" fishery.[38] Dry fishing required more attention and consequently a larger share, and apparently became important after 1550. Attempts on the part of the sailors to obtain more than a third were resisted. "A la grosse aventure," that is, "subject to every risk," was the chief basis on which loans were made; and risks were carried from the time of raising the anchor until the return home or forty-eight hours thereafter. Loans were repaid twelve to fifteen days after the return. Previous to 1550 interest rates were apparently omitted because of ecclesiastical objections, but from then on they varied from 20 to 40 per cent, depending on the scarcity of money, the solvency of the borrower, and other considerations. Basques and Bretons paid more than those in La Rochelle. Musset estimated that in 1537 the *Xpristofle* of 70 tons was herself worth 1,080 livres, her total returns in fish, 900 livres, and the profits were 27.75 per cent.

Bayonne apparently became interested in the Newfoundland fishery at about the same date as La Rochelle. On February 18, 1520, Pes de

36. The supplies for a 90-ton boat with 30 men engaged for five months—for example, for the *Laurens* of Saint Pol de Léon, which sailed December 2, 1545—consisted of 200 hundredweight of salt *mesure d'oleron,* 7 cables for boats, 300 pounds of tar to repair the vessel, 12 large knives, 4 *bydons,* 7 barrels of *aouillettes* and *seillaulx,* lamps, 20,000 nails, 8 *pelles,* 40 dozen *nettes,* 200 *gros bois,* 60 pounds of candles, 8 pieces of artillery; lines, harpoons, *clavières,* 2 boats with masts and sails, small anchors, 400 *étoupes pour boucher les voies d'eau* (tow or oakum); 30 pipes of biscuit, 66 pipes of wine—a pipe being equal to 2 barrels or 450 liters; 30 *coustes de lard,* that is, sides of bacon; 4 pints of vinegar; and 3 *barates* of butter. In old-time cod fishing the technical terms involved, whatever the language, as also the weights, measures, and monies, are now largely obsolete. But research work has shed at least some light.

37. For both the "dry" and the "green" fishery, see Index.

38. A quarter was sometimes also given in the dry fisheries. See Nicolas Denys, *The Description and Natural History of the Coasts of North America* (Champlain Society, Toronto, 1908).

Le Lande was granted permission to finish loading the *Senct Pe*, near the small port of Cap Breton, for a voyage to Newfoundland; and on March 6 two merchants, one also a farmer, were given permission to purchase half their cargo of cider, "forty butts, of the best that can be found," outside the city. The owner of the *Senct Pe* and one of the members of the partnership were given permission on the last day of March to unload a cargo of red lead from the *Marie* in order to take advantage of favorable weather to make a speedy voyage to Newfoundland, and to meet threats of the sailors to leave them. On February 6, 1527, permission was asked to load 24 butts of cider or even 20 butts outside the city on the ground that the petitioner had no money to buy them inside it. The documents are inadequate, but it would appear that at Bayonne certain substantial merchants, with vessels engaged in various activities, hired men to go to fish at Newfoundland.

Bayonne and St. Jean de Luz were linked to La Rochelle at a later date when the master played a more active part in financing expeditions. In 1537 the master and *bourgeois* of the *Marie* of St. Jean de Luz borrowed "subject to every risk" from a merchant of Bayonne 133 livres, 80 sols, and 6 deniers, payable fifteen days after the return of the vessel to La Rochelle or Bordeaux, and the master of the *Baptiste* of the same port borrowed sums from merchants of Bayonne, St. Jean de Luz, and La Rochelle. M. Doste of Bayonne was a party to loans to both ships, and to a loan of 266 livres, 13 sols, and 4 deniers to the master of the *Marie* of Ascaing. The *Marie* of Bayonne was loaded by merchants of La Rochelle; and in the following year the master of a ship of the same name and port borrowed 65 livres tournois. The master of the *Catherine*[39] of St. Jean de Luz borrowed 260 livres tournois "subject to every risk" from the same individual in La Rochelle. In 1541—to cite three vessels of St. Jean de Luz—the master of one, the *Trinité*, purchased his red wine from a La Rochelle merchant; the master of another, the *Marie*, sold his fish in advance to a prominent moneylender, Durand Buschet; and the master of the *Madeleine*[40] borrowed at La Rochelle 248 livres tournois from a Bordeaux merchant. In the same year three masters and *bourgeois* from Bayonne and one from St. Jean de Luz, each with a quarter interest in the *Charles* of Bayonne, borrowed 845 livres tournois from Buschet, and the master of the *Saint Esprit* borrowed from a La Rochelle merchant. In 1542 the *Baptiste* of St. Jean de Luz borrowed 128 livres from a Bayonne merchant. In 1548 the master of the *Sainte Anne* of St. Vincent borrowed 130 écus sol d'or

39. A ship of this name was mentioned in 1539. Her master borrowed 50 écus sol d'or from a Bordeaux citizen in 1546, and purchased salt for a voyage in 1547.

40. The master borrowed 50 écus sol d'or from a La Rochelle lender in 1544.

for provisions.[41] Bayonne was handicapped in the later years of the period by the shifting of the bed of the Adour to the north, and fishing vessels became more dependent for finances and markets on Bordeaux. On April 13, 1546, the *Marie de Cap Breton* bound for Newfoundland took on board a "canonnier," a cooper, and sailors at Bordeaux; and on April 19, 1548, a Cap Breton merchant living at Croisic borrowed 50 écus sol d'or at 30 per cent from a *bourgeois* of Bordeaux. A cooper was taken on by the same expedition. Two suppliers of Libourne borrowed money to outfit two vessels from Bordeaux.[42]

The Spanish Basque fisheries to Newfoundland apparently emerged in the last decade[43] of the period following a decline of the fishery off the coast of Ireland.

The destination of European fishermen is difficult to determine. It was claimed that Jehan Denys of Honfleur and a pilot, Gamart of Rouen, were on the coast north of Bonavista in 1506, and that the *Pensée* of Dieppe owned by Thomas Aubert opened up the northern fisheries to the Normans in 1508. As early as 1506 and again in 1539 it was stated that the coast from Cape Race to Bonavista had been occupied by the Portuguese, but Bretons and Normans in large numbers as well as Portuguese were fishing at St. John's in 1527. The voyages of Cartier[44] show that fishermen from Brittany were interested in the northern region. Cartier and members of his crew had obviously been engaged in the fishery prior to 1534. Bonavista was a landfall for him just as it had been on earlier voyages made by himself and others. St. Catherine's Harbor (Catalina) was well known and the Funk Islands were visited as a matter of course for supplies of fresh meat in the shape of great auks. Cartier proceeded along a well-known route to well-known places, Groais Island and Bell Island near Cape Rouge and Cape Dégrat on the Straits of Belle Isle. Grand Quirpon Harbor and Dégrat Harbor were well known, the name of the latter being in itself an indication of

41. See Musset, *op. cit.*, pp. 14 ff.

42. Francisque-Michel, *Histoire du commerce et de la navigation à Bordeaux* (Bordeaux, 1867), pp. 269–271.

43. "The son of Matias de Echevete [said] that he was the first Spaniard who went to the Newfoundland fishery. [He went] in a French vessel in 1545 and afterward made 28 voyages up to 1599, being the founder of the Basque fishery there." Prowse, *op. cit.*, p. 44; see also references to a lawsuit in 1561 suggesting that the fishery began between 1541 and 1545; *idem*, pp. 47 ff.; Vera Lee Brown, "Spanish Claims to a Share in the Newfoundland Fisheries in the Eighteenth Century," *Canadian Historical Association Report*, 1925, pp. 64 ff.; H. A. Innis, "The Rise and Fall of the Spanish Fishery in Newfoundland," *Transactions of the Royal Society of Canada*, 1931, sec. 2, pp. 51–70.

44. The account follows H. P. Biggar, *The Voyages of Jacques Cartier* (Ottawa, 1924).

the position of the fishery.[45] Harbors on the North Shore bore generally recognized names including Chateau Harbor, Hillcocks Harbor (Black Bay), Whale Harbor (Red Bay), Blanc Sablon,[46] Woody Island, Bird Island (Greenly Island). At "a harbour and passage called the Islets [Bradore Bay] which is better than Blanc Sablon . . . much fishing is carried on." Brest Harbor (Bonne Esperance) was regarded as a favorable harbor for securing wood and water and for fishing. At Shecatica Bay on June 12 they "saw a large ship from La Rochelle that in the night had run past the harbour of Brest where she intended to go and fish; and they did not know where they were." Little was known of the territory[47] beyond Brest other than that it offered possibilities of exploration for such voyages as that of Jacques Cartier. His expedition made available information on the resources of the Gulf, on the salmon in Shecatica Bay, the abundance of cod at Bear Head near Port au Port on the west coast of Newfoundland, the birds on Bird Rock, the walrus on the Magdalen Islands, and the mackerel in Gaspé Harbor. The importance of the area, especially along the north shore of the Straits of Belle Isle, was further made plain in the report received from Indians at Nataskwan Point—on their return on August 5—to the effect that "the ships had all set sail from the Grand Bay [the Straits of Belle Isle] laden with fish." Fishing vessels from La Rochelle and from the northern coast of France were established in this area before 1534 and in that year had secured their cargoes between June 12 and August 1. Such names as Belle Isle, Chateau, Bréhat, and Brest suggest Breton fishermen.

Cartier's voyage in the next year followed much the same route to the Straits; and in the Gulf and the River St. Lawrence he reported further resources, including whales near Anticosti Island, walrus on the Moisie River, white whales at the mouth of the Saguenay and in Baie St. Paul, and, throughout the St. Lawrence, a great wealth of fish of various kinds. On his return in 1536 by way of Cabot Strait he met, during his stay from June 11 to June 16 at the islands of St. Pierre and Miquelon, "several ships both from France and from Brittany." He

45. See Index.

46. O. W. Junek, *Isolated Communities* (New York, 1937), chap. iv.

47. According to the account of the voyage from Dieppe in 1539, the coast from Bonavista to the Gulf of "Castelli" was visited by a vessel from Honfleur with a pilot from Rouen in 1506. The Indians of this region were described as smaller, more kindly, and more tractable than those of the south coast. They lived in the region of the Gulf of "Castelli" in cabins and small houses during the summer and engaged in taking seals, porpoises, and birds. It is possible that the French learned of the Funk Islands from them. Cod fishing was regarded as excellent. The map suggests that the French north of Bonavista fished with the small boat and lines, and vessels and nets. A vessel in Belle Isle is shown hauling in a net. Ramusio, *op. cit.*, III, 423–424.

stopped at Renewse, a well-known harbor, for wood and fresh water. The English expedition of 1536 describes the capture of "a French shippe . . . well furnished with vittaile," probably somewhere along the northern coast of Newfoundland. In Cartier's expedition of 1541 the harbor of Grand Quirpon was chosen as a rendezvous.

The discoveries of Cartier in the Gulf and in the river opened the way to the fishery in new waters. The geographical information[48] was summarized in the *Cosmographie*, dated 1544, of Jean Alfonse of Saintonge, an "excellent pilot" in Roberval's expedition, and it included a survey of the coast line of Newfoundland and the Gulf of St. Lawrence with descriptions, directions, and the distances of the various courses. Cabot Strait is called "le destroict des Bretons" and we learn of "la terre des Bretons," or the portion of Cape Breton adjacent to Cabot Strait, and also of "le cap des Bretons."[49] Breton fishermen had extended their activities westward from St. Pierre and Miquelon. The excellent fishing grounds of Gaspé had been discovered at La Baie des Molues, that is, Mal Bay, and adjacent areas. It could be said that "in this region and at the Island of Ascension [Anticosti] there is a great fishery of cod and many other fish, much more than about Newfoundland; and the fish, too, are much better than the Newfoundland fish."[50]

Little is known regarding the discovery of the Banks. Cartier does not mention them, and the documents generally support the inference that fishing was confined to the coasts. References to the departure of ships are generally dated in March and April, that is, at a time that would allow of their engaging in the coast fishery and of their returning in August and later autumn months, or at times when ships returned from fishing on the Newfoundland coast. Cartier left St. Malo first on April 20, 1534, and then on May 19, 1535; and in both cases the fleet was delayed until he had a crew. It is probable that it was, comparatively, a long time before ships were able to equip themselves for a return voyage to the Banks without touching at the coast of Newfoundland for fresh supplies of water or wood. The bank fishery probably developed after the coast fishery had been established. The first reference to early departure appears later. From 1543 to 1545 the number

48. Biggar, *op. cit.*, pp. 278 ff.

49. According to the account of 1539 the Indians along the south coast were inhospitable, and they were not fishermen in spite of an excellent cod fishery which was prosecuted by Frenchmen and Bretons who had discovered it as early as 1504. According to an accompanying map, Sable Island was known and the fishery there was carried on by hook and line. Ramusio, *op. cit.*, III, 423–424.

50. "En cest coste et à l'isle de l'Ascension y a grand pescherie de molue et de plusieurs aultres poissons beaucoup plus que à la Terre Neufve; et si est ledict poisson bien meilleur que celluy de ladicte Terre Neufve."

of ships leaving Rouen, Havre, Dieppe, and Honfleur averaged, during January and February, about two a day.[51] At a later date the bank fishery was described as beginning in April and being over by July. The fish were large and "always wet, having no land neere to drie," and were called "core fish."[52]

In the first half of the sixteenth century Europeans discovered and prosecuted the fishery on the southern and eastern coasts of Newfoundland and even reached out to the Banks. European nations with supplies of cheap salt, such as Portugal and France, that were weak in agricultural development, limited of technique in the production of transportable supplies of meat products (protein) for the navy and other purposes, and had a large Catholic population became actively engaged in the Newfoundland fishery. The Portuguese, sailing from a relatively small number of ports, concentrated on the Avalon Peninsula; and they probably attempted to drive vessels from the scattered, relatively independent ports of France, and especially of Brittany, to the more distant portions of the coast line farther north, and to the south and west. Breton ports and others relied on Rouen and La Rochelle for financial support and for markets. The expansion of the fishery, first carried on for the home market, was followed by the growth of an export trade, particularly to England. The resources of the Gulf of St. Lawrence, Newfoundland, and the Banks had been explored and in part developed. The technique of the industry had been mastered in so far as the green fishery was concerned, with its reliance on abundant quantities of salt, and on animal life—for example, the birds[53] of Baccalieu and the Funk Islands—for bait and food. And a financial and marketing organization had been built up.

51. Edouard Gosselin, *Documents authentiques et inédits pour servir à l'histoire de la marine Normande et du commerce Rouennais pendant les XVI et XVIII siècles* (Rouen, 1876), p. 13. Jean Alfonse, Roberval's pilot, gave an inadequate account of the Banks, and in the map accompanying the voyage of 1539 they appear as a long thin strip. In spite of Harrisse it is possible that the first French vessel to fish on the Banks was the *Catherine* under Jacques Frasel in 1536. See Harrisse, *op. cit.*, p. xxxiii. In the accounts of Gilbert's voyage in 1583 the Banks were described as being located about fifty leagues east of Newfoundland and running north to 52° or 53° and south indefinitely, and as about ten leagues across. The depth varied from 25 to 30 fathoms. They were recognized by the large number of seafowl hovering over them.

52. *Voyages of the English Nation to America before the Year 1600,* I, 333.

53. "Frenchmen that fish neere the grand baie [Straits of Belle Isle] doe bring small store of flesh with them but victual themselves always with these birds [the great auks of Funk Island], Which the French men use to take without difficulty upon that Island and to barrell them up with salt." *Voyages of the English Nation to America before the Year 1600,* I, 303, 334. In view of what we now know, the wonder is not that the great auk became extinct but that it survived as long as it did. We do not know what its original numbers may have been. But for nearly 350 yeers it served as both food and bait for great numbers of cod fishermen, and did not wholly disappear until 1844.

Appendix

COINS AND MONEY VALUES[1]

THE various coins and money values appearing in this volume may well seem hopelessly puzzling. What are *sols* and *livres tournois,* or *rials, ryalls, réals,* and *royalls* of 8, or *escuiz, escus, écus sol, escuis d'or sols,* and *escus sol d'or?* What were they worth? The context gives us their value in their own time; but, of present-day coins, which would come nearest them? And how did they compare with the contemporary pound, shilling, or penny?

Questions answerable only approximately. For with the discovery of America, great quantities not only of cod but of new gold and silver began to go to Europe. The result, for almost a century and a half, was a great cheapening of the precious metals, and endless shiftings of their reciprocal values. At the same time, too, royal greed, official necessity, and private clipping and counterfeiting tended constantly to reduce the weight and fineness of many coins. Indeed, for most of our period[2] only the little brass scales of the money changers could say with any final accuracy how the monies of England and the Continent really "balanced."

But with all that there are things which clarify. Even as twenty English shillings made a pound and twelve pence made a shilling, so did twelve *sous* (or *sols*) make a French pound or *livre* and twelve *deniers* make a *sou* or *sol,* or at least a *sou* or *sol tournois.* All the *royalls, rials, réals,* and their like are simply one and the same *réal,* while all the *escuiz, escus sol d'or,* and *écus sol* are the same *écu sol.* The number of coins involved are in fact comparatively few, and some of them were of marked longevity and steadiness.

Taking English currency first, from 1489 to 1542 the sovereign or pound[3] weighed 240 grains troy, and was 23.3½ carats fine. That is, in present-day American gold it would be worth almost $10.00.[4] In "the great debasement" of the 1540's it was cut to 169 grains at 22 carats, and thereafter the figures were: 1552, 174; 1601, 171.9; 1604, 154.8; 1619, 140.5; 1663, 129.4; 1696–

1. By Arthur E. McFarlane.

2. To escape the monetary entanglements left by the first Great War we venture to set 1914 as our final date; and 1914 will be meant even when we speak of "the present," "today," etc.

3. Values are given as for the 20-shilling pound or the 21-shilling guinea, though the actual coins were of various names and weights.

4. Lack of space prevents our taking count of the downward trend in "fineness." Enough that while in 1552, 1604, and 1612 some English gold coins were still minted at 23.3½, their value was proportionate. In 1670 the 22-carat ratio became permanent, as also 925 parts fine per *M* for silver. Save for the *écu sol,* by 1641 French gold had dropped to a semipermanent 22 carats. French and American currency is now, of course, 900 parts fine. All weights are given in troy grains. For purposes of comparison, the half sovereign weighs 61.63; the shilling, 87.27; penny, 145.83. American gold: $10, 258; $5, 129; $1, 25.8; silver, 50 cents, 192.9; 25 cents, 96.45; dime, 38.58; nickel, 77.16; cent, 48. In carats, 24 c. = 1,000 fine; i.e., 23.3½ c. = 979.17, etc.

Dates refer to the most recent issue, and because of the many changes the date is all-important. See A. Blanchet et A. Dieudonné, *Numismatique française,* Vol. II.

1699, irregular but falling; from then till 1816, the 21-shilling guinea at 129.4; 1817 to 1870, 122.27; 1870 to 1914, 123.27.

Coming to the shilling and penny, in theory there was always a shilling in existence, which, at 925 fine, weighed 144 grains from 1500 to 1526, 128.40 to 1544, was then debased to as little as a third of its silver weight, but in 1560 was of good metal again, and weighed 96, then, in 1601, 92 and in 1670, 87.25, at which weight it remained to the present. As for the penny, silver till 1670, and always one twelfth of the weight of the shilling, in theory it likewise was continuously existent.[5] But, actually, for much of the time, both shilling and penny were driven into hiding by tradesmen's tokens, survivals of old coinages and clipped or counterfeit money; and, rightly, they should simply be thought of as one twentieth and one two hundred and fortieth of the gold pound, that is, merely as money of account, like our own mill and guinea.

On the Continent in 1500 three standard gold coins were being minted, the Spanish and the Venetian *ducat* and the French *écu sol*. All alike weighed 54.9 grains, and all were above 23 carats fine. If our fishermen sold their cod for so many thousands of *ducats* in the Atlantic ports of Spain, they were probably paid in Spanish *ducats*. worth at first about $2.30 today, and gradually decreasing to a value of some $1.95 in the two centuries that followed.[6] If the cod were sold in the Mediterranean, where the Venetian *ducat* had long been the great trading medium, early exchange would be about the same. While the sovereign still weighed 240 grains both *ducats* were current in England for at least 4*s*. 6*d*. And, as the sovereign lost weight, their exchange value rose accordingly. In 1618, for example, the Venetian *ducat* brought more than 7 shillings.

But the *écu sol*, or *écu d'or au soleil*—"the gold crown of the sun"—was preëminently the French gold piece of the fishery. It was so called because a tiny sunburst capped its escutcheon, and had nothing to do, of course, with the *"sol"* that was the old name for *sou*. All the *écu sol's* other names are merely archaic or colloquial. First minted in 1475, it virtually kept its value, with one bad but brief debasement, until it was officially withdrawn in 1692. Quoting it in modern *francs,* it was worth 11.60 till 1519, 11.35 till 1548, and 11.14 to the end, that is, in dollars, a decrease from some $2.24 to $2.15, or about 4 per cent. In 1641, reduced very slightly, and doubled in weight, it became the *"demi"* of the first *louis d'or*. As such under various names it lived on through the Revolution and the nineteenth century. And it is still the half *"louis"* or 10-*franc* gold piece of today. Suffice it that for more than two cen-

5. The ship money collected in 1639 by the sheriff of Monmouthshire was most of it in "such ragged pieces as broken groats, quarter-pieces of thirteen-pence half-pennies, ten-pence half-pennies, harpers and four-pence half-pennies." A. E. Feavearyear, *The Pound Sterling* (Oxford, 1931), p. 84. As late as 1816 Pascoe Grenfell could claim that "there was now nothing like a Tower," i.e., bona fide "shilling in the country." French 10-*sou* pieces were then circulating in England at a premium of 20 per cent. *Ibid.*, p. 196. See also H. A. Grueber, *British Museum Handbook of Coins*.

6. The "Spanish ducat of silver" was, about 1640, worth some $1.37.

turies it was the true gold standard of France. And even where it does not appear in the transactions of the French cod fishery it was the *écu sol* that was the key to all actual value.

For the *livre tournois*—the *livre of Tours*—and the other unspecified *livre*—really the *livre parisis,* or the *livre of Paris*—were not coins at all, but *monnaie de compte,* French money of account. The *sou* or *sol* was a coin, first silver and then copper. The *denier* was a tiny bit of minted silver or copper until, in the end, it too became only *monnaie de compte.* But no *livre* had ever been minted either in Tours or Paris. *Livre* was merely a term for what had once been a definite part of an 8-ounce block of gold. And, in 1364, the *livre tournois* was still held to be worth as much as a gold *franc,* or 18 *francs* ($3.47) today; that is, much more than any *écu sol.* But *livres* were only fiat money, unbacked even by stamped paper. The *gros,* the *teston,* and other coins of real silver or gold were being minted. People much preferred them. And the *livre tournois,* the *livre parisis* which had grown up with it, and all their coördinated *sols* and *deniers* continued to lose value. In 1500, 1 *livre tournois,* 6 *sols,* and 3 *deniers* (1 *l.t.* 6*s.* 3*d.*) were still worth an *écu sol,* but in 1519 2 *livres tournois* were demanded; in 1523, 2 *l.t.* 5*s.*; in 1561, 2 *l.t.* 10*s.;* and in 1573, 2 *l.t.* 14*s.* In 1577 a royal attempt—the first of many—was made, as we would say, to peg or tie the *livre tournois* to the gold *écu* at 2 *l.t.* 60*s.;* but by 1578 the price was 3 *l.t.;* by 1602, 3 *l.t.* 5*s.;* by 1640, 5 *l.t.* 4*s.;* by 1643, 5 *l.t.* 4*s.* 6*d.;* by 1669, 5 *l.t.* 10*s.;* and by 1689, 6 *l.t.* 5*s.* For the next century the drop continued, with the figures doubtful. But we know that the Revolution found the *écu sol* worth substantially ten times the *livre tournois.*

For present purposes, however, the record above has its worth for us in this: Because of the lasting staunchness of the *écu sol* we cannot merely translate its value, from 1500 to 1692, into contemporary shillings and present-day dollars. Linked as it is to the *livres,* it offers us France's detailed price chart for her whole Newfoundland fishery.

There still remains the silver coin of Spain. And if its account is easily rendered, it is a memorable one. One coin only enters in, the *réal,* originally an eighth of the ancient Moorish *piastre,* and, from the fourteenth to the eighteenth century, the Spanish monetary unit. When in the latter period of the cod fishery the *réal* began to appear in the record, it could pass for half a shilling, and indeed, was often called the Spanish sixpence. Two-, four-, and eight-*réal* pieces were also coined, the largest being the "pieces of eight" of the Spanish Main; and the "piece of eight" was to know its own changes. It became the Mexican dollar; but it still remained a dollar "of eight pieces," which in old California became "bits," and thus gave the *réal* a place in American currency which, if unofficial, bids fair to be lasting. Meanwhile, the eight-*réal* piece had also become the American dollar itself. And finally this one-time *piastre* of the Moors in old Spain became the "trade dollar," went around the Horn in a hundred Yankee clippers, and played the greatest of all parts in opening up age-old China and the East.

CHAPTER III

THE SPANISH AND ENGLISH FISHERIES 1550–1600

The English have had more absolute trade to Newfoundland since the year 1585 than ever before for in that year the war broke out betwixt Spain and us; whereupon the queen sent certain ships to take such Biscaines and Portuguese as fished there; a service of great consequence to take away the ships and victuals from our enemies subjects; and since that they have almost abandoned their fishery thereabouts. Out from these men came the great sickness that the judges and justices died of at Exeter.

SIR WILLIAM MONSON'S *Naval Tracts*

THE second half of the century was characterized by the rise and decline of the Spanish fishery, the expansion of the English fishery, and the adjustment of the French fishery to these major developments. Commercialism in England and France thrived on the rising prices which accompanied specie imports to Spain.[1]

The English advanced from a position of minor importance in the Newfoundland fishery at the beginning of the second half of the century to one of major importance at the end.[2] Legislation was[3] enacted to encourage the fishery,[4] but it apparently encouraged trade rather than industry. An act of 1548 (2 and 3 Edw. VI, c. 19) regarding Lent was extended in an act "for the better maintenance and encrease of the navy" (5 Eliz., c. 5), 1562–63, which added Wednesdays to Saturdays as fish days. The same act permitted any subject for the four years following April 1, 1564, to carry fish out of the realm in any British ship without paying customs dues. This provision was revived and continued for six years in 13 Eliz., c. 11, 1570. In 1579 the New-

1. Fish quoted in France at, respectively, 20 sols, 10 sols, and 4 sols in 1545 had increased in price to 30 sols, 20 sols, and 10 sols by 1562. H. A. Innis, "The Rise and Fall of the Spanish Fishery in Newfoundland," *op. cit.;* E. J. Hamilton, *American Treasure and the Price Revolution in Spain, 1501–1650* (Cambridge, 1934).

2. Whitbourne stated in 1621 that the English had been conducting thriving and profitable voyages to Newfoundland for more than four-score years. Richard Whitbourne, *A Discourse and Discovery of Newfoundland* (London, 1622). Mason wrote "60 years and upwards" in 1620. About 1610 the Charter of the Bristol Company speaks of "50 years and upwards."

3. R. G. Lounsbury, *The British Fishery at Newfoundland 1634–1763* (New Haven, 1934), p. 33.

4. A. L. Rowse, "The Dispute Concerning the Plymouth Pilchard Fishery, 1584–1591," *Economic History,* January, 1932, p. 461; and William Cunningham, *The Growth of English Industry and Commerce* (Cambridge, 1903), pp. 63 ff.

foundland fishery was declared exempt from an embargo on ships and mariners. In 1580–81 an act (23 Eliz., c. 7) prohibited imports of foreign-cured fish by Englishmen, but exempted such fish as came from Iceland, the Shetlands, and Newfoundland. The preamble of 39 Eliz., c. 10, 1597–98, which repealed it, stated that the

navigation of this land is no whit bettered by the means of that act, nor any mariners increased nor like to be increased by it; but, contrariwise, the natural subjects of this realm not being able to furnish the tenth part of the same with salted fish of their own taking, the chief provisions and victualling thereof with fish and herrings hath ever since the making of the same statute been in the power and disposition of aliens and strangers, who thereby have enriched themselves, greatly increased their navigation and (taking advantage of the time) have extremely enhanced the prices of that victual.

Wars with France and Spain hindered the fishing industry.[5] In spite of Anthony Parkhurst's enthusiastic report in 1578 that the English fishing fleet in Newfoundland, "since my first travell being but 4 yeeres, are increased from 30 sayle to 50,"[6] the fishery made slow progress until after the defeat of the Armada in 1588. Sir George Peckham wrote in 1584, "It is well knowen that in sundry places of this realme ships have been built and set forth of late dayes, for the trade of fishing onely: yet notwithstanding, the fish which is taken and brought into England by the English navy of fishermen will not suffice for the expense of this realme foure moneths, if there were none els brought of strangers."[7] Dried fish were in demand to victual the army and the navy and ships going below the line. Two merchants of Chester in an agreement

5. From 1542 to 1565 vessels from Brittany inflicted heavy losses on the English. M. Bronkhorst, *La Pêche à la morue* (Paris, 1927), p. 25. See reference to the losses of John Link, a merchant of London in the Newfoundland trade in 1563. A. M. Field, *The Development of Government at Newfoundland 1638–1713,* master's thesis, University of London, 1924; also C. B. Judah, *The North American Fisheries and British Policy to 1713* (Urbana, 1933), p. 27.

6. H. A. Innis, *Select Documents in Canadian Economic History 1497–1783* (Toronto, 1929), p. 9.

7. *Voyages of the English Nation to America before the Year 1600,* ed. Edmund Goldsmid (Edinburgh, 1889), II, 17. Robert Hitchcock speaks of the fishery of that time as something "the advantage and profite whereof this realme and subjects of late years for the most part have lost, and suffered strangers, the Fleminges and other nations, to take." The Flemings and other nations seem to have worked together in groups or companies. There was one at Dieppe, he says, "that serves and victualles all Pickardie; [another] at Newhaven that serves all base [lower] Normandie; [another] at the towne of Rone [Rouen] that serves all the hie countries of France"; and also, it would appear, it "served" La Rochelle, Bordeaux, and Spain, Portugal, Italy, "Barbary," and Africa. Moreover, "the further south and southweste that the fish well used [well cured] is caryed, the dearer it is and greatly desired." Robert Hitchcock, *A Pollitique Platt for the Honour of the Prince* (London, 1580). See also *Sir Walter Raleigh's Observations on the British Fishery,* reprinted (London, 1720).

made on September 26, 1580,[8] purchased from the master of a ship "34,000 Newland fish merchantable, at 10*s*. the 100 . . . also foure tonnes traine at £12 per tonne."[9] The war with Spain was responsible for high prices and a temporary decline in the English fishery. At Chester, in 1586, 20,000 Newfoundland fish were purchased at 20 shillings a hundred for the army in Ireland.[10] Trading was prohibited to Brittany, though allowed to Jersey and Guernsey, and in 1589 it was reported that only 10,000 Newfoundland fish came on the market.

After the defeat of Spain recovery was rapid. Fish were sold to France and through France to Spain. On July 20, 1590, Richard Seguiner, an English merchant at Weymouth, sold three merchants at Honfleur 5,000 or 6,000 green cod[11] "bonne et suffisante, loyalle et marchande à 66 poignées par chacun cent, moyennant le prix et somme de 6 escus sol chacun cent."[12] In 1593, only the Newfoundland fleet made profits; in 1594, it was estimated to number 100 sail. Plymouth alone had 50 sail in 1595. It was of these ships that Sir Walter Raleigh wrote to Sir Robert Cecil, "If these should be lost it would be the greatest blow ever given to England."[13]

In 1597 Dutch, Irish, and French ships were at Plymouth in September to purchase fish on the arrival of the Newfoundland fleet. A year later large cargoes of Newfoundland fish were brought to Southampton and Poole and were mostly sold by October for shipment to Spain by way of France. On September 17, 1599, a license was given to export from England 60,000 "Newland fishe, whereof there is at this

8. D. W. Prowse, *A History of Newfoundland* (London, 1896), pp. 84–85.

9. "Traine" or "trayne" was an abbreviation of "traine oil"—the oil of seal, whale, or cod.

10. Prowse, *op. cit.*, p. 76. In 1595 Sir G. Fenton wrote that he had "sent out again two ships to Newfoundland for fish and other provisions for the army in Ireland and the people." *Idem.*, p. 58 n.

11. Charles and Paul Bréard, *Documents relatifs à la marine Normand* (Rouen, 1889), p. 59. A document of about the last decade of the sixteenth century, "A Speciall Direction for Divers Trades," refers to the exporting of "a small quantitie drye newe land fyshe" to the Levant, of more "drye newland fishe" to the Canaries and the Madeiras, and also of the sending of "all kinds of course wares, waxe and tallowe, butter and chease, wheate, rye and beanes, byskye" [biscuit] to St. Jean de Luz "so that it be brought thither at Christmas or shortly after to sarve the newefoundland men. . . . This port [St. Jean de Luz] sarves when we have a restrainte between Spaine and us." On the other hand, from Bilbao they imported whale oil "that come from Newfoundland." The prices commanded by cod in St. Jean de Luz were, "wett newland fish," £1 per hundred, dry fish 10 shillings the hundred, and cod oil £9 a ton. N. S. B. Gras, *The Evolution of the English Corn Market* (Cambridge, 1915), pp. 429–439.

12. "Good and satisfying, reliable, and merchantable at 66 *poignées* a hundred; in other words, at an average price of 6 crowns a hundred."

13. Judah, *op. cit.*, p. 32.

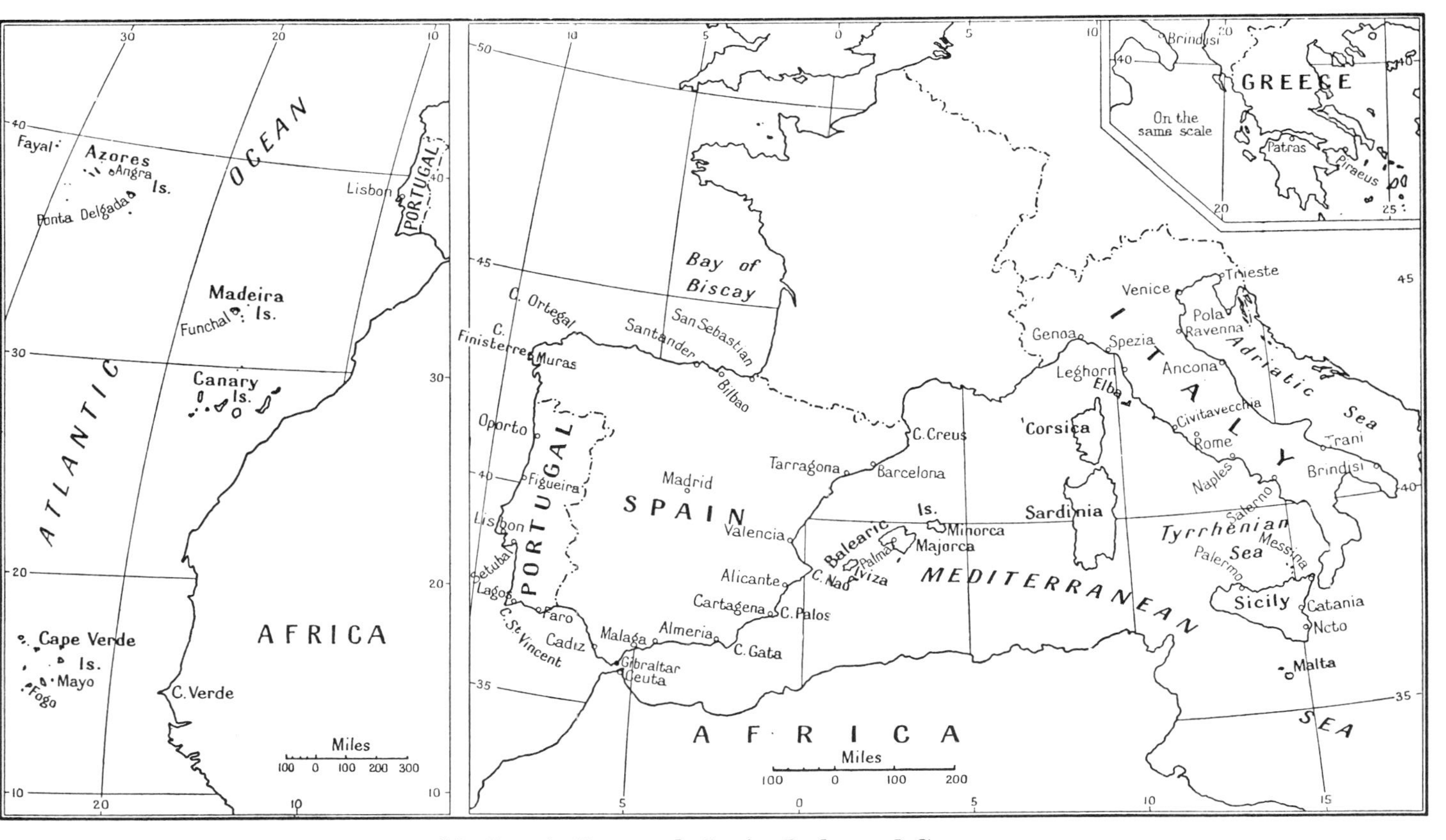

Markets in Portugal, Spain, Italy, and Greece

present good quantitie in the most partes."[14] The act passed in 1598 (39 Eliz., c. 10) permitted British subjects to carry foreign fish. A document of 1604 provided for the insurance of the *Hopewell* of London, 120 tons, "from the day and hower of the lading of the saide fishe aboorde the said shippe in the newe found land aforesaid, and so shall continewe and endure untill suche time as the same shippe with the same fishe shal be arived at Toulone and Marcelleze and the same their dischardged and laid on land in good safety . . . after the rate of seaven per cent." That is, London vessels were engaged in carrying fish from Newfoundland to France and to Marseilles, and moreover were able to secure insurance for the voyages.

With the rise of the industry there was built up an elaborate and elastic financial organization.

> In Englande in the West countrey . . . the fishermen conferres with the money man, who furnisheth them with money to provide victuells, salte and all other needefull thinges to be paied twentie five pounde at the shippes returne, upon the hundreth pounde in money lent. And some of the same money men doth borrowe money upon ten pounde in the hundreth and puts it forthe in this order to the fishermen, and for to be assured of the money ventured, they will have it assured gevying five pounde for the assuring of every hundreth pound to hym that abides the venture of the shippes returne.[15]

The development of the fishing industry and of the attendant trade contributed to the breakup of the control of trade to Spain and Portugal by chartered companies. The first charter[16] to trade with Spain and Andalusia was granted in 1530 and included ships from Southampton, Lyme, Exeter, Barnstaple, and Chester. An application by an exclusive company of merchants for a charter to trade with Spain and Portugal was opposed in 1574, but was conceded in 1577. Nevertheless, ships not belonging to the company were carrying corn and fish to Spain and Portugal in 1580. Their union in 1581 became a source of strength rather than of weakness to English trade, and it continued to be so even after the destruction of the Armada and the expulsion of English merchants from Lisbon in 1589. Imports of sugar, especially

14. Prowse, *op. cit.*, p. 84. In 1594, Dartmouth protested, apparently with success, against an embargo on English vessels. The following year regulations permitted exports after one fourth had been reserved for domestic consumption. In 1599 shipowners were required to have licenses to take fish out of the country. But a large catch made restrictions unnecessary, and the export market also became increasingly important with an increase in the consumption of meat in England and a decrease in the consumption of fish. Judah, *op. cit.*, pp. 40–42.

15. Hitchcock, *op. cit.*

16. V. M. Shillington and A. B. Chapman, *The Commercial Relations of England and Portugal* (London, 1907), pp. 129–176.

from the Indies, Barbary, and the Azores, were added to those of wine and salt.[17] The growing of sugar in Brazil resulted in direct trade thither from England, with the support of the English Council (1582), in spite of protests from the company trading to Spain and Portugal. An expansion of trade because of, and in spite of, both company control and the outbreak of hostilities was largely responsible for a favorable treaty in 1604.

The concentration of the English fishing interests in the West Country and the equipment of large English ships for the "shipping and furniture of munitions"[18] strengthened the hold of the English in Newfoundland as contrasted with that of other nations. The problem of cheap supplies of salt was partially solved by what was made available by foreigners in Newfoundland. In return for this the English seem to have conferred upon themselves in some manner a sort of general overlordship, at any rate as regarded the Portuguese. Parkhurst[19] refers to the English as "commonly lords of the harbours whence they fish." And, he says, they "doe use all strangers help in fishing if need require according to an old custom of the countreys, which they [the strangers or foreigners, offer to] do willingly so that you take nothing more than a boate or twaine of salt as a tax for protecting them." The English probably maintained their early affiliations with Portugal and assisted the Portuguese against the French—though Parkhurst purchased fish from "the Portugals and Frenchmen" both. He complained that the Portuguese had promised to deliver salt in return for protecting them from the French for two years, and that their failure had caused a loss of 600 livres, of which half was lost by his fishermen. He therefore asked for Hakluyt's advice as to whether he should attempt to collect in Portugal, or seize Portuguese goods in England, or salt in Newfoundland to that amount, or as much fish as 600 livres worth of salt would make. An increase in the number of "western men," that is, West Country men, meant increased demands for salt even when conditions were favorable for drying "light-salted" fish.[20]

17. *Idem,* Part II, pp. 6 ff.

18. It has been suggested that the heavy, strong ships essential to success in the struggles in Iceland were an important factor in giving the English an advantage in Newfoundland, but it was apparently the east coast which was chiefly concerned with Iceland. *The Cambridge History of the British Empire,* I, 60.

19. The account of Gilbert's expedition refers to English merchants "that were and alwaies be Admirals by turnes interchangeably over the fleetes of fishermen within the same harbour. . . . The maner is in their fishing, every weeke to choose their Admirall anew; or rather they succeede in orderly course." In St. John's the "English marchants . . . have accustomed walks unto a place they call the garden." *Voyages of the English Nation to America before the Year 1600,* I, 335–337.

20. In early August we learn the weather was "so hote this time of yeare except the

Any expansion of the industry and of trade depended on the accessibility of supplies of salt.[21] The Portuguese found it difficult to carry more than enough for their own green fishery and provisions, and consequently had far from an adequate supply to meet the demands of the English. Parkhurst hoped to find rich mines "more profitable for fishing than any yet we have used, where wee shall have not far from thence plentie of salt made undoubtedly, and very likely by the heate of the sunne, by reason I find salt kerned [dried] on the rocks in nine and fortie [49°], and better: these places may bee found for salte in three and fortie."[22] Sir George Peckham wrote:

> And the chiefest cause why our English men doe not goe so farre westerly as the especiall fishing places doe lie, both for plenty and greatnesse of fish, is for that they have no succour and knowen safe harbour in those parts; but [except] if one nation were once planted. . . . Whereas they now fish but for two moneths in the yeere, they might then fish as long as pleased themselves, or rather at their comming finde such plenty of fish ready taken, salted, and dried, as might be sufficient to fraught them home without long delay (God granting that salt may be found there) . . . and withall the climate doth give great hope that though there were none naturally growing, yet it might as well be made there by art, as it is both at Rochel and Bayon or elsewhere. Which being brought to passe shall increase the number of our shippes and mariners, were it but in respect of fishing onely.[23]

As a result of the salt problem the English fishery[24] tended to be concentrated in Newfoundland in regions suitable for "light-salting," and where trade could be carried on with the Portuguese; i.e., in the regions south of Trinity Bay, or roughly in the Avalon Peninsula, including Harbor Grace, Conception Bay, St. John's, Ferryland, Fermeuse, Renewse, Aquafort Tor Bay, Capelin Bay, Broyle, Trepassey, St. Marys, Placentia, and St. Pierre.

As early as 1579 Whitbourne went to Trinity Harbor in a 300-ton ship sent by Master Cotton of Southampton which "killed great store of fish" and returned to the home port. Four years later he was in command of a 220-ton ship sent by Master Crooke of the same port, on which voyage he witnessed the ceremony of the taking possession of St. John's by Sir Humphrey Gilbert. In that year Gilbert had planned to

very fish which is laid out to be dryed by the sunne be every day turned it cannot possibly be preserved from burning." Parkhurst refers to "four and twentie persons . . . turning of drie fish."

21. Edward Hughes, *Studies in Administration and Finance, 1558–1825, with Special Reference to the History of Salt Taxation in England* (Manchester, 1934), chap. i.

22. *Voyages of the English Nation to America before the Year 1600,* I, 304.

23. *Idem,* II, 17.

24. Whitbourne, *op. cit.,* Preface.

arrive in Newfoundland in August when, he felt, his fleet "should be relieved abundantly with many necessaries which after the fishing ended they [the fishermen] might well spare and freely impart unto us."[25] For a rendezvous Gilbert's ships had settled upon Cape Race, Renewse, or Fermeuse. They arrived at St. John's Harbor on August 3 and found "36 sailes of all nations" of which about 20 were Spanish and Portuguese and the remainder English.

There is little evidence that the English fishery had penetrated to the Gulf of St. Lawrence through Cabot Strait until after the defeat of the Spanish. The first vessels went to the Magdalen Islands in 1593 to participate in the walrus fishery. The *Grace*, of 35 tons, left Bristol on April 4 in the following year in search of whales, which, it was said, when wounded went to Anticosti to die; but the *Grace's* crew succeeded in finding only "wonderfull faire and great cod fish." On her return she started to collect a cargo of fish at Placentia but went on to Farillon (Ferryland) where she found 22 English vessels. Charles Leigh and Abraham Van Herwick of London sent two ships, one of 120 and the other of 70 tons, which left Falmouth on April 28, 1597, for the Magdalen Islands. They obtained a pinnace of seven or eight tons at Ferryland from "M. Wil Sayer of Dartmouth, Admiral." At an island called Menego(?) north of Cape Breton, in sixteen fathoms they "caught great store of cods which were larger and better fish than any in Newfoundland. . . . We fell to fishing where the cods did bite at least 20 fathomes above ground and almost as fast as we could hale them into the ship."[26] At the Magdalen Islands, with four hooks, they caught 250 cod in an hour. At that point and along the coast of Cape Breton and Newfoundland they carried out a series of raids on Spanish Basque vessels. The name "English Port," given for a time to the present Louisburg, was a tribute to English interests.

Nor is there much evidence of the English fishery in the north. Parkhurst in 1578 suggested the possibilities of fortifying the Straits of Belle Isle as a means of capturing the fishery "and from thence [to] send wood and cole with all necessaries to Labrador lately discovered." Davis in his voyage along the Labrador coast in 1586 noted an abundance of cod at 57°, and again at the entrance of a harbor at 56°, while at 54° 30′ there was "great abundance of cod so that the hooke was no sooner overboard but presently a fish was taken. It was the larg-

25. *Voyages of the English Nation to America before the Year 1600,* I, 335. "The shippes goeth forth from Englande and Irelande in March and comes home loden in August."

26. *Idem,* II, 70; also Marc Lescarbot, *The History of New France* (Toronto, 1911), II, 24.

est and best fed fish that ever I sawe, and divers fisher men that were with me sayd that they never saw a more suaule[?] or better skull of fish in their lives." He refers again "to certaine fish which we purposed to weather and therefore left it al night covered upon the island." On September 7, 1586, "wee saw an incredible number of birds; having divers fishermen aboord our barke, they all concluded that there was a great skull of fish; we, being unprovided of fishing furniture, with a long spike nayle made a hooke and fastened the same to one of our sounding lines. Before the baite was changed we took more than fortie great cods, the fish swimming so abundantly thicke about our barke as is incredible to bee reported of which, with a small portion of salt that we had, we preserved thirtie couple." So promising were the possibilities that in 1587 "two shippes were appointed for the fishing and one pinnasse for the discoverie." Between 54° and 55° "in sixteene days the two shippes had finished their voyage and so presently departed for England."[27]

The Portuguese fishery declined to relatively minor importance, and when Portugal became a part of Spain in 1581 she was affected by the aggressiveness of England. Parkhurst wrote in 1578 that the Portuguese, "not lightly above 50 saile, make all wet"—preserved their cod green or merely salted—and were engaged in the fishery from April to July. Their tonnage was estimated at 3,000, or an average of 60 tons to a vessel. They were engaged in the Gulf of St. Lawrence and Cape Breton fishery as well as in that of Newfoundland. Sir Humphrey Gilbert wrote in 1583, "Whereunto the Portugals (above other nations) did most willingly and liberally contribute . . . [and we] were presented above allowance with wines, marmalads, biskets, sweet oyls." After the defeat of the Armada the Portuguese were not in a position to give any aid to the English fishery, and England was compelled to secure supplies of salt from France.

The Spanish fishery acquired importance especially after the Peace of 1559, although it was said that in 1553 a Spanish fishing fleet which refused to accept a convoy to Newfoundland numbered as many as 200 ships and 6,000 men. In 1561 a dispute arose in Spain as to whether the church should be given 2 per cent of the proceeds of the Newfoundland fishery. In 1574 ships were described as leaving about the end of March and returning the latter part of September, while whalers left in the middle of June and returned in December or January. The fishery probably reached its peak in the decade from 1570 to 1580. Parkhurst gave an estimate in 1578 of "above 100 saile of Spaniards that come to take cod . . . besides 20 or 30 more that come from Biskaie to kill whale for

27. *Voyages of the English Nation to America before the Year 1600,* I, 245–246.

traine. These be better appoynted for shipping and furniture of munition than any nation saving the Englishmen." Parkhurst's total tonnage, one of 5,000 or 6,000, suggests an average ship of 50 or 60 tons. They "make all wet and do drie it when they come home."

The union of Portugal and Spain was followed by opposition from England. In August, 1582, a Mr. Ughtred of Southampton saw twenty Spanish and Portuguese vessels at St. John's, and in the same year he robbed more than that number. Sir Humphrey Gilbert took possession of Newfoundland in 1583, and provided an excuse for attacks on the Spanish fishery. In 1585 Sir Bernard Drake captured Spanish ships and sailors. An embargo was laid on the Spanish Basque fleet in 1586 and was not lifted until late in 1587. As has been indicated, the destruction of the Armada contributed to the marked decline of Spain's position in Newfoundland, and thereafter references to the Spanish fishery are scattered. English and French seized Spanish vessels as prizes during the difficulties of the latter part of the sixteenth century. In 1594, eight Spanish vessels were reported at Placentia, and in 1597 Spanish vessels were seen in the Bay of St. Lawrence on the south shore of Newfoundland, and at the Magdalen Islands and Cape Breton. In 1602 seven Basque vessels and several other cod-fishing vessels were noted in Newfoundland, and the Spanish Basques apparently maintained their position as whalers.[28]

The extensive French fishery,[29] with its dependence on divergent ports and relatively small ships, while aggressive toward English ships had itself been exposed to attacks from Spain. In 1554, 9 French fishing vessels, under convoy of a ship from Saint Pol de Léon, and another ship from Ile de Ré had been attacked by a Spanish squadron; and in the following year Spanish ships had captured 48 French vessels in Newfoundland and made a successful raid on St. John's. The Peace of Cateau-Cambrésis in 1559 had brought about renewed activity, and Jumièges, Vatteville, and La Bouille had 38 vessels, chiefly from 100 or 120 to 150 tons, which left in January and February, 1560, presumably for the Banks. In 1565 Croisic and St. Malo[30] each had 25 ships. Parkhurst described French and Breton shipping in 1578 as to-

28. Davis reported a Basque ship near the Straits of Belle Isle engaged in whaling on August 17, 1587. An English vessel discovered the wrecks of two Biscay ships in St. Georges Bay, Newfoundland, with seven or eight hundred whale fins. Lescarbot describes the Basques as engaged in whaling at Les Escoumins.

29. Charles de la Roncière, *Histoire de la marine française* (Paris, 1906), III, 589–595. In 1584 an edict was issued providing for the administration of the fisheries, indicating a growth of unity in control and increasing interest.

30. The strength of St. Malo was made manifest by its vigorous protests against monopoly in 1557.

taling 150 sail, mostly small ships not over 40 tons, with some larger, the whole tonnage probably not exceeding 7,000.

The financial organization of Rouen which was so close to the Paris market appears to have become extensive, and reflected the expansion of the industry, particularly the bank fishery. A list[31] of the loans on fishing vessels registered in 1564 in the archives at Rouen to protect the lender from fraud included about 50 vessels with a total tonnage of 5,500 or 6,000, of which Havre had 19, with 2,865 tons; Honfleur 4, with 460 to 480; Fécamp[32] 5, with 350; Vatteville 2, with 120; St. Valery en Caux 4, with 280; Barfleur 2, with 280; Jumièges one, of 70 tons. And a miscellaneous group of 14 ships, probably belonging to Rouen, totaled possibly 1,000 to 1,200 tons. The smallest ships averaged 50 to 60, a large number were listed at 80, several at 100 to 140, and two at 200. Financial support[33] was obtained mostly from Rouen.

31. Archives de la Seine-Inférieure, Fonds du Tabellionage de Rouen. Registre meubles. Registre 1564. 1er Janvier–21 Avril. Copied through the courtesy of H. P. Biggar. The ships are given as belonging to the port or harbor from which they were ready to leave for Newfoundland. "*Rouen* being to *Paris*," Sir William Petty tells us, "as that part of London which is below the bridge is to what is above it."

32. The sending of fishing ships to Newfoundland from this point began in 1561. See François Henry, "Fécamp, port de pêche," *Annales de Géographie*, March 15, 1930, pp. 181–184. As Havre became a commercial port Fécamp specialized in the fishery. M. A. Hérubel, *Pêches maritimes* (Paris, [1911]), p. 179.

33. For example, G. and N. Bongardz, merchants, lent 1,370 livres, or 770 and 600 livres respectively, at dates between January 8 and February 24. N. Bongardz and C. Bourdyneau lent 100 and 200 livres respectively to N. Daussi of St. Valery en Caux, master and *bourgeois* of the *Nicolas*, 80 tons. Generally loans were made through a large capitalist more conveniently located for the supervising of such ship business. N. Bongardz lent C. Cordier of Havre 500 livres and G. Bongardz 470 livres. Cordier was in turn *bourgeois* and "victuailler pour un tiers" of the *Barbe*, 80 tons, *bourgeois* "pour la totalité" of the *Don de Dieu*, 120 tons, and *bourgeois* "et victuailleur pour un quart" of the *Quanette Nefve*. He was chiefly interested in the large vessel, and interested in part in two smaller ships. Smaller loans were made to individuals as masters of single ships. G. Martel of Honfleur, master and *bourgeois* "pour la moitié," of the *Salamandre*, 160 tons, J. Maubert of Jumièges, master and *bourgeois* "pour un quart" of the *Vallentine*, 80 tons, and Pierre de la Fenestre of Bliquetuit, master and *bourgeois* "pour un quart" of the *Loise*, 140 tons, were each lent 100 livres by G. Bongardz. The loans in each case were "to purchase salt and other necessaries," and were apparently guaranteed by a lien on the ship with an interest charge of 35 per cent. A further illustration was the loan of 400 livres by François de St. Mesmin, merchant, to Nicolas Besnard *l'aîné*, *bourgeois* "pour la totalité" of the *Cerf-Volant*, 80 tons and *bourgeois* of the *Licorne*, 60 tons; and to Thomas Gueroult, *bourgeois* "pour la moitié" of the *Andrieu*, 50 tons, and of the *Marie*, 60 tons. Two *bourgeois*, of two small vessels each, joined to obtain a large loan from a merchant of Orleans. From February 1 to March 3, 1564, Pierre Lailler, merchant, lent 1,160 livres in eleven transactions involving fifteen vessels. On February 1, G. Atinguetz of Havre, master and *bourgeois* "pour deux tiers" of the *Esmerillon*, 55 tons, borrowed 100 livres; on February 10, C. Cordier of Havre, already heavily indebted to G. Bongardz, borrowed 150 livres; five days later, Pierre de la Fenestre (who, as we have seen, borrowed 100 livres from G. Bongardz) and J. Le-

The risks that the borrower ran made it necessary for him to depend on a large number of lenders.[34] The amount of capital required to finance a fishing vessel varied.[35] When the needed funds were secured the vessels were outfitted and left for such places as the Bay of Croisic in Brittany to obtain salt en route for the Banks. On their return, fish were discharged at Havre, Honfleur, and other points. The wide distribution of small loans, the high rate of interest, and the financial structure by which dependence was placed primarily on those directly concerned with the industry were possibly a result of heavy losses but more probably of the overwhelming importance of the vessel as a unit. The master was in many cases the *bourgeois*, but there were several cases in which the *bourgeois* financed the master. In some instances the

fevre, master, *bourgeois* and "victuailleur pour un quart" of the *Grace*, 80 tons, also of Bliquetuit, jointly borrowed 100 livres at 35 per cent; and on the following day J. Maubert (who had had a loan of 100 livres from Bongardz) borrowed 50 livres. A large loan of 300 livres was made on February 22 to N. Delisle, master and *bourgeois* "pour un quart" of the *Ysabeau,* 120 tons, to G. Rogniez, master and *bourgeois* "pour un quart" of the *Salamandre,* 100 tons, and to Olivier Delisle, master and *bourgeois* for three quarters of the *Petite Normande,* 60 tons, all of Vatteville. Four days later J. Anet, master and *bourgeois* "pour un demi-quart" of the *Rubis,* 120 tons, and J. Fichet, master and *bourgeois* "pour un demi-quart" of the *St. Jean,* 140 tons, both of Jumièges, borrowed 100 and 60 livres respectively at 35 and 33 per cent. A day later Jacques Duyn of the same place, master and *bourgeois* "pour deux quarts" of the *Louise,* 120 tons, and *bourgeois* for "un seizième" of the *Vincente,* 75 tons, Nicolas Duyn master, borrowed 100 livres. Early in March loans of 80 livres, with 28 livres interest, were made to E. Picquefeu, master and *bourgeois* of the *Bellenfant,* and to Jean Galanard of Caudebec, master and *bourgeois* of one half of the ship; 90 livres with 31 livres, 10 sols interest to R. Clerel of Jumièges, master and *bourgeois* of a quarter of the *Barbe,* 80 tons; and 30 livres with 10 livres interest to J. Vauquelin of Lenduit near Jumièges, master and owner of a quarter interest in the *Bonne Aventure,* 70 tons.

34. For example, J. Vauquelin, having borrowed 30 livres from P. Lailler, obtained on his new ship the *Bonne Aventure,* 70 tons, another 30 livres from Thomas Fontayne of Rouen; with his father, M. Vauquelin, he obtained 100 livres from Thierry Voisin, merchant of Rouen, and 165 livres with interest, raising the sum to 214 livres, from Mahier Hue, merchant. In addition to borrowing 325 livres he sold "un seizième du corps dudit navire" to Pierre Fossart.

35. On February 14, Etienne Canelet, merchant of Rouen, and master and *bourgeois* for one half of the *Geneviève,* 90 tons, purchased the other half from Cardin Chevremont, "moyennant" 300 livres tournois, involving a total outlay by the *bourgeois* of 600 livres tournois. On February 22, G. Clerice of Havre, master and *bourgeois* for three quarters of the *Licorne,* 90 tons, sold half the "victuailles et la moitié d'avant des compaignons dudit navire moyennant" 430 livres tournois (sold half the provisions and also parted with half his interest in the advance money made to the *compaignons* for, in all, a total of 430 livres tournois) to Pierre Lefevre and Jean Baudouin, merchant of Rouen. It is probable that for a ship of such a tonnage the interest of the *victuailleur* would be 600 livres tournois and the total advances to the *compaignons* 260 livres. On March 22 Clerice sold to Jean Beaudouyn "un quart des victuailles." On March 16 M. Fontaine of Fécamp sold "un quart des victuailles" of the *Georges* "moyennant" about 250 livres which would give the cost of provisions as 1,000 livres.

"victuailleur" is mentioned as well as the master and the *bourgeois*. The division of risk among *bourgeois*, master, *victuailleur*, and many small borrowers was characteristic of the industry. In France as in England it demanded the initiative of large numbers carrying on the industry from small vessels. The demands were reflected in aggressive commercialism in contrast with large-scale company organizations. It opened a breach in mercantilistic control which was progressively widened.

A list of ships registered at Honfleur[36] from 1574 to 1583 inclusive suggests more clearly in the dates of the majority of the documents—dates which run from January to March—the increasing importance of the Banks. During the above ten-year period references were made to about one hundred expeditions, the tonnage of the ships used varying from 60 or 70 to 100, 120, and 150. Allowing for similarities in names and a consequent difficulty of identification, about 50 ships were engaged in the fishery and about the same number of captains. The largest number of vessels recorded in one year was 17 in 1580, and the smallest number 3 in 1575. One of the captains, R. Baril, is recorded as having made 8 voyages from 1576 to 1583, 4 voyages in the *Esperance* and 4 in the *Michelle*, both vessels of 120 tons. Two captains, Jehan Poesson and Jehan Geffroy, made 6 voyages each, and 2 others made 5 voyages each. Seven captains made 3 voyages; 10 captains, 2; and 27, one voyage each. The ships showed even more conspicuous signs of specialization, 3 vessels making 6 voyages each; 6, 4 voyages; 6, 3 voyages; 13, 2 voyages; and 22, one voyage. Again, allowing for difficulties with names, the 2 ships making 6 voyages each had the same captain on each voyage. Four ships had the same captain 4 times; 5 ships the same captain 3 times; and 9 ships the same captain twice. Three of the captains went out 4 times in the same ship, 4 captains went 3 times, and 9 captains twice. One item would indicate 2 trips to the Banks or a second trip to the coast of Newfoundland. The *Esperance*, Nicolas Morin captain, referred to in February, 1582, is again referred to in the sale of a one-eighth share for 16 écus, 40 sols to François Péronne, *bourgeois*, on April 22, 1582.

As in the earlier period, the captains were largely a party to the outfitting of the vessels. A captain of a vessel may appear as a *bourgeois* or as a *bourgeois* "pour un quart" or "pour un seiziesme," or as borrowing 16 crowns (écus) to go to Newfoundland, or as selling a "demi-quart de son navire" (an eighth interest in his ship) in order to make the voyage. A company was formed in Rouen in 1570 to develop the

36. Bréard, *op. cit.*, pp. 53 ff.; also N. E. Dionne, *La Nouvelle France de Cartier à Champlain 1540–1603* (Quebec, 1891), pp. 291 ff.

fishery or perhaps to lend money to fishermen.[37] In February, 1577, Thomas Legendre, a merchant of Rouen associated with Fernand de Quintanadoine, sieur de Brétigny, supplied Jehan Morin, captain of the *Catherine*, with 60 cahizes of Spanish salt to go to the fishery in Newfoundland in return for a sixth share in all the proceeds of the voyage—"tout le rapport du navire." In February, 1582, Legendre lent 100 écus to Mathieu Le Tellier, captain of the *Saint Jehan*, at 35 per cent, and on March 21, 1582, he lent Guillaume Tuvache, *bourgeois* for an eighth of the *Jacques*, 33 écus for the Newfoundland voyage. It would appear that by 1582, Legendre's group were becoming more interested in the purely financial side of the expeditions. In January, 1576, two *bourgeois* are named for a 70-ton boat; in January, 1577, at least four *bourgeois* are named for the *Charles*. On February 2, 1580, Robert Godeffroy is given a loan of 100 écus at 40 per cent by "noble homme" Jacques de Courseulle on three vessels going to the fishery. On March 30, 1590, Jehan Courel, captain of the *Marie*, borrowed "16 escuz deux tiers" (16⅔ écus) at 35 per cent to repair his vessel. The rate of interest varied from 35 to 45 per cent, possibly because of a lack of insurance facilities. It appears probable that the captain of the ship was in a strong financial position, but that in many cases he relied for support on the *bourgeois* or on direct loans from financiers. A large proportion of the loans were, as has been said, "à la grosse aventure" (subject to every risk), with the vessel as security.

At La Rochelle the interests supporting the dry fishery began to be prominent, especially with the expansion of the French Basque fishery and the importance of the Spanish market. In 1556 the *Baptiste* of 100 tons had brought home 40,000 dry cod, 60,000 green cod, 20 barrels of whale oil, and 12 barrels of cod oil, the total being valued at 4,325 livres. In 1582 La Rochelle had about 10 ships which went to Newfoundland for dry cod, each ship carrying from 8 to 12 shallops. They all unloaded at La Rochelle. And in addition this port still had some 14 ships which went to the Banks.[38] In 1606 Lescarbot wrote, "When we were leaving La Rochelle there was a very forest of ships at Chef la Baie. . . . They set sail two days before us (May 11) and departed on the same tack for Newfoundland."

Bayonne continued to rely on La Rochelle for financial support. On March 26, 1556, B. Deperelongue of Bayonne, *bourgeois* of the *Saint*

37. Bréard, *op. cit.*, p. 51.

38. "Il y avait encore à La Rochelle environ 10 navires allant à Terre Neuve aux morues seches, de 12 a 8 chaloups chacun, qui toutes dechargeaient à La Rochelle et qu'il y avait en outre 14 navires pour le banc de Terre Neuve." Georges Musset, *Les Rochelais à Terre-Neuve 1500–1550* (Paris, 1893).

Esprit of Bayonne, 80 tons, and Etienne Depayne of Cap Breton borrowed 200 livres from André Blouet, a merchant of La Rochelle, to send a ship to fish for cod at Newfoundland. On October 7 of the same year two men of La Rochelle, Jehan Disnemartin and Pierre Johanneau, chose Janycot de Montguiel of Cap Breton as master of the *Jehan* of La Rochelle and outfitted the ship to fish for cod; and on December 17 Basset, a baker and *bourgeois* of La Rochelle, sold J. Raymond 50 quintals of biscuits for a voyage to Newfoundland. On April 1 Bertrand and Saulvat Debahongnes, respectively *bourgeois* and master of the *St. Esprit* of Cap Breton, 100 tons, along with Jacques le Roy of La Rochelle, borrowed 200 livres from Claude Furgon and Joseph Barbier to engage in the whale fishery. On April 10 Barbier made an additional loan.

Bordeaux occupied a position minor but similar to that of La Rochelle in financing the fishery. In March, 1552, the *Saint Esprit*, 140 tons and 40 men, of St. Jean de Luz, purchased provisions at Bordeaux on this basis: for the boat one quarter, for the men[39] one third divided into 34½ lots, and for the supplier the remainder. One of the two partners of St. Jean de Luz who owned the boat borrowed various sums, and by April 22 completed the equipping of the vessel with guns, arms, and ammunition, besides six shallops and a "batteau," or small boat. The outfit provided by the supplier included a ton of powder, 20 tons of wine, 120 quintals of biscuits, 10 of bacon, 2½ of olive oil, 22 barrels of vinegar, 120 pounds of candles, one barrel of beans, and 2 barrels of peas. In addition, he paid one half the cost of bringing the men to Bordeaux. In that year the take of a St. Jean de Luz vessel was valued respectively as follows: large green cod, 12 livres tournois a hundred; average green, 6 livres; small green, 3 livres; and "dressed cod also 3 livres [tournois] a hundred." In the same year (1552) six vessels from St. Jean de Luz, one from Bayonne, one from La Rochelle, and one from Olonne, for the Newfoundland whale fishing, were reported at Bordeaux. These vessels varied in value from about 400 livres tournois to 1,380. The loans carried from 30 to 45 per cent interest.[40] The cargoes were sold at La Rochelle or Bordeaux.

39. One third was the share for the men engaged in the dry fishery on vessels from Bordeaux; but in the case of Basque vessels there was a division on the basis of the value of the cargo of the ship. Nicolas Denys, *The Description and Natural History of the Coasts of North America* (Champlain Society, Toronto, 1908), p. 271.

40. The high rate was attributed to the war with Spain. Théophile Malvezin, *Histoire du commerce de Bordeaux* (Bordeaux, 1892), II, 164 ff. Two vessels were cited at Bordeaux in 1564 and 1565 as borrowing, respectively, 200 livres and 100 livres at 30 per cent, and 100 livres at 27½ per cent, to go to Newfoundland. "There goeth out of Fraunce commonly five hundreth saile of shippes yearely in Marche to Neuvefoundlande, to fishe for Newlande fish and comes home again in August. Amongest

French trade with Spain increased after the Peace of 1559, and especially as a result of the later hostilities between Spain and England. In 1577 the profits from the Newfoundland fishery as conducted from St. Jean de Luz, Zibura, Azcavin, and Urnia were estimated at 60,000 ducats. In 1584 it was reported that 50 large ships brought cargoes of fish and oil valued at over 140,000 ducats to San Sebastian, Bilbao, and Castrio. French Basque vessels became exposed to English attacks. A petition dated April 18, 1591, from Bayonne to the Privy Council of England asked for the release of a ship from Newfoundland with 108,000 dry fish, 4,000 green, and 14 hogsheads of train oil; and in September, 1591, English vessels were reported as having captured a ship from St. Jean de Luz that had a cargo of 15,000 dry fish and 60 hogsheads of train oil. The French Basque ports appear to have been largely concerned with the dry fishery, chiefly because of the nearness of the growing Spanish market and the decline of the Spanish fishery itself. By 1600 Newfoundland fish were selling in Bayonne at 14 to 18 royals (réals).

The diversity of markets for cod meant that the scattered French ports were interested in widely separated fishing grounds. Between 1600 and 1610 Rouen vessels were engaged in the fishery at Renewse, and St. Malo vessels at St. John's. One of Sir Humphrey Gilbert's ships seized two French vessels, one loaded with wines and the other with salt. There appear to have been no French ships at St. John's at the time of his visit although they were engaged in fishing in adjoining harbors. They were found along the south coast of Newfoundland, around Cape Breton, in the Gulf of St. Lawrence, and along the north shore of Newfoundland as far as Labrador. Fish "parés" (dressed) were sold at La Rochelle in 1596 at 4 sols if from the north and 5 sols if from the south. French Basque whalers had penetrated the Gulf from the south and left their imprint in the Basque place names of the western coast of Newfoundland.[41] In 1564 St. Jean de Luz, Cap Breton, and Biarritz vessels counted on obtaining, at Placentia, salt which had been brought from Spain. Sixteen vessels left Cap Breton, most of them before the beginning of February. Survivors of Gilbert's ship, the *Delight*, wrecked off Cape Breton, were rescued by Frenchmen on the south coast of Newfoundland and returned on a vessel sailing to St. Jean de Luz. Near the end of the century a vessel from St. Jean de Luz was reported in the Bay of Great St. Lawrence, and another from La Rochelle was seen at

many of them, this is the order, tenne or twelve marryners doeth conferre with a money man, who furnisheth them with money to buy shippes victualls, salte, lines and hookes, to be paied his money at the shippes returne either in fishe or in money with five and thirtie pounde upon the hundreth pounde in money lent." Hitchcock, *op. cit.*

41. De la Roncière, *op. cit.*, III, 315.

St. Marys Bay. In 1594 two ships from Sibiburo were reported fishing for cod off the Island of St. Pierre, and "3 score and odd sailes from Saint John de Luz, Sibiburo and Biscay" were seen at Placentia. Of these only eight were Spanish ships. Apparently the *Bonaventure* from St. Malo was the first French vessel to go to the Magdalen Islands,[42] where in 1591 it was successful in killing 1,500 walrus. The English captured a ship carrying 40 tons of train oil. In 1593 a ship from St. Malo, three-fourths loaded, was found by them in the same locality. In 1597 ships from Sibiburo and a large ship of 300 tons, a Biscay vessel from St. Vincent, were reported at ports in Cape Breton. At Louisburg English vessels found two ships from La Rochelle and two from Biscay, and at the Magdalen Islands two from St. Malo and two from Sibiburo. Lescarbot and Champlain told of meeting a Basque named Savalet, from St. Jean de Luz, at White Haven near Canso in August, 1607. He claimed that this was his forty-second voyage to those waters. By the end of the century most of the Basque fishermen were apparently French, and they had successfully established themselves in the dry fishery, walrus hunting, and whaling.

The rise of the English fishery in Newfoundland tended to force the French[43] to outlying areas, particularly in the Gulf and on the mainland; and the financial organization of the Channel ports gave support to an expansion to regions capable of meeting the demands of the Spanish market for dried fish, especially in the Mediterranean. Financiers such as Pierre de Chauvin[44] of Honfleur quickly appreciated the possi-

42. *Idem*, p. 304.

43. An attack by French vessels on St. John's in 1596 inflicted heavy damage upon the English.

44. Chauvin received, on November 3, 1596, 108 écus, 33 sols tournois from J. Mauduict on a ship the *Poste* engaged in the cod fishery. On March 2, 1597, Chauvin lent 8 "escuz" at 35 per cent to M. Faride to outfit a vessel under G. Duglas as master to go to the Canadas. The loan was repaid on October 13, 1597. Ships cited as going to Terre-Neuve in March, 1597, were the *Cygne*, master G. Premort; the *Catherine*, master R. Poesson; the *Perle*, master N. Tuvache; the *Francoise*, 80 tons; the *Georges*, captain S. Couillard; the *Fauron;* the *Isabeau*, 80 tons, captain N. Missent; the *Esperance*, captain S. Morin; and the *Bon Espoir*, captain G. Caresme. The following season the captain of the *Perle* planned, on December 20, 1597, to go to Spain for salt, and thence to the fishery; but changes were made and the *Perle* left for Peru in the following March. For January, February, and March, 1600, "armement [equipment] pour le voiage des Terres neuves pesches des morues des navires" included eleven ships, four of which had been noted in 1597, ranging in size from 60 to 150 tons. On September 30 the *Marguerite* was listed as having made two voyages. In November loans were made "subject to every risk" on a cod-fishing voyage to Newfoundland and the coast of Canada, to G. Le Chevallier, captain and *bourgeois* "au corps total du navire le *Don-de-Dieu*," 80 tons, by F. De Sarcilly and to four other ships; and between December 15 and 24 to two of these ships and four more. On January 19, 1601, M. de Sarcilly lent 50 écus at 35 per cent to G. Chefdhostel (master and owner of a half interest, and victualer of the ship *Jehan*) and other sums to ten other ships.

bilities of the new areas. Interest in Canada on the part of capitalists coincided with the ambitions of Troilus du Mesgouez,[45] Marquis de la Roche. Privileges granted in commissions dated March, 1577, and January, 1578, resulted in unsuccessful expeditions in 1578 and 1584, the latter having been arranged with the support of St. Malo and St. Jean de Luz. A commission dated February 16, 1597, was followed by an arrangement (March 4, 1597) with Thomas Chefdhostel[46] of Vatteville in which the *Catherine,* of 180 tons, was employed to take an expedition to Sable Island to engage in the fishery and in the capture of prizes. With new letters patent of January, 1598, an expedition in two vessels, the *Catherine* and the *Francoise,* both of 90 tons, was sent to establish a more permanent settlement on Sable Island, and apparently it maintained itself, with some help from France, for five years. Meanwhile, with the assistance of Pont-Gravé of St. Malo, Chauvin secured commissions dated November 22, 1599, and January 15, 1600, by which he became a lieutenant of De la Roche especially as applying to certain territory "dans cent lieues seulement dans La Baye au long de la rivière vers Cadossart ou Tadossart" (extending only one hundred leagues in the Gulf and along the river toward Tadoussac). Under this commission a post was established at Tadoussac in 1599.

The French pursued their advantages in the bank fishery. Lescarbot reported near the end of the century that some men came to the Banks as early as February. The French "settinge furthe in January, broughte their bancke fishe, which they tooke on the bancke forty or three score leagues from Newefoundelande, to Roan [Rouen] in greate quantitie by the ende of May and afterwarde returned this yere again to the fisshinge and are looked for at home towardes the fifte of November."[47] Whitbourne at almost the same date stated that about 100 French vessels went to the Banks, some of them making two voyages to them and a third elsewhere. "I have often . . . met French ships comming . . . deepe loden with fish" about the first of April that had gone out to fish in January, February, and March.[48] The fish were larger, thicker, and sweeter than on the Newfoundland coast, and, in France, Portugal, and Spain "that kind of fish usually sell at a great price." They were taken on the Grand Bank by Normans from Honfleur, Dieppe, Boulogne, and Calais, and by fishermen from Brittany, Olonne,[49] and the country about Aulais. A 100-ton ship had a crew of

45. See Gustave Lanctot, "L'Etablissement du Marquis de la Roche à l'Ile de Sable," *Canadian Historical Association Report* (Ottawa, 1933), pp. 33–42.

46. Bréard, *op. cit.,* pp. 75–78; also Innis, *op. cit.,* pp. 14–15.

47. *Voyages of the English Nation to America before the Year 1600,* II, 223.

48. Whitbourne, *op. cit.,* p. 97.

49. "The men of Olonne came here [to Louisburg] in old times to winter in order

15 to 18 men and provisions for six months. The vessels loaded their salt at Brouage, Oleron, Ile de Ré, or in Brittany, where it was purchased at 10 to 12 livres per hogshead of about 31 bushels. Salt made up "almost all the rest of that which the ship can carry." Each man[50] had from 8 to 12 lines and a larger number of hooks, 12 to 15 leads of 6 pounds each, and knives for heading, opening, and splitting the cod. Outboard staging was built along one side of the vessel, on which each fisherman placed a half hogshead reaching to the waist. A large leather apron from the neck to the knees projected over the edge of the half hogshead. The line was attached to it and the hook allowed to drop to about a fathom from the bottom and the lead to about two fathoms. With two lines, one cod was pulled up while the other was being put over the side. The tongue was taken out and kept as a means of counting the number caught by each man. Pieces of herring or cod entrails were used as bait. The catch might vary from nothing to between 25 and 200, or, exceptionally, 350 or 400 a day, about the limit of a fisherman's capacity. Boys took the fish to those who dressed them.[51] The salter made a layer on the bottom of the hold arranging head by tail, and covered the layer with salt. Successive layers were laid above it and similarly salted, until the end of the day. After the fish had been left for three or four days and the water had drained away, the surplus salt was removed and new layers were put down and covered with new salt as before. The last treatment was regarded as final and the fish were ready to be taken to France. With the approach of Lent, ships with as little as half or two thirds of a cargo set out with the fish for the Paris market, for the first arrivals got the best prices for the new cod. They might, by arriving early, return for a second voyage and still be in time for the Lenten sale. The bank fishery was usually completed by the end of May. Denys estimated that at least three quarters and sometimes

to be first upon the Grand Banc for the fishery of green cod and to be first back to France because the fish is sold much better when first brought in." Denys, *op. cit.*, p. 181. In 1596 it was stated that "les Olonnais ont envoyé sur le banc de Terre-Neuve, pour faire la pêche de la morue, cent navires ou environ, montés par plus de 1,500 hommes, marins du dit lieu, des Sables, d'Olonne et de la Chauline (the Olonnais have sent to the Newfoundland bank fishery a hundred ships or thereabouts, their crews being made up of more than 1,500 men, sailors of local origin from the Sables, Olonne, and Chauline).

50. Lescarbot gives an average crew of fishermen as some fifteen or twenty, each equipped with a line of 40 to 50 fathoms carrying a hook and a three-pound lead. Lescarbot, *op. cit.*

51. According to Lescarbot the fish were placed on narrow tables, where one man cut off the head and threw it into the sea, a second cut open the bellies and disemboweled them, and a third cut out the backbone. The fish were put in a salting tub for twenty hours and then packed away. The men worked for three months or until they got a full load or, failing that, proceeded to better fishing grounds.

practically all of the catch was for Paris consumption. A ship of 100 tons would bring back from 20,000 to 25,000 fish; and a fisherman could make from 35 to 40 écus. To carry on the bank fishery with an eye to the Paris market during the Lenten season, ships had to start very much earlier. In many cases two trips were made to the Banks, and a third to the coastal areas.

In the first half of the sixteenth century, the fishing fleets in the New World had belonged chiefly to France and to a lesser extent to Portugal. In France, the industry had first been centered in Channel ports which sold what they caught to the Paris markets. Later the industry had extended itself to La Rochelle and the Biscay ports. Toward the end of the period the growing market of the Channel area had made possible another extension of the fishery, and had sent large vessels to the Banks. Vessels from Rouen and other ports of Normandy concentrated on the southern coast of Newfoundland. Brittany ports such as St. Malo were engaged in the fishery of Petit Nord and along the mainland of the Labrador, as well as on the south coast of Newfoundland and Cape Breton. La Rochelle gave strong financial support to the vessels from adjacent ports and, since the local demand was not great, also gave its support to the production of a surplus of dry fish for the English market.

In the second half of the century the Spanish fishery had become important and involved competition first with the French and then with the English. The fishing industry in England had begun to shift its activities from the east-coast ports with their interest in Iceland to the West Country ports which were concentrating upon Newfoundland. In 1580 Denmark enforced the paying of license fees in the Iceland fishery, and accelerated a migration of the English fishermen to the new grounds. This probably necessitated a change in their technique. "Englishmen have not the use of barrelling up of cod, and if it be not barrelled it is not vendible in France, neither can they make haberdine, for if they could it would be well sold in Spain and Portugal."[52] What they had learned from their experience with Iceland stockfish, and also learned when limited to Newfoundland's smaller fish and scant supplies of salt, could be of value in meeting the demands of warm countries. It helped England to produce a hard, dry cure for the Mediterranean. English vessels had obtained salt from the Portuguese in Newfoundland and in alliance with them had forced the smaller French vessels to betake themselves to regions less favorable. During the war with Spain, which for the time included Portugal, England relied on France for sup-

52. Henry Harrisse, *Découverte et évolution cartographique de Terre-Neuve* (Paris, 1900), pp. xii–xiii.

plies of salt. Finally, with the decline of Spain, London ships carried fish produced by West Country fishermen from Newfoundland to the Mediterranean. With this development the struggle between the carrying trade allied with London and corporate interests and the fishing ships became more active. In the case of France, ships from her Channel ports and those from La Rochelle and the Bay of Biscay had the support of separate financial organizations. When they were forced to leave the Avalon Peninsula, the ships belonging to the French Basque ports moved to waters off the mainland—to the Canso regions, for example. As for Channel-port ships, they moved to the Gulf, particularly Gaspé; and there they produced dry fish for direct export to Spanish and Mediterranean markets.

The larger size of the English ships, the centering of England's fishing interests in her West Country ports, and their concentration upon the dry fishery with its smaller demand for salt, as also their looking to foreign countries both for their markets and for supplies of solar salt, were outstanding factors in the competition which grew up between England and France. The French fishery possessed an extensive home market, and in particular a demand for the large fish produced and salted green on the Banks. Its fishing ships came from many diverse ports which were interested in both the dry and green fisheries. It was a fishery which had that flexibility of organization incidental to smaller ships and varied technique. The bank fishery had become important to the French Channel ports and the demands of the Paris market; and the dry-fishing industry had expanded, particularly as carried on by the fishermen of the Biscay ports for the Spanish markets. With the decline of the Spanish fishery, the market for dry fish grew in Spain and the Mediterranean, and consequently the French Channel ports and the English fishery both set themselves to produce it. West Country fishermen went to Newfoundland, cured the fish, and at first brought the product home to England for domestic consumption. Later they exported it from England. And, finally, ships from London went to Newfoundland and purchased fish to be carried directly to the Mediterranean. France brought back the heavier-salted green fish, dried it at home, and thereby produced a poorer cure.* At St. Malo, fish brought back in August and September were spread out over large areas in the vicinity of the town. For France, cheaper supplies of salt, a large home market, and the distance between the various important centers of the industry—that is, Rouen, the Channel ports, La Rochelle, and ports on the Bay of Biscay—all this made it the more necessary for the French fishery to depend upon the Banks and on the more distant home areas for the drying process.

*See Notes to Revised Edition, p. 509.

England's relative scarcity of salt, her more limited home market, and her concentration of fishing interests in the West Country forced her to depend on areas suited to drying and adjacent to the fishing grounds. She acquired a foothold on Newfoundland for the dry fishery, while France had less interest in acquiring and occupying the land for such a purpose. France might expand to Gaspé and extend her interests in the fur trade of the St. Lawrence valley. But by the end of the century England was established on the Avalon Peninsula. The expansion of the markets for light-salted dry fish enabled her to increase her shipping, her direct trade in fish, and her indirect trade, by using ships released in the long closed season. Dried cod was food[53] of excellent keeping qualities, which gave it exceptional value for sailors on ships that went south of the line.* Ships and sailors[54] were essential to the fishery. Moreover, it meant an expansion of trade, if only because the English fishery depended on solar salt produced in southern continental countries. The demand for cod became a demand for English ships and men. Directly and indirectly the fishing industry hastened the growth of shipping and trade which followed the decline of Spain and Portugal, and the opening of markets beyond the powers of France to occupy. The contribution of the industry to flexibility of organization apparent in the inroads on company trade, and in a lowering of interest rates through the development of insurance, became more conspicuous in the next century. The rise of prices in Spain, following the inflow of specie from the New World, had its part in the consolidation of the West Country fishing industry in Newfoundland and its expansion to New England.

53. See the provisions for Frobisher's expedition in 1577 which included 2½ tons of stockfish. Innis, *op. cit.,* p. 12.

54. Lounsbury, *op. cit.,* chap. i; Judah, *op. cit.,* chap. ii.

CHAPTER IV

THE STRUGGLE AGAINST MONOPOLY 1600–1650

NEWFOUNDLAND

Ships called sacks being commonly in great number every year . . . carry fish from Newfoundland into the Straits, France, Portugal and Spain and . . . bring in their return into England bullion and other native commodities of those countries. . . . The trade of fishing upon that coast is of so great concernment to France, Spain, Portugal, the Straits and other parts that they cannot well have or be without that yearly supply which they receive in fish which comes from that island. Neither can the Hollands, Spaniards or Portugals well get any ship to the Indies without Newfoundland fish, there being no fish that will endure to pass the line sound and untainted but the fish of that country salted and dried there. . . . Above 200 sail do trade there yearly there to fish.

SIR DAVID KIRKE, *c.* 1650

THE growth of the power of Holland after 1581 and her importance in the North Sea fishery accentuated the development of the West Country fishery interests in Newfoundland.[1] A three-cornered trade from England to Newfoundland, Spain, and the Mediterranean provided a basis for expansion, and gave England an industry with an abundance of shipping, an outlet for manufactured goods and provisions, a supply of semitropical products and specie, substantial profits, and ideal possibilities for the development of a mercantile policy. England was able, in part because of her relatively shorter distance from Newfoundland and in part because of the nature of fish as a foodstuff, to secure a strong and continuous hold on a product by which she obtained a share of Spanish specie and the products of the Mediterranean. Cod from Newfoundland was the lever by which she wrested her share of the riches of the New World from Spain.

1. For an account of the efforts of Great Britain to develop the North Sea fishery see J. R. Elder, *The Royal Fishery Companies of the Seventeenth Century* (Glasgow, 1912); also "John Keymor's Observations made upon the Dutch Fishing about the Year 1601," *Phenix Reprints* (2d ed., 1721). On the decline of Spanish shipping partly as a result of the improved technique in Dutch shipping, see A. P. Usher, "Spanish Ships and Shipping in the Sixteenth and Seventeenth Centuries," *Facts and Factors in Economic History* (Cambridge, 1932). Dutch ships apparently proceeded directly to Newfoundland as early as 1593, and English ships began to make direct voyages from Newfoundland to Spain, certainly by 1604. R. G. Lounsbury, *The British Fishery at Newfoundland, 1634–1763* (New Haven, 1934).

The expansion of the fishery gave rise to a conflict which characterized the history of the fishing industry. It was limited to definite seasons and was prosecuted in favorable sites along the coast with a relatively small amount of fixed capital. The increase in the size of the English fishing fleet led to problems of space. Sir Thomas Hampshire attempted to solve these problems when he went to Newfoundland in 1582 by decreeing that "whatever room or space of foreshore a master of a vessel selected he could retain it so long as he kept up his buildings on it and employed it for the use of the fishing."[2] Sir Humphrey Gilbert, in the following year,

> granted in fee farme divers parcels of land lying by the waterside both in this harbour of St. John and elsewhere . . . of grounds convenient to dresse and to drie their fish whereof many times before they did faile, being prevented by them that came first into the harbour. . . . After this divers Englishmen made sute unto Sir Humphrey to have of him by inheritance their accustomed stages, standings and drying places in sundry places of that land for their fish as a thing they doe make great accompt of, which he granted unto them in fee farme.[3]

These rights were eagerly sought for and later refused,[4] ostensibly because Sir Humphrey Gilbert believed that such land might contain minerals, but probably because of the protests of West Country fishing ships. The struggle of the fishing ships against government interference began at an early date.

The rise in value of the fixed equipment consequent upon the scarcity of lumber, the abuses which characterized seasonal occupancy, the advantages to be gained by monopoly control over favorable sites, and the character of the three-cornered trade were factors responsible for the struggle between fishermen and settlers. The number of men employed on fishing expeditions to Newfoundland was in excess of the number necessary to man the ship, Whitbourne's estimate being ten extra per vessel. Trade with Spain and the Mediterranean involved either carrying the extra crew to these markets at an added expense, leaving them in Newfoundland until the following year, or employing ships to carry the fish to such markets and return directly to the West Country. Whitbourne stated that the number of ships in Newfoundland for the purpose of purchasing fish was not less than 40 in some years. According to Lewes Roberts (born in 1596), who had been in Newfoundland "in my younger days," about 500 vessels "great and small"

2. D. W. Prowse, *A History of Newfoundland* (London, 1896), pp. 60–61.
3. *Voyages of the English Nation to America before the Year 1600,* ed. Edmund Goldsmid (Edinburgh, 1889), I, 338, 369; also Lounsbury, *op. cit.,* pp. 26–27.
4. *Voyages of the English Nation to America before the Year 1600,* I, 353.

left England for Newfoundland annually, sailing about the end of April and returning in September. "And in this time," Roberts wrote, "[they] doe not onely catch as many fish as will lade their shippes but also as many as will lade vessels of greater burthens that in the summer come hither from England and other parts to buy up the same and purposely to transport it for Spaine, Italy and other countries."[5] The sack ships became an integral part of the trade,[6] and fish were sold from Newfoundland rather than on return to the West Country. Such ships made for an increase in the production of fish in Newfoundland and an increase in settlements,[7] whereas the West Country fishing ships were competitors in the carrying trade and objected to settlements as an infringement of their rights on the coast.

The struggle between the trading and carrying interests and the fishing interests became acute when the former attempted to establish settlements under the charter (May 2, 1610)[8] of the London and Bristol Company. The fishermen were elaborately protected by instructions, plans, and, in particular, by the following clause:

> Nevertheles our will and pleasure is, and we doe by theis p'sentes [presents] exp'sse and declare, that there be saved and reserved unto all manner of p'sons of what nation soever, and alsoe to all and everie our loveing subjects, which doe att this p'sent or hereafter shall trade or voiag to the parts aforesaid for fishing, all and singuler liberties, powers easementes and all other benefitt whatsoever, as well concerning their saide ffishing as all other circumstances and incidentes thereunto in as large and ample manner as they have heretofore used and enjoyed the same without anye ympeachment, disturbance, or exaccion, any thing in theis p'sentes to the contrarie nothwithstanding.[9]

5. Lewes Roberts, *The Merchants Mappe of Commerce* (London, 1638). It was customary "to sell the said *fish* either by *tale* or by the *hundredweight* in *England* by *contract* before they depart their homes, or before the said fish be caught, at profitable rates." The "price of fish once generally cut [set] at their fishing stales [stalls?] doth afterward, in lieu of coine by way of commutation, all that yeare passe currant for all needful commodities." See Adam Smith, *Wealth of Nations* (New York, 1937), p. 23. See p. 61 *infra.*

6. Sack—*sec,* dry, from the French—was used in referring to white wines from Spain and the Canaries, and it is probable that this return cargo gave the name to these ships.

7. Mason states that planters could be brought out at 10 shillings a man, and 20 shillings "to find him victuals." The freight rates on commodities sent out "by shippes that goe sackes" were 10 shillings a ton outward-bound and 30 shillings homeward.

8. See *The Cambridge History of the British Empire,* I, 90.

9. *In the Privy Council, In the Matter of the Boundary between the Dominion of Canada and the Colony of Newfoundland in the Labrador Peninsula,* Volume IV of Joint Appendix, p. 1703. Hereafter the matter in the citation above will be abbreviated to *P.C.* The legal jargon quoted reminds one forcibly of certain lines by Adam Smith. "It has been the custom in modern Europe to regulate, upon most occasions,

The instructions given in 1610 to John Guy, who was placed in charge of operations, provided for the exporting of "twelve months victualles with munition, nets and with all manner of tooles and implements," as well as domestic animals.[10] And lest these measures should raise apprehensions, this was added:

> Upon your first arrival there the sooner to operate our patent and to prevent ye murmuring of suspicious and jealous persons that perhaps will not [fail] to spread abroad that this enterprize wilbe to the prejudice of ye fishermen as well of our nation as others, we do hould it expedient that you call an assembly of all the fishermen that shall be nere thereabouts, and there in presence openlie and distinctlie cause to be read the graunt under the King's Majesties great seal which you shall have along with you, that by the tenour of it they may be satisfied that there is no intent of depriving them of their former right of fishing.[11]

The company planned at the end of the fishing season to send the fish over and above their own needs on the ships returning to England, and Guy was advised to purchase cod oil if it was obtainable at £8 a ton, and to send it to Bristol or to keep it in warehouses until a ship was sent to get it in January or February. Similarly, "a shipps lading of masts sparres and deal boards" was to be prepared to load any ship sent out with salt

> which you shall unloade and lay it in your warehouses to be readie there for our use to be used in fishing or to be sold to ye fishermen. By employing of shipping of great burden the trade between Bristoll and Newfoundland may be profitable. . . . If you can buy there 60,000 of good dry fish reasonable you may likewise do it and charge us home by exchange and place it in our warehouses until we send a bark thither to take it in and to go with it there home to Spain, which coming there alone may sell better than that which came first, the great glut maring oftentimes that market.[12]

The company expected to sell pine boards for making or mending fish-

the payment of the attornies and clerks of court, according to the number of pages which they had occasion to write; the court, however, requiring that each page should contain so many lines, and each line so many words. In order to increase their payment, the attornies and clerks have contrived to multiply words beyond all necessity, to the corruption of the law language of, I believe, every court of justice in Europe. A like temptation might perhaps occasion a like corruption in the form of law proceedings." Adam Smith, *op. cit.*, p. 680.

10. *Idem*, pp. 1712–1713. See also J. W. Damer Powell, "John Guy, Founder of Newfoundland," *United Empire*, June, 1933, pp. 323–327; and "The Explorations of John Guy in Newfoundland (1612)," *Geographical Journal*, December, 1935, pp. 512–518; and C. M. Andrews, *The Colonial Period of American History* (New Haven, 1934), Vol. I, chap. xv.

11. *P.C.*, IV, 1712.

12. *Idem*, p. 1713

ing boats and it was hoped that timber could be found for the manufacture of hoops and staves.

"Certain orders for the fishermen to observe and keep in the Newfoundland," published on August 13, 1611, on the other hand, involved conflict with the fishermen. "Ballast, press stones or anything else hurtful to the harbours" were not to be thrown out in the harbor but were to be carried ashore. No one was allowed to "destroy deface or any way work any spoil or detriment to any stage, cook room, flakes, spikes, nails or anything else that belongeth to the said stages," and they were to use only the stages they needed and to repair them with timber fetched from the woods and "not with the taking down of other stages." The "admiral" of each harbor was allowed only the beach and flakes needed for the number of boats "that he shall use with an overplus, only for one boat more than he hath," and everyone coming afterward was allowed only what he had use for. Changing the marks on the boats was prohibited. No one was allowed to use the boats of others except in case of necessity and after notifying the "admiral." No one was to destroy at the end of the voyage the stage, cookroom, or flakes he had used during the year. No one was allowed to set fire to the woods. All of these prohibitions carried penalties for their violation.[13]

Whitbourne, who was favorable to the establishing of settlements, undertook an investigation in 1615, "under the broade seale of the Admiralty," with "a barke victualled and manned at my owne expense." He found general disobedience. "It is well knowne that they which adventure to *New-found-land* a-fishing begin to dresse and provide their ships ready commonly in the moneths of December, January and February and are ready to set foorth at sea in those voyages neere the end of February, being commonly the foulest time in the yeere." Every captain made all speed to reach the Newfoundland harbor first, which would make him that year's "admiral." For the "admirals" had "the chiefest place to make their fish on where they may doe it with the greatest ease and have the choyce of divers other necessaries in the harbors which do them little stead; but taking of them wrongs many others . . . which arrive there after the first." He said that the men fished with hook and line on Sundays; that they dumped very large stones, used to press dry fish, into the harbors, thus spoiling the anchorage and endangering the ships and cables; and that the harbor at Renewse was in danger of being ruined.

Many men yeerely . . . unlawfully convey away other men's fishing boats from the harbour and place where they were left the yeere before; and some

13. *Idem,* pp. 1715–1716.

cut out the markes of them; and some others rippe and carry away the pieces of them, to the great prejudice and hindrance of the voyages of such ships that depend on such fishing boates and also to the true owners of such boats. . . . Some arriving there first, rippe and pull downe stages for the splitting and salting of fish . . . other stages [are] set on fire. . . . Some who arriving first . . . take away other men's salt . . . left the yeere before, rip and spoile the vats. . . . Some men likewise steale away the bait out of other men's nets by night and also out of their fishing boates by their ships side, whereby their fishing . . . is overthrowne for the next day. . . . Some men take up more room than they need or is fitting to dry their fish on. . . . Some men rip and take away timber and rayles from stages and other necessary roomes . . . fastened with nailes, spike or trey naile, and some take away the rindes and turfe [used as roofing]. . . . Some yeerely take away other men's trayne oyle there by night.[14]

Late arrivals were sometimes occupied for twenty days securing boards and timbers to fit boats for fishing and rooms for salting and drying fish, all of which involved a heavy expense in provisions. It necessitated an early start and delay after arrival, as well as the danger of complete loss in ice and bad weather. Whitbourne argued in favor of opening Newfoundland to settlement not only to check irregularities but also as an important halfway station to the West Indies, Bermuda, Virginia, or New England. "In the yeare 1615 . . . three ships returning from the West Indies did arrive there, purposely to refresh themselves with water, wood, fish and fowle and so have divers others done at other times to my knowledge." Newfoundland would also serve as an important base in case of war with France, Portugal, or Spain; and the expansion of the fishery would increase shipping, mariners, and wealth, besides providing relief for unemployment and making it unnecessary to rely on the Dutch for supplies. He elaborated on the advantages due to the expansion, which would follow the development of plantations and the suppression thereby of the difficulties with the fishing ships which he had described. Taking a 100-ton ship with 40 men, he estimated the cost of provisioning and outfitting her for a year to be £420 1*s*. 4*d*.[15] With 8 three-man fishing boats catching an average of 25,000

14. Richard Whitbourne, *A Discourse and Discovery of Newfoundland* (London, 1622), pp. 21 ff., 64 ff.; see also Anthony Parkhurst's memorandum of about 1578. Lounsbury, *op. cit.*, pp. 24–25.

15. The 40 men included 24 fishermen, 7 skilled headers and splitters, 2 boys to lay the fish on the table, 3 to salt fish, 3 to pitch salt on land and to wash and dry fish.

The outfit and provisions comprised, according to Whitbourne:

"11,000 wt [weight] bisket bred 15/ [shillings] per hundred	£ 82. 10. 0
26 tun of beere and sider 53/4 the tun	69. 6. 8
2 hhds of very good English beef	10. 0. 0

(*continued on next page*)

15 (*continued*).

2 hhds. of Irish beef	5. 0. 0
10 fat hogs salted, caske and salt	10. 10. 0
30 bus. peas	6. 0. 0
2 firkins butter	2. 10. 0
2 cwt. cheese	0. 6. 0
one bus. mustard seed	1. 5. 0
1 hhd vinegar	1. 0. 0
wood to dresse meate withall	2. 0. 0
one great copper kettle	2. 0. 0
2 small kettles	2. 0. 0
2 frying pans	0. 3. 4
Platters, ladles and cans for beere	1. 0. 0
1 pr. bellowes for the cooke	0. 2. 0
Locks for bread roomes	0. 2. 6
Tap, boriers and funnels	0. 2. 0
1 cwt candles	2. 10. 0
130 quarters of salt at 2/ the bus. (15 gals to a bus.) is 16/ the quarter	104. 0. 0
Mats and dynnage to lye under the salt	2. 10. 0
Salt shovels	0. 10. 0
More for repairing 8 fishing boats, 500 ft. of elme boords of 1" thickness at 8/ the hundred	2. 0. 0
2000 nails for said boats and stages at 13/4 the thousand	1. 6. 8
4000 nails at 6/8 per 1000	1. 6. 8
2000 nails 5 d per hundred	0. 8. 0
5 cwt. pitch	2. 0. 0
1 barrel of tar	0. 10. 0
2 cwt. of clacke ocome [oakum]	1. 0. 0
Thrummes for Pitch mabs [mops]	0. 1. 6
bolles, buckets and funnels	1. 0. 0
2 brazen crocks	2. 0. 0
canvas to make boat sails and small ropes fitting for them at 25/ each sail	12. 10. 0
10 boats, anchors, ropes	10. 0. 0
12 dz. fishing lines	6. 0. 0
24 dz. fishing hooks	2. 0. 0
squid hooks and squid line	0. 5. 0
for pots and liver mands	0. 18. 0
Iron works for 10 fishing boats	2. 0. 0
10 keipnet irons	0. 10. 0
Twine to make keipnets etc.	0. 6. 0
10 good nets at 26/ a net	13. 0. 0
2 saines, a greater and a less	12. 0. 0
2 cwt. lead	1. 0. 0
small ropes for seines	1. 0. 0
dry-vats for nets and seines	0. 6. 0
flaskets and bread boxes	0. 15. 0
twine for store	0. 5. 0
so much hair cloth as may cost	10. 0. 0
3-tun vineger caske for water	1. 6. 8
2 barrels otemeal	1. 6. 0
1 doz. deale boords	0. 10. 0
1 cwt. spikes	2. 5. 0
heading and splitting knives	1. 5. 0
2 good axes, 4 hand hatchets, 4 short wood hooks	

(*continued on next page*)

fish per boat,[16] the season's total catch would be some 200,000 cod. Assuming this to be a load for a 100-ton ship, allowance being made for the needed space for water, wood, victuals, and provisions for the men, the fish could be taken directly to Marseilles or Toulon "where the customes upon fish are but little and the kentall lesse than 90 pounds. . . . Such fish . . . I have not knowne to be sold for lesse than twelve shillings every kentall and commonly a farre greater price, and there speedy sales are usually made . . . and good returns had: and if any man will returne his money from thence hee may have sure bills of exchange, for payment thereof heere in London, upon sight of such bills." Two hundred thousand fish sold at Marseilles, Whitbourne calculated, would weigh more than 2,200 "kentalls," i.e., quintals, which at 12 shillings a quintal would bring £1,320; or, at 16 shillings, £1,760. If the ship had been chartered by the month she could be discharged at Marseilles, or chartered anew to go on to Spain for a cargo for England, and get to England in time for the next year's voyage. In addition to the fish he put down in his estimate 12 tons of cod oil which could be sold in Newfoundland at £10 a ton, or £120 in all. In England it would sell at £18 or £20. To this he added 10,000 large green fish. They sold in New-

15 (*continued*).

2 drawing irons, 2 adizes	0. 16. 0
3 yds. good woolle cloth	0. 10. 0
8 yds. good canvas	0. 10. 0
a grinding stone or two	0. 9. 0
an iron pitch pot and hooks	0. 6. 0
1500 dry fish to sped thitherward	6. 1. 0
1 hhd. aqua vitae	4. 0. 0
2000 good Orlop nails	2. 5. 0
4 arm saws, 4 hand saws, 4 thwart saws	
3 augers, 2 crowes of iron, 2 sledges, 4 iron shovels	
2 pick axes, 4 mattocks, 4 cloe hammers	5. 0. 0
other necessaries	3. 0. 4

"If 10 men winter they will require of the above, 5 cwt. biscuit bread, 5 hhds. beere or cider, ½ hhd. beef, 4 whole sides bacon, 4 bush. pease, ½ firkin butter, ½ cwt. cheese, 1 pecke mustard seed, 1 barrel vinegar, 12 lbs. candles, 2 pecks oatmeal, ½ hhd. aqua vitae, 2 copper kettles, 1 brasse crock, 1 frying pan, 1 grind stone and all axes, hatchets, wood hooks, augers, saws, crowes of iron, sledges, hammers, mattocks, pickaxes, shovels, drawing irons, splitting knives, hair cloth, pinnaces sails, pinnace anchor, ropes, seine, some nets, the 8 fishing boats and their iron works . . . also pikes, nails etc. to build houses." Whitbourne, *op. cit.*, pp. 81 ff.

16. "Three men to sea in a boate with some on shoare to dresse and dry them in 30 dayes will kill commonlie betwixt 25 and thirty thousand, worth with oyle arising from them 100 or 120 pound." John Mason, *A Briefe Discourse of the New-found-land*, Prince Society Publications (Boston, 1887). Whitbourne estimated that three men would take 1,200 fish a day, reckoning six score to the hundred. Each fish with its oil might be valued at a penny, and the total, when split, salted, and dried, worth £6 sterling. "A single man may take in that employment (hooke and line) above forty shillings of fish per day." William Vaughan, *The Golden Fleece* (London, 1626).

foundland at £5 a thousand, but in England they would bring twice that, and make the grand total £2,250. Of this the master and the ship's company were entitled to one third, allowing a "small" sum for victualing, ship's expenses, and charges. The owners were entitled to one third after deducting for the master's allowance and for bonuses, over and above their share in the first third, to those men who had proven themselves much better fishermen than the others; and the victualers were entitled to one third. This, according to Whitbourne's estimate, would leave £750, after deducting £420 1*s.* 4*d.* for supplying the 40 men, or "a profit of £331 11s." (*sic*). During years of high prices for salt, bread, and beer the victualers were allowed one half and the crew and the ship one half. A ship chartered by the month for nine months at £40 a month or less would cost accordingly.

Salt was brought to England from Spain, Portugal, or France, and the ships leaving for Newfoundland "each yeare take neere halfe their lading of salt," or not less than 7,000 tons. This, purchased at 20 shillings a ton, with an additional 20 shillings a ton for freight, involved an outlay of £14,000. In England, supplies[17] such as nets, leads, hooks, lines, bread, beer, beef, and pork represented the work of "great numbers of people [such] as bakers, brewers, coopers, ship carpenters, smiths, net makers, rope makers, line makers, hooke makers, pully makers and many other trades which with their families have their best means of maintenance." Those engaged in financing the fishery were in a position to hire ships to load cod and take it to France or Spain; but the difficulties involved in the rigid terms of the charter party, in which the merchant must "discharge and also relade" though the commodities were "much dearer than at some other place not farre from it," made it advisable to purchase outright. Whitbourne suggests that a ship of 100 tons should be purchased, and that two small barques of 30 tons each and two fishermen's boats for each should be hired by the month. After

17. "From this island our English transport [fish] worth £20,000. . . . First this trade of fishing multiplyeth *shipping* and mariners, the principal proppes of this Kingdome. It yearely maintaineth 8,000 persons for 6 months in Newfoundland. . . . It releeves after their returne home with the labour of their hands yearely their wives and children, and many thousand families within this kingdom besides, which adventured with them or were employed in preparing of nets, caskes, victualls, etc. or in repayring of ships for that voyage." Vaughan, *op. cit.* "But of all other plantations that of Newfound Land may deserve to be furthered . . . for the more secure and commodious prosecution of our fishing trade on those banks in which imployment a dozen of men only in a few moneth's time are able to improve their labour to farre greater advantage, than by a whole year's toile in tilling of the ground or any handicrafts mysterie whatsoever." Henry Robinson, *England's Safety in Trades Increase* (London, 1641), p. 13. "Of all fabricks a ship is the most excellent, requiring more art in building, rigging, sayling, trimming, defending, and moaring." *Travels and Works of Captain John Smith, ed. Edward Arber* (Edinburgh, 1910), II, 950.

acquiring the ship it was important to get as master a good man who understood fishing and who would hire the best fishermen. The barques should sail earlier than the large ship, which would leave about the end of March. The barques could sell to merchants 100,000 or more fish to be delivered in Newfoundland, and in return receive money in London on bills of exchange. Finally one barque might load with train oil and the other with green fish, and proceed to the most promising market. If the ship were chartered, fish could be contracted for in England:

> They are to bargaine for their fish heere in England with such as doe set forth ships in the fishing trade which fish may be bought beforehand of them, to be delivered there at eight shillings the hundred-waight, or neere that price and to pay for the same within 40 days more or lesse after such times heere in England, that there comes from thence the sight of any bills of exchange from those that receive the fish there in that maner; and the ship so hired being there loaden may saile from thence unto France, Portugall, Spaine, or any other port within the straights of Gibraltar. I suppose the fraight of every tun of fish so to be transported there will be neere foure pound the tun . . . which freight and hire for the ship men and victuals in all that time, it may be agreed to bee paid there where the fish is sold, so that for the hire of the ship, men and victuals there will be no occasion to disburse any money before the ship safely arrive to either of the places aforesaid, where, by Gods assistance, anyone shall so intend to make sales.

The fish could be sold in Marseilles, Spain, Portugal, the Biscay ports, in "Nance [Nantes], Bordeaux, Rochell, Bayonne, Rouen," or in the British Isles.

The fishermen opposed these elaborate arguments[18] for the establishing of settlements as a means of forwarding the expansion of the fishery. The western ports in December, 1618, launched an attack upon the settlers or "planters" brought out by the company, charging that they usurped "the chiefest places of ffishinge there" and disposed of the same to such as pleased them, that they took salt, casks, boats, stages, and other provisions left by the petitioners, that they prohibited the use of birds taken on Baccalieu Island for bait before the ordinary bait came on the coast, that they encouraged pirates, and that they summoned a court of admiralty "in the chiefest tyme of ffishinge" and "exacted ffees of trayne and ffishe for not apperinge."[19] These charges were in the main denied but it was conceded that, "in regard of theire chargeable maintenance of a colonie on land, there all the yeare, it is conceivable to be lawful for them the inhabitants to make choice of their fishing place and not to leave the benefit thereof to the uncertayne commers thither."

18. Vaughan, *op. cit., passim.*
19. *P.C.*, IV, 1717–1718; but especially Prowse, *op. cit.*, p. 100.

The fishing interests replied that "no privilege [is] given by charter to planters for fishing before others; if choice of places is admitted, contrary to common usage, the peticioners contend that they ought rather to have it. [They] desire that the liberties reserved to them by charter may be confirmed."

The general confusion arising from the struggle encouraged piracy.[20] In 1612[21] a pirate named Easton took four ships from the Isle of May and later five more ships from other parts, about one hundred pieces of ordnance, "victualles and munition," while from the English he took as much as £20,400, as well as 500 fishermen "taken from their honest trade of fishing." Two years later Sir Henry Mainwaring arrived with "8 sails of warlike ships," one of which had been taken from the Banks and another from the coast, and exacted from the fishing fleet a sixth of its men and "the one first [?] part of all their victuals." On September 14 he seized almost 400 sailors and fishermen. In 1616 pirates took the ordnance from two ships, one from Bristol and the other from Guernsey. Part of Sir Walter Raleigh's fleet on its return from Orinoco in 1618 taxed fishermen in all the harbors of Newfoundland for powder, shot, and the like to a value of £2,000 besides "one hundred and thirty men they took away." In 1619 the Mayor of Poole demanded the suppression of piracy in Newfoundland.[22]

Vaughan claimed that plantations would check piracy both in Newfoundland and in the Mediterranean. "Those petty merchants, which were led with desire of gaine, not willing to enranke themselves into an orderly societie" might be saved by "a couple of good ships on the charge of the fishermen, which yeerely frequented that coast, continually to assist them against the invasions of pirats, who had a few years before pillaged them to the damage of fortie thousand pounds, besides a hundred peeces of ordnance, and had taken away above fifteene hundred mariners to the great hindrance of navigation and terrour of the planters." A check would also be imposed on "the misgoverned and stragling courses of the Westerne merchants which either of foolehardinesse, carelessenesse, or of a griping humour to save a little charge, adventured in their returne from Newfoundland, without fleets, or wafters to guard them, or any politicke order to passe through the straights of Gibraltar to the Dominions of the King of Spaine, to Marseilles, or Italy, where yeerly they met with Moorish pirates."[23]

20. Lounsbury, *op. cit.*, pp. 41–42.

21. Prowse, *op. cit.*, pp. 102–103.

22. Henry Harrisse, *Découverte et évolution cartographique de Terre-Neuve* (Paris, 1900), p. xii. M. Oppenheim, *History of the Administration of the Royal Navy* (London, 1896), p. 199.

23. Vaughan, *op. cit.* It was claimed that the company lost £40,000 as a result of piracy, and that Algerian vessels captured 466 English ships between 1609 and 1616;

It was also claimed that settlements would stop the destructive activities of the fishing ships. Captain Edward Wynne wrote from Ferryland on August 17, 1622: "For there hath been rinded this year not so few as 50,000 trees and they heave out ballast in the harbours though I look on." At the same time, one "N.H." wrote describing a fire which "began between Fermouse and Aquaforte, [and] it burned a week." Daniel Powell, in a letter dated July 28, 1622, wrote: "But the woods along the coasts are so spoyled by the fishermen that it is a great pity to behold them, and without redresse undoubtedly [it] will be the ruine of this good land. For they wastfully barke, fell and leave more wood behinde them to rot then they use about their stages although they imploy a world of wood upon them."[24] The best trees were cut down to build stages and rooms, i.e., sheds for various purposes, and many others were destroyed by cutting and barking, the bark with turf laid above it being used as indicated to cover the stages and roof, with the result that within a mile of the sea the woods were destroyed. The list of damages in 1620 included

> eight stages in several harbours worth at least in labour and cost £180 maliciously burned by certain English fishers, besides many more in the harbours of the country, greatly to the prejudice of fishing trade. . . . Great damage done by certain English fishers to a saw mill and a grist mill built by the plantacion, not to be repaired for forty pounds. The woods daily spoiled by fishers in taking the rind and bark of the trees, and 5000 acres of wood burned maliciously by the fishers in the bay of Conception anno 1619 with many more thousands of acres burned and destroyed by them within these 20 years. Harbours frequented by English near 40 in number almost spoiled by casting out their balast, and presse stones into them. Portugals, French and all other nations frequenting that trade are more conformable to good orders than the English fishers.[25]

The company had failed to form settlements in the face of such confusion. The first establishment at Cupid's Cove, Conception Bay, was followed by an offshoot at Harbor Grace, but continued difficulties led to the disposal of large tracts in grants.[26] Trepassey was settled by Sir

in October, 1617, 30 Turkish frigates captured 7 English fishing ships on their way to Italy. See Astrid Friis, *Alderman Cockayne's Project and the Cloth Trade* (London, 1927), pp. 150, 183. A supporter of the home fishery stated that "Newfoundland employeth some 150 saile, from all parts, of small ships, but with a great hazard, and therefore [they] that voyage feared to be spoiled by heathen and savage as also by pirates. . . . New found land may breed and employ some fifteene hundred: but seeing what discouragements they have, what casualties they are subject to, we may judge of their incertainty." [Robert Keale(?)], *The Trades Increase* (London, 1615).

24. Prowse, *op. cit.*, p. 130.

25. *Idem*, p. 103.

26. *Idem*, pp. 110 ff.; also Lounsbury, *op. cit.*, pp. 46–48; also J. D. Rogers, *Newfoundland* (Oxford, 1911), pp. 56 ff.

William Vaughan in 1617, Ferryland was established under Lord Baltimore[27] in 1621, and the area between Renewse and Aquaforte was granted to Lord Falkland but never developed. Lord Baltimore was held to have had thirty-two boats in the fishery at Ferryland,[28] but he probably lost money. John Mason succeeded John Guy in 1615 but returned to England in 1621, where he was attracted to the possibilities of New England in 1622. In 1632 he became treasurer of an association organized to compete with the Dutch in Scottish waters.

West Country merchants not only carried on a struggle against monopoly control through the fishing ships in Newfoundland but also engaged in a struggle against centralized control in England. Expansion by the sea to Newfoundland contributed to the breakdown of control over external development by company organization. The control of external trade by chartered-company organization had reached the peak of effectiveness at the end of the sixteenth century,[29] and the end of the war with Spain in 1604 brought about a renewal of the struggle of the outports, particularly those of the West Country,[30] against London. The Spanish charter of 1577 was confirmed in 1604, and, although of a more democratic character than the original, was the object of a determined attack. West Country demands for freedom of trade in cloth were supported by those of small traders in grain and, particularly, in fish. In 1604 bills for free trade, supported by four representatives from Devon and Cornwall and by others led by Sir Edwin Sandys, formerly a representative of Plympton in Devonshire, constituted one evidence of the attitude of the West Country. Another, in 1605, was an attack on the Spanish charter by Sir George Somers of Lyme Regis. Fishermen were sending fish direct from Newfoundland to Spain, Portugal, and Italy and at prices less by one half than those obtainable by

27. Prowse, *op. cit.*, pp. 128 ff. A patent to Avalon granted to Baltimore on April 7, 1623, uses the words "saving always and every [*sic*] reserved unto all our subjects free liberty of fishing as well in the sea as in the ports of the province and the priveledge of drying and salting their fish as heretofore they have reasonably enjoyed." *Idem*, p. 132.

28. L. D. Scisco, "Testimony Taken in Newfoundland in 1652," *Canadian Historical Review*, September, 1928, pp. 239–251.

29. As to the attacks of the outports upon the Merchant Adventurers, see John Wheeler, *A Treatise of Commerce*, 1601, ed. G. B. Hotchkiss (New York, 1931), p. 61.

30. The Merchant Adventurers had little trade with Spain and France "by reason that our English merchants have had a great trade in France and Spain and so serve England directly from thence with the commodities of these countries," Wheeler, *op. cit.*, p. 23. See V. M. Shillington and A. B. Chapman, *The Commercial Relations of England and Portugal* (London, 1907), I, chaps. ii, iii; and Friis, *op. cit.*, Part II, chap. iii, especially for an illuminating account of the constitutional problem, and particularly considering the influence of the West Country in the House of Commons and its ability to exercise its influence on the Privy Council as contrasted with the eastern ports.

the merchants. Even the company conceded that "if the fishermen persist in trading themselves, at least they make their homeward lading with money, salt or oranges and not in frivolous commodities."[31] Legislation in 1606 granted liberty of trade with Spain, Portugal, and France. Attempts of merchants trading with Spain to secure incorporation in 1619 and 1620 were defeated in turn by western ports[32] which drove "a greate, and ample trade into the dominions" of Spain. In 1609 the West Country insisted on a democratic charter for trade with France: that all fishermen should "have liberty to transport and carry their fish into France and the Dominions thereof and to return their monies in any kind of merchandise and bring over the same"[33] to England. Correspondingly, the Levant Company encountered difficulties when it attempted to prevent West Country merchants from returning with Levant goods after disposing of their fish in Italy. A navigation proclamation of 1622 which imposed mercantilistic restrictions on shipping was held responsible for a sharp increase in the cost of shipbuilding materials, and concessions were made by the Privy Council which required Eastland (Baltic) commodities to be sold at West Country ports at the same price as in London. The influence of the West Country was also evident in customs regulations.[34] A decline in the cloth trade was offset by the increasing importance of exports of Newfound-

31. Shillington and Chapman, *op. cit.*, II, 44.

32. The traders in Devonshire cloth aided in opposing the charter. "The fishing trade . . . hath a reference to the want of money, or to speak ingeniously, is a chiefe cause of the want of money which might be procured thereby, whereby both the trade of cloth and fishing might flourish together contrary to the opinion of the severall societies of merchants before alleadged; for although they be of severall companies yet such orders may be designed by the corporation to be made of fishing and shall not infringe their severall priviledges anyway." Gerrard Malynes, *The Maintenance of Free Trade* (London, 1622), pp. 100–101. "There hath been a continuall agitation above 30 yeeres to make busses [fishing vessels] and fisher boats but the action is still interrupted because other nations doe finde too great favour and friends here to divert all the good intentions and endevours of such as (with the author of this discourse) have employed their time and good meanes therein; for the merchants adventurours, the companie of merchants trading in Russia; and the East land merchants, did also oppose themselves against it at the Councell table. . . . So that in conclusion England (by their saying) cannot maintaine the sea trade and the land trade." Malynes assailed the "admitting of forraine nations to fish in his Majesties streames and dominions without paying anything for the same, whereby their navigation is wonderfully increased, their mariners multiplied, and their countries enriched with the continuall labour of the people of all sorts both impotent and lame, which are set on work and get their living." *Idem.*

33. Cited by Friis, *op. cit.*, p. 166.

34. *Commons Debates 1621,* ed. Wallace Notestein, F. H. Relf, H. Simpson (New Haven, 1935), II, 490; VII, 258–259. Regulations freed the industry from customs, and from tithes which imposed a burden of 10 shillings a man, whereas the "clear gains" were not above £5 a man. Such customs and tithes had contributed to the difficulties of the fishery.

land fish to Spain. By 1625 the government was forced to yield to the strategic position of the West Country. An attempt of merchants to form a company in 1630 aroused new protests from Dartmouth and Cornwall; and a similar proposal in 1635 brought opposition from Exeter, Plymouth, Dartmouth, Totnes, Barnstaple, Southampton, Poole, Weymouth, and Lyme. The failure of the London and Bristol Company in Newfoundland was followed by the failure of other attempts to control the carrying trade.

The failure of company organization to control West Country trade and the West Country fishing industry led to the substitution of regulations which found an extension in the navigation system. In 1626 an order in council required fish caught by Englishmen in English, Newfoundland, or New England waters to be exported in English ships. Another order in council in 1630 required that this regulation should be enforced; but it was relaxed in 1631 as a result[35] of protests from the West Country and was disregarded in subsequent years. Trinity House complained in 1633 that foreign ships were buying fish in Newfoundland for Spain and Italy while English shipping was idle, although fish was apparently shipped to a greater extent in London vessels. West Country insistence on the importance of competition between London and foreign purchasers and on freedom of the fishery in Newfoundland culminated in 1634 in the granting of the first Western Charter, as it is known, in which freedom of the fishery was guaranteed. The regulations embodied in its letters patent, issued on February 10, 1634, stated that

> our people have many yeares resorted to those partes where, and in the coasts adjoyninge, they imployed themselves in fishing whereby a greate number of or [our] people have been set on worke and the navigation, and mariners of or realme hath been much increased. And or subjects resorting thither . . . and the natives of those parts, were orderlie and gentlie intreated untill of late some of or subjectes of the realme of England plantinge themselves in that country, and there residinge, and inhabitinge, upon conceipt that for wrong or injuries done there, either on the shoare or in the sea adjoyninge they cannot be here impeached . . . by that example or subjectes resortinge thither injure one another and use all manner of excesse, to the greate hinderance of the voyage, and common damage of this realme.[36]

The difficulties involved in a free fishery, as pointed out by the planters, were recognized, however, and the regulations followed closely those published under the hand of John Guy;[37] but they were more stringent

35. Lounsbury, *op. cit.*, chap. ii.
36. *P.C.*, IV, 1719.
37. Similar regulations had apparently also been published in 1621.

in dealing with stealing of bait or other property, the rights of admirals, the disposal of ballast, the protection of stages, and the obliteration of owners' marks on boats. No one was allowed to use the bark of trees "either for the seelinge of shippes houldes, or for roomes on shoare, or for any other uses, except for the coveringe of the roofes for cookeroomes to dress their meate in, and those roomes not to extend above sixteene foote in length at most." Anchoring or otherwise interfering with the hauling "of seanes for baite in places accustomed thereunto" was penalized as was also the stealing of bait from nets or boats; and no person was permitted to set up taverns to sell "wine, beer or stronge waters, cyder or tobacco to entertayne the ffishermen." Provision was likewise made for worship on Sundays. In order that penalties should be speedily enforced, authority was vested in the mayors of Southampton, Weymouth, Melcombe-Regis, Lyme, Plymouth, Dartmouth, Eastlowe, Ffoye (Fowey), and Barnestaple, and in the vice-admirals of the counties of Southampton, Dorset, Devon, and Cornwall.[38]

The Western Charter of 1634 was a compromise with the fishing interests, and it was followed by further efforts on the part of London interests to secure control over the carrying trade and plantations by the grant of a patent in 1637 to Sir David Kirke. West Country fishermen were guaranteed ample supplies of shipping[39] but Kirke was permitted to tax all foreigners "making fish" in Newfoundland five fish out of every hundred, and 5 per cent of the oil. He could demand five fish out of every hundred and twenty in the case of foreign sack ships. The grant protected the fishermen by stating that the patentee

shall not fell, cutt downe, roote up, waste or destroie anie trees or wood whatsoever nor make erecte or builde any house or houses whatsoever or plant or inhabitt within sixe miles of the sea shore of any parte of Newfoundland aforesaide between the Cape de Race . . . and Cape Bonavista . . . save only that the planters and inhabitantes shall have like libertie of fishinge there and takeinge and cuttinge of wood for their use aboute fishinge as other our subjectes have and enjoye, and also shall have full power and libertie to builde any fforte or ffortes att any place or places within the saide lymitt for the defence of the saide countrie and fishing, and shall have and take convenient tymber and wood where it may bee spared with leaste prejudice to the fisheinge for the makeing of such ffortes . . . and also . . . shall not att any tyme hereafter appropriate to themselves or any of them or ppossesse or take upp before the arrivall of the ffishermen aforesaide the beste or moste convenient beaches or places for fishing within the capes aforesaide nor take

38. *P.C.*, IV, 1719–1722; also Prowse, *op. cit.*, pp. 154–155; and C. B. Judah, *The North American Fisheries and British Policy to 1713* (Urbana, 1933), pp. 80–82.
39. Lounsbury, *op. cit.*, pp. 80, 82.

away, burne, spoile, waste or destroie any stages saltboates nettes or any necessaries whatsoever of any ffishermen cominge thither.

The right of free fishing was elaborately protected.

Nowe and for the tyme beeing for ever hereafter [they] shall and may from tyme to tyme and att all tymes for ever hereafter peaceably and quietlie have hould use and enjoye the freedome of fishinge . . . as fullie freelie liberallie effectuallie and beneficially as att any tyme heretofore hath beene used and accustomed with full power and authoritie to goe on the shoare or land in or upon any place of the saide contynent of Newfoundland aforesaide as well for dryeinge, saltinge and husbandinge of their fishe on the shoares thereof, cuttinge off all manner of trees and woods for makeinge of stages shippes and boats, and making all manner of provisions for themselves theire servantes mariners shipps and voyages and for doeinge all other thinges necessaire or usefulle to or for them-selves or theire trade of ffishinge or marchandize as att any tyme heretofore hath beene had used or enjoyed.[40]

Under this patent a settlement was formed at Ferryland in 1638, and Sir David Kirke exacted a yearly license fee from English fishermen in the vicinity. The chief profits arose, however, from the sale of cargoes sent out from London, the purchase of fish, and the export of cargoes of dried cod to southern Europe.[41] But the West Country fishermen complained of the aggressiveness of Kirke and, in 1651, finally succeeded in securing his recall to England.[42] The victory of the fishermen over monopoly was again complete.

The compromise in the Charter of 1634 and the agreements under Kirke's patent of 1637, which emerged from the struggle between the fishing interests, the settlements, and the carrying trade, were supplemented by navigation legislation designed to defeat the Dutch.[43] The

40. *P.C.*, IV, 1723–1737.

41. *The Cambridge History of the British Empire,* VI, 130–131.

42. See L. D. Scisco, "Calvert's Proceedings against Kirke," *Canadian Historical Review,* June, 1927, pp. 132–136. See the petition of merchants, owners of shipping, seamen, and fishermen of Plymouth (March 24, 1646) against the "insolencies and oppressions" of Sir David Kirke, "a notorious malignant, and others the planters in the said Newfoundland." L. F. Stock (ed.), *Proceedings and Debates of the British Parliament Respecting North America* (Washington, 1924), I, 177.

43. "This is their mine and the sea the source of those silvered streames of all their vertue, which hath made them now the very miracle of industrie, the patterne of perfection for these affaires; and the benefit of fishing is that *primum mobile* that turns all their spheres to this height of plentie, strength, honour and admiration." Captain John Smith, *op. cit.,* I, 194. About 1613 Yarmouth sent only two or three ships to Bordeaux and a similar number to Rouen and Nantes whereas the Dutch loaded twelve ships at Yarmouth, chiefly from English merchants, for Leghorn, Genoa and Marseilles. The rise of the Newfoundland fishery enabled the English to develop shipping and, with their cod, to compete in these markets. Tobias Gentleman, Fisherman and Mariner, *Englands Way to Win Wealth, and to Employ Ships and Mariners* (London, 1614). Judah, *op. cit.,* pp. 87–89; Lounsbury, *op. cit.,* pp. 85–86; Violet Bar-

building of faster ships in the latter part of the sixteenth century and the peace with Spain in 1609 had enabled them to dominate Spanish and Mediterranean shipping. In 1612 the pirate Easton captured a great Flemish ship in Newfoundland valued at a thousand pounds. In spite of proclamations in 1615 and 1622 demanding the enforcement of acts requiring all English commodities to be exported directly to the country where they were consumed and foreign commodities to be imported directly from their place of production, Dutch shipping increased. A memorandum of about 1640 protested against the presence in Newfoundland of Dutch traders who purchased fish "in great abundance to the hurt of English merchants, taking the prime of the market." Difficulties led eventually to the Navigation Acts of 1650 and 1651, and the outbreak of war with Holland.

The position of the English fishing ships in Newfoundland was strengthened throughout the first half of the century in spite of numerous vicissitudes.

> There were then [1615] on that coast of your majesties subjects above 250 saile of ships great and small. The burthens and tunnage of them al one with another . . . allowing every ship to bee at least threescore tunne (for as some of them contained lesse, so many of them held more) amounted to more than 15,000 tunnes. Now for every three score tunne burthen, according to the usual manning of ships in those voyages . . . there are to be set downe twenty men and boys, by which computation in 250 saile there were no lesse than five thousand persons. Now everyone of these ships, so neere as I could ghesse had about 120,000 fish and five tun of traine oyle one with another [selling at 12 pounds per ton £15,000] so that the totall of the fish in 250 saile . . . (being sold after the rate of four pound for every thousand of fish, sixe score fishes to the hundred, which is not a penny a fish and if it yield less it was ill sold) amounted in money to 120,000 pound.[44]

According to Mason "the fish and traine in one harbour called Sainct Johns is yearly in the summer worth 17 or 18 thousand pounds."[45]

bour, "Dutch and English Merchant Shipping in the Seventeenth Century," *Economic History Review,* January, 1930, pp. 261–290.

44. Whitbourne, *op. cit.,* pp. 11–12. Other estimates state that the English had 150 ships in Newfoundland in 1615 and 120 in Iceland. Oppenheim, *op. cit.,* p. 200. The merchants of Lyme Regis, in the reign of James I, "being engaged in trade to Newfoundland acquired large fortunes and raised the town considerably." *The Victoria History of the Counties of England: Dorset* (London, 1908), Vol. II.

45. Mason, *op. cit.* Vaughan stated that Devonshire sent 150 ships annually to Newfoundland, that the fishery employed 8,000 people for six months every year, and that 500 to 600 ships "doe yearly resort thither by which means they augment their princes' customes." Including fishing ships and cargo vessels other estimates give England between 300 and 400 vessels. Prowse, *op. cit.,* p. 136; L. D. Scisco, "Kirke's Memorial on Newfoundland," *Canadian Historical Review,* March, 1926, pp. 46–51; *The Cambridge History of the British Empire,* VI, 129.

After living in Newfoundland from 1615 to 1621 he estimated that there were 300 English ships and 3,000 seamen, and 20,000 employed in England. A temporary decline followed the difficulties with pirates and the early settlements, and in 1621 the fishery was apparently a failure. By 1625 piracy had been suppressed and the number of Plymouth and Dartmouth ships sent out increased from 2 or 3 a year to 60 or 80. It was the same with other ports such as Barnstaple and Topsham. In 1634 it was stated that 27,000 tons of shipping and, in spite of an impressment of "splitters" by the navy, 18,680 men were regularly engaged in the Newfoundland fishery,[46] and that it brought in a gross income of £178,880. According to one estimate, the total number of ships had increased to about 500 in 1637. Another outbreak[47] of piracy "whereby Bristol and the western ports that cannot have so great shipping as London, are beaten out of trade and fishing" (1640) may have caused a decline as did the Civil War in England. The total was given as less than 200 sail in 1652.

In the territory between Cape Bonavista and Trepassey, where the English were in sole possession, there were about 500 permanent residents, including 350 women and children, scattered between 30 and 40 settlements of which St. John's, Bonavista, and Conception were the most important. In addition, about 1,000 boatmen and servants were employed for short periods. The settlers took, roughly, one third of the catch, and were outnumbered three to one by English fishermen during the summer. Settlements had developed in spite of the colonization companies rather than because of them. The companies were not a financial success, this being due to the hostility of the fishermen, lack of experience on the part of the company managers, and difficulties in adjusting colonization to the fishing industry. By the middle of the century the dominance of the West Country fishermen had been established.

NEW ENGLAND

Going southward from Newfoundland the English have had a new plantation by the favour of the sea, that yields them great store of better and a larger sort of fish than the other coast does; only it is too thick to dry and therefore not to be vended in the straights or southern-most part of Spain.

SIR WILLIAM MONSON'S *Naval Tracts*

THE rapid growth of the English fishery in Newfoundland and the opening of the Spanish market early in the seventeenth century were

46. The capital employed was estimated at £300,000 yielding 12 per cent. W. R. Scott, *The Constitution and Finance of . . . Joint Stock Companies to 1720* (Cambridge, 1910), II, 317 ff.

47. Stock, *op. cit.,* I, 177. There were 270 ships in 1644, averaging say 80 tons, and

accompanied by an expansion to new areas, and particularly to the rich fishing grounds of the Gulf of Maine and New England. In March, 1602, Bartholomew Gosnold, who made a voyage to New England shores to trade and to search for sassafras, reported that "in the moneths of March, April and May, there is upon this coast, better fishing, and in as great plentie, as in Newfoundland. . . . And, besides, the places . . . were but in seven faddome water and within less than a league of the shore; where, in New-found-land they fish in fortie or fiftie fadome water and farre off."[48] As a result of his favorable reports Martin Pring was sent out by Bristol merchants with two vessels in 1603, and returned with similar reports. "Wee found," he wrote, "an excellent fishing for cod which are better than those of New-found-land and withall we saw good and rockie ground fit to drie them upon."[49] George Waymouth, after a visit to the Maine coast, wrote, in his published account, of prospects

in a short voyage with good fishers to make a more profitable returne from hence than from Newfoundland; the fish being so much greater, better fed and abundante with traine. . . . We were so delighted to see them catch so great fish, so fast as the hooke came down, some with playing with the hooke they tooke by the backe, and one of the mates with two hookes at a lead at five draughts together haled up tenne fishes. All were generally very great, some they measured to be five foot long and three foot about.[50]

In the expedition[51] supported by Sir Ferdinando Gorges and Sir

for each 100 tons 50 men and 10 boats. Total, 21,600 tons, 10,800 seamen, and 2,160 boats. Each boat meant, roughly, 5 men and generally from 200 to 300 quintals, which sold at 14 to 16 reals or 7 or 8 shillings per quintal. Prowse, *op. cit.*, p. 190.

48. *Early English and French Voyages, Chiefly from Hakluyt 1534–1658,* ed. H. S. Burrage (New York, 1906), pp. 331–332.

49. *Idem,* pp. 345 ff.

50. *Idem,* pp. 390–391.

51. *Idem,* pp. 399 ff. Smith wrote that whereas "New found land doth yearely fraught neere 800 sayle of ships with a sillie leane skinny poore John and codfish; which at least yearely amounts to 3 or 400,000 pound . . . yet all is so overlaide with fishers as the fishing decayeth and many are constrained to return with a small fraught." Cod was to be found in New England in abundance in March, April, May, and half of June; and "in the end of August, September, October and November you have cod againe to make cor fish or poore John and each hundred is as good as two or three hundred in the Newfound Land: so that halfe the labour in hooking, splitting and turning is saved. And you may have your fish at what market you will before they can have it in Newfoundland, where their fishing is chiefly but in June and July. . . . Your corfish you may in like manner transport, as you see cause, to serve the ports in Portugall, (as Lisbon, Avera, Porta Port,) (or what market you please) before your Ilanders [Icelanders] return. They being tied to the season in the open sea; you having a double season." He estimated that £2,000 would fit out two ships of 100 and 200 tons, respectively. The dry fish, sold in Spain at 10 shillings a quintal—and generally at 15 to 20 shillings on first arrival—would bring £2,000 and in addition the money gained by exchange; the freight of vintage taken home was "cleere gaine . . .

John Popham in 1607–8 cod were caught on the Sable Island Bank "very great and large fyshe, bigger and larger fyshe than that wch coms from the bancke of the New Found land; hear we myght have loaden our shipe in lesse time than a month." Near La Have "we took great stor of cod fyshes the bigeste and largest that I ever saw or any man in our ship." Although the expedition failed to establish a colony, its failure provided the seeds of the fishing industry. "There was no more speech of setling any other plantation in those parts for a long time after: only Sir Francis Popham having the ships and provisions, which remayned of the Company and supplying what was necessary for his purpose, sent divers times to the coasts for trade and fishing."[52] Captain John Smith went with four merchants from London to Monhegan Island in 1614 for whales; but failing in this took furs, train oil, and core fish to England. The best of 7,000 core (green) fish was sold for £5 per hundred and the remainder for between £3 and 50 shillings. The dry fish (40,000) were sold in Malaga at 40 reals (20 shillings) per quintal, 100 fish being held to make 2½ quintals.[53] He met one of Sir Francis Popham's ships "having many yeares used onely that porte." In 1616 Captain John Smith reported four or five sail leaving Plymouth and as many leaving London. Expeditions sent back ships loaded with dry fish for Spain, the Canaries, and Bilbao, and other ships with furs, train oil, and green fish for England. In 1619 a 200-ton ship left Plymouth and earned for each sailor £16 10*s.* for seven months' labor. The following year three ships reported £20 a share.[54] In 1621 the number of ships for New England increased to 10, in 1622 to 37, in 1623 to 40,[55] and in 1624 to 50. Piracy and difficulties be-

with your shippe of a 100 tons of traine and [cod?] oyle . . . besides the bevers and other commodities." "New Englands fishings (arc) neare land where is helpe of wood, water, fruites, fowles, corne or other refreshments needeful." Captain John Smith, *op. cit.*, II, 712 ff.

52. Hakluytus Posthumus, *Purchas His Pilgrims* (Glasgow, 1906), XIX, 270–271. Captain Hobson and Captain Herley were engaged in fishing and capturing Indians in 1611. *Idem*, p. 272.

53. Captain John Smith, *op. cit.*, I, 187 ff.

54. From 1618 to 1620 there were many ships that made good voyages, some of six months, others five. One ship of 200 tons manned by 38 men and boys sold her freight "at the first penny for one and twenty hundred pounds besides her furres." Six or seven more "went out of the West, and some sailors that had but a single share had twenty pounds and at home again in seven months, which was more than such a one would have got in twenty moneths had he gone for wages anywhere." For detailed accounts of expeditions during 1614–20 see Captain John Smith, *op. cit.*, I, 240–242; also *The Cambridge History of the British Empire*, I, 88–89.

55. In 1622: "From the west to fish five and thirty saile, two from London with sixty passengers for them at New-Plymouth, and all made good voyages." In 1623 "went from England onely to fish, five and forty saile and have all made a better voyage than ever." Captain John Smith, *op. cit.*, II, 941–943. Vessels from Weymouth

tween the fishing interests and the company "in regard that the Newfoundland fishing hath fayled of late years" compelled fishermen to turn to New England.

The fishing industry in New England prosecuted by West Country fishermen was, as in Newfoundland, subject to difficulties with monopolies.[56] The struggle moved to the mainland. In 1620 Sir Ferdinando Gorges obtained a charter covering the territory between the parallels of 40° and 48°. There no one was allowed to visit the coast without obtaining a license from the New England Council, and fishermen were forbidden to land or procure wood to build stages on which to dry their fish. Each fishing vessel was required to secure a license.[57] When John Mason returned to England from Newfoundland in 1621 he was consulted by Sir William Alexander regarding the settlement in Nova Scotia and secured grants of land through Gorges in New England. In the struggle of the colonizers with the fishermen, Captain Francis West was dispatched in 1622 to enforce the regulations,[58] but found that "he could doe no good of them; for they were to stronge for him and he found ye fisher men to be stuberne fellows." The following year licenses were granted to five ships from Plymouth, and two from Dartmouth were seized. Gorges' charter was attacked by Sir Edwin Sandys[59] and

were fishing at Casco Bay in 1624. Forty or fifty sail went to New England "from England yeerely, and all that have gone thither have made advantageous voyages." Sir William Alexander, *The Mapp and Description of New England* (London, 1630). "Another ship for all this exclamation of want is returned with 10,000 corfish and fourescore kegs of sturgeon, which they did take and save when the season was neare past and in the very heat of summer yet as good as can be." Captain John Smith, *op. cit.*, II, 956.

56. Dartmouth complained of serious losses as a result of Gorges' restrictions. *Commons Debates 1621*, III, 408, 441–442. See H. L. Osgood, "The Colonial Corporation," *Political Science Quarterly*, XI, 259–278, 502–534; also E. R. Johnson, *The History of Domestic and Foreign Commerce of the United States* (Washington, 1915), I, chap. ii; Raymond McFarland, *A History of the New England Fisheries* (New York, 1911), chaps. ii–iv; W. R. Scott, *op. cit.*, II, 299 ff.

57. It has been stated that they were issued for £6 13*s*. 4*d*.

58. Prowse, *op. cit.*, p. 121; also *The Cambridge History of the British Empire*, I, 147–149.

59. Judah, *op. cit.*, pp. 55–66. In 1625 Sir Edward Coke argued: "Your patent contains many particulars contrary to law and the liberty of the subject; it is a monopoly and the ends of private gain are concealed under color of planting a colony; to prevent our fishermen from visiting the sea coast for fishing is to make a monopoly upon the seas which are wont to be free; if you alone are to pack and dry fish you attempt a monopoly of the wind and sun." Cited in Prowse, *op. cit.*, p. 121. It was urged against the bill "for freer liberty of fishing," that these lands were "not yet annexed to the Crown," and therefore not under the jurisdiction of the Commons. See Thomas Chandler Haliburton, *An Historical and Statistical Account of Nova Scotia* (Halifax, 1829), II, 322–323; also Lounsbury, *op. cit.*, p. 72. As for control over the fishing industry of New England, the demands of Parliament, made in the debates of the House of Commons, beginning in 1621, in opposition to the rights of patents un-

others in support of a bill for "freer liberty of fishing," partly on the ground that expansion and increased sales in Spain would increase imports of specie to England. It was an industry "of 200 ships and 10,000 men, 8,000 marryners, and bringes in great store of bullion from Spaine."[60] "It is a gaine and sewerly retournes money, ex lege Hispania."[61] Sandys took an active part in the Virginia Company. He made plans to produce salt for the fishery which were opposed to Gorges' monopoly, and whereas Sandys formerly had opposed the interests of the West Country fishermen, because of his interest in the Spanish company, he now supported them. By insisting on expanding the New England fishery[62] he brought out the importance of specie imports and the danger that lay in importing tobacco from European countries.

> The fishinge at Monhigen exceedeth New Foundland fishinge caryed into Spayne intercepted by the merchants of France and to the value of £100,000 per annum now brought home in tobacco. . . . The fish is carryed into Spayne where for Royalls of 8 they buy what is allowed for victualls to be exported. This money received for fish and which may come into England is intercepted by merchants of other companies. . . . Money drawne out of Spaine for the overplus of our commodities, increased by a newe fishinge discovered on the north coast of Virgynia [New England] havinge ever constantly beene a £100,000 by yeere, is intercepted, by the way, by merchants tradinge into other countries.[63]

In 1627 tobacco imports from Spain were excluded and a monopoly was conceded to Virginia.

But in spite of determined protests by representatives of the Virginia Company, of Parliament, and of West Country interests, Gorges continued to hold his charter.[64] "Notwithstanding the fishing ships made

der the Crown were a reflection of the growing importance of the West Country with its wool trade, its contributions to shipping, imports of treasure from Spain, and assistance to the navy. The impact of commercial expansion was of far-reaching significance economically, politically, and constitutionally. See Stock, *op. cit.*, I, xi–xii; G. L. Beer, *The Origins of the British Colonial System, 1578–1660* (New York, 1908), pp. 266 ff.

60. *Commons Debates 1621,* IV, 367. 61. *Idem,* III, 81.

62. New England fish were carried to the Straits, Spain, and Italy, "from where they made noe returne but in treasure, for the Levant Company would not suffer them to deale in any commodityes. . . . Nothing was expended in this trade but onely employment of men and salt, whereof ther was hope to be supplyed from Virginia." *Idem,* IV, 255–256. "The trade of fishing is a most beneficiall trade to this realme for the increase of shipping navigation and mariners and the bringing in of bullion and victual to a very great yearly valewe and supplie." May 28, 1624. Stock, *op. cit.,* I, 70. Sir Dudley Diggs, March 17, 1624: "Fishermen are chiefly to be cherished for they bring in muche wealth and carry out nothing." *Idem,* p. 60.

63. *Commons Debates 1621,* II, 139–140.

64. See M. Christy, "Attempts toward Colonization: the Council for New England and the Merchant Ventures of Bristol, 1621–1623," *American Historical Review,* IV,

such good returnes, at last it [New England] was ingrossed by twenty pattenties, that divided my map into twenty parts and cast lots for their shares."[65] Captain John Smith charged that as a result of the impositions of the New England Council the fishery "hath ever since beene little frequented to any purpose." Settlements were gradually established and the fishery developed along lines different from those of Newfoundland. Monopoly restricted West Country fishing in New England, but helped in the settlement of the country and in the local fishery which eventually destroyed monopoly control. It was in part the possibilities of the industry that attracted the Pilgrims to Plymouth. They secured assistance from the fishermen at Monhegan,[66] but the settlements established at Weymouth by Weston, an English fishing merchant, and at Cape Ann by "the Dorchester Company" ended in failure. The experience gained in these failures led to success at Salem after 1628. By 1630 the settlements were growing and were carrying on the fisheries not only there but also in the vicinity of the Island of Monhegan, on the mainland,[67] and eastward to Penobscot Bay. Fishing stations were established at Dorchester, Marblehead, and Scituate by 1633, and the industry[68] had by 1635 begun to recover from such handicaps as the lack of capital, small numbers, the distances between settlements, and the uncertain returns. In 1639 vessels participating in the fishery were made exempt from taxation for seven years. Men were dispatched from Boston to engage in the walrus and seal fishing on Sable Island, and in 1645 vessels went to Bay Bulls in Newfoundland.

The increase of settlements was accompanied by the development of a winter fishery, especially after 1630.

678–702. For an account of the protection of fishing rights in later charters, see Stock, *op. cit.*, III, 423, 425. A clause in Gorges' charter in 1639 provided "liberty of fishing." Andrews, *op. cit.*, I, 325.

65. Captain John Smith, *op. cit.*, II, 892. See J. M. Morse, "Captain John Smith, Marc Lescarbot, and the Division of Land by the Council for New England in 1623," *New England Quarterly*, September, 1935, pp. 399–404.

66. Judah, *op. cit.*, pp. 61–63.

67. Robert Trelawney of Plymouth and others were given a patent to land including Richmond Island in 1631, and John Winter as agent conducted a fishery. Ships were sent out with supplies and provisions and returned with fish to Europe, especially to Bilbao, or in case of an incomplete cargo to Casco Bay or Newfoundland. The station was handicapped by the desertion of fishermen. "This country needs, as the fishinge prooves, good pliable boats masters and good fisher men." "They think they cann do them selves more benyffit to be (masters) of them selves for fishinge or any thinge els which is heare to be donn in the country and for selling their fish at a greater prize." Poorer grades of fish were sent to Boston. Ships also came out, for example from Barnstaple, to sell goods and to purchase and take fish. "The best fishinge is heare in January." See the "Trelawney Papers," *Documentary History of the State of Maine* (Portland, 1884), Vol. III, *passim*.

68. See F. X. Moloney, *The Fur Trade in New England, 1620–1676* (Cambridge, 1931) for general references to the fishery; also Andrews, *op. cit.*, I, 329, 500.

It was the winter fishery that placed on our coasts a class of permanent consumers, and gave to agriculture the possibility of flourishing. The lumber trade marched beside it. In these pursuits, they who tilled the land during the short summer could find profitable employment in the winter on the ocean or in the forest near their homes. The elements for supporting a family were thus united together. It was the winter fishery, prosecuted in boats from the shore, as it usually was, that furnished, not merely a supply of food to the fisherman's family, but an article which was a medium of exchange that was in demand with the traders on land, or the fishing smacks which came in fleets to fill up a cargo, and sure to command goods or money, as their necessities demanded. It secured employment all the year round to the industrious and made a residence profitable. It thus also gave to the industrious the great boon of independence, the foundation of character in the individual, and in the State. Agriculture followed with halting steps where it led the way. There was no crop that the land produced for export, like the tobacco of Virginia or the indigo and sugar of the West Indies; no great prairie range for pasturage of either cattle or sheep. . . . The discovery that the cod approach these shores to spawn in the winter, whilst late in the spring and summer they are found at greater distances from the coast, and notably on Georges, the Grand Banks, Jeffries, etc., completed a fisherman's round, giving him a home fishery for the months when the dangers on the Banks are greatest, and perfecting an economical employment of his time. . . .

The continuous employment a residence on these coasts afforded to the fisherman, gave him great advantages over the European and those who had no winter fishery at their doors, and the fishing population rapidly increased in numbers and prosperity, bringing with it commerce and an agricultural population. Let me be clear, neither Pilgrims nor Puritans were its pioneers; neither the axe, the plough, nor the hoe led it to these shores; neither the devices of the chartered companies nor the commands of royalty. It was the discovery of the winter fishery on its shores that led New England to civilization, and fed alike the churchmen and the strange emigrants who came with the romance of their faith in their hearts, and the *lex talionis* in their souls to persecute because they had been persecuted.[69]

A market for poor grades of fish emerged with the growth of slavery in Virginia and more tropical regions. The monopoly of the English tobacco market which Virginia obtained led to attempts to form plantations in Guiana and the West Indies. The failures of Guiana plantations preceded the success which crowned the efforts of colonists in the West Indies. In 1623 St. Christopher (St. Kitts) was settled, and in 1629 this island and Barbados were exporting tobacco to England.[70] By 1630 colonists had also been settled in Nevis, and by the middle of the

69. C. L. Woodbury, *The Relation of the Fisheries to the Discovery and Settlement of North America* (Boston, 1880), pp. 23–26.

70. Beer, *op. cit.*, chap. iv.

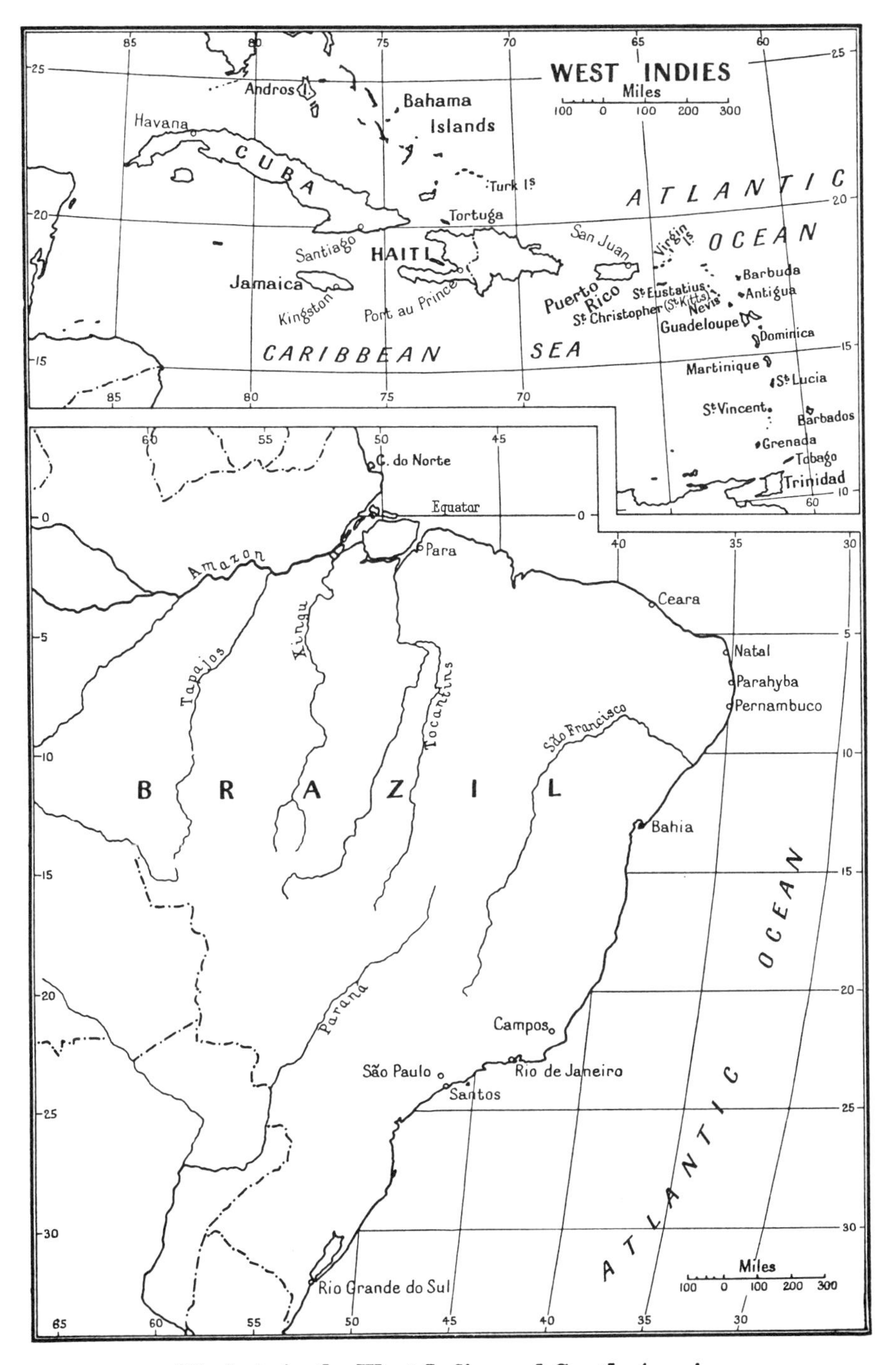

Markets in the West Indies and South America

decade colonies had been planted in Antigua and Montserrat. The West Indies, especially Barbados and St. Christopher, shared with the northern colonies the marked emigration from England to the New World after 1630.[71] The decline of Spain was evident in the foothold which had been obtained by England and other countries in the West Indies. By 1650 England was in possession of several important islands that were chiefly engaged in the production of tobacco.[72]

The disappearance of exports of tobacco from Spain was made good by exports of sugar, both from Spain and from the Azores, Madeira, and Brazil. The separation of Spain and Portugal in 1640 was followed in 1642 by a treaty between England and Portugal in which greater freedom was conceded to England in Africa and the East Indies. English ships became increasingly important in the Brazil trade; and in 1654, after the disruption due to the Civil War, a treaty made possible an English trade with Brazil that included everything but fish, wine, oil, and Brazil wood, which remained under the control of the Brazil Company. The decline of the Spanish and Portuguese fishery in Newfoundland was also accompanied by an increase in exports of English fish not only to Portugal but also to Brazil via Portugal. The increase in the sugar trade with Spain and Portugal in turn led England to encourage the production of sugar in the British West Indies. "Men are so intent upon planting sugar that they had rather buy foode at very deare rates than produce it by labour, so infinite is the profitt of sugar workes after once accomplished."[73]

Not only was there a market in Spain for the product of the winter fishery and a return of specie because of the check to imports of tobacco from Spain, but the production of tobacco and sugar in Virginia and the West Indies also provided a market for poorer grades of fish. Trade began in a small way between New England and Virginia, and from Virginia to Barbados. In 1635 a Dutch ship brought tobacco and 140 tons of salt from St. Christopher to Boston.[74] On February 26, 1638, a vessel which left Salem seven months before brought back cotton, tobacco, and negroes from the Isle of Providence and salt from the Tortugas.[75] On August 27, 1639, a vessel arrived from the West Indies

71. C. S. S. Higham, *The Development of the Leeward Islands under the Restoration 1660–1688* (Cambridge, 1921), chap. viii; A. P. Newton, *The European Nations in the West Indies 1493–1688* (London, 1933), chaps. xi, xiii–xiv.

72. See C. M. MacInnes, *The Early English Tobacco Trade* (London, 1926), chap. vi.

73. Richard Vines to Governor Winthrop (1647). Thomas Hutchinson, *A Collection of Original Papers Relative to the History of Massachusetts Bay* (Boston, 1769), p. 222.

74. *Winthrop's Journal 1630–1649,* ed. J. K. Hosmer (New York, 1908), I, 151.

75. *Idem,* p. 260.

with indigo and sugar and returned with commodities from New England.[76] Trade expanded rapidly and with it shipping and shipbuilding.[77] In 1641 a ship of 300 tons was built at Salem and one of 160 tons at Boston; and in the same year 300,000 dry fish were sent to market.[78] In 1642 a Dutch ship arrived with salt from the West Indies and took on a cargo of planks and pipe staves; and a small vessel arrived from the Madeiras and returned with a cargo of pipe staves and other products.[79] In March, 1643, a Boston-built ship returned from a voyage to Fayal where it had sold pipe staves and fish, bought wine and sugar, sailed for St. Christopher, and there exchanged some of the wine for cotton, tobacco, and iron. Small vessels were going to the West Indies for cotton, and larger ships were taking clapboards, pipe staves, fish, and other products to the Canaries. On December 3, 1644, a Boston ship of 60 tons arrived from Palma Island in the Canaries with wine, pitch, sugar, and gum. The *Trial*, 160 tons, built in Boston, left in March of the same year for Bilbao and Malaga with fish, and returned with wine, fruit, oil, iron, and coal. In 1645 a ship returned after taking a load of pipe staves to the Canaries, slaves from the Cape Verde Islands, and wine, sugar, salt, and tobacco from Barbados. In addition to this tropical trade, New England vessels were leaving for England and Holland with wheat, rye, and furs in return for linen, woolens, shoes, stockings, and dry goods. The difficulty of obtaining furs to send to England had encouraged the opening of trade with the West Indies; and cattle and provisions were exported in return for sugar, cotton, tobacco, and indigo, which were in turn exported to England in exchange for English goods. By 1650 trade was being carried on with Virginia, Barbados, England, Portugal, Spain, Holland, and France.[80]

Dry fish were sent from fishing areas to Boston, and from thence the merchantable grades were dispatched to Lisbon, Bilbao, Marseilles, Bordeaux, Toulon, and other French ports, the refuse grades going to the West Indies. Supplies were purchased from the southern colonies. Relations between Newfoundland and New England were closely interwoven. Sack ships carried freight and passengers to New England,[81]

76. *Idem*, pp. 309–310.

77. For a valuable account see Andrews, *op. cit.*, I, 513–518; also M. J. Lanier, "The Earlier Development of Boston as a Commercial Centre," doctor's thesis, University of Chicago, 1924.

78. *Winthrop's Journal*, II, 42.

79. *Idem*, p. 89.

80. *Johnson's Wonder Working Providence 1628–1651*, ed. J. F. Jamieson (New York, 1910), p. 247.

81. In 1634 Winthrop referred to a "petition of shipmasters (attending how beneficial this plantation was to England) in regard to the Newfoundland fishery; which they took in their way homeward." In January, 1639, merchants trading to Spain, Portugal, and the Straits complained of their being forbidden to send their ships to

and merchants in the Newfoundland fishery were engaged in the New England trade and fishery. New England traders sold corn and cattle in Newfoundland, purchased fish and oil, and carried back to New England fishermen brought out from the West Country.[82] The base of supplies for Newfoundland was beginning to shift from England to New England. The growth of the settlements in New England and settlement difficulties in Newfoundland left the fishing ships in control of Newfoundland. The 50 vessels which had sailed from the West Country to New England in 1624 had dwindled to 15 in 1637. The outbreak of civil war in England weakened West Country fishermen in New England and hastened the growth of self-sufficiency and economic organization.[83] The cessation of immigration and a collapse of prices necessitated a development of trade.

In the Newfoundland trade, division within the industry complicated its problems. The importance of competition in the carrying trade, high interest rates, and the character of the financial support tended to stress the value of short-term credit and rapid turnover, and to strengthen the opposition of the fishing interests to the attempts of the carrying trade to gain control through either monopolies or regulations. Seasonal limitations in Newfoundland and agricultural difficulties militated against the establishing of settlements, whereas in New England the possibility of developing the winter fishery and wider markets made for their increase. Settlers were brought out to New England, and ballast to Newfoundland. The problem of regulating ballast thrown into its harbors could be contrasted with the growth of settlements in New England. Ships carried settlers, "who for their passage will spare the charge of double manning their ships, which they must doe in Newfoundland to get their fraught [freight]; but one third part of that companie are onely but proper to serve a stage, carry a barrow and turne poor john."

Commercialism based on the fishing industry in the West Country restricted settlements in Newfoundland, and commercialism based on the

New England as it would "deprive the Kingdome of much trade, (of) the importation of much money, his Majestie of much Custome and many ships and seamen of employment, and therefore humbly besought the Boord to grant them liberty to send their shipping intended for Newfoundland and other places and that, by the way, they may take in such helpe of fraight by passengers and goodes for New England as shalbee presented to them, so that his Majesty's custome, navigation and merchants may be cherished and increased."

82. Prowse, *op. cit.*, pp. 151–153. "New England having had of late great traffic with Newfoundland where they vent the growth of their plantation." B. M. Egerton MSS. 2395, No. 259, cited by A. M. Field, *The Development of Government at Newfoundland 1638–1713*, master's thesis, University of London, 1924.

83. Judah, *op. cit.*, pp. 67–68.

fishing industry in New England made for settlements. The latter meant the expansion of the fishery, of shipbuilding, and of trade. The struggle against monopoly was less acute but eventually effective. Long-term credit[84] became more important with the growth of settlements, and companies were more efficient organizations to provide it. Metropolitan organizations, centered in London, Bristol, or elsewhere, were less of a failure on the mainland than in Newfoundland. On the other hand, the expansion of the fishery in New England implied independent organization; and the growing compromises in Newfoundland between the shipping interests of metropolitan centers, such as London, and the West Country fishing interests called for an increasingly close relationship among English activities. The effective representation of London and West Country interests contrasted sharply with the conflict between the widely separated ports of France, inherent in the scattered regions in which the French prosecuted the fishery in general and the relatively more important green fishery.

FRENCH, SPANISH, AND PORTUGUESE FISHERIES

Our nations tradinge by meanes of the imposition paid by strangers are greatly advanced in their marketts, being able by their freedome to undersell the stranger, by which means our english trade will be encouraged and increased and strangers discouraged.

DAVID KIRKE'S *Memorial on Newfoundland*

THE expansion of the French fishing industry to the north and to the mainland on the south was hastened by the contraction of their Newfoundland fishing grounds as a result of English aggression.[85] The pat-

84. "All trades settled in joynt-stocks must restrain the trade to London; from thence all ships for the carrying on of such trades must have their egress and thither must return, which as well as the grievances before mentioned will occasion complaints from the rest of the trading towns, the city of Exon and Bristol and others being as well seated and accomodated for carrying on a trade for Africa as London." John Pollexfen, *A Discourse of Trade, Coyn and Paper Credit and of Ways and Means to Gain and Retain Riches* (London, 1697). The position of companies in England concerned with the wool trade and dominated by London meant weak development of companies in the provinces and hostility to London. The problem of securing long-term credit essential to colonization in North America was more acute in the provinces and explained in part the difficulties of companies in New England. See E. F. Heckscher, *Mercantilism* (London, 1935), I, 375, 402, 424, 428, 429, 432; also E. A. J. Johnson, *Predecessors of Adam Smith* (New York, 1937), chaps. iii, iv. The Hudson's Bay Company succeeded, having London support and limited demands for long-term capital. The land trade and the sea trade were developed under widely varying conditions and the possibility of linking them involved greater difficulties than were generally appreciated, as France and England learned to their sorrow.

85. In 1612, the pirate Easton robbed and spoiled 25 French ships fishing about

ent to Sir David Kirke in 1637, as we have seen, entitled him to impose a tax on foreigners fishing, drying, or purchasing fish in Newfoundland, but an attempt to collect it along the southern coast in 1639 was only partially successful.[86]

The French fishery on the south coast was responsive to the demands of the Spanish market. Champlain took note of the fishery at Placentia, Trepassey, and St. Pierre, "where many vessels go to fish and where they dry the fish." About 1613, 30 ships from St. Jean de Luz brought 16,000 cargas (about 64,000 bushels) of codfish to Passages and 150 cargas of oil to Bilbao. Partly as a result of the profitable character of Spanish trade, Sibiburo was reported as having grown from 30 houses to 500 between 1600 and 1630. In the early 'forties the French Basque region of Bayonne had a fleet of 60 ships which employed 3,000 seamen and annually took from Spain 400,000 ducats of silver. It was significant that by 1650 San Sebastian had become an important financial center, as indicated by its loans to St. Jean de Luz.[87]

In the Petit Nord and on the Labrador the migration of ships from the Avalon Peninsula was in part a source of irregularities. On October 10, 1610, the town of St. Malo asked permission to send two armed boats to make war on the Indians who had killed "a number of shipmasters and sailors while fishing for cod."[88] Letters patent in 1615 prohibited the tearing down of staging in Newfoundland. An *arrêt*[89] issued at Rennes, July 31, 1640, provided that the first arrival at the harbor called Le Havre du Petit Maître (Petty Master's) was entitled to first choice of the harbor and ground necessary for his fishery. After he had made his choice other masters were entitled to make their selections in the order of their arrival. In 1647 a vessel was dispatched at the ex-

Newfoundland, doing damage to the extent of £6,000. In 1614 pirates robbed a French ship at Harbor Grace of 10,000 fish and took another at Carbonear. In 1616 a third French ship was captured; and, two years later, Captain Wollaston and others of Raleigh's fleet robbed four French ships of their dry fish, and sold them at Leghorn for £3,000. They did damage to other French fishermen to the extent of £500, and caused a loss of an equal amount to another French ship.

86. Lounsbury, *op. cit.*, pp. 80–85; also Prowse, *op. cit.*, p. 137. L. D. Scisco, "Kirke's Memorial on Newfoundland," *Canadian Historical Review*, March, 1926, p. 50.

87. See H. A. Innis, "The Rise and Fall of the Spanish Fishery in Newfoundland," *Transactions of the Royal Society of Canada*, 1931, sec. 2, pp. 51–70. See H. P. Biggar, *Early Trading Companies of New France* (Toronto, 1901), p. 65.

88. Marc Lescarbot, *The History of New France* (Toronto, 1907), I, 59; also on the French fishery to the north about 1640 see Henry Kirke, *The First English Conquest of Canada* (2d ed., London, 1908), p. 39.

89. *P.C.*, V, 2174. See Ferdinand Louis-Legasse, *Evolution économique des Iles Saint-Pierre et Miquelon* (Paris, 1935), pp. 39–40, and Harrisse, *op. cit.*, p. xxxvi. See also Prowse, *op. cit.*, pp. 139–140; Biggar, *op. cit.*, p. 102; Hippolyte Harvut, "Les Malouins à Terre-Neuve," *Annales de la Bretagne*, November, 1893, pp. 20–21.

pense of St. Malo and Binic to protect 4,000 men engaged in the fishery between Dégrat and Cape St. John.

The expansion of Canada was in part a result of the difficulties in Newfoundland. In the financial records of French Channel ports, the name of Canada appears more frequently after the turn of the century and generally in connection with the Mediterranean market. On March 16, 1602, a loan was made to H. Gohorel on the *Esperance*, 100 tons, G. Dieres master, of 10 écus at 35 per cent if he returned directly to Honfleur, or 40 per cent if he returned by way of Spain. Dieres himself received a loan on April 2 of 25 livres at 40 per cent to go to Canada, return to La Rochelle and Bordeaux and thence home. On December 1, 1604, Jehan Desamaison, master, employed[90] P. Gadoys to go to Leghorn and from there to Canada. On February 9, 1607, the same Jehan Desamaison borrowed 100 livres from P. du Sausay (who married the widow of Chauvin) to go to Leghorn with a cargo of wheat, thence to Spain for salt, and finally to Canada or Newfoundland for cod. On February 12, Dieres, master of the *Esperance*, borrowed money to undertake a voyage to "Malleque" (Malaga) or Cartagena and thence to Canada to fish for cod. The rate of interest was given as 27 per cent in a document of April 4, 1608, and as 25 per cent in one of March 20, 1609, and was apparently declining with the result that capitalists[91] had become more important; individual names such as Andrieu and Du Sausay appeared more frequently, and it became possible to undertake more extensive expeditions to the mainland and to the Mediterranean. By 1604 the fishery was carried on at Cape Breton, especially Ingonish and Louisburg, and Canso, but not beyond Mahone Bay, Sambro, and Port Mouton, although Gosnold[92] reported in 1602 that Basques or French from St. Jean de Luz had been trading along the coast of Maine. At Canso, "fishing for both green and dry fish is here carried on."[93] In 1603 Champlain probably followed the usual

90. For this service he was to receive "le demi-tiers et la moitié du demi-tiers à ladite navigation aux us et coutumes de la mer, avec 105 livres de pot de vin, douce cents morues seiches, un demi poinson d'huile et un poinson de morue verte" (a one-sixth and a one-twelfth interest, respectively, in the said venture and, as additional payment, according to the usages of the sea, 105 livres' worth of wine, 1,200 dry cod, a quarter cask of oil, a half cask of green cod, and 21 livres a month).

91. According to Harrisse not less than five companies were formed in 1612 to support fishing vessels from Rouen, Dieppe, and other ports. Associations of "armateurs" (equippers) were formed to establish the dry fishery in Newfoundland, one such association in La Rochelle in 1608 having twenty-seven proprietors of vessels.

92. *Early English and French Voyages, op. cit.*, pp. 330–331.

93. *The Works of Samuel de Champlain*, ed. H. P. Biggar (Toronto, 1922), I, 466. "The merchants who were partners of M. De Monts, not knowing that the fishing extended beyond this spot. . . ." Lescarbot, *op. cit.*, II, 350; III, 4. On Cape Breton there were "two harbours where the fishing is carried on namely English Harbour

route. Leaving Honfleur on March 15 he proceeded to Gaspé, Codfish Bay, and Isle Percée, where there was both green and dry fishing, and on August 24 he returned from Gaspé to Havre de Grace. Smaller vessels were able to avoid the ice which blocked Cabot Strait until the end of May by going through the Gut of Canso,[94] but the late season was a handicap. An arrival at Tadoussac on May 19, 1609, was said to have been the earliest in sixty years. Chauvin had large interests in the fishing industry and in the company of 1599 which had established a post at Tadoussac. Protests from other fishing interests led in 1603 to the inclusion in this company of merchants of Rouen and St. Malo, and, in 1604, of La Rochelle and St. Jean de Luz. On Champlain's expedition of 1604 a vessel was found trading in furs at Liverpool Bay on May 12, and Basque vessels were seized at Canso for encroaching on the company's trading privileges.[95] In 1606 the company learned of several vessels about Cape Breton which were engaged in the fur trade. With the extensive fishing industry, monopoly of the fur trade proved difficult to enforce; and according to Champlain the hostility of the Basques and Bretons was responsible for the difficulties of the company in 1607 and 1608. Although by 1607 the experiment of colonization in Acadia had failed, the *Jonas* which brought back the colonists[96] left Canso with 100,000 cod, green and dry. It sailed on September 3 and arrived at St. Malo the end of the month.

The disappearance of the company, the revival of competition, and attacks by the English in Newfoundland probably contributed to the wider dispersal of fishing vessels. On August 13, 1609, the first year of competition, Champlain found a number of vessels catching and curing fish at Isle Percée. The fishery was divided chiefly between the Normans and the French Basques,[97] but Spanish Basque vessels were noted at

[Louisburg] . . . and Ingonish [Niganiche]." *The Works of Samuel de Champlain,* I, 467. A Basque, Savalet, from Saint Jean de Luz, was reported at Whitehaven near Canso in 1607 with a ship of 80 tons and 16 men. Lescarbot put the profits on his voyage at 10,000 francs; and at the date of his (Lescarbot's) visit he was taking 50 crowns' worth of cod a day.

94. *The Works of Samuel de Champlain,* I, 466–467.

95. *Idem,* pp. 27–29, 231.

96. Lescarbot, like Captain John Smith, noted the relation between fishing and colonization. "But here I must stop to consider that those who in these last voyages have crossed to those shores have had an advantage over those who wished to colonise Florida, in this recourse which we have mentioned, to the French ships which frequent the Newfoundlands, without having the trouble to build large vessels, or to abide the extremities of famine as have done those in Florida." Lescarbot, *op. cit.,* II, 284. But the dry fishery was not sufficiently important to the French to support a sustained development of settlements, and it is significant that colonists returned by a *fishing* vessel.

97. *Idem,* II, 348–349.

Tadoussac.[98] In 1611 men from St. Malo and La Rochelle were engaged in trading at Port Royal,[99] and in 1613 two ships from St. Malo were reported near Port Mouton. In the same year Poutrincourt had a magazine of salt on Long Island in the Bay of Fundy, but it was looted by the English. When Captain John Smith arrived on the coast of Maine in 1614 he was obliged to meet competition[100] from French ships that had already established themselves there. The importance of the fishery had seriously affected the fortunes of the monopoly in the fur trade, and it was not until 1613 that Champlain succeeded in arranging an association which limited the number of vessels to three from Rouen and one from La Rochelle. Merchants of St. Malo protested against their exclusion and at the end of the season they also were included.[101] In 1614 the new company shares were divided between Rouen, St. Malo, and La Rochelle; but the latter failed to participate. A reorganization in 1620 added more shareholders from Rouen. But, while monopoly had established itself in the fur trade of the St. Lawrence Valley, the difficulties of competition in the fishing industry of the maritime region[102] proved insuperable.

The industry was competitive because of the difference of interest between the various French ports as well as between the French and the English. In 1613 Captain Samuel Argall from Virginia destroyed the French settlements in Acadia and took the first steps in the developments which gradually narrowed the range of the French fishery. In 1622 and 1623 Sir William Alexander dispatched vessels to establish a settlement in Nova Scotia under a charter granted to him, but it came to little. In the latter year his vessel sailed along the Acadian coast to Port Mouton[103] where it met a French captain who "in a very short

98. *Idem,* III, 6; also *The Works of Samuel de Champlain,* II, 11–13.

99. Lescarbot, *op. cit.,* III, 52; on the difficulties with La Rochelle see Biggar, *op. cit.,* p. 111.

100. Hayman wrote in 1630 that "the French and Biscays do yearly in great number fish at the Mayne and dispossess us."

101. *The Works of Samuel de Champlain,* II, 246.

102. La Rochelle and Basque vessels become notorious for participating in smuggling, and were effective in trade because they carried cheaper goods. Company organization continued to have an interest in the fishery at Gaspé and adjacent points, Miscou and Prince Edward Island especially, under the direction of the Sieur de la Ralde. *Idem,* Vol. V, *passim.*

103. George Patterson, "Sir William Alexander and the Scottish Attempt to Colonize Acadia," *Transactions of the Royal Society of Canada,* 1892, sec. 2, pp. 79–107. In 1622 Alexander's vessel reached St. Peters Island, Cape Breton, but wintered at St. John's, Newfoundland. The second expedition reached St. John's on June 1, 1623, and found that members of the first expedition "had engaged themselves to serve fishermen, by which meanes they gained their maintenance and some means beside so that they could hardly be gathered together again." Finally they left St. John's for points "where their ship was to receive her loading of fishes." However, the ship was

time had made a great voyage, for though he had furnished one ship away with a great number of fishes there were neere so many ready as to loade himself and others." In 1629 Sir William Alexander's squadron discovered three French ships at anchor off the coast of Cape Breton, one of which was a barque of 60 tons, probably the *Marie* of St. Jean de Luz. On August 26, the French captain Daniel from Dieppe found a Bordeaux vessel anchored in Baleine Cove. A final attempt to include the maritime area under monopoly control by the formation, in 1627, of the Company of New France[104] with its headquarters in Paris was responsible for a temporary success gained by Alexander. The charter of the company was marked by distinct centralizing features[105] to offset the decentralizing tendencies of the fishing regions. Among those excluded by the French Company were the Kirkes of Dieppe; and their coöperation with the English, and later with Alexander, in capturing Quebec and holding it until 1632 was a further indication of the failure to control the fishing regions by monopoly, particularly with the ever-present danger of conflict with the English.[106] Sir David Kirke was, for his part, destined to fail in his ambitions to maintain a monopoly in New France, as later he was to fail, as we have seen, in Newfoundland.

The difficulties met by monopoly in the Maritimes became evident in the bitter contest between the French occupants. In 1633 Denys, in partnership with Razilly and a merchant of Auray, established a depot for the sedentary fishery at Port Rossignol near Liverpool. A shipload of fish was sold in Brittany under favorable conditions; but a larger venture in a 200-ton ship and attempts to export to Oporto failed because of difficulties between Spain and France in 1634–35.[107] Denys established a post at Miscou in 1645, but this was seized two years later by D'Aulnay. The vicissitudes of the Denys venture and of other French settlements were a result of the close relationship between the fur trade and the fishery in the maritime regions, of inherent difficulties in the fishery which made monopolies impossible, and of the disappearance of limited markets upon the outbreak of war.

apparently left in Conception Bay, and the members "dispatched themselves in several ships that belonged to the west part of England."

104. The monopoly "reserves all rights in the cod and whale fishery, which His Majesty wished to see made free to all his subjects." *Edits, ordonnances royaux, déclarations et arrêts du conceil du roi concernant le Canada* (Quebec, 1854), I, 8.

105. *Idem,* pp. 5–6.

106. Biggar, *op. cit.,* chap. viii; G. P. Insh, *Scottish Colonial Schemes 1620–1686* (Glasgow, 1922); Kirke, *op. cit.;* and J. B. Brebner, *New England's Outpost Acadia before the Conquest of Canada* (New York, 1927), chaps. i, ii. In 1630 Sir William Alexander was concerned with the development of the fishery in Scotland as a means of opposing the Dutch. Elder, *op. cit.,* chap. iii.

107. Nicolas Denys, *The Description and Natural History of the Coasts of North America* (Champlain Society, Toronto, 1908), pp. 142–144.

The growth of settlements in Nova Scotia was consequently slight. Denys's description of the coast line, written about 1669 from the experience of a long stay in the region, indicated points of development. De Lomeron was noted as the first place along the coast which offered good shelter and abundant cod close to shore, the fish coming in earlier than anywhere else; but an establishment was destroyed by the English about 1628. Seals were taken at Seal Islands, and Cape Sable provided a good harbor and an abundant cod fishery. La Tour had a fort at Barrington Bay. Negro Harbour had an excellent fishery, but Denys reported never having seen a ship there. He had seen ships at Port Mouton where the fish were taken from two to two and one half leagues from shore and dried in an unsatisfactory way on hurdles (flakes) "on which one is obliged to dry the fish when there occur at the place of the fishery only sand and grass." Razilly occupied an establishment at La Have, but his death in 1635 was followed by its disappearance. At Prospect men often engaged in the fishery and they could dry the cod on the rocks. The beginnings of a settlement were made at St. Marys River some time after 1650. Canso was an organized and well-known fishing area. Both harbors had gravel beaches, but flakes were also necessary. The "admiral," or first arrival, chose the better harbor, leaving the other for the "vice-admiral." At Green, Goose, and Harbor Islands one or two vessels anchored and dried the fish on the islands. At Chedabucto Harbor (Guysborough), Denys had established a sedentary fishery but as a result of attacks had been forced to move to St. Peters in Cape Breton. The Michaux Islands provided a good fishery and Louisburg was a famous locality with a good harbor and good fishing. Farillon was a point at which the Bretons had fished from an early period and had an excellent fishery. The harbor of St. Annes provided good shelter. Ingonish (Niganiche), according to Denys, was one of the first points to be occupied because "the fishing there is good and early." Cape North was suitable for one vessel to "make fish." At Cheticamp, cod were abundant but shelter was lacking. Along the coast of Cape Breton "that which makes it valued are the ports and roadsteads which the ships use to make their fishery. Mackerel and herring are very abundant around the island and the fishermen make their boitte or bait of them for catching the cod, which is very fond of them."[108]

108. *Idem,* p. 186. "The cod fishing begins very early in the spring near Louisburg . . . and continues good till July when that fish proceeds towards Cape North and the Gulf of St. Lawrence; but in August when the mackerel, herring and capling [capelin] make their appearance the cod returns in pursuit of them in such numbers and with so much eagerness that many are caught in very shallow water. The end of the fall is the usual time for leaving off fishing." D. C. Harvey, *Holland's Description of Cape Breton Island and Other Documents* (Halifax, 1935), p. 124.

In the Gulf, the Magdalen Islands provided walrus, seals, and good fishing for cod. Fishing vessels on their way through Cabot Strait stopped at Bird Island for eggs and birds. Prince Edward Island was generally neglected because of the difficulty of getting into the harbors, although Denys said that he had seen three large Basque vessels at Cascumpecque Bay. At Miscou, vessels were afforded valuable protection. "I have seen as many as five or six ships here making their fishery. They make flakes upon this point of sand, for there is no gravel on it." The Caraquet Islands provided shelter for vessels, good fishing, and gravel beaches, but "flakes" were also needed. Vessels anchored at Paspebiac, which was a very important fishing center where cod were generally to be found even when they failed at Bonaventure Point. Fish were dried on the gravel beach. Along the coast to Isle Percée cod were abundant but shelter was not adequate. At the latter point the Normans left their ships and went out to the Orphan Banks where the fish were as large as those on the Grand Banks and were put down green. This fishery served as the point of contact by which the Normans became interested in the fur trade of the St. Lawrence. Denys had seen eleven vessels there loaded with fish. Capelin were plentiful for bait and, coming on shore, attracted the cod. Pebbles were spread on the beach on which fish could be dried and flakes were set up on the level meadows. The fir trees had been cut down for some distance to make stages, and building had become expensive. The fishermen had made gardens and planted cabbages, peas, and beans. Bonaventure Island, although possessed of an excellent fishery, had a gravel beach suitable for only one vessel; and the remainder were forced to build flakes to which it was necessary to make a road and carry the fish up a hill. Denys reported having seen three vessels in the cove. At Gaspé fishing vessels were able to anchor and there was a gravel beach sufficient for two large vessels. When the fishing failed at this point the men went to Farillon to make a "dégrat." The coast suffered from a lack of adequate shelter, drying facilities, and fish. Localities with an abundance of fish were of little value without harbors. The scarcity of gravel beaches and the heavy expenditure of effort in erecting flakes were additional handicaps. The more favorable localities had consequently a high rental value and were subject to conflict under the "admiral" rule, and the extent of the fishery was limited.

According to Denys, those principally engaged in the French dry fishery were Basques, men from La Rochelle and its adjoining islands of Ré and Oleron, and also fishermen of Brittany and Bordeaux, and they had a total of from 100 to 150 vessels. Less salt by half was needed than was called for by the bank fishery. For a 200-ton ship engaged in the dry fishery about 50 men were required, with provisions for eight or nine months, and the catch would average 200,000 fish. Generally the skill they displayed varied directly with the distance of their home ports from Paris and inversely with their distance from Spain and the Mediterranean, the chief markets for dried fish. "The Basques are the most skilled. Those of La Rochelle have the first rank after them and the Islanders who are in the vicinity, then the Bourdelois and then the Bretons." The business agreements made were based in part on the skill of the fishermen.

> The Basques make their agreements on the basis of the cargo of the ship. It is estimated that the ship can carry so many quintals of fish; the owners make an agreement with the crew, and make two or three hundred shares according to the number. They give to the captain a certain number of shares according to the reputation that he has in this business, to the beach master so many, to the pilot so many, to the dressers so many, to the masters of boats so many, to each of the stowers and boatswains so many, and to each of the boys so many parts.

The men from La Rochelle were less efficient. They were not so well equipped; for example, the Basques carried a duplicate outfit of clothing, so that in case of rain one suit could be left to dry. Instead of the precise division based on the estimated cargo of the ship, the men of La Rochelle were given one quarter of the fish. The crew shared equally, but 100 écus (500 livres) were provided in addition for each boat and its five men, which sum was divided between them in proportion to their worth. The ships from Bordeaux gave one third of the cargo to the crew, who agreed among themselves as to the disposal of their share. The captain decided on the arrangements of the crew, and he hired boys by direct payment and took their shares. The extent of the wage system varied inversely with the skill of the fishermen.

The dry fishery was carried on chiefly by the ports more closely associated with the Bay of Biscay and the green fishery by the Channel ports. Denys claimed that it gained at the expense of the green fishery, and larger numbers of vessels from the Channel ports became engaged in it. The number of French vessels fishing on the Banks was estimated by him at from 200 to 250. A 200-ton ship employing 25 men would produce 45,000 to 50,000 fish. The great proportion were taken by Norman fishermen from Honfleur, Dieppe, Boulogne, and Calais; also by men from Brittany, Olonne, and the country of Aulnes; and three quarters were sold in Paris. In 1603 it was estimated[109] that 80 vessels left Havre alone for the Banks, although some of them were probably engaged in the dry fishery. In 1617 St. Malo was reported to have sent

109. Biggar, *op. cit.*, p. 49.

120 vessels to the Banks and the harbors,[110] chiefly in the Petit Nord. In 1611 the French were held to have 600 ships;[111] but in 1615 Whitbourne's estimate was 400, including those of the French Basques and the Portuguese and those going to Canada. Vaughan[112] put the total at from 200 to 300 ships other than English in Newfoundland; and Hayman[113] in 1630 estimated that there were 400 French ships on the Banks. Fishermen made voyages "into these countries in search of codfish wherewith they feed nearly all Europe and supply all sea-going vessels."[114]

The expansion of the French to the more distant regions was hindered by the dominant importance of the bank fishery in the domestic market. The technique of the green fishery involved possibilities of development over a wide area without the handicap of climate, whereas the dry fishery tended to be restricted to definite areas like the Petit Nord and select portions of the mainland such as Canso, Cape Breton, and Gaspé. The French were forced to concentrate on the more distant restricted areas in their expansion of the dry fishery. Competition from the English became increasingly effective; New England was a competitor at Bilbao, and Newfoundland in the Mediterranean.

Vessels extended their activities to the coast of Nova Scotia and the Gulf of St. Lawrence, and became engaged in the fur trade for the home market and the dry fishery for those of Spain and the Mediterranean. The financial structure that had been established in order to send ships to the Banks early in the year, and have them return without delay with green fish for the Paris market, was further broadened to aid both in the export of dry fish and other products to the Mediterranean and in the penetration to the Gaspé fishing grounds, which admitted of the production of both dry and green fish. When vessels sailed to Spain, and then to Newfoundland and Canada for furs and dry and green fish, there was a more complete occupation of the land areas throughout the year. The later arrival of vessels facilitated the profitable development of the Gulf fishery and the fur trade. The superiority of the Channel ports in the fur trade was related to the importance of the Paris market, to the increasing complexity of the fishing industry as conducted from these ports, and to the profits of the green fishery as conducted on the Banks. The decline of the Spanish fishery was responsible, with the French as with the English, for the development of a three-cornered trade. The French brought green and dry fish to Honfleur and the Channel ports, dry fish to the Mediterranean, solar salt

110. *Idem*, p. 106. In 1628, St. Malo was reported to have a fleet of 112 ships.

111. *Idem*, p. 25.

112. Prowse, *op. cit.*, p. 136.

113. *Idem*, p. 138.

114. Lescarbot, *op. cit.*, II, 22.

from Spanish or French ports, and French provisions and supplies to the New World; whereas the English took dry fish from Newfoundland to Spain and the Mediterranean, returned with salt and other products to England, and then carried supplies and provisions from England to Newfoundland. Paris was the important market for furs and green fish, and Spain and the Mediterranean for dry fish.

The expansion of the fishery to the Gulf and along the coast led to contacts with the Indians and the development of the fur trade. This meant at first exporting trading goods of small bulk but, to the Indians, of great value. These goods served to round out the outward-bound cargo of fishing vessels; and the highly valuable small-bulk furs could be easily carried with the cargo of fish returning to Europe. With the expansion of the fur trade, the increasing demand for diversified products contributed to the increasing complexity of trade in relation to fishing. The fur trade developed in close connection with both green and dry fishing, as prosecuted by vessels from the Channel ports, and in response to the metropolitan demands of Paris.

The development of the fur trade and the characteristics of this trade provided a setting for the development of control by company organization and later by the state. The success of monopoly in the fur trade of the continent was in sharp contrast with the struggle against it in the fishing industry. The concession to De la Roche and its numerous successors in the maritime regions failed, while in the fur trade monopoly struck deeper roots. The atomistic, or small-unit, commercialism of the fishing industry was entrenched in a century of development. England was less handicapped because of her relative freedom from the friction between the demands of such widely different interests as the fur trade and the fishing industry. She was compelled to divide concessions over an extended Atlantic seaboard in which separate areas involved separate problems; but from Newfoundland to New England she was forced to concentrate on the fishing industry. The success of the atomistic commercialism of the West Country in Newfoundland and in England was partly a result of the retreat of company organization to New England and the mainland. In New England commercialism was less conspicuously successful in the beginning, but eventually triumphed. The fishing industry meant conditions of cumulative corrosiveness which destroyed company control of trade.[115] The Navigation Acts,

115. The strictures of Adam Smith on companies were supported by experience in the fishing industry. "Of all the expedients that can well be contrived to stunt the natural growth of a new colony, that of an exclusive company is undoubtedly the most effectual." Adam Smith, *op. cit.*, p. 542. "These companies, though they may, perhaps, have been useful for the first introduction of some branches of commerce,

the colonial system, and the increasing use of treaties emerged in the place of the company.

The Portuguese and Spanish fisheries declined as the English and French expanded. The Anglo-Spanish treaty of 1604, which made it permissible to import fish into Spain, recognized the marketing advantages of the English, and their aggressiveness in Newfoundland and New England weakened the position of competitors. The Portuguese, like the French, suffered severely from English piracy and confusion in Newfoundland.[116] In 1615 Whitbourne wrote that "sundry Portugall ships have also come thither purposely to loade fish from the English and have given them a good price for the same and sailed from thence with it to *Brasile* where that kinde of fish is in great request and they have made great profit thereby." It was probably difficult, however, to compete with traders carrying fish from England or Portugal.

The Spanish fishery declined in spite of numerous attempts to aid in its recovery.[117] In 1625 San Sebastian was said to have 41 ships with 295 shallops and 1,475 men engaged in the fishery; but by the end of the first half century the decline had been marked. The increase in prices incidental to the importation of specie in Spain continued to be a powerful factor, weakening Spanish industry and leading to the

by making, at their own expense, an experiment which the state might not think it prudent to make, have in the long-run proved, universally, either burdensome or useless, and have either mismanaged or confined the trade." *Idem,* pp. 691, 700, 704–705. "Without a monopoly, however, a joint stock company, it would appear from experience, cannot long carry on any branch of foreign trade. To buy in one market, in order to sell, with profit, in another, when there are many competitors in both; to watch over, not only the occasional variations in the demand, but the much greater and more frequent variations in the competition, or in the supply which that demand is likely to get from other people, and to suit with dexterity and judgment both the quantity and quality of each assortment of goods to all these circumstances, is a species of warfare of which the operations are continually changing, and which can scarce ever be conducted successfully, without such an unremitting exertion of vigilance and attention, as cannot long be expected from the directors of a joint stock company." *Idem,* pp. 712–713.

116. In 1612 the pirate Easton was reported as responsible for "spoiling [the] voyage of 12 Portugal ships £3000." Two years later Mainwaring "with divers other captains . . . from the Portugal ships took all their wine and other provisions save their bread." In 1616 pirates took "a French and Portuguese ship." In 1620 "certain English fishermen entered aboard a Portugal ship in the night in St. John's Harbour with swords and axes wherewith they cut many of his ropes." Again there was "a great combat betweene some insolent English and certain Portugals in Petyte Harbour." Mason, in *A Briefe Discourse* (1620), noted the arrival "in the midest of May some Portingalls." Prowse, *op. cit.,* pp. 102–103.

117. *The Cambridge History of the British Empire,* I, 90; Vera Lee Brown, "Spanish Claims to a Share in the Newfoundland Fisheries in the Eighteenth Century," *Canadian Historical Association Report,* 1925, p. 67. In 1631 Spain imposed a prohibitive export duty on salt and compelled England to depend on France.

growth of trade from France and England. Trade with Spain hastened the development of the dry fishery in the New World, the beginning of English settlements in Newfoundland and New England, and of those of the French in Acadia and New France.

The elimination of the Spanish and the Portuguese left the fishery to be divided between the English and the French. By virtue of their reliance on dry fishing and the geographical advantage of distance, the English had succeeded in obtaining a secure hold on the Newfoundland coast from Cape Bonavista to Cape Race, and from this solidly built steppingstone they strode to New England on the mainland. There monopoly control, crushed in Newfoundland, found a scarcely less pleasant resting place. Lord Baltimore moved to Maryland and John Mason to New England; but the fishing industry in New England, if hindered by monopoly as carried on from the West Country, gained by the planting of settlements and by the local fishery which provided a base for New England commercialism. By the end of the first half of the seventeenth century the English had developed an important fishery in these two widely separated areas. In both, and in the fishing regions of New France, control through monopolies was exercised with difficulty and finally disappeared. The fishing interests at home and abroad served as a spearhead in the struggle for freedom of trade beyond the confines of England. In the fifteenth century English fishermen and merchants from Lynn, Hull, and Bristol had ignored Norwegian attempts to control the Iceland fishery from Bergen; in the sixteenth century they had gained the ascendancy over France and Spain in Newfoundland; and in the seventeenth they fought against settlements established by companies in Newfoundland. A monopoly of land trade proved inadequate to control sea trade. The struggle in Newfoundland for the right of free fishing reflected the growing importance of the sea. Commercialism, built on the fishing industry, drove a wedge into monopoly control over external trade as exercised from Great Britain. Hostility to Sir Humphrey Gilbert's monopoly in Newfoundland was intensified in the struggle against the Newfoundland Company, and was followed by compromises in the case of both the Charter of 1634 and Kirke's limited rights in 1637. Competition characterized the English and the French fishing industries. It characterized the industries individually and collectively. The breakdown of monopoly control in New France in the fishing regions contributed first to the collapse of the monopoly exercised by the English under Alexander in Nova Scotia, and then in turn to that of Kirke in New France and Newfoundland.

An industry prosecuted in small vessels subject to limited control, and from European ports scattered over a wide area divided between French and English and between green and dry fishing, conducted on a profit-sharing basis, and selling to different markets was essentially competitive and sharply divisive.

CHAPTER V

THE EXPANSION OF THE WEST INDIES AND NEW ENGLAND, 1650–1713

NEWFOUNDLAND

During the course of years [1670–97], our fishing trades to Newfoundland, Greenland and the northern seas . . . have decreased, occasioned we presume by the increase of other trades which have been found more beneficial to the traders and easy to the seamen, which has drawn off our people from those trades and given opportunities to foreigners more used to hard labour and diet to get a great share of them.

Parliamentary Debates (1699)

THE decline of Holland and the influence of the Navigation Acts gave support to the English carrying trade and contributed to the expansion of the French fishery. Competition from French fish weakened the West Country and, throughout the period, led to energetic efforts to dominate the trade of Newfoundland. West Country fishing interests, faced with the growing settlements in New England, concentrated on Newfoundland. The Navigation Acts were tempered to meet the demands of West Country interests. An act passed in 1647, which for three years exempted Newfoundland from the advantages of the removal of all duties except excise applying to American plantations, was followed by the Navigation Act of 1651 which forbade imports of fish unless taken by English fishermen and carried in English ships, and exports of fish in other than English ships with English masters and a majority of English in the crews. From 1656 to 1659 the West Country[1] was favored by a waiving of this legislation, which permitted the entrance of fish into the Spanish market in foreign bottoms during the war with Spain; and the providing of convoys for ships en route to Spain also favored it. Aliens were allowed to purchase fish in Newfoundland and New England free of duty, and Newfoundland fish in England on the payment of three pence a quintal. An act of 1660 admitted, free of duty, fish caught by Englishmen and brought in English ships, and the

1. In spite of the demands of the navy in the war with the Dutch in 1653–54, fishermen proceeded to Newfoundland. See R. G. Lounsbury, *The British Fishery at Newfoundland, 1634–1763* (New Haven, 1934), pp. 92–96; also C. B. Judah, *The North American Fisheries and British Policy to 1713* (Urbana, 1933), pp. 92 ff.; C. M. Andrews, *The Colonial Period of American History* (New Haven, 1938), IV, 38 ff.

Navigation Act of that year permitted the export of salt fish in foreign bottoms. In the same year the duty on salt[2] to be used in the fishery was remitted, other materials were exempted, and a stricter observance of Lent[3] was required. In 1662 English vessels carrying exports of fish to the Mediterranean were likewise exempted from special regulations. In 1663 the Newfoundland fishery was relieved of all taxes, and duties were imposed on all fish and products imported into the kingdom by foreigners and aliens,[4] and in 1667 the importing of fish taken by foreigners was prohibited. Thus, advantages gained by the carrying trade through the Navigation Acts were adapted to meet the peculiar position of the West Country fishing ships in Newfoundland as "a nursery for seamen." In 15 Chas. II, c. 16, following 2 and 3 Edw. VI, c. 6, it was provided "that no person or persons whatsoever do collect, buy, or cause to be levied, or taken in Newfoundland any toll or other duty of or for any cod or poor john or other fish of English catching."

In 1652 the appointment as commissioner of John Treworgie, who had been engaged in the fisheries of Maine and Newfoundland, probably followed recommendations intended to check the proprietary interests of Kirke. According to his instructions in 1653 he was "to collect the imposition of fish, due from and paid by strangers, and likewise the imposition of oil for the use of this Commonwealth." Regulations were repeated regarding ballast stones, the destruction of stages, the limitation of the admiral's rights, the alteration of marks on property, stealing, the destruction of trees except for cookrooms "not to extend above 30 foote in length at the most" (instead of 16), the hindering of the hauling of seines for nets, the stealing of bait, and the setting up of taverns. Treworgie was instructed to see that no planter was permitted

to keep any more stage room than he hath fishing men in possession for the managing of it, and that every planter in each harbour may take their stages and fishing room together in one part of the harbour and not scattering as they now do, wasting as much room for one or two boats as may serve 6 or 8 boats. . . . That no planter be permitted to build any dwelling house, storehouse, court-ledge, or garden or keep any pigs or other cattle upon or near the ground where fish is saved or dried. . . . That all provisions imported for sale necessary for fishing be free for any person to buy for his own present occasion so it be demanded within six days after its arrival, and not to be ingrossed by a few to make benefit on others thereby.[5]

2. Salt was obtained for Newfoundland and New England chiefly from the Isle of Maio, one of the Cape Verde Islands, and the Tortugas.

3. On political Lent see E. F. Heckscher, *Mercantilism* (London, 1935), II, 37–38.

4. Lounsbury, *op. cit.*, p. 113.

5. *P.C.* (*Privy Council*), IV, 1740–1743; D. W. Prowse, *A History of Newfoundland* (London, 1896), pp. 167–168.

In spite of encouragement under the Navigation Acts and the support that was given to the West Country, the handicaps due to the Civil War, the wars with the Dutch, and competition from the French were accompanied by a marked decline. These difficulties apparently led to the emergence of the byeboatkeeper who carried on the fishery independently of the fishing ships or the planters. Rules were enacted on January 26, 1661, which practically renewed the Charter of 1634 and reinforced what Rogers has called "the ten commandments of the fisheries."[6] The threat of the byeboatkeeper as a competitor was met by the regulation that

> for the benefit of the said trade there we do hereby straightly charge, prohibit and forbid all . . . and every the owners of ships trading in the said Newfoundland fishery that [neither] they nor any of them do carry or transport or permit or suffer any person or persons to be carried or transported in their or any of their ships to the said Newfoundland other than such as are of his or their own or other ships company or belonging thereunto, and are upon the said ships hire and employment or such as are to plant and do intend to settle there.[7]

The duty of prosecuting for offenses against these regulations was entrusted to the mayors of Southton (Southampton), Weymouth, Melcombe, Lyme Regis, Plymouth, Dartmouth, Eastlowe, Foye (Fowey), and Barnstaple, and to the vice-admirals of Southton, Dorset, Devon, and Cornwall. On November 27, 1663, the towns of Dartmouth, Totnes, Plymouth, and Barnstaple, apparently the towns most directly interested, complained that the clause prohibiting the carrying of passengers was being violated and that as a result "the trade is so reduced that men can only be found for a quarter of the ships formerly sent out, whereby both the trade and his Majesty's service suffer great hurt." It was held that the enforcement of the clause would increase the number of ships and seamen for the navy, benefit the handicrafts, and stop "the loss through keepers of private boats drawing away able seamen."[8] On December 4, in response to the petition, an order was passed requiring closer observance of the regulations.

A failure of the fishery, the establishment of the French at Placentia in 1662,[9] and the raids by the Dutch on St. John's, Bay Bulls, and

6. See Lounsbury, *op. cit.*, p. 112, n. 45.

7. *P.C.*, IV, 1746.

8. *P.C.*, IV, 1747.

9. *P.C.*, IV, 1, 768. A petition of merchants in 1659 stated that the English fishery formerly sent from 50 to 60 ships to Bilbao and St. Sebastian, 40 to Cadiz and St. Lucan, 20 to Malaga and Velez, from 20 to 30 to Alicante, Valencia, Cattagua, and other Spanish towns, and the remainder to Majorca, Minorca, Sicily, Sardinia, Naples, and Milan; but that the French usurped the trade by obtaining permission to enter the ports of Bilbao, St. Sebastian, and Passages, "where they do not only furnish the

Petty Harbor in 1665[10] increased the difficulties of the fishing ships and led to complaints against the byeboatkeepers and settlers. On August 28, 1667, "merchants, owners of ships, and others, inhabitants" of Totnes, Plymouth, Dartmouth, and other places presented petitions,[11] and a commission was appointed to examine the whole question. Bristol and London carrying interests urged that the byeboatmen and those who had settled in Newfoundland be given more favorable consideration. In a statement[12] of January 8, 1668, they estimated that the revenue arising out of the fishery totaled £40,000 a year and the returns from the trade £300,000; and "the merchants of London, Bristol, Hampton and Weymouth and other parts petitioning and consenting to the settlement and securing of Newfoundland are concerned three-quarter parts in carrying on the fishing-trade." In 1669 Captain Robert Robinson,[13] in a strong case in favor of the establishment of settlements and the extension of government, argued the importance of the fishery for revenue, for seamen, and as a means of resisting the French, urged correction of the abuses incidental to the lack of a resident government, and the necessity of reviving trade, which had sunk from £500,000 to less than one third of that. A small tax of 1 per cent on the fishery, he said, would serve to protect 300 sail and 15,000 seamen. On the other hand the West Country argued that "for many years past few have made 10 per cent on this fishery and last year both Dartmouth and Plymouth lost considerably," chiefly because of the settlers. They should be removed "so that the trade in provisions etc., now mostly supplied from New England may be carried on by fishing ships from England and the seamen augmented." If the fishery continued to be carried on by byeboatkeepers and residents under a governor "the trade in a few years will be removed from this kingdom and become as that fishery of New England which at first was maintained from these parts, but is now managed altogether by the inhabitants of New England." A governor at St. John's would be of little value in the 48 fishing places along a 300-mile coast line from Bonavista to Trepassey and would mean a heavy tax.[14] Further protests were made on December 23, 1670, by the mayors of Exeter, Dartmouth, Plymouth, Lyme Regis, Barn-

men of war that take us but the fish is carried three or four hundred miles up in the countries of Spain." The French had 250 ships. St. Jean de Luz vessels had increased from 8 to 50, a good part English prizes. John Collins, *A Discourse of Salt and Fishery* (London, 1682).

10. More than a thousand men stayed at home in 1665 through fear of being impressed. *P.C.*, IV, 1752.

11. *Idem*, p. 1749.

12. *Idem*, p. 1751.

13. *Idem*, pp. 1753–1754, 1764–1765.

14. *Idem*, pp. 1755–1756.

staple, Weymouth, and Poole that "private boatkeepers still continue to fish in Newfoundland and great number of passengers still go there."[15] Ships "lie by the wall for want of men."

The success of the West Country protests was evident in the recommendations of March 10, 1671, made by His Majesty's Council for Foreign Plantations, that the fishery should be protected against the residents, that "masters of ships be required to bring back all seamen, fishermen and others, and none to be suffered to remain in Newfoundland," and that "the inhabitants of Newfoundland were to be encouraged to go to Jamaica or other foreign plantations." They recommended that no alien or stranger be allowed to fish between Cape Race and Bonavista, that no resident be allowed to "inhabit or plant within six miles of the shore," that "no planter or inhabitant . . . do take up or possess any of the stages, cookrooms, etc., beaches or places for taking bait or fishing before the arrival of the fishermen out of England and that they be all provided." The regulation of 1661 prohibiting the transport of passengers was narrowed. The clause "such as are to plant and do intend to settle there" was omitted, and in its place there was inserted "such as are engaged in the voyage and share, or shares, or hire of the said ship." No master or owner of any fishing ship was allowed to take more than 60 persons to 100 tons of burden. It was commanded that every fifth man taken out must be a "green man." The masters and owners were required "to provide in England victuals, and other necessaries (salt only excepted) for the whole voyage, or fishing season, for themselves and companies, and to put the same on shipboard before the going out of port here." No fishing ship or company was allowed to leave England directly for Newfoundland before March 1, or for Cape Verde Islands and thence to Newfoundland before January 15. It was further ordered "that no fisherman or seaman, carried out as aforesaid, be suffered to remain in Newfoundland in the winter after the fishing voyage or season is ended," and "that no master of any fishing ship or others do take up or use any stage already built . . . with a less number of men than twenty-five, who are to be of one entire company." Men were required to leave Newfoundland before the last day of October.[16]

These severe regulations against settlements were followed by serious disturbances in 1671 and bitter controversy. In the first year after their introduction Captain Davis of the convoy ship *Success* arrived at Bay Bulls, and dispatched the first convoy of twenty-three vessels on August 28. He reported that the fishery on the average was poor, yielding only 140 quintals of fish per boat with the exception of Conception

15. *Idem*, pp. 1757–1758.

16. *Idem*, pp. 1759–1763, also p. 1767.

Bay[17] which yielded over 200. He complained that the West Country men were at fault for various abuses. "The West Country owners at the end of the year send their men to New England to save their passage home, by which fishermen are made scarce, and many serviceable seamen lost. By reason of a late act for turning the planters six miles into the country the chiefest have gone to New England." Bad years, and losses resulting from the war with the Dutch from 1672 to 1674 strengthened these arguments and others advanced[18] by the planters and London interests. West Country representatives urged, however, that the French had practically excluded the English from their domestic markets and that they were competing in foreign markets. New England took "great quantities of large fish, about sixty thousand kintalls a yeare, and by increasing the trade there, bring much detriment to that of Newfoundland"; ships and men had been lost in the wars, especially with Spain; settlers and planters continued to destroy "whatever the adventurers leave yearely behinde," to take the best places, and to sell brandy and wine to the seamen. Settlements would increase the consumption of products from New England, "the shipping of which country furnish them with French brandy and Madeira wines in exchange for their fish, without depending for any supply from hence." To limit the establishment of settlements in Newfoundland would indirectly weaken New England. An order in council of May 5, 1675, approved the regulations of 1671 and provided for more rigid enforcement and "that inhabitants be advised to move from the island and future habitation discouraged."[19] The expansion of New England was an argument for limiting the growth of settlements in Newfoundland.

To carry out these regulations Sir John Berry, as commander of the convoy, arrived in Newfoundland on July 11, 1675. He found forty ships in St. John's Harbor, most of which were going to market. However, he conducted an investigation and protested against regulations which, when laboring men could earn twenty pounds in a summer, would remove them to England and put them on the parish. Instead of finding the planters guilty of destroying the stages, he alleged that three fourths of the admirals and the commanders of 45 vessels were in favor of tearing them down at the end of the season, using them for fuel in their ships, and selling the remainder to sack ships. Seamen were persuaded to stay behind by the fishing vessels, and not by the planters, in order to save the passage money of 30 or 40 shillings. Instead of New England ships selling wine and brandy he found that they had "taken good quantities of those goods from hence, the product of which is

17. *Idem*, p. 1766.
18. Lounsbury, *op. cit.*, pp. 135–139.
19. *P.C.*, IV, 1768–1777.

shipped in English vessels for a market."[20] He reported that 175 ships with 4,309 men and 688 boats had caught 250 quintals a boat, worth, at 12 shillings a quintal, £103,200; 20 quintals of core fish at 5 shillings per quintal, £3,440; and made 7 hogsheads of train oil a boat worth, at 40 shillings per hogshead, £9,816; or a total of £116,272. And he went on to say that the planters numbered 1,655 men, who used 277 boats and cured 69,250 quintals of merchantable fish, "most of it shipped in English vessels, which with their core fish and oil will amount to £46,813, upwards of one third [of the value] of the fish taken by the merchant adventurers."[21] In spite of these facts the privileges conferred on the West Country men in 1660, in 1671, and in 1675 were reaffirmed on January 27, 1676.[22]

Nevertheless, determined protests from John Downing, the planters' agent, reinforced by Berry's report, brought temporary relief. The restrictive regulations were suspended in 1677, notwithstanding arguments of West Country merchants in March "that the renewal of the western charter two years ago" had been followed by a revival of trade, and that, in 1676, 7,500 men had gone out with fishing ships besides those on the sack ships. Downing held that a colony of 150 families was of great importance to the fishery since they kept the stores for the fishermen over the winter and offered some slight protection from the French.[23] Further investigations in 1678 by Sir William Poole and in 1679 by Charles Talbot resulted in a report,[24] made by the latter, that there were 1,700 people in the colony, that they were of great assistance to the Adventurers, sawing boards, building boats, and making oars in the winter, and that everything that pertained to the fishery in addition to their bread, clothing, malt, salt meat, and peas came from England, with the exception of such provisions and clothing as came from Ireland; and that only tobacco, sugar, molasses, rum, fresh meat, bread, and flour came from New England. The latter traded for fish but did little fishing, the chief complaint being that New England's fisheries were on the increase, for "they steal fishermen every year." Brandy, wine, and salt were obtained from France, Spain, and Portugal, but only in English ships. The planters had not more than three boats each and owned about a fourth of them. Boatkeepers were left behind by the Adventurers but "there are many that pay their passages out and home and fish the season." They did not regard as important the "pernicious practice lately introduced by the masters and owners of the fishing ships of carrying to Newfoundland byeboatkeepers, and their crews and

20. *Idem*, pp. 1772–1773.
21. *Idem*, pp. 1774–1775; also p. 1784.
22. *Idem*, pp. 1777–1782.
23. *Idem*, pp. 1785–1786.
24. *Idem*, pp. 1788–1790.

servants, to assist the inhabitants in their fishery." In answer to claims that the fishery had increased, it was said that the number of fishing ships was only 125 in 1676, and the season was very bad in 1677. Only 159,000 quintals had been taken in 1679 and this amount was disposed of to the sack ships, of which there had been 140. In 1680 fishing ships totaled 97 and the catch, averaging merely 170 or 180 quintals a boat, was valued at £126,000, of which the sack ships took about 60 per cent and carried it to market.[25] In a memorial of February, 1680, the planters said that "by the violence of the Western Adventurers they have been forced to disperse to twenty different places," and asked that the number should be reduced to four: Trinity Harbor, St. John's, Ferryland, and Trepassey.[26] Because of these investigations new regulations were introduced in February, 1680,[27] which were favorable to residents. They were allowed to keep taverns and public houses, to live near the shore,[28] and to retain possession of their stages. But they were not allowed to build more until after "the Adventurers be all arrived," nor to hire servants in England and transport them to Newfoundland. The regulation of 1671 which prohibited the departure of ships before March 1 was rescinded.

The regulations resulted in encroachments by the residents and became the object of protests from the fishing ships. On October 27, 1684, Captain Francis Wheler[29] reported that because of the scarcity of fuel at St. John's, where the residents were forced to go two miles for it, they had pulled down the Adventurers' stages and that, while the fishing ships had sufficient room for their stages in the ports, warehouses had been built which took up "good room." Fishing ships declined to 32 in 1682 and rose to 43 in 1684, with seamen numbering 1,012 and 1,489, and boats 189 and 294 respectively. In the same years the planters' or residents' boats numbered respectively 299 and 304, and took nearly as much cod as the fishing ships.

The expansion of New England began to have its effect in lending support to the settlements. She was a competitor for labor and for trade. Provisions, molasses, and rum were exported[30] to Newfoundland

25. Lounsbury, *op. cit.*, pp. 162–163. Other accounts show that the industry varied respectively for the years 1675, 1676, and 1677 as follows: fishing ships, 105, 120, 109; seamen, 3,278, 4,556, 4,475; boats 677, 894, and 892. Planters numbered 138 and 152 in 1676 and 1677 and servants 943 and 1,355; boats totaled 206 and 337.

26. *P.C.*, IV, 1792.

27. *Idem*, p. 1791.

28. On February 26 the regulations were modified and planters were forbidden to have buildings, other than those necessary for the fishery, less than a quarter of a mile from the shore. *Idem*, pp. 1793–1794.

29. *Idem*, pp. 1806–1809.

30. See R. G. Lounsbury, "Yankee Trade at Newfoundland," *New England Quarterly*, III (1930), 607–626. See a list of American vessels going to Newfoundland in

to secure bills for an expanding New England trade or European goods.* Labor was attracted to the higher wages of New England. According to Wheler, clothing, tackle, brandy, and Fayal wines were brought from England, but salt, liquor, and provisions came from France and New England. "Considerable quantities of rumm and molasses are brought hither from New England with which the fisher's grow debauch'd and run in debt so that they are obliged to hire themselves to planters for payment thereof."[31] "It would be impossible to continue the trade, for ten hours in the boats every day in the summer and the intolerable cold of the winter makes living hard, without strong drink. But the worst thing is that the New England men carry away many of the fishermen and seamen, who marry in New England and make it their home."[32] It was reported on January 12, 1687, that great quantities of European goods were being imported into New England "under colour of trade to Newfoundland for fish." "The island is become a kind of magazine of contraband goods."[33] "They have their agents in most harbours in the land . . . and so drive an indirect trade, and supply the plantations with several commodities which they ought to have directly from England." "Their vessels generally make two or three trips in a year with bread, flower, porke, tobacco, molasses, sugar, lime-juice and rum." "They sell their provisions some small matter cheaper to the inhabitants, but then they oblige them to take a quantity of rum."[34] Provisions

1698; Prowse, *op. cit.*, p. 200, and C. P. Nettels, *The Money Supply of the American Colonies before 1720* (Madison, 1934), pp. 76–79.

31. *C.O.* (*Colonial Office*) 194: 2. "Debts were never wont to be paid in Newfoundland till the 20th of August but for these two or three years past the rocks have been stript by night, and the fish carryed off in June and July without weighing, a second hath come and taken it from the first and perhaps the planter [resident] hath had twenty or thirty quintalls of fish spoyled in the scuffle, and the rest of his creditors are forced to go without any satisfaction; nay the poor fishermen who helped to take the fish have not one penny wages; salt provisions and craft are all payable here before wages and considering how poor fishermen are used I admire how the planters and inhabitants procure hands from England to fish for them." *P.C.*, IV, 1813. The fishing ships became trading ships and West Country admirals were able to act as tyrants over the population and became known as "kings." Fishing ships disposing of salt to the planters were charged with compelling them to purchase a butt of wine and a quarter cask of brandy with every ten hogsheads. Every house became a tavern. "Drunkenness abounds exceedingly. I have often seen from 100 to 200 men drunk of a sabath day." *C.O.* 194: 5. The problem was more acute in districts remote from the center of trade at St. John's, and explained demands that the price of fish and oil and of goods at St. John's should prevail throughout the country, and that standard measures for oil and other products should be introduced. Topsham, Dartmouth, Plymouth, and Bristol ships sold salt at 62 wine gallons a hogshead, shoveling it in like flour and not allowing it to settle, whereas Poole and Barnstaple sold at 63 beer gallons a hogshead. *C.O.* 194: 3.

32. *P.C.*, IV, 1807.

33. *Idem*, p. 1809.

34. *Idem*, p. 1812.

*See Notes to Revised Edition, p. 509.

and supplies from New England were sold more cheaply than those brought from England in order that she might obtain bills.[35] When fish were abundant and bills scarce the former were purchased by New England and sold to British ships for a "rial" (royal) or two less, with the result that the market for merchants' fish was weakened and they were forced to store it and "rendered uncapable of paying their men's wages or buying the necessarys in England for the next season."[36]

The New England men never carry their fish, which they receive in exchange from the inhabitants and planters for their cargo, to market, but either sell the same upon bill in England by which they gain five or six and thirty pounds per cent, or else for wine, brandy, dowlas and other sorts of linen cloth, silks, alamode and lustring, sarcenets and paper from France. . . . I am told that the New England vessels last year carryed out of Conception Bay upwards of 500 men some of which were headed up in casks because they should not be discovered.[37]

Direct trade between Newfoundland and Europe followed the expansion of New England trade, and was an outgrowth of the restrictions of the Navigation Act and the decrease of fishing ships. Newfoundland offered an escape from the restrictions of the colonial system.* Various ships, although claiming to be English-owned, brought their supplies from Spain and Portugal and took on men from foreign ports. Competition from those selling rum necessitated more direct trade and cheaper supplies of liquor from Europe, and involved a loss of trade to the fishing ships. New England was an important market for European liquors.[38] "At least one fourth of the ships here come from Spain and

35. They "generally sell their cargo for money and bills which makes 25 per cent to them in New England; but if they cannot get them they buy refuse fish and go to the West Indies." November 13, 1698. *P.C.*, IV, 1802; also p. 1805. In 1712 New England sold tobacco and rum or bought fish and sold it to British ships for bills which brought 40 to 45 per cent advance. Only one New England vessel was reported to be fishing on the coast at Ferryland.

36. *C.O.* 194: 5.

37. *P.C.*, IV, 1812.

38. An act for preventing frauds and abuses in the plantation trade (7 and 8 Wm. III, c. 22, 1695) was intended to check smuggling which had grown up under 15 Ch. II, c. 17, 1663, an act that permitted salt to be carried directly from Europe to Newfoundland and New England; but it was evaded on the ground that Newfoundland was not a colony. They "carry considerable quantities of wines and brandy from home which is brought here by the merchant ships from France, Portugall and Spaine, the duties of all which His Majesty is defrauded of." Large quantities came in from France in 1698. Captain Norris, November 13, 1698, *idem*, pp. 1801 ff. It was claimed that this direct trade was responsible for a revival of the Spanish and Portuguese fishery by shipowners with English crews. "The enumerated plantation commodities especially tobacco and sugar which ought not to be transported to any foreign parts without having been first landed in England are frequently carried thither, and sold to the fishing ships who [*sic*] carry them directly to Italy, Spain and Portugal to the great prejudice of his Majesty's revenue in the customs, which ought to be paid here;

Portugal, supplying the land with all manner of necessaries, selling them cheaper than our owners can afford and returning with their cargoes of fish and glutting the market abroad. Their merchants living on the spot can generally undersell our merchants-adventurers, whereby our owners are much discouraged." It was recommended that "all ships making voyages to Newfoundland should be obliged to clear from England, and not bring more liquor than be thought fit to allow for each ship."[39]

The expansion of New England trade and the decline of the fishing ships facilitated the growth of the French fishery, again narrowed the market for Newfoundland fish, and increased the troubles of the fishing ships. French salt was weaker than the Lisbon or Spanish salt used by the English "soe they [the French] use more and their fish weighs the heavier and its [the salt] being weak does not easily burne."[40] They fished "as they drive, and split it alive which makes it very white as does the abundance of salt they use for want of flat room to dry it, and cure it upon ships deck and carry it to market." It was claimed that French duties had destroyed a market for 500,000 quintals of English fish, and that the remission in 1675 by the English of the 5 per cent tax on fish caught in Newfoundland contributed to French success. The French[41] profited by many things, including their aggressiveness in reducing the losses caused by Algerian pirates to ships going to Mediterranean markets, and their fish were said to have the preference in Alicante, Barcelona, Genoa, Leghorn, and Naples.

Increased trade with New England and Europe, a competition from French and New England fish, the growth of settlements and the English carrying trade,[42] together with the drain of labor to the New Eng-

and on the other side European commodities are carried thither from foreign parts and sold to the plantation-ships trading there; who disperse them at their return in the several plantations whither they are bound, contrary to law, and to the great prejudice of this kingdom." April 24, 1701. L. F. Stock (ed.), *Proceedings and Debates of the British Parliament Respecting North America* (Washington, 1924), II, 398. See p. 113 *infra*.

39. *P.C.,* IV, 1805.

40. *C.O.* 194: 2. The quality of English Newfoundland fish, on the other hand, deteriorated. "We have lately received letters from your Majesty's Consuls and the merchants residing in Spain, Portugal and Italy, that the fish brought to those markets from Newfoundland for some years past, has been for the most part so very ill cured that the consumption thereof is greatly abated and that the trade is in danger of being thereby lost."

41. See G. L. Beer, *The Old Colonial System 1660–1754* (New York, 1912), II, 227–229.

42. The settlers and residents were at a disadvantage in the matter of the cost of building, and of food during their eight idle months. They "build suttling houses, gardens and meadows for their cows adding storehouses to their rooms to receive the remainder of this country cargoe; and as few or any can themselves occupy all the

land winter fishery, weakened the position of the fishing ships and contributed to the expansion of the byeboat system as an efficient method of prosecuting the Newfoundland fishery.

> The small boatkeepers of our parts fitting out for Newfoundland have the advantage of the adventurers of taking the choice of the ablest fishermen and shoremen . . . the reason being that they live somewhat easier with them [the fishermen and shoremen], not being obliged to do any ship-work, but only to do the labours of the voyage and so rest in time when it is not weather to work about the fishery. . . . These byeboatkeepers in England generally choose the best sailing ships so as to gain their passage sooner, and if they reach the country early they place themselves in the best and most convenient places by the water side, whereby later ships are often obliged to hire both stage and room from them.[43]

Fishing ships with boats were unable to get a sufficient number of men "by which means instead of catching our lading we are obliged to make use of our owner's credit to buy our lading from the boatkeepers [byeboatkeepers], otherwise [we] must go dead-freighted." The Adventurers claimed they were placed at a disadvantage, what with having their stages pulled down, their boats staved in during the winter, the most convenient room taken up, in some cases, and their being compelled to work behind the byeboatkeepers "which puts them to the expense of one man in five," or to hire stage and room. Captain Norris recommended (November 13, 1698) that the merchant adventurers should "have the preferable encouragement before the planters [and] boatkeepers." The boatkeepers (byeboatkeepers), it was charged, to save paying the return passage home for their men, hid them or even sent them to New England and were able practically to control the land on which their stages were built, and were even known to sell it. The renting of stages to the fishing ships became a practice and increased. "We have known from £5 to £15 given this year for a boat's room in this harbour."[44]

fishing room they have taken possession to . . . they let [them] out to boatkeepers or latter [later] ships who thereby become their tennants." Byeboatkeepers were more efficient than the residents because of their ability to attract the best men by good wages and the share system. Passengers paid £3 for the passage out and a third or a half less for the passage home. "The byeboatkeeper employing himselfe wholly on the fishing account and generally in partnership there [is] seldom or never any dilligence wanting either in catching or curing it." *C.O.* 194: 2.

43. *P.C.*, IV, 1804.

44. *P.C.*, IV, 1804–1805. A boat's room was described as "as much flake as will spread 70 quintals of wet fish," or 80 feet square for 70 quintals of large fish, and 100 feet square for small fish. "The broad flakes spread one third more than the narrow kind." The rule for beaches as at Placentia "where there were ships enough to occupy the whole, was [a frontage] the length of the ships main yard and so extending back inland." In 1727 the governor at Placentia leased the beach to fishing ships for £130

Sir Josiah Child, interested in ship chandlery in the West Country, attributed the decrease of fishing ships "to the growing liberty which is every year more and more used in the Romish countries, as well as others, of eating flesh in Lent and on fish days," to the increase of the French fishery at Placentia and other parts of Newfoundland, to the wars which had impoverished the western merchants and "reduced them to carry on a great part of that trade at bottomry viz., money taken upon adventure of the ship at twenty per cent per annum," and

to a later abuse crept into that trade, which has much abated the expense within these twenty years of the commodity, of sending over private boat keepers, which has much diminished the number of fishing ships. . . . By the building, fitting, victualling and repairing of fishing ships, multitudes of English tradesmen and artificers besides the owners and seamen, gain their subsistence; whereas by the boats which the planters and boat-keepers build or use at New-found-land, England gets nothing.[45]

The outbreak of war with France in 1689 meant immediate losses to the fishing ships because fishermen were seized by the press gangs, Ferryland was attacked in 1694, and St. John's and Ferryland were captured in 1696.[46] On January 13, 1697, it was recommended that convoys from Plymouth and Milford should accompany the salt ships leaving at the end of January, the fishing ships leaving at the end of February, and again, the salt ships sailing at the beginning of June.[47] Land defense assumed greater importance, and forts were recommended for Ferryland, St. John's, and Fermeuse.[48] On the other hand the policy of encouraging the fishing-ship industry as a basis of naval support remained in force. After the restoration of peace by the Treaty of Ryswick in 1697 it was recommended that the number of residents "during the winter should be limited to 1,000 lest by the increase of their numbers they engross the fishery to themselves to the prejudice of our navigation."[49] Following the Revolution of 1688, the fisheries of Newfoundland came more definitely under the control of the House of Commons

sterling, each boat being allowed 9 flakes of 60 feet by 8 feet "and 4 feet between," i.e., 5,400 square feet. Other reports give 5 flakes of 100 feet by 6 feet, that is, 40 by 20 yards front, or 7,200 square feet. "In well settled harbours the ancient custom is strictly adhered to and in case of dispute is ever the standard, forty feet front being esteemed one boat's room without limitation backward."

45. Sir Josiah Child, *A New Discourse on Trade* (4th ed., London, n.d.). See Prowse, *op. cit.,* pp. 188–189.

46. Conception Bay, Trinity Bay, and Bonavista "all suffered from French raids and only Carbonear escaped." Report of Colonel Gibson, June 28, 1697. *P.C.,* IV, 1798–1799. See petitions from Dartmouth, Exeter, Poole, and Bristol in November, 1696, Stock, *op. cit.,* II, 178–181.

47. *P.C.,* IV, 1797. Convoys were likewise sent in 1695 and 1696.

48. See also report of Captain Norris, March 17, 1698, *idem,* p. 1800.

49. *Idem,* p. 1797.

and legislation was introduced to support them. "An Act to Encourage the Trade to Newfoundland" (10 and 11 Wm. III, c. 25) was passed in 1699,[50] and went into effect on March 25, 1700. Its preamble emphasized the importance of the Newfoundland trade and fishery from the standpoint of seamen and ships, the consumption of provisions and exports "whereby many tradesmen and poor artificers are kept at work," and the returns from other countries in the form of "great quantities of wine, oil, plate, iron, wooll, and sundry other useful commodities." The regulations in the charters, such as "the ten commandments," were repeated and elaborated; and, "whereas several inhabitants in *Newfoundland* and other persons, have, since . . . [1685] ingrossed and detained in their own hands . . . several stages, cookrooms, beaches, and other places . . . (which before that time belonged to fishing ships, for taking of bait, and fishing and curing their fish) to the great prejudice of the fishing ships," such individuals were required to give up "to the publick use of the fishing ships . . . all and every the said stages, cookrooms, beaches and other places." After March 25 no one was allowed to take up stages, cookrooms, beaches, or other places "before the arrival of the fishing ships . . . and until all such ships shall be provided with stages, cook-rooms, beaches and other places." Buildings which had been put up since March 25, 1685, and had not belonged to fishing ships since that date were to be held in peaceful possession. Byeboatkeepers were strictly forbidden to disturb buildings "that did belong to fishing ships" since 1685 or that might have been built since March 25, 1700. The byeboatmen and residents were placed under further handicaps by the requirement that every master should hire two "fresh" men in every six, one who had made only one voyage and one who had never been to sea, and "every inhabitant should employ two such fresh men . . . for every boat kept by them." This was intended to offset the regulation requiring every master or owner of a fishing ship to take out one green man in every five.

The act had little effect in reversing the prevailing trends of the fishery.[51] Because the fishing ships had fallen off so much in the years before 1685 they could not occupy more than a third of the stages and rooms, and the residents built and occupied the other two thirds. In some cases the residents relinquished their stages to the fishing ships, but used them before their arrival and after their departure. The fishing ships were "deprived by the Act of the right they had to the said places . . . which must be assigned for one of the reasons why they quitted the fishing trade so soon after the Act passed and why they have declined it ever since."[52] A report by George Larkin, made on August 20, 1701,[53]

50. *P.C.*, I, 250–256.
51. For an extended account see *P.C.*, IV, 1815 ff.
52. *P.C.*, IV, 1835.
53. *P.C.*, IV, 1811 ff.

stated that "these byeboatkeepers can afford to sell their fish cheaper than the Adventurers which must lessen the number of fishing ships," and it paid aggressive fishing-ship masters to secure large numbers of passengers to further a competing development. The admiral of St. John's Harbor in 1701, Captain Arthur Holdsworth from Dartmouth, had brought 236 passengers, "all or great part of which are byeboatkeepers and under a pretence of being freighters aboard his ship, which is only for some few provisions for their necessary use, he hath put and continued them in the most convenient stages, etc. in this harbour, which all along since the yeare '85 have belonged to fishing ships." He had brought out no fresh men nor green men and with others had searched the market towns in the West of England for passengers, and made "an agreement with them that, in case they shall happen to be Admirals of any of the harbours, they will put and continue them in fishing-ships roome." The number of byeboats increased as a result. The number of residents rose from 2,159 in 1701 to 3,153 in 1715.[54] The boats belonging to them had become practically as many as were attached to the fishing ships at the end of the century.[55]

54. Ferryland had 30 houses and families, Cape Broyle 12, Bay Bulls 20, Brigus 6, Bell Inn 3, Toads Cove 2, Munemables Bay 6, Petty Harbor 6, St. John's 60, Quidi Vidi 20, Torbay 4, Holyrood, Salmon Cove, and Harbor Grace, each 12, Carbonear 30, Bay de Verde 10, Old Perlican 6, Trinity 12, Bonavista 25, Greenspond Island, which had apparently been settled in 1698, 3. Total: 267 families. At St. John's "the houses were built on the northern shoar and every family had a sort of a wharfe before their houses to dry their fish on." John Oldmixon, *The British Empire in America* (London, 1708), Vol. I.

55. The total number of men employed, also total residents, boats, shallops, and the catch of cod in quintals, all for about the year 1696, are given in La Potherie, *Histoire de l'Amérique septentrionale* (Paris, 1753), Vol. I. See *Documents Relating to the Early History of Hudson Bay,* ed. J. B. Tyrrell (Toronto, 1931), pp. 181–183. The following table for 1700 is taken from *C.O.* records, and is printed verbatim.

	Residents	Stages	Boats	Byeboat-keepers	Boats of Bye-boatkeepers
Renewse	112	14	20		
Fermouse	104	19	16		
Aquafort	37	9	6		
Ferryland	166	29	24	25	5
Caplin Bay	..	4	..		
Cape Broyle	13	11	2		
Brigus	56	7	10		
Toads Cove	124	11	18		
Whitless Bay	37	6	6	6	1
Bay Bulls	..	27	31	51	9
Petty Harbour	..	..	..		
St. John's	256	74	50	136	45
Quidi Vidi	75	7	26	5	1
CONCEPTION BAY AND TRINITY					
Torbay	30	5	6		

(*continued on next page*)

55 (*continued*). CONCEPTION BAY NORTHWEST

	Residents	Stages	Boats	Byeboat-keepers	Boats of Bye-boatkeepers
Portugal Cove	16	5	5		
Harbour Main	..	..	..		
Baron Cove	..	..	..		
Brigus	56	12	10	6	1
Port de Grave	216	31	37	5	2
Hailins Cove	24	3	..		
Bay Roberts	60	6	10		
Bryants Cove	34	3	3		
Harbour Grace	151	23	20	28	5
Mosquito Cove	..	..	..		
Carbonear	345	44	56	42	6
Croques Cove	46	4	6		
Capelin Cove	..	..	..		
Bay de Verde	11	14	14		
Galleys Cove	17	9	6		
Clown's Cove	51	17	13		
		TRINITY BAY SOUTH			
Old Perlican	225	21	33	13	2
Lance Cove	104	11	18	10	2
Scilly Cove	110	11	10		
New Perlican	97	14	12		
Heart's Content	35	5	7		
		AU NORD			
Heart's Ease	41	6	7		
Trinity	277	39	29		

Total catch of cod in quintals, 188,800.

In 1705 the fishery was roughly distributed as follows:

	Inhabitants' Boats	Cod in Quintals	Train Oil, Tons	Men
Bonavista	24	7,200	42	120
Trinity Bay	116	4,800	28	80
Conception Bay	40	12,000	70	200
St. John's and South	80	24,000	140	400

Stock, *op. cit.*, II, 433–434.

The fishery extended to the north. In 1698 William Wyng was reported as having fished for some years 14° north of Bonavista and "this year one Nevill has been that way and has more fish for his two-men-boats than those here for shallops so that next summer several inhabitants of this harbour design to remove thither as also the masters of ships that have fished here this year." *P.C.*, IV, 1802–1803.

	Number of Residents	Stages	Boats
English Harbour	40	4	7
Salmon Cove	70	7	13*
Green Island	50	3	7
Bonavista	350	32	52†
Bails Cove	42	7	14
Keells	52	4	8
Salvage	43	3	7
Greenspond	51	3	7

* Also 12 byeboatkeepers and 2 boats. † Also 20 byeboatkeepers and 3 boats.
See Appendix, p. 140.

The Newfoundland fishery as conducted by the English had been handicapped by competition from France, New England, and the carrying trade, by wars, and—scarcely less serious—rumors of wars. It had borne grudgingly the tax of impressed sailors levied on an area favored by legislation as "the nursery for seamen." An absence of military defense because of a policy imposed in the interests of the navy involved serious burdens incidental to the destruction wrought by wars. The support given by the fishing ships[56] to the navy weakened the position of settlements and the development of military defenses. The protection of the settlements by land defense was neglected because of reliance on naval defense, and because additional taxes would sap the strength of the English fishery in competition with the French. On the other hand the growth of settlements in spite of restrictions involved exposing them to attack from the French and disastrous losses. Settlements and the growing importance of the sack-ship trade, and of London interests[57] fostered by New England support, necessitated greater attention to land defense which called for a type of government other than that provided by the admirals of the fishing ships. The burden of naval defense strengthened New England and weakened the fishing ships of the West Country. Defense was built up at the expense of "opulence" and the ultimate withdrawal of New England suggests that "the act of navigation" was perhaps not "the wisest of all the commercial regulations of England."

NEW ENGLAND

New England is the most prejudicial plantation to this Kingdom. . . . New England produces generally the same we have here viz. corn, and cattle, some quantity of fish they likewise do kill but that is taken and saved altogether by their own inhabitants which prejudices our Newfoundland trade, where, as has been said, very few are, or ought according to prejudice, to be employed in those fisheries but the inhabitants of Old England. The other commodities we have from them, are some few great masts, furs, and train-oil, of which the yearly value amounts to very little, the much greater value of returns from thence being made in sugar, cotton, wool, tobacco, and such like commodities, which they first receive from some other of his Majesty's Plan-

56. "The security of a fishing colony must ever depend upon a naval force (to the support of which such a colony is supposed greatly to contribute). To do it by fortifications and inhabitants would be impracticable in such a country as Newfoundland which abounds with more harbours than any known country of equal extent, all which in that case should be fortified and require more expense than the whole charge of the navy." 1764. *C.O.* 194: 17.

57. "The merchants of London were induced to set up a fishery in New England, which has been ever since carried on to the great discouragement of the adventurers." *P.C.*, IV, 1819.

tations, in barter for dry cod-fish, salt mackerel, beef, pork, bread, beer, flower, pease, etc., which they supply Barbadoes, Jamaica etc. with, to the diminution of those commodities from this Kingdom. . . . Of all the American plantations, his Majesty has none so apt for the building of shipping as New England nor more comparably so qualified for the breeding of seamen, not only by reason of the natural industry of that people but principally by reason of their cod and mackerel fisheries; and in my opinion there is nothing more prejudicial and in prospect more dangerous to any mother Kingdom than the increase of shipping in her colonies, plantations, or provinces. . . . Of ten men that issue from us to New England and Ireland what we send to or receive from them does not employ one man in England. . . . I must confess that though we lose by their unlimited trade with our foreign plantations yet we are very great gainers by their direct trade to and from Old England. Our yearly exportations of English manufactures, malt and other goods from hence thither amounting in my opinion to ten times the value of what is imported from thence.

SIR JOSIAH CHILD, *A New Discourse of Trade*

THE expansion of New England trade, through the cod fishery, to the West Indies and Europe was supported by the Navigation Acts.[58] They restricted Dutch shipping and extended the field for shipping in the colonies and the motherland. The success of Dutch shipping based on the herring fishery was offset by the success of English shipping based on the cod. The Navigation Act of 1650 prohibited all foreign ships from going to the English colonies and the handling of exports or imports other than in English vessels. The act of 1651 forbade the bringing of goods produced in Asia, Africa, or America into England, Ireland, or the colonies, and the bringing of European goods into England, Ireland, or the colonies, except in English shipping or in such as belonged to the place of production or to the port whence the said goods were usually shipped for transportation. None but vessels entirely owned and manned by Englishmen could engage in the coasting trade. The act of 1660 forbade the shipping of certain enumerated articles,[59]

58. G. L. Beer, *The Origins of the British Colonial System, 1578–1660* (New York, 1908), chap. xii; also H. C. Hunter, *How England Got its Merchant Marine, 1066–1776* (New York, 1935), pp. 102–226. For a statement of the significance of the fishery to the expansion of Dutch trade and an appreciation of the basic importance of the fishery see Thomas Mun, *England's Treasure by Forraign Trade* (London, 1664) (reprint, Oxford, 1928), pp. 74 ff.; also G. N. Clark, "The Navigation Act of 1651," *History*, VII (January, 1923), pp. 282–286.

59. "In the exportation of their own surplus produce too, it is only with regard to certain commodities that the colonies of Great Britain are confined to the market of the mother country. These commodities having been enumerated in the act of navigation and in some other subsequent acts, have upon that account been called *enumerated commodities*. The rest are called *non-enumerated,* and may be exported di-

such as sugar, tobacco, cotton, wood, indigo, ginger, fustic or other dyewoods produced in English colonies, elsewhere than to England, Ireland, or another English colony. Trade between the colonies and the mother country was restricted to English-built or English-owned ships. Following complaints in 1661, New England was granted an exemption which permitted her vessels to sell timber, staves, fish, and other commodities to European countries other than England and to purchase commodities for sale in England instead of merely bringing specie. Another act in 1663 permitted salt to be carried directly from Europe to Newfoundland and New England, and became the basis of the smuggling trade in European and American goods. It prohibited importations into the colonies of any European commodities that had not been loaded and shipped in England,[60] and imports in the colonies were to be carried only in English-built shipping.[61] The effectiveness of these acts

rectly to other countries, provided it is in British or Plantation ships, of which the owners and three-fourths of the mariners are British subjects." Adam Smith, *Wealth of Nations* (New York, 1937), p. 543; also 429–431. An act of 1706 (3 and 4 Anne, c. 5) added naval stores, rice, and molasses to the enumerated articles. G. L. Beer, *The Old Colonial System 1660–1754* (New York, 1912), Vol. I, chap. ii. For a discussion of the significance of weak staple development in England to the development of colonial policy, which was a staple policy applied to the whole, see Heckscher, *op. cit.,* II, 69–71.

60. John Hull wrote from Boston on December 22, 1677: "If that we send our fish to Bilboa and carry the produce thereof into the Straits, at great charge and hazard, and procure fruits, oil, soap, wine, and salt (the bulk of our loadings salt, because that most necessary for us, and always ready to be had at Cadiz); and because we have little other goods, for our necessity calls not for much—we must go to England to pay his majesty's customs; which is as the cutting off our hands and feet to our trade; we must neither do nor walk any more; but this orphan plantation will be crushed. If we carry our provisions which we have raised with great difficulty, because of long winters, etc., to the West Indies, we pay custom for our cotton, wool, and sugar there; and the bulk of them are sent to England again from hence, and pay custom there a second time. If we might have liberty for our vessels only to trade into the Straits, or a certain number of them every year, though it were but two or three ships in a year; to supply the country with such necessaries as those parts afford; but, for this so remote plantation to be punctually bound up to the acts of trade relating to England, methinks, if represented to a gracious sovereign and compassionate parliament such a poor orphan plantation might have some exemption from the severity of those acts of trade." "Diary of John Hull," *Transactions and Collections of the American Antiquarian Society,* III, 130.

61. "The land was good and of great extent, and the cultivators having plenty of good ground to work upon, and being for some time at liberty to sell their produce where they pleased, became in the course of little more than thirty or forty years (between 1620 and 1660) so numerous and thriving a people, that the shopkeepers and other traders of England wished to secure to themselves the monopoly of their custom. Without pretending, therefore, that they had paid any part, either of the original purchase-money, or of the subsequent expence of improvement, they petitioned the parliament that the cultivators of America might for the future be confined to their shop; first, for buying all the goods which they wanted from Europe; and, secondly, for selling all such parts of their own produce as those traders might find it convenient to buy. For they did not find it convenient to buy every part of

in North America was enhanced after the capture of New York and the Peace of Breda by the restriction of Dutch shipping. In 1673 an act levied export duties on enumerated products where shipped from one colony to another. An act in 1696 required the carrying of both imports and exports in English-built ships. The colonial system had its advantages for the colonies, not the least being the protection of the mother country, and its disadvantages were overcome in part by violation of the laws and by the development of trade through Newfoundland.[62]

The expansion of New England was furthered under the Navigation Acts by the opening of the West Indies. The attempts of the Brazil Company of Portugal to keep English merchants from carrying on trade directly with Brazil were followed by efforts to encourage sugar production in the British West Indies. The first sugar cane was planted in Barbados in 1641, and production increased rapidly after 1650. Sugar, and later indigo, cotton, and other products, displaced tobacco. The rapid spread of the drinking of chocolate[63] from 1650 to 1660 and of coffee and tea at a later date was responsible for a marked increase in the consumption of sugar. The production of sugar in Jamaica, which came into English possession in 1658, increased, after the first cane was planted in 1660, to an annual production of 1,710,000 pounds by 1670. From 1680 to 1684 Jamaica exported 31,647 hogsheads to England, from 1686 to 1691 more than 57,000, and from 1698 to 1700 nearly 20,000. Between 1697 and 1700 exports from the West Indies to England totaled about 70,000 hogsheads of Muscovado and over 2,000 hogsheads of white sugar. Quantities were also exported to the American colonies. In 1677 and 1678 Jamaica was sending sugar to Virginia, to the Bay of Campeachy, and to Carolina; Nevis to Virginia and to New England; St. Christopher to New England and to Virginia; and Barbados to Bermuda, Carolina, Virginia, New York, and New England. From 1697 to 1700 Barbados exported to the American colonies 6,875 hogsheads, 6,403 tierces, 6,837 barrels, and 3,452 small casks of sugar. Quantities were reëxported to England. Rum was manufactured in the American colonies, which offered the chief market. In 1666 Eng-

it. Some parts of it imported into England might have interfered with some of the trades which they themselves carried on at home. Those particular parts of it, therefore, they were willing that the colonists should sell where they could; the farther off the better; and upon that account proposed that their market should be confined to the countries south of Cape Finisterre. A clause in the famous act of navigation established this truly shopkeeper proposal into a law." Adam Smith, *op. cit.*, p. 580.

62. Heckscher, *op. cit.*, Vol. II, chap. x, contains a study of the importance of Newfoundland; but see Lounsbury, *op. cit.;* also Child, *op. cit.*, chap. iv.

63. Ellen D. Ellis, *An Introduction to the History of Sugar as a Commodity*, Bryn Mawr College Monographs, IV (Philadelphia, 1905), pp. 88 ff.

lish plantations in the West Indies were giving employment to 400 sail of English ships and 10,000 seamen.

The rapid expansion of the sugar trade meant increased demands for cheap labor, and white labor was rapidly displaced by the importation of slaves. By 1670 negroes had become an important source of labor in Barbados, Jamaica,[64] and the Leeward Islands. During the period of monopoly held by the Royal African Company, or from 1672 to 1688,[65] 46,396 slaves were imported by the British colonies. In the free-trade period from 1698 to 1708, Jamaica, Barbados, and Antigua alone imported nearly 90,000, in addition to nearly 20,000 brought in by the company.[66] Ships left Birmingham, Sheffield, and other centers with goods such as woolens, firearms and ammunition, iron, brass, malt spirits, tallow, tobacco, pipes, Manchester goods, glass beads, linens, cutlery, and East Indian goods and traded them in Africa for slaves. Having delivered the slaves in the West Indies they returned to England with sugar, the excess in value of slaves over sugar being paid in specie or bills.

The expansion of sugar production involved not only demands for labor but also demands for such commodities as staves, and lumber for building. Horses were also needed to work the mills that were used to grind the cane. When agriculture was almost confined to the production of sugar, a demand arose for protein foodstuffs, especially beef and fish. The demand in the tropics for fish and general supplies and the demand in New England for tropical products contributed to the rapid expansion of trade.

The middle of the seventeenth century was a peak in the history of the fishing industry. The period of marked emigration to the colonies came to an end[67] with the outbreak of England's Civil War. As a result of it, English vessels left the home ports and adopted New England as a base for the prosecution of the fishery in Newfoundland; and the number of ships from the west coast of England to Newfoundland[68] declined from 275 to 100. The fishing ships from England lost the income derived from the fares paid by settlers outward-bound for New England and from the freight charges on the chattels they took with them.

64. On the significance of Jamaica as a center for the distribution of slaves to the Spanish colonies see Nettels, *op. cit.*, chap. i.

65. *The Cambridge History of the British Empire*, I, 440 ff.

66. F. W. Pitman, *The Development of the British West Indies, 1700–1763* (New Haven, 1917), chap. ii; also Beer, *op. cit.*, Vol. I, chap. v.

67. *The Cambridge History of the British Empire*, Vol. I, chap. v; Lorenzo Sabine, *Report on the Principal Fisheries of the American Seas* (Washington, 1853), pp. 93 ff.; and S. E. Morison, *Builders of the Bay Colony* (Boston, 1921), pp. 25 ff.

68. Prowse, *op. cit.*, p. 159.

The last English ship to engage in the New England fishery was said to have sailed in 1661.[69] The depression which followed the Civil War, the stoppage of emigration, and the loss of markets because of a lack of protection[70] were also in part responsible for a growth of the fishery in New England.

As a result of the expansion of markets in the West Indies and Spain, Villebon, speaking of the Isle of Shoals about the end of the century, described them as places

> where great quantities of fish are dried. . . . Stationed there, are some sixty vessels each with a crew of four men besides the beach-masters and the women who take charge of the fish on shore. Altogether perhaps 280 men, but . . . from Monday to Saturday they are all away fishing on the banks in the open sea. . . . In Ipswich bay there were about six hundred men, at Cape Anne some forty fishermen's houses, at Salem four hundred houses, the inhabitants all fishermen and sailors.

Marblehead was less important than Salem. Barques from Boston enjoyed a virtual monopoly of trade in the Bay of Fundy. New England had had as many as 30 shallops at Port Rossignol in Nova Scotia since 1670, and a small warehouse existed at that point in 1686.[71] At La Have 12 to 15 large vessels were engaged in the fishery.[72] In May, 1699, Villebon noted that the English had ketches of 40 tons, and that they had secured one load of fish and had returned for a second. Vessels of from 20 to 30 tons with five men including the beachmaster were used and were able to take from 900 to 1,000 quintals in a summer. In 1699 fishermen from Salem and Marblehead offered to pay a license fee for permission to fish on the banks of Nova Scotia and to get wood and water.[73] In 1708 it was claimed that 300 fishing vessels from New England had been on the coast of Acadia.

The fishermen of New England not only participated in the shore fishery on an extended coast line but also solved the problem of drying

69. *P.C.*, IV, 1756. Sir Josiah Child wrote, late in the period, that English ships had been displaced in the carrying trade of New England except for "the liberty of carrying now and then by courtesy, or purchase, a shipload of fish to Bilboa when their own New English shipping are better employed or not at leisure to do it."

70. Prowse, *op. cit.*, p. 163. "It [the Civil War] caused all men to stay in England in expectation of a new world, so as few are coming to us, all foreign commodities grow scarce, and our own of no price. . . . These straits set our people on work to provide fish [and] clapboards." Winthrop's Journal, cited by Judah, *op. cit.*, p. 103.

71. H. A. Innis, *Select Documents in Canadian Economic History 1497–1783* (Toronto, 1929), pp. 55–56.

72. *Idem*, p. 60.

73. *Idem*, pp. 52–53. The Sieur de Dièreville found ten English vessels fishing along the Nova Scotia shore. *Relation of the Voyage to Port Royal in Acadia or New France by the Sieur de Dièreville*, ed. J. C. Webster (Toronto, 1933), p. 73.

fish from the offshore banks. A limited number of men in small ships were employed on these banks; and having caught and salted down a load of fish, they brought them in to shore, washed off the salt, and dried them. The large vessels coming from France and England to engage in the dry fishery anchored in the harbors and were more dependent on boats which fished directly offshore. New England, being able by virtue of its nearness to the base to build small and yet seaworthy ships, could exploit offshore banks for the dry fishery, an advantage not possessed by the large Atlantic ships from Europe that were engaged in the shore fishery. The French lamented their lack of settlements in Nova Scotia and their inability either to build small ships or to fish offshore with large ships.[74] The element of distance from the base was an important factor in the size of the ship and its adaptability to fishing.

The development of the offshore dry fishery in Nova Scotia coincided with the demands for the poorer grades of fish bought for the slaves. Small two-masted ketches suited to fishing in summer made two voyages a year to the West Indies during the winter season when cargoes of sugar were available. Sugar, molasses, indigo, and cotton were imported by New England and reëxported to England in return for dry goods and supplies for the fishery. In 1661 New England was "the key to the Indies without which Jamaica, Barbadoes and ye Charibby Islands are not able to subsist, there being many thousand tunns of provisions as beefe, pork, pease, biskett, butter, fish, carried to Spain, Portugall and the Indies every year."[75]

74. Innis, *op. cit.*, p. 52.

75. *Documents Relative to the Colonial History of the State of New York*, ed. E. B. O'Callaghan (Albany, 1856–87), III, 40; also C. S. S. Higham, *The Development of the Leeward Islands under the Restoration 1660–1688* (Cambridge, 1921), ch. viii. On the coast of Maine "shop-keepers there are none, being supplied by the Massachusets merchants with all things they stand in need of; keeping here and there fair magazines stored with English goods; but they set excessive prices on them; if they do not gain cent per cent they cry out that they are losers. The fishermen take yearly upon the coasts many hundred kentals [quintals] of cod, hake, polluck, etc., which they split, salt and dry at their stages, making three voyages in a year. When they share their fish (which is at the end of every voyage) they separate the best from the worst, the first they call merchantable fish, being sound, full grown fish and well made up; which is known when it is clear like a Lanthorn horn and without spots, the second sort they call refuse fish, that is such as is salt burnt, spotted, rotten and carelessly ordered; these they put off to the Massachusetts merchants: the merchantable for thirty and two and thirty ryals [reals] a kental; the merchants sends the merchantable fish to Lisbonne, Bilbo, Burdeaux, Marsiles, Talloon [Toulon], Rochel, Roan and other cities of France, to the Canaries with claw-board and pipe-staves which is there and at the Charibs a prime commodity; the refuse fish they put off at the Charib-islands, Barbadoes, Jamaica, etc., who feed their negroes with it. To every shallop belong four fishermen, a master or steersman, a midshipman, and a foremast-man and a shore man who washes it out of the salt and dries it upon hurdles pitcht

In the last quarter of the century large New England ships carried fish and lumber to the West Indies, fish to Spain, Portugal, the Mediterranean, and England, rum to Africa, and slaves from Guinea and Madagascar. Freights might be picked up in European markets, in which the fish were sold, and taken to England, where the ship and cargo would be disposed of in return for bills on London, or for cordage, iron, hemp, fishing tackle, and manufactured products[76] to be taken to the colonies. Bills were sold to Boston merchants for supplies for the fishery. Specie was brought from the West Indies and Portugal. Masts[77] were sent direct to England in return for dry goods and supplies. Direct trade from England and Europe to the West Indies made the Barbados and the West Indies, as well as Newfoundland, a clearing-house through which New England ships imported European goods.

The market for the large winter-cured New England fish—"great merchantable" fish—was chiefly to Bilbao. About 1700, Boston exported some 50,000 quintals of dry fish of which three fourths went to that market. New England cod caught throughout the winter and cured in cold weather brought from fifty cents to a dollar more per quintal in Bilbao than Newfoundland fish. Merchants expected profits of 50 per cent on fish sold, and 100 per cent on goods bought in London with specie taken from Spain. "Little merchantable" fish were sold in Lisbon and Oporto; middling fish and medium in the Canaries, Madeira, Fayal, and Jamaica; and refuse in Barbados and the Leeward Islands.

In 1662, it was estimated, Boston had a population of 14,300, and possessed 300 vessels engaged in trading with Nova Scotia, Virginia,

upon stakes breast high and tends their cookery; they often get in one voyage eight or nine pound a man for their shares; but it doth some of them little good; for the merchant, to increase his gains by putting off his commodity in the midst of their voyages and at the end thereof, comes in with a walking tavern, a bark laden with the legitimate blood of the rich grape which they bring from Phial [Fayal], Madera, Canaries, with brandy, rhum, the Barbadoes strong water and tobacco; coming ashore he gives them a taste or two, which so charms them that for no persuasions that their employers can use will they go out to sea, although fair and seasonable weather for two or three days, nay sometimes a whole week, till they are wearied with drinking, taking ashore two or three hogsheads of wine and rum to drink off when the merchant is gone. . . . When the day of payment comes they may justly complain of this costly sin of drunkenness for their shares will do no more than pay the reckoning; if they save a kental or two to buy shooes and stockins, shirts and wastcoats with, tis well, other-wages they must enter into the merchants books for such things as they stand in need off, becoming thereby the merchants slaves and when it riseth to a big sum are constrained to mortgage their plantation if they have any" (1663). John Josselyn, "An Account of Two Voyages to New England," *Collections of the Massachusetts Historical Society,* Series 3, III, 350–352.

76. "As yet (1657) our chief supply in respect of clothes is from England." From the "Diary of John Hull," *Transactions and Collections of the American Antiquarian Society,* III, 180.

77. R. G. Albion, *Forests and Sea Power* (Cambridge, 1926), chap. vi.

the West Indies, and Madeira. Toward the end of the century and especially after 1690 the Revocation of the Edict of Nantes, the wars with France, and the struggle with the Indians at home known as King Philip's War were responsible for depression. But in spite of this Boston had 174 ships in 1700, and there were 70 in other ports. Boston and Charleston cleared a thousand vessels a year. In 1709 Massachusetts cleared 200 large ships and 120 of lighter burden in the West Indian trade. In or at the end of the period, it was estimated that 300 sail of some 30,000 tons were in the trade between England, Newfoundland, Nova Scotia, New England, and in lesser degree other parts of the mainland. Twenty-seven hundred men were employed, and in money the annual revenue to England amounted to about £260,000.[78]

FRANCE

The French by this trade had so far encreased their riches and naval power at that time, as to make all Europe stand in fear of them; which plainly shows that twenty years of quiet possession is capable of making any prince that has it the most formidable both by sea and land by the yearly encrease of men, ships, bullion etc. . . . The whole encrease of the naval greatness of France had its foundation from this trade, and for the nature of it is such that about one fourth of the men employed in it are green men, that were never before at sea, and the climate being very healthy scarce one man in fifty dies on a voyage. . . . [They] have quite eaten the English out of this trade.

Considerations on the Trade to Newfoundland.
A Collection of Voyages and Travels (London, 1745)

The increase in the production of sugar in the British West Indies and the growth of trade under the Navigation Acts attracted shipping and facilitated the expansion of the French fisheries. New England played a direct part in the falling off of fishing ships in Newfoundland and the development of the byeboat and the resident fishery. The aggressive commercialism of New England thrived on the fishing industry and showed itself in the extension of the fishery to Nova Scotia, in the growth of trade to the West Indies, Newfoundland, and other regions in the Atlantic basin, and in urgent demands for currency. The contribution of the fishing industry to the integration of the British Empire was in striking contrast to its contribution to the French.*

France, like England, began the second half of the seventeenth century with possessions in the West Indies and on the coast and mainland

78. W. B. Weeden, *The Economic and Social History of New England, 1620–1789* (Boston, 1890), Vol. I, chaps. v, ix; also J. D. Phillips, *Salem in the Eighteenth Century* (Boston, 1937), chaps. ii, iii, iv.

of the north-temperate regions of North America. From 1663 on, Colbert pursued an aggressive policy of unification alike in New France, the West Indies, and the fishing industry. But limited development in the St. Lawrence as a result of the demands of the fur trade and of severe competition from New York and from Hudson Bay, the importance of the bank fishery, the scattered character of dry-fishing regions, and the pressure of the Dutch in the West Indian trade, following their exclusion from British trade by wars and the Navigation Acts, were factors which, for France, limited the possibilities of coördinated growth.

France succeeded in establishing settlements in the West Indies—in St. Christopher by 1630, in Martinique and Guadeloupe in 1635,[79] the Tortugas in 1640, St. Martin, the Saints, Marie Galante, and St. Lucia in 1648, St. Croix and Grenada in 1650, and St. Bartholomew in 1659. The French population, chiefly from Normandy and Brittany, totaled 7,000 in 1642, and had increased to 15,000 in 1655. As in the British West Indies, the planters turned from tobacco to sugar, especially subsequent to 1640,[80] and the number of slaves had increased to 12,000 by 1655. The trade which followed this increase in population[81] and in the production of tropical products attracted the Dutch, particularly after the passing of the English Navigation Acts and the loss of New York.[82] To check this trade, Cayenne was settled, and in 1664 the French West India Company was organized and given a monopoly for forty years.[83] By 1683 it was estimated that more than 200 French ships were engaged in the West Indian trade. Sugar consumption had increased to 17,700,000 pounds, and 29 refineries had been established in France. It was held that by 1680 Martinique, the most profitable island, every year required from 1,000 to 1,200 new slaves, while the remaining islands needed from 1,500 to 1,800, or a total of 2,500 or 3,000. And even these were not adequate, as planters complained of the restrictions they were under because of the monopoly possessed by the company. The increase in population meant demands for foodstuffs. In return for sugar, cotton, indigo, ginger, tobacco, and hides, the southern ports—Bordeaux, La Rochelle, and Nantes, which had not suffered from the wars that handicapped the Channel ports—sent out wines, brandy, salt, flour, pork and beef, staves, and headings.[84] But these supplies were inade-

79. See Maurice Satineau, *Histoire de la Guadeloupe sous l'ancient régime 1635–1789* (Paris, 1928), chaps. i, ii; L. P. May, *Histoire économique de la Martinique 1635–1763* (Paris, 1930), pp. 1–14.

80. S. L. Mims, *Colbert's West India Policy* (New Haven, 1912), chaps. xi, xii.

81. May, *op. cit.*, chap. ii.

82. Mims, *op. cit.*, chap. i.

83. *Idem*, chaps. ii, iii; see also Satineau, *op. cit.*, chap. iii and Arthur Girault, *The Colonial Tariff Policy of France* (Oxford, 1916), pp. 17 ff.

84. Mims, *op. cit.*, pp. 236 ff.; May, *op. cit.*, pp. 136–149.

quate; and salt beef, more than 30,000 barrels of which were consumed annually, was purchased in Ireland[85] and reëxported from France to the West Indies. France alone was not able to furnish her sugar plantations even with foodstuffs.

Newfoundland offered possibilities as a basis of supply for the West Indies and the navy.[86] "Without fish from that place [Newfoundland]," it has been written, "that nation [France] cannot be supplied nor the King of France's navy furnished with fish." Driven by the English from the territory between Cape Race and Cape Bonavista, France attempted to consolidate the fishery to the north and south. As a result of difficulties with the English fishery she established a settlement at Placentia in 1662.[87] To encourage the industry she provided a protected market in 1664 by imposing import duties on cod.[88] Placentia was free from field ice in the spring, could be used for fishing operations earlier than the eastern bays, and was close to the Banks and to Cape St. Mary with its abundant supply of spring herring for bait.[89] Lahontan, describing it, said, " 'Tis a place of refuge to the ships that are obliged to put into a harbour where they go or come from Canada and even to those which come from South America when they want to take in fresh water or provisions and have sprung their mast or been dammag'd in a storm."[90] The establishment of a garrison permitted fishermen to pursue their activities with greater security in small neighboring harbors, as at Great Burin, St. Lawrence, Mortier, and Chapeau Rouge.[91] A memorial dated January 8, 1668, noted that "the French planters being now more than the English, and thus fortified, make dry fish where they please and load therewith at least 100 great ships whereas last year there were not above 10 or 12."[92] Attacks by the Dutch were re-

85. Léon Vignols, "L'Importation en France au XVIIIe siècle du bœuf salé d'Irlande," *Revue Historique*, September, 1928, pp. 78–95.

86. The importance of the fishery to the navy was emphasized continually. Innis, *Select Documents,* pp. 146–147.

87. D. W. Prowse, *A History of Newfoundland* (London, 1896), pp. 178 ff.

88. From 8 to 10 livres, 10 sols a thousand on dry fish and 3 livres a thousand on green.

89. R. G. Lounsbury, *The British Fishery at Newfoundland, 1634–1763* (New Haven, 1934), pp. 231–233; also Prowse, *op. cit.,* pp. 182–183: "They get to work six weeks before us and take such catches that they are generally gone before the end of July" (1684). *P.C.,* IV, 1807. La Potherie described the harbor as rather difficult to approach because of the tides. *Documents Relating to the Early History of Hudson Bay,* ed. J. B. Tyrrell (Toronto, 1931), p. 157.

90. *New Voyages to North America by the Baron de Lahontan,* ed. R. G. Thwaites (Chicago, 1905), I, 335.

91. *Idem,* p. 338. For an account of the regulations introduced in the South for the St. Malo fishery see Hippolyte Harvut, "Les Malouins à Terre-Neuve," *Annales de la Bretagne,* November, 1893, pp. 23–26.

92. *P.C.,* IV, 1751–1752.

sponsible for a partial decrease,[93] but ten years later the French fishery still employed 60 "great ships." In 1675 "the St. Malo fleet of 20 sail go without convoy, being all considerable ships, and about 40 or 50 with convoy" from a rendezvous at Trepassey.[94] In 1681 French ships of 200 to 400 tons from Bordeaux and St. Jean de Luz and 15 or 16 "Biscainers" were at Placentia. It was estimated that there were 100 ships fishing from St. Mary's to St. Pierre, those from St. Malo being at the latter point.[95] With settlements to go to, fishermen arrived in June and stayed until October, catching an average of 200 quintals for every boat and five men. In 1684 there were 30 vessels, and the French

catch 300 quintals to a boat of four men while we catch but 100 quintals. The usual price is six *livres*. The French catch more, victual cheaper, finish earlier, and get the first of the market, so they profit more by the trade than we. The French markets are France, Spain, Portugal and Italy. Their trade generally increases except during war with Spain. . . . They are supplied with salt provisions from France and with rum and molasses from New England.[96]

In 1698 the fishery had been profitable,[97] yielding 300 quintals and upward a boat, and there were about 50 sail of merchantmen of which 20 had sailed by the beginning of September. The number of fishing vessels comprised 14 at St. Pierre Island, 3 at Little Placentia, 5 at St. Mary's, and 4 at Trepassey. Although vessels came out from year to year with their crews from France and returned at the end of the season, Placentia[98] had become important as a depot for fish.[99] It was

93. Prowse, *op. cit.*, p. 183.[See Notes to Revised Edition.]
94. *P.C.*, IV, 1775.
95. Prowse, *op. cit.*, pp. 185–186. In 1680 Englishmen were severely punished for the destruction of boats and property which had been left over the winter by a French fisherman at Colinet Island. *Idem*, pp. 173–174; see also J. D. Rogers, *Newfoundland* (Oxford, 1911), pp. 282–290.
96. *P.C.*, IV, 1807.
97. *Idem*, p. 1803; also, for 1701, *idem*, p. 1814.
98. "Ships at Placentia keep their boats at St. Mary's where they cure their fish but dry it at Placentia."
99. Although it was stated that "their fishery is managed all by fishing ships, without making use of sack-ships or by-boats," Lahontan wrote that there were generally from 30 to 40 ships and sometimes 60 from France at Placentia, some of which were engaged in the fishery, while others "have no other design than to truck with the inhabitants." Both the inhabitants and the fishermen sent sloops about two leagues from the port to fish, the fishery lasting from June to the middle of August. Lahontan, *op. cit.*, pp. 335–337. "They manage their trade of fishery as our merchants do. . . . They usually bring from Europe some merchandize to support the inhabitants." Their trade decreased in 1700, "occasioned by the great quantity of capelin (which the fish don't take as usual when scarce)." In that year it was stated that they had 30 sail at Placentia, 24 at St. Pierre, 4 at Grand Burin and St. Lawrence, 2 at Mainclon, and 1 at Fortune Bay. Ships from 80 to 300 tons carried 4 to 20 boats each, and their men received a fifth part of the fish. In 1701 the *St. Louis* of 200 tons with 16 boats took 7,000 quintals between June 10 and the end of August. The total number of boats belonging to ships, residents, and byeboatkeepers was 1,010. In general three

stated that 29 families or some 160 people and, with the soldiers, about 300 in all wintered in the colony, 12 families being in Little Placentia. By 1713 it was estimated that there were only 180 French on the island.[100]

Placentia, as an encroachment on English territory, suffered from raids,[101] from the interference of the government, and from difficulties in preparing large bank fish. Control under a governor involved heavy taxes and apparently corruption, in contrast with English Newfoundland which was alleged to suffer because of being without a governor. Both Lahontan and La Potherie referred to the superiorities of other fishing grounds, the former[102] noting that the fish at Ile Percée were "more proper for drying"; and both described the advantages of Cape Breton.[103] The handicaps included a deficiency of gravel on the beaches and a lack of space. Demands for space were, as ever in the dry fishery, a basic consideration.

St. Malo ships were particularly concerned with the ports of the Petit Nord. The *arrêt* of 1640 was extended to apply to all France in 1671, and in 1681 to other ports of this fishery.[104] In 1706 it was esti-

men would work in the boat and two ashore; and the catch brought 24 reals a quintal. Residents numbered: Placentia 200, St. Pierre 100, Mainclon 86, Bay of Fortune 150, and St. Mary's 7. *Colonial Office,* 194:2. See also Ferdinand Louis-Legasse, *Evolution économique des Iles Saint-Pierre et Miquelon* (Paris, 1935), pp. 37–38. In 1701 it was stated that the French plantations did not increase "nor are they of any other use than preserving the boats, craft and goods left by the merchant ships for the succeeding voyage." The fishermen never baited their own boats but had boats supply them twice a day. "They make their voyages as we do, and some pay by the share and some by the voyage." *P.C.,* IV, 1804. The residents employed their fishermen chiefly by shares, and apparently paid them by allowing them one third of the fish and the first right to purchase fish and oil "au prix courant de la coste" (at prices current on the coast). Innis, *op. cit.,* pp. 87–88. The technique and organization of the industry at Placentia entered largely into the later problems of Cape Breton.

100. Prowse, *op. cit.,* p. 185. For the number of settlers, see Rogers, *op. cit.,* pp. 88–89. He gives as the number at Placentia 123 in 1687, 108 in 1691, 126 in 1693, and 213 in 1710. He makes the total number of French in Newfoundland 640 in 1687, 355 in 1693, 555 in 1704, and 579 in 1710.

101. Rogers, *op. cit.,* chap. v.

102. Lahontan, *op. cit.,* p. 305.

103. *Idem,* p. 324; *Documents Relating to the Early History of Hudson Bay,* ed. J. B. Tyrrell (Toronto, 1931), p. 160. "The throater, header, splitter and salter (*piqueur, decoleur, trancheur* and *saleur*) dress and salt the cod. A bed of cod nine to ten feet long and 3 feet high is salted one row of cod on top of the other with flesh side uppermost." The cod was left in "salt bulk" five or six days, then washed in the sea to remove the salt, and piled to allow the water to drain off. "Then they heap them together. They remain there two days and, after St. John's Day (June 29) for one day only on account of the heat." Next they were spread on the beach and turned flesh up during the day and flesh down at night. After they were dried, five or six cod were piled head on tail "en mouton" for three days and three nights and finally put in a large pile, sometimes of 300 quintals, and left for a month to sweat before loading in the vessels. *Idem,* pp. 158–159.

104. *P.C.,* V, 2178–2180. Louis-Legasse, *op. cit.,* pp. 40–41; also Lahontan, *op. cit.,*

mated that 30 French ships with 60 to 150 men each were engaged north of Bonavista. They arrived the latter end of May and completed the fishery in six weeks or two months. In 1707, 14 vessels arrived from St. Malo of which 12 were for Marseilles; 3 from Granville (2 for Marseilles and one for Bordeaux); 2 from Brest (one for Brest and one for Marseilles); 2 from Binic to return there; and one from Norleigh(?) for Marseilles. The men were said to have been paid 200 livres each. North from Cape St. John they had an average of 40 sail carrying 7,000 quintals each, chiefly to the Straits.

The French fishery increased in the Gulf of St. Lawrence as on the south coast of Newfoundland. An *arrêt* of 1669 permitted exports of cod to France by the inhabitants of Canada, who paid the same duties as were paid by residents of Havre. The ordonnance of 1681 fixed the limits of the free fishery at Cap l'Espoir and Cap du Rosier. The handicap of lack of shelter and drying space at Ile Percée[105] grew more serious when the fishery expanded. By 1686 difficulties had become acute and, according to De Meulles, open warfare between the fishermen from Bayonne and La Rochelle and the Normans from Honfleur was avoided only by regulations. Whereas the struggle in Newfoundland was between the West Country and the settlers, in the French fishing regions it was between ships from France.[106] The fishing ships coming each year from France were given the rights to "graves, galets et vignaux" (curing beaches, landings, and drying frames or flakes) in preference to the residents. Space was definitely allocated on the basis of three *vignaux* 40 fathoms long—the fathom was one of five feet—by 5½ feet wide, per boat with an addition of 3 feet for a road between *vignaux*. The ships were allowed one boat per 20 tons, and, as the ships averaged from 200 to 300 tons, an allowance per ship of at least 11,000 square feet was necessary. The Normans, or those ships making green fish as well as dry, were limited to two *vignaux* per boat. As in Newfoundland, captains and others were forbidden to destroy their buildings, but were allowed to take down their "échafaux" (stages) because of the dan-

p. 308; and Harvut, *op. cit.*, pp. 27–28. The use of the "jigger" was forbidden in 1684. *Idem*, p. 28. A "jigger" is a plummet of lead, with two or three hooks stuck at the bottom, projecting on every side and quite bare. This is let down to the proper depth; then a man, taking a hitch of the line in his hand, jerks it smartly the full length of his arm, and having let it down slowly, jerks it again.

105. Lahontan, *op. cit.*, p. 306; also, on the scarcity of "galets" and the need of building flakes, see Innis, *op. cit.*, p. 423.

106. *Idem*, pp. 47–48. In 1673 Simon Baston, a merchant of La Rochelle, was murdered at Percé. J. M. Clarke, *L'Ile Percée* (New Haven, 1923) (reprint title, 1935, *The Gaspé*), p. 138. Father le Clercq speaks of there being 400 or 500 French at Ile Percée. *New Relations of Gaspésia* (Toronto, 1910), p. 81. The English under Phips captured six vessels in 1690.

gers from ice during the winter and to pile them out of the way; and this property was given freedom from molestation for three years. Again, captains or others were forbidden to take the boats of those absent; but they were permitted to leave people over the winter. Finally the residents were given the right of ownership in "vignaux ou galets" on their own property. A road was marked by De Meulles from the shore to the houses. In 1706 it was reported that there were ordinarily seven or eight vessels in peacetime, chiefly Basque and Norman. A system of exchange had grown up by which Normans traded their small fish suitable for drying to the Basques, who in turn traded their large fish, suitable for the green fishery and the Paris market, to the Normans, the trade being conducted at the rate of two small for one large.[107] The possibilities of expansion, however, were limited. There appears to have been little development in the Magdalen Islands.

In the area outside the free fishing zone the influence of monopoly persisted. On the North Shore of the Gulf of St. Lawrence concessions of fishing rights were probably adapted to the seal fishery and hastened the latter's development. By 1713 concessions had been granted on the Island of Anticosti, at Mingan, and near the Straits of Belle Isle; but losses due to the outbreak of hostilities with the English, competition by the fur trade for labor and capital, and the high price of salt restricted development.[108]

The industry was handicapped in Nova Scotia, as we have seen, by encroachments from New England, by the limitations put upon company control by the fishing industry, and by inadequate technique. According to a memoir[109] of Villebon in 1699, where at Placentia the fishery was limited to the period from May to August, the fishery in Acadia began, in the east, in March and continued without diminution to Christmas, the cod migrating from Cape Sable along the coast, generally in May. In July the fish stopped biting from eight or nine o'clock in the morning until four or five in the afternoon; but fishing at night was so successful that with two hooks to a line two cod were often caught. Unfortunately large vessels such as the *biscayennes* and *charois* used at Placentia could not be employed because of the length of time necessary to get back to the harbor from banks offshore, as also because of the difficulty of leaving with an onshore breeze and the dangers of

107. Innis, *op. cit.*, p. 50.

108. Quebec merchants were complaining of the monopoly prices of goods from La Rochelle in 1692. *Idem*, p. 325. See J. N. Fauteux, *Essai sur l'industrie au Canada sous le régime français* (Quebec, 1927), chap. xi; also E. T. D. Chambers, *The Fisheries of the Province of Quebec* (Quebec, 1912), Part I, pp. 59 ff.

109. Innis, *op. cit.*, pp. 52–53.

night fishing. As has been said, the small New England vessels of 15 to 30 tons were in their success a conspicuous contrast.

In February, 1682, a grant of the fishery along the coast of Nova Scotia as far as Saint John was made to a company formed by a Huguenot, M. Bergier of La Rochelle, and MM. Gautier, Boucher, and de Mantes of Paris. The policy of granting licenses to English fishermen ceased in 1684.[110] In 1686 M. Gautier secured fishery concessions for twenty years in Cape Breton, Prince Edward Island, and the Magdalens. A fishing station at Chedabouctou, Fort St. Louis, was attacked by the English in 1688 and captured with Port Royal in 1690. After the Treaty of Ryswick in 1697, the company established posts at La Have[111] and at Chebuctou (Halifax) in 1698.[112] As a result of continued difficulties the company's concession was canceled in 1708. Fishing was limited to small scattered settlements such as Port Roseway, Whitehaven ("flakes are used for drying because there is no beach"), Canso, Baleine, Louisburg, and Ingonish. De Dièreville[113] claimed to have started a fishery at Port Royal in 1700. There the people built 20 shallops and in the spring and summer of that year put down more

110. Eight English vessels were seized, six of which were confiscated, in 1684. After the treaty of peace in 1686 the French attempted to regulate the fishery along the coast to the Kennebec River and continuous raids followed. The English settled on both banks of the Kennebec, and continued to fish off the coast of Nova Scotia. J. C. Webster, *Acadia at the End of the Seventeenth Century* (Saint John, 1934), Part I; and *The Ancient Right of the English Nation to the American Fishery and its Various Diminutions; Examined and Stated* (London, 1764). The entire region was then known as Nova Scotia.

111. "There is not much beach available for a large fishing industry but . . . flakes could be used and they without question produce the finest quality of fish." Webster, *op. cit.*, p. 135.

112. Innis, *op. cit.*, p. 56. The Sieur de Dièreville reported in 1699 on his voyage from La Rochelle to Port Royal that he found a deserted fishing establishment at Chebuctou. "It was half as long and quite as wide as the Mall in Paris, built on a fine beach along the river at a distance which permitted the water to pass under it at high tide and carry away the refuse of the cod. Imagine a wooden bridge, built over the land, of large piles driven well in on the side facing the water; at their extremities other pieces of wood placed crosswise and securely clamped; imagine the same construction on the land side but not so high because it was on a slope. Over all this the trunks of young fir trees long enough to rest on either edge laid evenly one alongside the other and well nailed at both ends to the wooden supports and you will then know what this contrivance which is called by the fishermen a *degras* [dégrat] is like. The cod carefully split are spread upon it during the summer and turned and returned continually so that they may dry and attain the proper state, the one familiar in a thousand places in the world to which they are easily transported. This station was uninhabited; it had been made before the last war by French fishermen who had established themselves there in the name of a company which did not prove profitable." De Dièreville, *Relation of the Voyage to Port Royal,* pp. 73–74. The settlement was composed of Huguenots who deserted to Boston. See Webster, *op. cit.*, pp. 124, 206–208.

113. De Dièreville, *op. cit.*, p. 97.

than 30,000 green fish. But little more was heard of it, and he appears scarcely to have earned his title of "père des pêcheurs" (father of the fishermen). Nova Scotia failed to provide the conditions by which France might have solved the problem of combining the demands of the fur trade, sugar production, and the fishing industry, and consequently accentuated the difficulties. New France was handicapped by her limited development of the dry fishery and the continued struggle against monopoly.

The French fishery on the North Atlantic apparently reached its peak between 1678 and 1689, coinciding with the difficult period of the English fishery, which declined from about 250 ships with 20,000 men in 1640 to less than 100 in 1680. It was estimated that the French had 300 vessels and 20,000 men in the fishery in 1678, but[114] it was claimed that by 1700 the number had dropped, chiefly as a result of wars, to about 100 ships of 50 to 150 tons on the Banks and 20 on the Canadian coast.

In spite of the energies of Colbert and his officials and the expansion of the fishing industry, attempts to link up New France and the West Indies failed. The fishery was scattered over various areas and this diminished the possibilities of specialization for diverse markets. Again, limited space in such areas meant congestion. The small fish of the Petit Nord were taken by St. Malo fishermen to Marseilles and the Levant, the large green cod by the Normans of Honfleur to Paris, the large dry cod, particularly those of the Basques, to Spain and Portugal, and the average dry cod to the home market. Settlements under these circumstances developed to slight extent.[115] The larger crews neces-

114. *The Cambridge History of the British Empire,* VI, 133–135. Other accounts give larger estimates. Lounsbury, *op. cit.,* p. 231. In 1710 the bank fishery had dropped to 50 or 60 sail. Bank ships belonged chiefly to La Rochelle, Normandy, and Bordeaux. It was claimed that the English, by virtue of their ability to secure a return cargo, succeeded in forcing the Malouins from the Italian markets as early as 1669, but there is much evidence against this. In 1696 the English were said to have sent 60 large ships from Newfoundland to the Mediterranean and that by 1700 they had become entrenched in the southern Spanish markets. Albert Girard, *Le Commerce Français à Séville et Cadix au temps des Habsbourg* (Paris, 1932), pp. 388–390.

115. Cod that are to be cured or dried "can only be taken on the coasts: which requires great attendance and much experience. M. Denys asserts that in order to carry on this fishery there to advantage the fishers must be persons residing in the country. . . . It can only be carried on from the beginning of the month of May till the end of August. Now if you bring sailors from France, either you must pay them for the whole year, in which case your expenses will swallow up the profits, or you must pay them for the fishing season only, in which they can never find their account. . . . But if they are inhabitants of the place, the undertakers will . . . be better served. . . . By this means they will take their own time to begin the fishery, they will make choice of proper places, they will make great profits for the space of four months; and the rest of the year they may employ in working for themselves at home." P. F. X. Charlevoix, *Journal of a Voyage to North America* (Chicago, 1923), I, 78.

sary for fishing rather than for manning the ship could be taken back to the home ports without the necessity of traversing such long distances as were involved in the English fishery and trade to Spain and the Mediterranean. The regions and technique of the French fishery were less suitable for settlements than those of the English. Dependence on home markets and on ships fishing in scattered areas and coming from many different ports of France continued to maintain a situation unsuited to settlements. France could take advantage of European markets because she was increasing her production, but she was unable to coördinate the fishing settlements at Placentia with New France and the West Indies. Placentia was even dependent to a slight extent on New England. Her competition with England in Europe contributed to the difficulties of the English fishery in Newfoundland and facilitated the development of the New England fishery for the West Indies market. But the French expansion in Newfoundland was related to Europe, and neither New France nor France proper provided adequate supplies for the sugar plantations of the French West Indies.

Talon in Canada[116] attempted in 1670 to develop a three-cornered trade by sending to the West Indies three vessels loaded with planks, peas, and Indian corn; and in 1672 two ships made the voyage, but the trade was of slight importance. An *arrêt* of April 31, 1685,[117] exempting trade between the West Indies and Canada from duties had little effect. In 1708 trade was reported as increasing gradually and the *Afriquain* took a small cargo of flour, oil, butter, suet, and planks. The *Biche* of some 60 or 80 tons attempted to engage in direct trade with the West Indies in the same year, bringing 1,000 minots of salt, some *cassonade*, and specie (1,200 piasters) in exchange for barrels of fine flour,[118] salmon, eels, and dried cod. But not only was it difficult to develop a direct trade with the West Indies; it was also difficult to develop a direct trade with the fishery. Lahontan reported ships returning via Placentia for a cargo of codfish, but that there was "more lost than got by that way of trading."[119] The export of wheat as flour and biscuit to Placentia and the purchase of dried cod for France had been thought to promise large profits, but actual trade appears to have been slight.[120] In 1708 there was little flour for export because of the bad

116. See Thomas Chapais, *The Great Intendant* (Toronto, 1914), pp. 50–51; *Lettres de la Révérende Mère Marie de l'Incarnation* (Tournai, 1876), II, 447. According to Lahontan, French vessels returning from New France picked up coal at Cape Breton and carried it to Martinique to be used in the refineries. Lahontan, *op. cit.*, I, p. 374.

117. Innis, *op. cit.*, pp. 340–341.

118. It became necessary to pass an ordonnance prohibiting the fraud of putting good flour at both ends of the barrel and poor flour in the middle. *Idem*, pp. 328–329.

119. Lahontan, *op. cit.*, p. 374.

120. Innis, *op. cit.*, p. 326.

harvest, and in 1709 there was a surplus of 5,000 *quarts* (barrels) which could not be exported for lack of vessels. In the latter year Canada exported 7,637 quintals of flour, 1,775 quintals of biscuit, and 3,250 quintals of peas, as well as butter, cheese, dried cod, and other provisions to the West Indies, Placentia, and Nova Scotia. The demands of the fur trade upon French Canadian man power had much to do with the country's limitations in agriculture.

The difficulties of a three-cornered trade between New France, the West Indies, and France itself were brought out very clearly in a *Mémoire touchant le commerce du Canada aux Isles Antilles françaises de l'Amérique*[121] of December 15, 1670. The scarcity of harbors made it necessary for ships to go to Martinique, Grenada, or to the Saints near Guadeloupe during the hurricane period from about July 15 to October 15.[122] During October, November, and December the sugar cane began to sprout and no cargoes of sugar were available. Under these circumstances a ship would have to leave Canada between November 1 and November 11 and arrive in the West Indies about the end of December or the beginning of January. In about six weeks, or by the end of February, the ship could load and leave with sugar and other products for a French port. Arriving in France within another six weeks, or between April 15 and 30 at the latest, the next two months, May and June, were available for refitting, and for reloading the vessel with goods for Canada. Alternatives to this arrangement included the departure of ships from Canada in the spring; but they would be able to leave only about the end of April at the earliest because of the ice, which did not allow sufficient time to go to the West Indies and load before the hurricane period. Again, ships might leave Canada toward the end of August and arrive in the West Indies about October 15, and then, after disposing of their cargoes of food products and loading the sugar, leave the West Indies the end of November and arrive in France the middle of January; but this period was not satisfactory for the loading of fish in New France and the selling of it in the West Indies.

Numerous suggestions were made for overcoming these difficulties. Lahontan proposed the development of a fishery in Cape Breton and advised its settlement and that of Prince Edward Island. The people might use sloops to engage in the fishery, to provide a supply of fish which would be purchased by ships about the latter part of August, and

121. *Idem*, pp. 320 ff. Talon's first vessels in 1670 arrived in the West Indies in December.

122. The general sailing routes followed the northeast trade wind from Europe to the West Indies and, working north, returned to Europe by the antitrades. August to October was practically a closed season. See Higham, *op. cit.*, pp. ix–xiii.

thus overcome the handicaps of an absence of harbors in Prince Edward Island and the small number in Cape Breton. He foreshadowed the plan that was followed after 1713. In 1706 an anonymous writer[123] pointed to the rise of the English fishery in Newfoundland from 1650 to 1706, and estimated that it was sending out a hundred vessels laden with dry cod, and could point to another hundred vessels in the New England fishery. The latter was carried on in barques on the Banks, and the fish were dried on the home shores. The French were chiefly engaged in the green fishery.

It is an established fact that most of the ships from France come for the green fishery on the Banc a Verd, the Banc de St. Pierre, and the Sable Island Bank, and even in the Gulf of St. Lawrence. They divide their fishing for four sizes of cod, the large, the medium-sized, the small, and the smallest of all which they call "raquet," and they do it for this reason: When they get back to France they send out four classes of fish, each having its price in the market. But, for all that, it is difficult to set these prices satisfactorily. And, more, it often happens that there are so many more of the small and the "raquet" size, that the two larger and of better quality do not sell at a proper profit.[124]

These difficulties would be overcome if the fishermen landed their fish, and dried the small ones and green-salted the large ones as was done at Ile Percée. The losses, and the dangerous voyages absorbing four months' time which characterized French fishing voyages such as those of ships leaving France in March would be avoided and they would be able to come at the end of the fishing season and leave with a uniform cargo of fish for definite markets. It might be possible to make two voyages in a year and to develop a three-cornered trade. Finally, French ships leaving the West Indies which took advantage of southerly winds and returned to France by the Grand Banks would in Cape Breton have a place to outfit and obtain a fresh supply of provisions.

As a result of the difficulty of receiving adequate supplies from France and from New France, the French West Indies were forced to depend to an increasing extent on the areas beyond French jurisdiction.[125] Horses for operating the mills were obtained from the Dutch colony of Curaçao, and later from the Spanish colonies. Staves, hoops, and headings for barrels were obtained from New England.

The failure of the French Empire to coördinate the activities of its

123. J. S. McLennan, *Louisbourg from its Foundation to its Fall, 1713–1758* (London, 1918), pp. 22 ff.

124. *Idem*, p. 24.

125. May, *op. cit.*, pp. 149–154, 162–165; Mims, *op. cit.*, pp. 318–319; Innis, *op. cit.*, pp. 320 ff.

territory was in part a result of the increasing strength of the English Empire which flourished in spite of, as well as because of, attempts to coördinate the activities of its various parts. The dependence of the French West Indies on the English colonies was an indication of the character of support given to the British West Indies.

The weakness of the fishing ships from the West Country was a result of the expanding trade from New England to Newfoundland and to the southern colonies, the West Indies, Africa, and elsewhere, which prevailed in spite of legislation, protests, and all the arguments that the fishing industry was a nursery for seamen.[126] French expansion in the fishing industry, on the other hand, was an indication of the limitations of France in the carrying trade.[127] Says the author of *Britannia Languens*, "Our fishing for white herring and cod was deserted for this trade," that is, for plantation commodities. "Our fishing trade hath decayed continually of late years; we formerly supplied France, Spain, Muscovy, Portugal and Italy with great quantities of white herring, ling and codfish which trade is now lost to the Dutch, French etc. We have only the trade of red herrings which we retain." In Iceland, he declared,

> we have not a fourth part of the trade we had twenty or thirty years since; the like may be said of our Newfound-land fishing; and our Greenland where we had the sole trade is quite lost; the Dutch had far beaten us out of these trades but the French of later years have struck into a good share of the whole, beating out the English more and more.[128]

126. "A fishing trade is one great and certain nursery of seamen and brings wealth and comfort to sea towns." (William Petyt?), *Britannia Languens, or a Discourse of Trade* (London, 1680). "The fishery is of an absolute and indispensable necessity to the well-being both of King and people." R. l'Estrange, *A Discourse of the Fishery* (London, 1674). "No trade is more likely to increase seamen, than our fishing trade, because great numbers (as well as some landmen which thereby become seamen) are imployed in the taking and making as well as in carrying it to foreign markets; and no trade can produce more clear profit to the nation, because the whole value ariseth from the labours of our people, excepting the salt." John Pollexfen, *A Discourse of Trade, Coyn and Paper Credit and of Ways and Means to Gain and Retain Riches* (London, 1697).

127. Some of the advantages of the English over the French in shipping were these: the French could not victual so cheaply nor sail with so few hands; with lack of good coasts and harbors they could not keep ships in port except at twice the cost of the English; the scarcity and distance of ports from one another; seamen and tradesmen could not correspond with and assist one another so easily, cheaply, and advantageously as in other places. See Sir William Petty, *Political Arithmetick* (London, 1699).

128. (William Petyt?), *Britannia Languens;* (London, 1680); see also Roger Coke, *A Treatise . . . that the Church and State of England are in Equal Danger with the Trade of it* (London, 1671). A petition of London merchants trading to New York and New England stated that these colonies "take off vast quantities of the manufac-

> Newfoundland trade much diminished and northern fishing trades disused. By which we have suffered two great inconveniences, the loss of the greatest nurseries we have for seamen and the use our neighbours have made of it to increase theirs. By the northern fishing the Dutch have made their greatest numbers of seamen and by the banks of Newfoundland the French, and thereby make those trades difficult to be retrieved; for as long as we have not a number of seamen over and above what may be imployed in our other trades, [it is] difficult to be found that they will go to the fishing trades, in any great abundance, because [they] are attended with great labour and hardship.[129]

The limitations of the fishing industry as a support to the navy intensified the need for ways of securing specie. The fishery had been stressed by those insisting on a monopoly of tobacco in Virginia and of the sugar industry in the West Indies. But it had been weakened by competition from the French. Attempts were made to check imports from France.* The expansion of sugar production in the British West Indies brought about a marked decline in prices in England. Wines displaced sugar in exports from Portugal to England. Average imports of 8,500 tons of wine from France from 1675 to 1678 and minor quantities of Portuguese wine were followed by average imports of 7,000 tons from Portugal from 1679 to 1685 and minor quantities of French wine. The outbreak of war with France in 1689 intensified the change. The Methuen Treaty of 1703 was designed to increase this trade with Portugal.[130] English woolens were by it admitted free to Portugal, and the duty on Portuguese wines was made a third lower than on the French. A petition of merchants trading with Spain, Portugal, and Italy in 1709 stated that

> since the war great quantities of wines have been imported into this kingdom from Portugal, Italy and Spain which has encouraged those nations to take off much greater quantities of woollen manufactures, and fish from Newfoundland and New England and other the product of this kingdom than formerly; whereby much more shipping is employed and the Portuguese (who had set up the manufacture of cloth and well nigh brought it to perfection and thereupon prohibited our woollen cloth) have, a few years past, taken off the said prohibition to encourage our continuing the consumption of the wines.[131]

torie of this kingdome . . . the chief return for which is beaver and other furs. That the trade of these colonies doth imploy a great quantity of shipping and next to the Newcastle trade (now the French have in effect beat us out of our Newfoundland fishery) is the greatest nursery for seamen this kingdom hath left." May 14, 1690. Stock, *op. cit.*, II, 26.

129. Pollexfen, *op. cit.*

130. Adam Smith, *op. cit.*, pp. 512–513, 625–626.

131. Stock, *op. cit.*, III, 211. "The importation of French wines will very much

In 1707 the value of fish exported from Newfoundland to Spain and Portugal was estimated at £130,000, and returns were taken in wine, brandy, salt, oil, and linen for England and for New England through Newfoundland. Cod from Newfoundland was exported to the Madeiras in exchange for wines, and from New England to the West Indies for sugar and rum.

The aggressive commercialism of New England thrived on, and contributed to, the increase of settlements and the decrease of the fishing ships in Newfoundland. Similarly, the expansion of sugar production in the British West Indies provided increasing markets for the fishing industry of New England. The varying size of cod, its wide range, its adaptability to varying technique in preparation for various markets, and the existence of a central exchange area in which cargoes were available for divergent ports contributed to the rapid growth of trade and industry. The fishery meant a demand for skilled labor,[132] for carpenters, carvers, blacksmiths, and blockmakers, for ships, boats, and provisions; and supplies for a shipbuilding industry such as iron, rope, rigging, sailcloth, planks, and pulleys. With year-round navigation and large numbers of small ships, flexibility of capital was given free play by the adventure system.[133] A constant supply of fish in winter on the

prejudice the woollen manufactures and the fisheries of Great Britain and Newfoundland, the greatest part being sold in Portugal, Spain and Italy and the products thereof returned mostly in wines." *Idem,* pp. 247, 249.

Imports from Portugal doubled between 1662 and 1700 and exports increased one and a half times. The balance of trade was sharply in favor of England, particularly after the Methuen Treaty when exports to Portugal totaled £780,664 and imports from Portugal £330,689. Exports of bullion to England continued throughout the period. See V. M. Shillington and A. B. Chapman, *The Commercial Relations of England and Portugal* (London, 1907), chap. iv.

132. "In the highest place in the scale of labor is the seaman." Petty, *op. cit.* The importance of the fishing industry to labor gave it a significant place in mercantile theory. See E. A. J. Johnson, *Predecessors of Adam Smith* (New York, 1937), pp. 240–242, and chap. xv; also E. D. Furniss, *The Position of the Laborer in a System of Nationalism* (Boston, 1920). "It is therefore a general maxim to discourage the importation of work and to encourage the exportation of it." James Stewart, cited by Johnson, *idem,* p. 308.

133. March 27, 1664: "The Lord brought in a small vessel sent out by myself and others last winter for Alicant." This was in spite of its having been boarded by the Turks. April 6: "The Lord brought in safe the several vessels that I had adventures in." October 30, 1666: "I sent to England a considerable adventure in sundry ships." "Diary of John Hull," *op. cit.,* III, 154, 156. "It is not to be doubted but those who have the trade of shipping and fishing will secure themselves of the trade of timber for ships, boats, masts, and cask, of hemp for cordage, sails and nets, of salt, of iron, as also of pitch, tar, rosin, brimstone, oil and tallow, as necessary appurtenances to shipping and fishing. Those who predominate in shipping and fishing, have more occasions than others to frequent all parts of the world and to observe what is wanting or redundant everywhere; and what each people can do, and what they desire; and consequently to be the factors and carriers, for the whole world of trade. Upon

coast of the Gulf of Maine, and in summer off Nova Scotia, and an expanding market in the Catholic countries of Europe and in the sugar-producing areas of the tropics gave elasticity, variety, stability, and continuity to the economic organization of New England. Expansion on the sea facilitated development on the land, with an exploitation of forests and increase in agriculture. The land and the sea were joined to support shipping, trade, and industry. An increasing population which accompanied an expanding fishery, industry, and trade meant increased demands for manufactured products from England. The importance of shipping, the fishing industry, and trade brought demands for bills. The geographical location of New England in relation to a wide range of producing regions and a variety of products for purchase and sale meant substantial profits from shipbuilding, shipowning, and marketing. "The Bostoners may be said to be the carriers to most of the other plantations."[134] The ports of the northern colonies became extensive distributing centers.[135] As Thomas Mun wrote regarding England, "The value of our exportations likewise may be much advanced when we perform it ourselves in our own ships, for then we get not only the price of our wares as they are worth here, but also the merchants' gains, the charges of insurance and freight to carry them beyond the seas." So too, might he have written of New England. English merchants in 1676 complained that

which ground they bring all native commodities to be manufactured at home; and carry the same back, even to that country in which they grew. All which we see. For do they not work the sugars of West Indies? the timber and iron of the Baltick? the hemp of Russia? the lead, tin and wool of England? the quick silver and silk of Italy? the yarns and dying stuffs of Turkey? . . . Husbandmen, seamen, soldiers, artizans and merchants are the very pillars of any common-wealth. . . . Now the seaman is three of these four. . . . The employment . . . of seamen is free to the whole world. . . . It is certain that somewhere or other in the world, trade is always quick enough and provisions are always plentiful, the benefit whereof, those who command the shipping enjoy, and they only. The labour of seamen and freight of ships is always of the nature of an exported commodity, the overplus whereof, above what is imported brings home money etc. (which is) wealth at all times and in all places whereas abundance of wine, corn, fowls, flesh etc. are riches but hic et nunc." Petty, *op. cit.*

134. Letter from J. Higginson at Salem, August, 1700, cited by Nettels, *op. cit.*, p. 94.

135. Nettels, *op. cit., passim,* especially pp. 67–72, and a criticism of Beer for neglecting invisible items of trade, pp. 128–132. "Our principal commodities are dry merchandise, cod-fish fit for the markets of Spain, Portugal, the Straits, also refuse dry fish, mackerel, lumber, horses and provisions for the West Indies; the effects whereof mostly return for England. . . . The making of returns for England by way of Barbados, Leeward Islands, Bilboa, Oporto, Cadiz and Isle of Wight would be more easy and safe than direct from home; and it's probable more advantageous. . . . A man may sell more goods and better get in his debts more speedily and certainly for barter of goods for those markets than direct." *Idem,* p. 94.

all sorts of merchandise of the produce of Europe are imported directly into New England, and thence carried to all of the other King's dominions in America and sold at far cheaper rates than any that can be sent from home, and . . . they take in exchange the commodities of the plantations which are transported to Europe without coming to England, so that New England is become the great mart and staple by which means the navigation of the kingdom is greatly prejudiced, the King's revenue unexpressibly impaired, the price of home and foreign commodities lessened, trade decreased and the King's subjects much impoverished.[136]

New England ships became competitors with English ships and New England trade began to compete with English trade.[137] The advantages of the Navigation Acts to New England contributed to the emergence of a menace to England.[138]

The increasing strength of New England meant pressure on the French and contributed to the withdrawal of the latter from Nova Scotia and Newfoundland as agreed to in the Treaty of Utrecht. Newfoundland more and more became marginal territory which served as an overflow and a spillway for growing quantities of goods from Europe rather than from England. Struggles between the planters and the West Country were growing pains incidental to the expansion of New England trade and the metropolitan growth of London, by which the West Country was persistently weakened. The competition of the French with a commodity*for which the market was comparatively restricted and probably contracting, and on which a wide range of industries was dependent, contributed to the effectiveness of New England trade as a support to settlements and to the necessity for more efficient types of the industry. Attempts to enforce impressment or to force the fishing industry to support the navy, in so far as such attempts tended to be successful, contributed to the success of New England by encouraging the migration of labor. The demand for exchange in an expanding commercial area was met by increasing trade to the West Indies, to Europe, and to Africa, by an extensive participation in tropical trade, and by the sale of products to Newfoundland for bills. The commercialism of New England was a powerful factor militating against the effectiveness of regulation, and called repeatedly for the making of new regulations.

The expansion of New England and the plantation trade and the support of the Newfoundland resident fishery eventually meant encroachments on the French markets in spite of English statements to

136. *Idem,* pp. 132, 279–280. For a valuable account of the clash between the free-trade policies of New England and the British commercial system see V. F. Barnes, *The Dominion of New England* (New Haven, 1923), chap. vii.

137. Nettels, *op. cit.,* pp. 140–141. 138. *Idem,* pp. 44–46.

the contrary. "Since the English consume almost no dry fish in Europe they take it to Spain and Portugal and even to the Levant where they sell it in competition with the French, who should rightly be the masters in that species of trade."[139] The scattered character of the French dry fishery and the difficulty of developing a central market in which the product could be sorted and graded for particular markets seriously affected its competition with the English product. Lack of coördination in the fishery and in the regions of the New World corresponded with lack of coördination[140] in the ports of France. The Bay of Biscay ports such as La Rochelle, Bordeaux, and Bayonne tended to displace Havre, Dieppe, and the Norman ports; and wine, brandy, tobacco, and iron were sent to Canada, Newfoundland, and the West Indies. La Rochelle became increasingly prominent, but suffered a serious blow through the revocation of the Edict of Nantes. A strong, powerful, metropolitan development such as was seen in the expansion of London, the concentration of West Country ports, and of Newfoundland, New England, and the colonies, and an increase in shipping under the Navigation Acts, was impossible for France.

The French Empire was handicapped by the development of a continental economy on the St. Lawrence deeply affected by the fluctuating character of the fur trade and by difficulties in linking up with the tropical products of the West Indies. Duplessis wrote in 1704:

> It is easy to judge, from what is put before us by the evidence, of the gain it would be to New France if a little help could be given to her sea trade rather than, as always, to the fur trade of her forests. For in that fur trade there has ever been a lack of stability which, today, is bringing Canada to ruin. For her Canadian colonists have never interested themselves in anything but fur-trading, and they are now falling into bankruptcy because of the low prices that furs are bringing. In the mere diversity of the various kinds of commerce that can be carried on by sea there is the saving factor. For if one fails to bring returns, another makes up for it and prevents any general ruin such as that which has now overtaken the trade in furs.[141]

An able writer, writing anonymously in 1706, brought into contrast the position of the English.

> If anyone gives considered attention to the progress the English have made in the case of their New England colonies, he will have good reason to tremble for our colony in Canada. There is no single year but sees more children born in New England than there are men in the whole of Canada. In a few years we shall be facing a redoubtable people, one to be feared. As for Can-

139. McLennan, *op. cit.*, p. 29.
140. Lahontan, *op. cit.*, p. 374.
141. Innis, *op. cit.*, pp. 326–327.

> ada, her people will not number many more than they do today. Whether we must seek the reason in that mildness of climate which is so favorable to agriculture, stock-raising and all-the-year-round navigation, or whether we must seek it in the demesne of specialized industry, this is certain: On those shores the colonies of England have become as solidly established as England herself.[142]

The relative absence of development in Nova Scotia was in contrast to the expansion of New England. Attempts to establish company control in the fishing regions involved difficulties with the more aggressive New England type of enterprise. In a memoir of October 27, 1699, Villebon wrote regarding Acadia:

> He [Villebon] believes the English should not be excluded completely from the country until His Majesty has had forts built and in condition to withstand all attacks, because he is convinced that, although they are at present under a strict government, all New England is concerned in the fishing industry, and there is danger that they might secretly instigate some freebooter, as they have done in the past, to harry our young settlements without appearing to have had anything to do with the matter.[143]

The expansion of the shipping both of New England and of England herself resulted, indirectly and directly, in encroachments on the fishing ships of Newfoundland. The exporting of fish by sack ships involved competition with the fishing ships and led to an increase in settlements and in the numbers of byeboatmen. The fishing ships contributed to the general trend by participating in the carrying of passengers for the byeboatkeepers. The expansion of the carrying trade caused regrets for the decrease of the fishing ships; but it meant the development of trade in more profitable lines of shipping, in spite of protests over the loss of the nursery for seamen. New England shipping and trade served to knit together the diversities of the empire in North America, enhanced the difficulties of coördinating the French Empire, and was an aid to the carrying trade of England. Such a happy arrangement was not destined to last. The fishing industry was essentially competitive and meant a struggle between New England and the West Country of a kind that was vital for the problems of government in the British Empire.

The effectiveness of the Navigation Acts and the defeat of the Dutch in the 'sixties of the seventeenth century made more rapid the expansion of shipping and the carrying trade in England and New England. The growth of the English carrying trade was accompanied by a temporary

142. McLennan, *op. cit.*, p. 29.
143. Webster, *op. cit.*, p. 139. On the position of companies in France see Heckscher, *op. cit.*, I, 350–351.

decline in the Newfoundland fishery and by the temporary rise of the French in that industry. But French naval power based on the fishery was less effective than English naval power based on the carrying trade; and England was to pay for the expansion of carrying trade under the Navigation Acts by the ultimate restrictions they put upon New England, and the loss of the colonies.

The Treaty of Utrecht[144] was signed on March 31, 1713, and according to its provisions the French were forced to cede Acadia, Newfoundland, and Hudson Bay to the English. Placentia was evacuated. Article XIII, however, provided that, while

> it shall not be lawful for the subjects of France to fortify . . . or to erect any buildings there besides stages made of boards and huts necessary for fishing and drying of fish . . . it shall be allowed to the subjects of France to catch fish and to dry them on land in that part only and in no other besides that, of the said island of Newfoundland which stretches from the place called Cape Bonavista to the northern point of the said island and from thence running down by the western side reaches as far as the place called Point Riche.

Cape Breton[145] remained in French possession. The Assiento monopoly of the Spanish slave trade held by France since 1701 was ceded to Great Britain in 1713 and gave her the right to participate[146] in the Spanish slave trade to the extent of 4,800 slaves a year for thirty years.

144. *P.C.*, V, 2181. See Judah, *op. cit.*, chap. viii.

145. For an extended argument against the ceding of Cape Breton to France see Stock, *op. cit.*, III, 317–319, 321–323; and pp. 102–109 for an account of French competition in Newfoundland in 1705 and 1706.

146. This privilege was given to the South Sea Company which collapsed in the South Sea Bubble. See W. E. B. Dubois, *The Suppression of the African Slave Trade from Africa to the United States of America 1638–1870* (New York, 1896), chap. 1; G. Scelle, "The Slave Trade in the Spanish Colonies of America, The Assiento," *American Journal of International Law,* IV, 612–661.

Appendix

The extent of the Newfoundland fishery in 1700, as conducted from various ports, is indicated in the following table.

Sailing from Newfoundland Port	*Number of Ships*	*Ports of Origin*				
		Plymouth	*Biddeford*	*Barnstable*	*London*	*Topsham*
Renewse	9		6 (1 for Oporto)	2 (1 for Lisbon)		
Fermeuse	15		11 (2 for Alicante; 1 for Lisbon)			
Aquaforte	5		1	2 (1 for Lisbon)	1	
Ferryland	12	2 (for the Straits*)	4 (for the Straits*)	3 (for Lisbon) 1 (for England)	1 (for Lisbon)	1 (for the Straits*)
Capelin Bay	4	4 (Cadiz, San Sebastian, Virginia, and Biscay)			1 (for Cadiz)	1
Cape Broyle	7		4 (3 for the Straits;* 1 for Biddeford)	2 (for the Straits*)		
Brigus South	2			2		
Toad's Cove	8	1	4 (1 for Lisbon; 1 for Oporto)		1 (for the Straits*)	
Witless Bay	3					3 (2 for Topsham)
Bay of Bulls	14				4 (2 for the Straits;* 1 for Oporto; 1 for Bilbao)	9 (4 for England, 1 for the Straits,* 1 for Oporto)

Petty Harbor	5	1 (for Plymouth)		3 (1 for the Straits*)
St. John's	44	1 (for Cadiz)	22 (5 for the Straits.* 2 each for Lisbon and Leghorn; 2 each for Alicante, Malaga, Seville and Cadiz; and 1 for Barcelona)	7 (3 for England; 1 for Bilbao; 1 for Cadiz; 2 for the Straits*)
Torbay	1	1 (for the Straits*)		
Portugal Cove	1	1 (for Plymouth)		
Belle Isle	1		1 (for Bilbao)	
Brigus North	2			
Port de Grace	5			1 (for England)

Additional sailings from Newfoundland ports completing the above table were: Renewse, 1 each from Tinsmouth (Teignmouth?), Torrington, Baseliton(?), and Weymouth; Fermeuse, 1 each from Dartmouth and Chester, the latter ship for Cadiz; Aquaforte, 1 from Lyme; Capelin Bay, 1 from Waterford for Leghorn; Cape Broyle, 1 from Tinsmouth for the Straits, Bay of Bulls, 1 from Dartmouth to England; St. John's, 1 from Tinsmouth for England, and 5 from Dartmouth, of which 3 returned to England; Belle Isle, 1 from Dartmouth; Port de Grace, 2 from Bristol; St. John's, 1 from Tinsmouth for England, 5 from Dartmouth, 3 returning to England, and 1 each from Limington for England, from Jersey for Bilbao, from Barbados for New England (in ballast), from Bridgewater for the Straits,* from Lancaster for Lisbon, from Guernsey for Boston, and 1 for Belfast.

* By "Straits" are meant the Straits of Gibraltar and the Mediterranean.

Sailing from Newfoundland Port	*Number of Ships*	*Ports of Origin*			
		London	*Bristol*	*Poole*	*Guernsey*
Harbor Grace	6	1 (for Oporto)			
Carbonear	25	5 (2 for the Straits;* 1 for Lisbon; 1 for Cadiz)	1 (for Bristol)		
Bay de Verde			1 (for Bristol)	1 (for Poole)	3 (1 for Guernsey; 1 for Canaries)
Old Perlican	5	1 (for London)		3 (2 for Poole)	
Aunt's Harbor	2			2 (1 for Spain; 1 for France)	
Scilly Cove	2		1 (for Spain)	1 (for Poole)	
Heart's Content	1			1	
New Perlican	3	2			1 (for Guernsey)
Trinity	14	1	1		1
English Harbor	1	1 (for Bilbao)			
Salmon Cove	4	1	1		
Bona Vista	8	2 (1 for Oporto; 1 for Alicante)	1 (for Leghorn)	3 (2 for England; 1 for Cadiz)	

Additional sailings from Newfoundland ports, completing the table, were: Harbor Grace, 2 from Dartmouth, 1 from Jersey, 1 from Lar pool (?) for Cadiz, 1 from Dublin for Cadiz; Carbonear, 1 from Dartmouth for Genoa, 1 from Jersey, 4 from Topsham, of which 3 were for Cadiz; 2 from Lynne of which 1 was for England, 1 from Plymouth, 1 from Southampton for Lisbon, 1 from Liverpool for the Straits;* Old Perlican, 1 from Liverpool; English Harbor, Trinity, 3 from Weymouth, of which 2 were for Malaga, 2 from Jersey, of which 1 was for Bilbao, and 1 from Southampton; Salmon Cove, 1 from Jersey and 1 from Liverpool for Cadiz; Bona Vista, 1 from Cadiz for Cadiz, and 1 from Lymington; Trinity, 1 from Portsmouth and 1 from St. Lucan; Kerles (?), 1 from Hampton.

* By "Straits" are meant the Straits of Gibraltar and the Mediterranean.

CHAPTER VI

THE SUPREMACY OF THE BRITISH EMPIRE 1713–1763

NEWFOUNDLAND

A spirit of contention and law appears to be too powerful amongst the people of that country for the prosperity of the fishery. For more than a century after the first discovery of Newfoundland and the establishment of its fisheries the opinions of government as to the most advantageous plan of carrying it on for the national benefit appear to have been very unsettled, wavering between two different and in some measure adverse propositions viz. either planting the island and establishing a civil government, and thereby encouraging a promiscuous fishery, or discouraging inhabitancy and thereby conforming the fishery entirely to ships fitted out from these Kingdoms; and by sometimes adopting and pursuing the one, and sometimes the other, as different interests prevailed, the nation lost many advantages which would have been derived to it, had either one or the other of the propositions been firmly and uniformly pursued.

Privy Council, IV, 1849 (1765)

THE growth of the New England fishery in response to the demands of the West Indies and Europe and competition in Europe by the French fishery had serious consequences for Newfoundland. It intensified the struggle between the fishing ships and the settlers. Labor was attracted to New England and the market for Newfoundland fish was restricted, especially during the war between England and Spain. The settlements were fed by New England provisions. The complaints by West Country ports of the New England trading in Newfoundland were an indication of its extent. The merchants of Bideford claimed in 1715 that

the inhabitants are supplied with provisions, tobacco, rum, sugar, rice, etc., from New England and the colonies of America and what profit they make by catching and curing fish is spent in Newfoundland; besides, inhabitants on the least encouragement will so increase in number as soon to carry on the whole fishery by themselves and the whole employ of this island being fishermen, there can be no fresh men to be bred up sailors, or if they were, Great Britain would gain nothing by having sailors bred for its plantations abroad.[1]

1. *C.O.* (*Colonial Office*) 194:5. See R. G. Lounsbury, *The British Fishery at Newfoundland, 1634–1763* (New Haven, 1934), chaps. viii, ix.

In the same year merchants of Poole and other ports petitioned that the selling of tobacco and liquors in Newfoundland by New England traders should be prohibited. "A nest of little pedlars," they said, "who go under the denomination of merchant factors have small storehouses, sell rum, tobacco and sugar by retail."[2] The New Englanders, after getting the planters into debt, compelled them and their servants to go to New England. From there it was estimated, about 1720, that Newfoundland imported some 600 hogsheads of rum a year, and in addition bread, flour, pork, molasses, tobacco, black cattle, and sheep, in value amounting to about £10,000. The buyers paid either directly in bills of exchange or indirectly in fish. The fish were sold to sack ships in return for bills,[3] or, if the fish were refuse fish, they were sold in the Madeiras and the West Indies. In 1733 imports from American plantations had an estimated value of £16,000. Five years later imports from New England included 300 cows and oxen, 600 sheep, and 300 swine. Trading ships from the colonies increased from 31 in 1716 to 66 in 1749, when their total tonnage amounted to 6,400, and their crews to 737 men. In 1750 there were 75 such ships. In 1751 there were 103, of 7,011 tons. Vessels bringing salt for New England still brought other goods from Europe; and New England sent back enumerated goods in return.[4] The importance of New England trade[5] was enhanced during the Seven Years' War. The colonies had improved their position in the carrying trade and in the supplying of provisions. Vessels from New York, Philadelphia,[6] Rhode Island, and Boston brought beef, pork, rice, peas, Indian corn, flour, bread, onions, molasses, rum, staves, and the like in

2. *C.O.* 194:7.

3. "Their principal view is to get bills of exchange for their remittances to Great Britain; they seldom load fish for the foreign markets but sell them to the sack ships from Europe" (1751). *C.O.* 194:13.

4. "Notwithstanding all the endeavours and methods that can be used few ships that arrive but what import goods prohibited by an Act of the 13th King Charles II, specially wines from France, Portugal, and Italy, soap, candles and tallow from Ireland, concealed in barrels, and [they] pass as beef or pork and [are] transported in coasters from harbour to harbour, as also rum and molasses from St. Eustatia and the French islands which are generally landed in our harbours and transported by coasters into others." *C.O.* 194:11, 1742. "It is inconceivable what quantities of French rum, molasses etc. they bartered with the Newfoundland traders. . . . I have known them bring ships even into several of the Bays of Newfoundland, and boats and ships have met them without ever coming to an anchor and exchanged with them the amount of their whole cargo." Griffith Williams, *An Account of the Island of Newfoundland,* cited by Janet Paterson, "The History of Newfoundland 1713–1763," master's thesis, University of London.

5. See Appendix A, p. 180.

6. Merchants were able to specialize in colonial products. In 1763 at least two cargoes from Philadelphia were consigned to one R. Bulley, of St. John's.

large quantities. The value of Newfoundland imports from the West Indies and North America in 1763 was put at £30,000.[7]

Colonial shipping grew in importance with colonial trade. In 1763 St. John's shipped fish totaling 56,365 quintals. Of this, vessels from Philadelphia carried 2,967 quintals; from New York, 4,100; and from other colonial ports, 8,630. Vessels from Dartmouth took 9,100; ships from Exeter, 3,600; from Teignmouth, 1,600; and from London, 6,386. Of a total of 59,596 quintals from Conception Bay, London ships took 13,920 quintals; ships from Jersey, 17,560; from Poole, 8,770; and from the colonies, 7,130. Increasing trade and the development of a trading organization in St. John's and Conception Bay went with the development of bank fishing and the migration of fishing ships from the West Country to the smaller outports, with the result that colonial shipping was of less importance. Poole and Teignmouth traded with Bonavista; Poole, Dartmouth, and Topsham with Bay Bulls. However, of the total of 20,300 quintals from Bay Bulls, vessels from Cadiz, Bristol, and London carried 5,600. While Waterford and Dartmouth ships took the bulk of 10,900 quintals of Ferryland cod, vessels from Philadelphia and Boston took 3,500, and even one from Ferryland, 2,000. Dartmouth and Teignmouth took fish from Renewse and Fermeuse, and trading vessels from Poole and Waterford carried most of the cod from Old Perlican and Trinity. But again other overseas ports and ships had their share. Of a total of 46,992 quintals from these ports, vessels from Cadiz took 3,800, from London 14,936, and from the colonies 3,067.

The position of the fishing ships was weakened by the increase in trade between New England and Newfoundland, by what trade did for the growth of settlements, and by the competition for labor resultant

7. IMPORTS TO NEWFOUNDLAND FROM NORTH AMERICA AND THE WEST INDIES (1763)*

Rum	128,000 gals.	Hams	6 tierces and 30 bbls.
Molasses	34,400 gals.	Pine boards	130,373 ft.
Bread	5,211,000 lbs.	Planks	19,385 ft.
Flour	2,909 bbls. (2 cwt. each)	Lime	20 hhds.
Butter	122 firkins (½ cwt. each)	Shingles	98,000
Pease	60 hhds.	Chocolate	59 boxes, 56 lb. each
Turpentine	121 bbls.	Bricks	21,000
Pitch	322 bbls.	Rice	63 tierces 45 bbls.
Indian corn	598 bus.	Beef	221 bbls.
Staves	171,758	Pork	635 bbls.
Tar	212 bbls.	Soap	113 boxes 25 bbls.
Candles	3,000 lbs.	Loaf sugar	27 hhds. 14 tierces 16 bbls.
Spirits	72 kegs	Brown sugar	3 hhds. 9 tierces 475 bbls.
Masts and spars	112	Tobacco	21,993 lbs.
Madeira wine	1 hhd. 60 gr. casks.	Coffee	65 bbls. 7 bags.

* *C.O.* 194:15.

upon the expansion of the shipping, fishing, and trading interests of New England. The migration of labor from Newfoundland to New England was felt to be one cause of an increase in wages, which, from £12 or £14 a season in 1708 rose to £20 or £30 in 1715. To such a labor migration was also ascribed a rise in the cost of taking cod, and it was held that to produce a profit there should be a price increase from 25 or 28 reals a quintal to 35 or 39. It was estimated that 1,300 fishermen migrated in 1717. Newfoundland was becoming "a cradle or nursery for seamen" not for England but for New England. "Whereby this fishery which in its first institution was wisely intended to be a nursery of seamen for the service of Great Britain, far from answering that end, [it] is become a dangerous drain from the Mother kingdom to increase the shipping of a colony negligent of the laws of navigation, frequently encroaching upon your Majesty's Royal prerogative and too much inclined to independence" (1728).[8] Although New England masters were required to give bonds not to take men, the regulation was difficult to enforce, and "spiriting," that is, smuggling men out of the country, generally to New England itself, became serious. "Their voyage is lost if they go without them." In 1729 New England men were charged with assisting byeboatkeepers. "I am informed that the masters of fishing ships and byeboatkeepers do connive at their servants going to New England or remaining in the country purely to save the charge of their passage home."[9] Palliser wrote in 1764: "there is a constant current of seamen, artificers and fishermen through this country into America."[10] Not only were wages raised in Newfoundland by such competition but fish were produced more cheaply in New England than in Newfoundland. In 1717 it was claimed that New England fish were competing with Newfoundland fish in the European market and were selling at a dollar a quintal cheaper.

Competition from New England and the demand of the latter for bills of exchange also kept down the price of provisions and encouraged the formation of settlements. "There are at present (1714) about five hundred families in Newfoundland but their condition . . . is more to be pitied than that of slaves and negroes."[11] The population[12] remained

8. *C.O.* 195:7.
9. *C.O.* 194:9.
10. *C.O.* 194:17.
11. *C.O.* 194:5.
12. Planters increased from 346 men in 1736 to 690 in 1749, and to 1,250 in 1764; servants from 3,727 in 1749 to 8,976 in 1764; and from 612, the number of women and children in 1710, there was an increase to 1,356 in 1738, and to 2,508 in 1754. As for the fishery, the number of boats owned by residents, which had averaged 381 from 1713 to 1716, rose to 654 in 1749 and to 1,236 in 1764. The catch by resident fishermen increased from 27,420 quintals in 1720 to 124,395 in 1738, to 200,960 in 1746, to 293,106 in 1749, and to 352,690 in 1764. J. D. Rogers, *Newfoundland* (Oxford, 1911), pp. 122–123.

fairly stationary at between 2,000 and 3,000 from 1713 to 1733, but increased to nearly 7,000 in 1750, to 10,000 in 1758, and to 16,000 in 1764.

*Scheme of the Fishery of Newfoundland at a Medium yearly, from 1736 to 1739.**

	Number of Boats		
St. John's and adjacent Harbour	240		
Bay Bulls, Breakers etc.	100		
Ferryland	40		
Renewse & Fermeuse	150		
Trepassey & St. Mary's	33		
Placentia	110		
Carbonear Bay, the sevl. Harbours	100		
Trinity Bay, the sevl. Harbours	100		
Bonavista	45		
	——	918	
Fogo & Twillingate, the Northern Fisheries		80	
Burin, St. Lawrence, Marteens, St. Peters, Oderan etc. the Western fisheries		120	
		1118	
Say 300 quintals per boat 335,400 qtls.			
Fish caught by fishing Ships on the Banks 45,000 Quin.			
Value of 380,400 quintals at 10/ per quintal			£190,200
Train oil 2 hhds. per 100 qtls. 1902 tons			
Seal and Whale Oyl about 350 tons			
1000 Tierces of Salmon, value at Market £3 p. Tierce			£3,000
Value of Furrs brought to England, about			5,000
Produce of Newfoundland (yearly)			£227,476
About 8000 People employed in the Fishery, and Ships to carry it off.			
21,452 Tons of Shipping required to carry off the whole produce.			
Craft, Cloathing, Provisions, Sail Cloth, Cordage, Ironwork, and other fishing Utensils necessary for fitting out the Fishing Boats as also the Outsets of the several fishing Ships & others employed in that Trade & Fishery, amounting to about			£80,000.

* Verbatim from *C.O.* 194:10.

Following the withdrawal of the French under the Treaty of Utrecht and later as a result of the Seven Years' War, the English began to move northward. In 1765, according to Griffith Williams, more than one third of the population, then 5,260,[13] owners of 496 boats, were concentrated about Conception Bay; and some settlements extended to Twillingate, Exploits, and Fogo Island. With this northward expansion,[14] seal and salmon fishing and the fur trade became of greater importance. Another migration, to the south, followed the withdrawal of the French from Placentia, but was limited because piracy was an obstacle.[15] The fishery, as formerly conducted by the French from eight to ten leagues from shore, required larger boats than those used by the English. The region had been placed under the control of Nova Scotia from 1713 to 1729. Many of the French had sold their "rooms," and by such sales introduced more directly the principle of ownership. The English officials[16] engaged in the fishery "in like manner as other traders and merchants immediately therein concerned." In 1728 it was reckoned that 30 planters with 5 boats each took an average of 200 quintals.

The expansion of the resident fishery had a part in the development of the bank fishery. A series of bad fishing years had begun with the severe and prolonged winter of 1713–14. It had chilled the water along the coast, and it had been followed by "the worst season for many years." Up to 1720 the average yearly catch was only 90,000 quintals;

13. *C.O.* 195:9, April 29, 1765; *P.C.* (*Privy Council*), IV, 1854.

14. Fish produced in the north sold at a lower price than southern fish—13 shillings a quintal in contrast with 16*s.* 9*d.* Where in the south there were six or seven men to a boat, in the north "where the fish [are] more plenty as the season is shorter, they allow eight men." On the other hand every year staging was taken down in the north because of danger from ice and in the south it was only repaired. In 1738 Fogo was reported as having 7 fishing ships, 4 sack ships, 70 passengers, 14 fishing ships' boats, 24 residents' boats, 135 byeboatmen, 21 families, and 215 residents, of whom 143 remained over the winter. It also produced 19,000 quintals of fish and seal oil and furs valued at £770 and £300 respectively. Twillingate had 2 fishing ships, 3 sack ships, 50 passengers, 8 fishing ships' boats, 16 residents' boats, 130 byeboatmen, 16 families, and 184 residents, of whom 152 remained over the winter. Twillingate produced 12,000 quintals of fish and seal oil and furs valued at £440 and £100 respectively. The fur trade north of Bonavista increased to £3,890 by 1757. Rogers, *op. cit.*, pp. 121–122. The value of seal oil produced increased from £1,016 in 1749 to £12,664 in 1768. D. W. Prowse, *A History of Newfoundland* (London, 1896), p. 298. George Skeffington, financed by New England enterprise under William Keen, had developed the salmon fishery north of Cape Bonavista. He apparently started about 1708, and was given a monopoly for 21 years of the fishery at Fresh Water Bay, Ragged Harbour, Gander Bay, and Dog Creek. *P.C.*, III, 2008–2010; IV, 1961–1963. Salted salmon was exported to Italy and Spain, more than 1,000 tierces going in 1743 and 4,848 tierces in 1757. Prowse, *op. cit.*, p. 285.

15. See Prowse, *op. cit.*, pp. 276–277; Rogers, *op. cit.*, pp. 116–117; *P.C.*, IV, 1850; and David MacPherson, *Annals of Commerce* (London, 1805), III, 166.

16. Instructions were issued on May 13, 1715, and later, with a view to extending to Placentia regulations favorable to the fishing ships in Newfoundland. *Royal Instructions to British Colonial Governors 1670–1776*, ed. L. W. Labaree (New York, n.d.), II, 690–692.

and this led to the development of the bank fishery by fishing ships. In 1714 and 1715, small ships were sent to the Banks

> to catch and load and then come in, put their fish ashoare to drŷ, and immediately out again leaving people ashore to cure it; and this manner of working has turned to prodigious account. . . . Ten hands employed in a barke this way where the whole amount of weare and teare of the vessel and wages of servants included has not been computed at above 70 pounds has catch'd 600 quintals and upwards whereas the service of seven boats employing in all 35 hands and wages to the amount of near 400 pounds has not caught an equall quantity. . . . Every fish brings its own bait with it to catch another with for by opening the maw you are always stock'd with fresh bait to proceed upon new purchase.[17]

The poorer residents were taken as fishermen and allowed one third of the catch, delivered and cured for market. "The fish taken by the vessels employed on the bank fishery (of which there are a great number) are split and salted the same day they are taken."[18] But there were complaints that the fish produced were of poor quality, "chiefly owing to the fishing ships who have left off keeping of shallops and fishing near the shore but send their ships and vessels on the banks for a month or five weeks, then bring the fish to land to cure; such fish as are caught at the beginning of the season are good, if rightly salted, but in the height of summer and latter end of the year very bad."[19] In 1723 fish "made" by residents commanded "generally a real per quintal and this year in most places twas three reals dearer"—that is, than the bank fish.[20] In 1726 it was two reals higher in price and in most places five reals. The market was weakened because of the northern product and the bank fish; and because of the failure to grade, fish were sold "tal qual," or just as they came. The price of merchantable fish was forced down, and very little refuse fish was available.

> All the fish buyers have the liberty of culling for themselves; it must be their own faults if they take what is bad and it often happens for cheapness they load green fish not thoroughly cured which proves of very great prejudice by heating in the hold and damaging that which was realy good and proves a great discreditt as well as loss in foreign markets.[21]

> It appears very true that the French fish has sold of late years at the price of one dollar per quintal, at least in the Italian markets, more than the English fish.[22]

17. *C.O.* 194:6.

18. *C.O.* 194:13.

19. *C.O.* 194:7 (1720). It required 50 quintals of bank fish to produce a hogshead of oil, and 40 quintals of shore fish.

20. *Idem.*

21. *C.O.* 194:10 (1739).

22. *Idem.*

Of late years the consume of Newfoundland dry'd cod fish called Baccalao has [been] greatly lessened in this province (Catalina) by the fisheries of the same kind of fish that are at present [1765] carried on with success on the coast of Norway and at Knall in Russia.[23]

Similar reports were made from Venice and Leghorn. Newfoundland fish were further handicapped in the Portuguese market by restrictions such as fees and duties amounting to as much as 22 per cent. Further duties of 4 per cent were added in 1756, and another 2 per cent in 1761.

Residents encroached "on the fishing ships rooms because the masters of the fishing ships dont keep up their stages nor employ any shallops but send their ships out on the Banks to fish."[24] As early as 1731 "few ships come purely on account of catching and curing of fish except it be from Bideford and Barnstaple."[25] "Of late [in 1749] few fishing ships bring more than a common sailing crew to manage the ship, their voyage depending chiefly on the freight and passengers out and home, and not on the profits in taking and curing of fish, few keeping more than three boats and many one boat only."[26] By 1763:

There is no such thing as getting an account of the ships rooms in any port, but if a person applys for a grant of a place any number, nay all the inhabitants, will readily certifie or even swear it never was a ships room, thus almost all stages and ships rooms are become private property; the ship-fishing is in a manner dropt or excluded. . . . All rooms and conveniences now used for the fishery to the southward of Cape St John are constantly possessed and kept by the same people for their own private benefit and are become private property. . . . Yct all ships arriving from Britain directly call themselves fishing ships because they clear out as such; though they have no more men engaged to them than is necessary for their navigation nor more boats than one each employed in the fishery.[27]

As ships had ceased to bring out men directly interested in the fishery, wages had for a long time largely displaced the share system. "Of later days [1715] they have given their men monthly wages which did not answer so well as when they went by the thirds; then every man made it his business and took more care for the good of the voyage having a more particular interest therein for the more fish was taken the greater his

23. Cited in Paterson, *op. cit.*

24. *C.O.* 194:7.

25. *C.O.* 194:9. Fishing ships indicated in the statistics apparently included sack ships. Their number fluctuated from 85 (8,000 tons) in 1714 to 80 (10,280 tons, 1,451 men) in 1749; 93 (11,450 tons, 1,597 men) in 1750; 122 (14,580 tons, 2,514 men) in 1751; and 141 (14,819 tons, 1,933 men) in 1764. Boats owned by fishing ships decreased from an average of 324 between 1713 and 1716 to 171 in 1749, increased to 199 in 1750, and to 295 in 1751, but dropped to 210 in 1764. In 1723 ships of 100 tons were outfitted at from £850 to £1,300.

26. *C.O.* 194:12. (1749.)

27. *C.O.* 194:17.

share."[28] In 1723, we learn, "One or two ships from Barnstaple and Bideford continue to allow their company's [crew's] shares."[29] But, by 1749, this could be said: "All ships and boats . . . are upon certain wages and not upon shares. . . . Some give a premium upon every thousand of fish to encourage their men to industry, who keep an account of every fish they catch."[30]

"A fishery is that branch of commerce which not only requires every attention and encouragement but will not admit of the smallest impediment or obstacle." For this reason the byeboatmen continued to have advantages over both residents and fishing ships. "They are the only support of the fishery in this country [and] ought greatly to be encouraged for their indefatigable industry and hard labour."[31] Three partners to a boat and two servants took 100 quintals more in a season than the residents who allowed six and seven men to a boat. They obtained rooms from the residents by leases, "which is found to be cheaper than to build," rentals varying from £6 and £7 during years of peace to £10 and £12 in wartime. The number of byeboatmen increased from 286 in 1716 to 421 in 1749, and to 554 in 1751; their servants increased from 1,538 in 1716 to 4,930 in 1749, but fell to 3,848 in 1751. Their boats numbered on an average 177 from 1713 to 1716, 349 in 1749, and 542 in 1751. Their catch increased from 30,480 quintals in 1721 to about 140,000 in 1750. But their position was seriously affected by the Seven Years' War.

The competition of both New England and the byeboatmen for skilled labor compelled the residents and fishing ships to depend on the relatively unskilled. New England was a low-pressure economic area to which labor and capital were drawn, leaving high-pressure areas such as Newfoundland to draw on grades of labor that were poorer and accustomed to lower standards of living. Fishing ships from Bideford and Barnstaple and other ports had brought out unskilled Irish laborers.[32] A large unskilled resident population depended on goods from England and New England, and wages were reduced by a resort to the truck system.

The inhabitants usually trust their servants for more than their wages and by that means [their servants] are obliged to serve them the next year; or

28. *C.O.* 194:5.

29. *C.O.* 194:7. The men were given one fourth of the catch. In a crew of 30 men, the master had two shares, the mate one and a quarter, the boatmaster one, midshipmen three quarters of a share, foreshipmen half a share, and seamen one quarter.

30. *C.O.* 194:12.

31. *C.O.* 194:7. (1720.)

32. These vessels supplied southern ports such as Renewse and Fermeuse, "to which places they return on account of the utensils they have there." The proportion

[they] dispose of them to whom they please. . . .[33] The trusting the fishermen with such quantitys of strong liquors is very prejudicial to the fishery, and that [is] greatly owing to the masters themselves, for they consider that the more the servant spends in liquor, by which he [the master] gains one half by the profit he makes, he lessens the bills the servants would otherwise receive; this encourages the excess by which many disorders are committed very prejudicial to the peace and quiet of trade in general.[34]

Rum was imported in large quantities, especially in New England ships. In 1723 Newfoundland had 74 taverns, 50 being in St. John's, and in

of Irish was larger in the south, although in the north it had increased somewhat. The following figures are from *C.O.* 194:13.

1753	English	Irish
St. John's, Quidi Vidi, Torbay	454	669
Bay Bulls, Willeys Bay, Toad's Cove	206	395
Renewse	82	100
Fermeuse	50	68
Ferryland	120	130
Old Perlican	100	166
Trinity Bay and Bonavista	513	700
Carbonear and Musquito	222	400
Bay de Verde	69	59
	1,816	2,687

According to Governor Edwards' letter of October 28, 1757, to the Board of Trade, "It having been a custom for some time past for the fishing ships not to bring with them from England their complete number of green-men, and a breach of the laudable custom of allowing shares of what they make in their fishing voyages, instead of wages, they have had recourse to getting over a great number of Irishmen, who, being generally Roman Catholics, they use them as they think proper and seldom pay them any wages, by which many of them are left on the Island, to the great terror and distress of the inhabitants." *Third Report from the Committee Appointed to Enquire into the State of Trade to Newfoundland 1793,* p. 77. In 1765, the total population was given as 9,976 men, 1,645 women, and 3,863 children. "Of these people full 9/10 ths of them are of no use to that country and are lost to this during six months of the year; for during that time they are perfectly idle, abandoned to every sort of debauchery and wickedness, become perfect savages, are strangers to all good order, government and religion, by habitual idleness and debaucheries they are averse to and unfit for labour, never becomeing either industrious fishermen or usefull seamen; or if they were either they are never of use for manning our fleets or for defense of the mother country, have no attachment to it and are always out of reach of it; they are subsisted with the produce of the plantations and use a great deal of foreign manufactorys; they as all inhabitants of Newfoundland ever did, always will carry on a trade prejudicial to the mother country; they claim and hold as property all the old and best fishing conveniencys which by law belongs to ship fishers; by such claims a great deal lies waste and on such as are occupyd they do not employ half so many or so good men as ship fishers would; in my humble opinion such inhabitants instead of being of benefit or security to the country and the fisherys are dangerous to both, for they always did and always will join an invading enemy." Governor Palliser, December 18, 1765. *C.O.* 194:16.

33. *C.O.* 194:7.

34. *C.O.* 194:13.

1750, 122. "The poorer sort of people live very hard and often die in wintertime for want."[35] Joseph Banks described St. John's as the most disagreeable town he ever met with. "For dirt and filth of all kinds, St. John's may in my opinion reign unequalled." In 1753 many thefts and disorders were committed "very prejudicial to the trade as well as to the peace and quiet of the inhabitants occasioned by the numbers of Irish that remain the winter in the country, who have nothing to subsist on but what they steal from the inhabitants and traders."[36] "Seven months in the year," wrote Palliser in 1764, "there is not employment for a tenth part of these inhabitants and . . . consequently they spend that time in idleness and subsist for the greatest part by robbery, theft and every species of violence and wickedness."[37]

Merchants were forced to increase prices to residents to offset losses from debt, and a vicious circle was the result. Goods were sold at 100 and 200 per cent advance, and "some articles almost at three hundred." In this system of engrossing or "cornering the market" to increase prices, monopoly soon had the largest part. "I did all in my power to hinder the engrossing [of] commodities but believe tis here as in most places, the richest people will take their opportunitys of advantaging that way."[38] At St. John's in 1741 three or four merchants engrossed the supply of goods: "The whole island is a monopoly."[39] In 1758 the situation was worse. "Engrossing . . . has increased to a very great degree of late years in Newfoundland."[40] Beef, pork, and butter from Ireland, sugar and molasses from the plantations, rum and other imports were

> engrossed by a few opulent merchants, store keepers, and considerable boat keepers who retail them to the rest of the inhabitants and to those they employ under them in the fishery, at exorbitant prices; by which means they

35. *C.O.* 194:10. Attempts to develop local agriculture were limited. The cows, pigs, and poultry all ate fish. Every planter had a small garden and a potato patch. In 1741, 500 acres were cleared and the colony had some 300 cattle, 200 sheep, and 300 swine. "Under all the severity of the climate, they are not without some horned cattle, but these are preserved with no less care and difficulty than at Louisburg. The inhabitants have also their little kitchen gardens for summer herbs; but all the other species of provisions as flour, salt, meat, etc. they are supplied from Boston, Pennsylvania and other colonies to the southward. With regard to the goods of other kinds, they are brought from England." Antonio de Ulloa, *A Voyage to South America* (London, 1758), II, 404. "They feed their men in the summer season mostly with fresh cod with some salt pork and a little beef and biscuit." The fishing industry was in some sense similar to agriculture in that people literally ate their way into the development of the industry in contrast to the fur trade and the lumber industry, in which food production involved more direct competition for labor.

36. *C.O.* 194:13.

37. *P.C.*, IV, 1852.

38. *C.O.* 194:10. (1735.)

39. *C.O.* 194:11.

40. *C.O.* 194:14.

keep them poor and in debt, and dependent upon them. . . . These merchants, store-keepers and boat keepers in order to secure the produce of the labour of the poor inhabitants to themselves, press their goods upon them in advance for that produce, so that they contract debts without a possibility of paying them; and thus mortgaging the fish before it is caught, their only study is how to defraud their creditors, contract fresh debts with other merchants, and so become indifferent about prosecuting their fishery; and if they do prosecute it, it is only to sell their fish clandestinely to others for immediate supplies or to the French. . . . The inhabitants under these conditions of oppression and deprived of every view of bettering their condition, become abandoned to that dissolute way of life . . . and remain under a slavish servitude to the merchant supplyers, store keepers, and boat keepers whose object it is by every method to induce and compel such as come out passengers from England and Ireland to remain in the country, exercising every scandalous act to defraud and cheat those servants whom they cannot persuade to run out their wages in truck or liquors. The entire dispaire of ever freeing himself has made many a laborious man desperately resolve upon laziness.[41]

The steady hard-working people who engaged in the byeboat fishery were the gainers. They escaped debt and were able to return to England. The high prices had

taught the servants through self defence to raise their wages in proportion and to receive the balance in bills of exchange at par by which means the sober and industrious servant who carries out in his chest the few necessaries he may want receives an extraordinary price for his labour, and at a certainty whether his master takes fish or not without having other discount against him than the passage out, which is very moderate and the master always engages for. . . . The difference between fish pay, and bill or cash is from twenty-five to thirty per cent.[42]

With the rise of the wage and the truck systems went the growth of a resident population, the importation and distribution of provisions and supplies, the purchase of fish, and their collection and sale to sack ships. The increasing importance of the residents favored the development of the merchant class which handled provisions and supplies and received dried cod in return. The large number of servants hired by the master residents who did not average one boat each reflected, however, the part played by individual initiative, the importance of the boat as a unit, the development of a wage system, and the growth of a financial and marketing organization to mobilize supplies from the colonies and from England and to market the finished product of large numbers of men over wide areas. The fishing industry depended upon the initiative

41. *P.C.*, IV, 1852–1853. (1765.) 42. *C.O.* 194:15.

of the individual fisherman, and for the old incentive of profit sharing in the fishing ships, new devices were found. The fishing boat became the basic unit, and the wage or truck system was in part an adaptation to this unit.

The lack of competition, with large ships able to supply the demands of an outport, the low standard of living, the general lack of cash, the dependence on a single staple such as cod, and wide variations in both catch and price were factors that gave support to the truck system. The congestion and lack of space[43] for the dry-fishing industry, resulting from the growing up of settlements, forced the fishing ships into the bank fishery and to outlying ports. The dependence of one, or more than one, small outport on an individual ship, particularly considering the increasing size of ships, meant that the ship alone could exercise control over supplies.[44] The fishing ships had become trading ships and sold merchandise to the fishermen in exchange for fish, or supplied their employers. The expansion of the dry-fishing industry called for movement outward along the shore line, and cheap navigation to the outports made it easier to escape from the control of a central trading organization; but it meant facing monopoly control in the guise of the ship.

The development of the bank fishery, in which "the admirals are chiefly employed in their own fishing and frequently are absent a month at a time on the banks,"[45] and increasing difficulties with the settlements, particularly in the winter seasons, made the demand for an improvement of government machinery inevitable. In winter there was "a sort of respite from all observance of law and government. . . . Theft, murder, rape, or disorders of any kind may be committed without control." And in summer "both the fishing admirals and the commander of His Majesty's ships exercise (1715) a most absolute and tyrannical power over their plantations, carrying away their fish by force and violence and leaving them to starve."[46] We read in a petition from Petty Harbor in 1720: "Your poor petitioners . . . labour under severe difficulties for the want of the administration of justice amongst us and in the winter season especially are in danger of our lives from our servants whose debauched principles lead them to commit wilful and open

43. The enormous amount of space required for drying, the problems of climate, and limited harbor facilities persisted as factors responsible for the migration of English fishermen from Newfoundland to the mainland and for the serious difficulties of the French who were compelled to concentrate on such areas as Gaspé and Canso. New England encroached upon the space of the French in Nova Scotia and Cape Breton.

44. Rogers, *op. cit.,* pp. 116–119. Compare Appendix A and B, pp. 181–182.

45. *C.O.* 194:10. (1736.)

46. *C.O.* 194:5.

murder upon their masters, an instance of which has lately happened in this place." But still the Lords Commissioners for Trade and Plantations could report in December, 1718, that

the fishery at Newfoundland from its first establishment has either flourished or languished according as the inhabitants have been discouraged or encouraged; that the principal obstructions which have attended the trade since the reign of Charles I when it was at its greatest height . . . are entirely owing to the project for carrying on the said trade by a colony of fishermen in opposition to the fishing ships belonging to the adventurers . . . and that the most effective method to remove all the aforementioned obstructions and to restrain the irregularities and disorders of the fishermen as well as to encourage the adventurers to return to their employment would be to remove the inhabitants or planters to Nova Scotia or to some of Your Majesty's plantations in America.[47]

Attempts were made to encourage migration to Nova Scotia the following year.[48] Finally in desperation fifty-one principal merchants and householders on November 26, 1723, made a social contract, quoting Locke's second essay, and three men administered justice from that date to February 25, 1724. On May 31, 1729, a governor was appointed,[49] after the settlement of various controversies as to his jurisdiction, particularly as it affected the rights of the admirals in the fishing vessels.

Disputes increased over property rights as a result of the decline in the control exercised by fishing ships over shore rights. In 1751, we learn, "The admirals in their respective harbours do not concern themselves in preserving the peace nor anything else but their own fishery."[50] Palliser wrote in 1764: "The admirals scarce ever act. . . . For the most part they are ignorant, illiterate men, and themselves the greatest offenders against the rules of the Act."

As ever, the outbreak of wars strengthened the trend toward the formation of settlements and the introduction of government. Convoys were not adequate; in 1757 the enemy captured several vessels. The risk which sailors ran of being impressed for the navy forced the fishing ships to pay higher wages. Bideford complained in 1759 that for many years the port had sent a good 25 ships and 1,000 men, but that trade had declined. Barnstaple had 20 ships and 1,000 men in the 'forties but in 1759 "not one ship." Not only did wars involve drains on men but they also narrowed the market for fish. Again, the growth of settle-

47. *P.C.*, IV, 1831, 1836.

48. Labaree, *Royal Instructions to British Colonial Governors, op. cit.*, II, 624, 692.

49. *P.C.*, IV, 1838; also Attorney-General Yorke's opinion on the powers of the several officers at Newfoundland, December 29, 1730. *Idem*, pp. 1841–1842.

50. *C.O.* 194:13.

ments involved an increasing reliance on land defense.[51] The shift in the character of defense measures necessitated the introduction of taxes in Newfoundland wherewith to build forts.[52] The institutions of government concerned with the problems of settlements gradually displaced institutions peculiar to the fishing industry as dominated by the fishing ships. The adaptation of government suited to the fishing ships to a government suited to settlements was slow but persistent. Oldmixon's comment, "There's no need of much law for the inhabitants have not much land and no money," was no longer applicable.

The increase of settlements and the growth of trade brought an increase in the number of sack ships. The total production of fish grew from 88,469 quintals in 1716 to 506,406 in 1749, and to 561,310 in 1764. Between 1716 and 1749 the number of sack ships increased from 30 to 125 (18,750 tons and 1,809 men), but dropped to 97 (11,924 tons and 1,039 men) in 1764.[53]

> The number of ships employed in this trade and the quantity of fish cured and carried to market, are, independent of many other peculiar advantages which would not attend a mere ship fishery, as great now [in 1765] as are stated to have been employed and caught in the most flourishing time of this fishery under the antient establishment, whilst the value of our exports to this island is five times as great as what it is stated to have been at this period . . . and though it be true as is represented, that the value of what they take from the colonies is double what they take from this Kingdom, yet that must not be accounted for loss, since whatever profits are gained by them finally center in this kingdom.[54]

51. By 1715 the divorce between ships of war and commerce was quite complete. R. G. Albion, *Forests and Sea Power* (Cambridge, 1926), p. 76.

52. See Captain Crowe's laws, 1711, Prowse, *op. cit.*, pp. 271–272.

53. It is difficult to differentiate sack ships from fishing ships.

54. *P.C.*, IV, 1854. In 1765 the men engaged in the fishery, excluding the residents, totaled 9,152. This total was made up as follows: those on ships from England and Ireland 2,211, "passengers" from England 1,993 and Ireland 2,753; on 26 ships from Jersey 310 men composing the crews, and 633 passengers; and 1,252 on sack ships. About 1,000 men migrated annually to the plantations. Of the total engaged in the fishery, 3,492 made up the crews of 253 ships which sailed to foreign markets. Two thirds returned to England, Ireland, and Jersey before the next voyage, and one third went directly back to Newfoundland; and 4,660 in 40 ships returned directly to England after the fishery. The total fishery comprised 293 ships or a tonnage of 31,621, 17,876 men, 1,823 boats, 1,005 stages, and 806 train-oil vats. The fishery produced 532,512 quintals of fish, 2,384 tons of oil, 1,172 tierces of salmon, £5,109 of seal oil, and in addition furs to the value of £980. *C.O.* 194:16. For an excellent criticism of Newfoundland statistics see Williams, *op. cit.* He estimated that Conception Bay produced a quarter of the total, Bay de Verde, Carbonear, and Harbor Grace a second quarter, Torbay, Quidi Vidi, St. John's, and Petty Harbor a third, and Bay Bulls, Ferryland, Fermeuse, Trepassey, and Placentia the remainder. The total reached 1,032,000 quintals and 5,160 tons of oil.

From 1720 to 1750 the price of fish rose from 9 shillings to 12 shillings, and in the 'fifties and during the Seven Years' War from 14 shillings to 16*s*. 9*d*. Ships to Lisbon and Alicante, carrying fish which they sold at six dollars a quintal, could stow only a quantity of wine equal to half the sale price of the fish, and were compelled to take the remainder in money. The industry therefore continued to support the objects of the mercantile policy by increasing control over money, and, in turn, over production and exchange, and by enhancing the flexibility of the trading organization.[55]

Control by the West Country through the fishing ships shifted to control through trade. In 1763 "the merchants and traders in the West of England being long skilled therein have been found by experience to carry on this trade to advantage while those trading from other parts of the kingdom have been losers thereby."[56] Defoe noted at Poole that "especially here were a good number of ships fitted out every year to the Newfoundland fishing, in which the Poole men were said to have been particularly successful for many years past." At Weymouth "merchants carry on now, in time of peace, a trade with France; but besides this they trade also to Portugal, Spain, Newfoundland and the Straights." Dartmouth had "some very flourishing merchants who trade very prosperously and to the most considerable trading ports of Spain, Portugal, Italy, and the plantations; but especially they are great traders to Newfoundland, and from thence to Spain and Italy with fish, and they drive a good trade also in their own fishing of pilchards which is hereabouts carried on with the greatest number of vessels of any port in the west except Falmouth."[57] Saltash merchants had some ships in the Newfoundland fishery, "but" says Defoe, "I could not hear of any thing considerable they do in it." "The chief commerce of these towns . . . is the pilchards and Newfoundland fishing, which is very profitable to them all."[58] In 1732, Weymouth and Melcombe-Regis had "80 sail of ships and barks engaged in the Newfoundland industry." In 1761

the principal branch of the foreign commerce of Poole's inhabitants is the Newfoundland fishery to which they send every spring in time of peace up-

55. "Of all the commodities, therefore, which are bought in one foreign country, for no other purpose but to be sold or exchanged again for some other goods in another, there are none so convenient as gold and silver. In facilitating all the different round-about foreign trades of consumption which are carried on in Great Britain, consists the principal advantage of the Portugal trade; and though it is not a capital advantage, it is, no doubt, a considerable one." Adam Smith, *Wealth of Nations* (New York, 1937), p. 516.

56. *C.O.* 194:15.

57. Daniel Defoe, *A Tour Thro' the Whole Island of Great Britain* (London, 1724), I, 60, 66, 72, 90–91, 99, 111.

58. Defoe, *op. cit.*

wards of seventy sail of vessels from the burden of 100 to 150 tons, laden with provisions, nets, cordage, sailcloth, and all sorts of wearing apparel with variety of other commodities for the consumption of the inhabitants and their servants. The smaller vessels fish on the Banks and make two or three trips every season. Their returns are in cod, oil, skins and furs, and in autumn they export their fish to Spain, Italy and Portugal. This is a trade not more profitable to those concerned than beneficial in general to the kingdom, as it subsists a prodigious number of hands, occasions a great export of our commodities and manufactures and breeds excellent seamen. . . . In time of war they have hitherto suffered extremely and as this is so exceedingly detrimental to a trade which is so apparently serviceable to the Royal Navy it deserves notice.[59]

Competition with London ships became of less importance because England was emerging as a cosmopolitan area with cosmopolitan interests far wider than those limited to a single center.

NEW ENGLAND

A great many galleys [from New England] go to Newfoundland, there purchase a loading of fish for the Streights where they deliver their cargo, and take a loading for England, Holland or the Baltick etc and so return home. The subsistance of the colonies is the supplying our sugar plantations with flower, bisket, pipe-staves, fish and other provisions . . . it being supposed that not less than eight hundred vessels belonging to the province of New England are employed in that coasting and fishing trade. They go to New England and the northern colonies with a cargo of goods, which they there sell at a very great profit, and with the produce build a ship, and purchase a loading of lumber and sail for Portugal or the Straights etc., and after disposing of their cargoes there, frequently ply from port to port in the Mediterranean, till they have cleared so much money as will in a good part pay for the first cost of the cargo carried out by them and then perhaps sell their ship, come home, take up another cargo from their employers and so go back and build another ship.

JOSHUA GEE, *The Trade and Navigation of Great Britain Considered*

FOLLOWING the Treaty of Utrecht, New England regained the position she had lost in the fishery as a result of the wars with France from 1698 to 1713, and she rapidly extended it.[60] In 1721 New Hampshire

59. *The Victoria History of the Counties of England: Dorset* (London, 1908), II, 357.

60. Raymond McFarland, *A History of the New England Fisheries* (New York, 1911), p. 81, also chap. v, *passim;* J. B. Brebner, *New England's Outpost, Acadia before the Conquest of Canada* (New York, 1927), chap. ii; and Lorenzo Sabine, *Report on the Principal Fisheries of the American Seas* (Washington, 1853), pp. 93 ff. See also L. H. Gipson, *The British Empire before the American Revolution* (Cald-

had about 100 fishing vessels. Marblehead had 120 schooners of an average of 50 tons and employing about 1,000 men, and 160 schooners twenty years later. Gloucester had several schooners in 1720 and about 70 on the Grand Banks in 1741. In the latter year Massachusetts had some 400 fishing vessels and an "equal number of ketches, shallops, and undecked boats." The outbreak of war in 1744 and unsettled conditions caused a falling off, and in 1745 the production of New England was estimated at only 230,000 quintals.[61] The number of Marblehead schooners fell to 120 in 1747 and to 55 in 1748. The decline from 1743 to 1763 was one of the immediate effects on the fishery of the wars with France and Spain. Fishermen had been quickly transferred from the fishery to the fleet, and the buying power of the markets in the West Indies and Europe had been much reduced.

Until the outbreak of the war with Spain, New England winter-dry cod had continued to be sent to the Bilbao market as the grade best suited to stand inland transport to Madrid. In 1728 a 60-ton vessel was chartered to take cod to Bilbao, freight to Lisbon or Cadiz, and, from there, salt to New England or freight to Ireland, Holland, or England. Salt was brought from the Tortugas, Turks Island, Cape Verde Islands, Lisbon, and the Bay of Biscay. The poorer grade from the Tortugas reached New England in the middle of April and produced a lower grade of fish than the salt used in Newfoundland, which was chiefly from Lisbon and the Bay of Biscay. In addition to salt, vessels that had sailed with fish, staves, and heavy lumber for Portugal and Spain brought back wines from the Azores and Canaries. In 1722 Bristol and London financed agents to build ships in Boston, load them with fish for the Straits and freight for London or with fish and lumber for the West Indies, and return with freight to London and then New England.

The expansion of the New England fishery was chiefly in response to the demands of the British and foreign West Indies. By 1731 this market for oil and refuse fish was held to be bringing sufficient returns to purchase the required rum and molasses, and the salt, provisions, and equipment used in the entire fishery. This left the receipts from the European market as clear profit. One half of the catch of the bankers was refuse fish. In 1747–48 Salem exported, in 131 vessels, 32,000 quintals

well, 1936), III, chap. x; and J. D. Phillips, *Salem in the Eighteenth Century* (Boston, 1937), chaps. xi, xii, xix, xx.

61. In two Massachusetts Bay districts Douglass cites a decline of from 120,384 quintals in 1716 to 53,000 in 1748. He claims that the milder salt used in Newfoundland meant that fewer fish were "salt burnt." In Newfoundland the fish were "worked belly up" and in New England "belly down." W. B. Weeden, *Economic and Social History of New England, 1620–1789* (Boston, 1890), II, 650.

of merchantable[62] fish to Europe and about 20,000 to the West Indies; and in 1762, in 30 vessels, 11,177 quintals of merchantable fish and 17,498 of refuse fish. In 1750, of the whole catch of 400 vessels in the fishery, half went to the West Indies in addition to the take of 200 vessels engaged in fishing for mackerel, haddock, hake, pollack, and scale-fish. In 1763 Massachusetts, with 300 vessels, took 102,265 quintals of merchantable fish and 137,794 of refuse fish.

The growth of fish exports to the West Indies was due partly to the development of the plantation system, which called not only for supplies in general, for building lumber, draft horses, and oxen, but also for the labor of slaves in large numbers. The slaves were fed for the most part on the poorest grade of cod, or refuse fish. The demand for it grew constantly greater because more and more slaves had always to be brought in, and for two reasons, widely differing. One could be found in the British West Indies, the other in the French. In the British "sugar islands" there was a growing exhaustion of the soil. To hoe, to plant, to spread the extra manure that was needed, larger and larger numbers of field hands had to be used. As for the French islands, their soil was relatively unexploited. They required far fewer hands per acre. For example, in 1717 it was estimated that while, in Barbados, to cultivate 30 acres 150 negroes were needed with a dozen horses and 50 or 60 oxen, in the newly opened French islands the same area required only 30 or 40 hands. But, this making it possible to produce sugar much more cheaply, greater areas of French West Indian sugar land were yearly being opened. The French and English planters were alike in their mounting demand for slaves.[63] After 1724 about 3,000 blacks were brought in annually. Between 1712 and 1762 their number rose from 42,000 to about 70,000. For the Leeward Islands the increase from 1720 to 1755 was some 94,000. Of these, Antigua and St. Christopher each took some 1,500 a year, or a total of about 75,000; and between 350 and 400 went to Nevis and Montserrat. But even this was not enough, and probably more were smuggled in by the Dutch. From 1702 to 1775 Jamaica brought in 497,736, but "reëxported" 137,114. The total number of slaves brought to the British West Indies from 1680 to 1786 was estimated at 2,130,000. Between 1735 and 1763 about 13,000 a year were brought from the Gold Coast alone.

The more profitable character of the relatively unexhausted soil of the French islands, the greater efficiency of the French in the handling

62. The higher-grade fish suited to European consumption.

63. F. W. Pitman, *The Development of the British West Indies, 1700–1763* (New Haven, 1917), pp. 70–71.

of slaves,[64] and the increasing demand in the gold fields of Brazil led, in the Leeward Islands, to a rise in their average cost of from £18 in 1726 to £24 in 1739, and in Jamaica from £35 between 1739 and 1741 to £37 between 1741 and 1745. With this increase in the cost of slaves, and in the number required, and with the decline in exports of sugar went a fall in the price of sugar, especially from 1728 to 1739. This followed the expansion of production in the French islands, particularly from 1720 to 1733, and the competition of this sugar on the European market.[65] About 1730 and later, as a means of lowering costs, large planters began to intensify their control throughout the whole of the British West Indies. Small planters were compelled to migrate to the Dutch islands, Virginia, Carolina, and the northern colonies. By 1748 the white population of Barbados had fallen by 2,551. Large-scale production and the emigration of small planters tended to decrease the supply of foodstuffs produced in the islands, to raise their prices, and to make the islands increasingly dependent on the northern colonies for provisions. The mounting demand of the British West Indies for provisions was accompanied by lower prices of sugar—particularly with an export tax on British West Indian sugar. While the costs of British West Indian sugar increased, the price of sugar declined due to the French competition. Prices of rum and molasses also fell, since they were by-products of sugar.[66] They were excluded from France by a high tariff because they meant competition with French brandy and wines. Consequently flour, grain, and lumber were sold, for example in Jamaica, for specie which was used to purchase sugar and molasses from the French islands. By 1730 the northern English colonies were trading on a large scale with the French West Indies, that is, San Domingo, Guadeloupe,

64. "But, as the profit and success of the cultivation which is carried on by means of cattle, depend very much upon the good management of those cattle; so the profit and success of that which is carried on by slaves, must depend equally upon the good management of those slaves; and in the good management of their slaves the French planters, I think it is generally allowed, are superior to the English. The law, so far as it gives some weak protection to the slave against the violence of his master, is likely to be better executed in a colony where the government is in a great measure arbitrary, than in one where it is altogether free." Adam Smith, *op. cit.*, p. 553.

65. The French gained control of the sugar trade of the Portuguese from Brazil in most of the territory north of Gibraltar and forced down the price of Brazil sugar from £7 or £8 per hundred to as low as 6, 7, and 8 shillings. "[The] Island of Barbadoes is very much worn out. . . . [The] French are able to undersell us." Joshua Gee, *The Trade and Navigation of Great Britain Considered* (London, 1730).

66. "But in our sugar colonies the price of sugar bears no such proportion to that of the produce of a rice or corn field either in Europe or America. It is commonly said that a sugar planter expects that the rum and the molasses should defray the whole expense of his cultivation, and that his sugar should be all clear profit." Adam Smith, *op. cit.*, p. 157.

and Martinique. There was an increase in smuggling to Curaçao, Surinam, Dutch and Danish Guiana, and foreign West Indian ports. British planters complained that the demands of the foreign colonies in the West Indies raised the prices of the products of the northern colonies and made it possible for such foreign colonies to raise sugar more cheaply. High prices after the war and the Treaty of Utrecht provided a stimulus which by 1730 had become of serious consequence. In spite of an act passed in 1715 which prohibited the import of foreign sugar into the British West Indies, French sugar was smuggled in and re-exported to England. On the other hand, due to their decreasing supply of sugar,[67] molasses, and rum, the inability of the British West Indies to meet the growing demands of the northern colonies grew more marked, and the latter were forced to resort to the foreign colonies. The direction of shipping to these foreign islands raised the price of shipping to the British West Indies, and by helping to drain away specie brought about the development of inflation on a serious scale. Smuggling between the Leeward Islands and Barbados and Martinique, and between Jamaica and San Domingo, was inevitable.

A centralization of control because of the relatively few large planters gave the British West Indies sufficient influence in the House of Commons to lead them to attempt to secure a guaranteed British market, to confine the trade of the northern colonies to the British West Indies alone, and finally to secure the passage of the Molasses Act of 1733.[68] This act practically prohibited the importation of products of the foreign West Indies into the northern colonies.[69] New England pro-

67. The increasing cost of sugar growing meant a drop in sugar exports. They declined from 22,769 barrels in 1736 to an average of 13,948 from 1740 to 1748.

68. Lilian M. Penson, *The Colonial Agents of the British West Indies* (London, 1924), chap. vi. "Jamaica [is] the most valuable plantation belonging to the Crown." "*Negroes* are the *first* and most necessary *materials* for *planting*." "The labour of negroes is the principal foundation of our riches from the plantations—to speak of our trade to Africa, which is a trade of the greatest value to this kingdom, if we consider the number of ships annually employed in it, the great export of our manufactures, and other goods to that coast, and the value of the product of our plantations annually sent to that coast, and the value of the product of our plantations annually sent to Great Britain. . . . [The African trade] is the spring and parent whence the others flow and are dependent." "We are enabled by our manufactures and goods of all kind to trade the cheapest and most advantageously to Africa and have a superior strength of any nation to protect our trade on that coast. . . . [They are] so very advantageous to Great Britain, by conducing so much to the support of our tobacco colonies and sugar plantations; and since so great a part of our foreign trade ariseth from them they ought undoubtedly to have all due encouragement and to be supplied at the most easy and reasonable rates with negroes." William Wood, *A Survey of Trade* (London, 1722).

69. Pitman, *op. cit.*, chap. xi; also Richard Pares, *War and Trade in the West Indies 1739–1763* (Oxford, 1936), pp. 79 ff.

tested that the British West Indies offered too small a market for their produce and provided too small a supply of molasses for their distilleries. With cheaper foreign sugar, and with the difficulties of enforcing the act in an area containing numerous islands, cheap water transportation, and relatively small boats, smuggling increased. Neutral ports, especially St. Eustatia, were the scenes of extensive trading, and during the wars between England and France served as depots in which the northern colonies were able to sell their provisions to France and to purchase increased quantities of European manufactured goods. During the war, or from 1744 to 1748, as a result of a scarcity of sugar in the British West Indies, New England was compelled to turn to the French West Indies. The resulting trade became to an increasing extent a basic part of the economic development of the northern colonies. After the war it became even more important, and when the French settled and developed the previously neutral Windward Islands of Dominica, St. Lucia, St. Vincent, and Tobago, this was made easier by exports from the northern colonies.

New England complained that the higher cost of sugar incidental to the restriction of trade with the British West Indies weakened her capacity to purchase manufactured products from Great Britain.[70] The Molasses Act virtually guaranteed a monopoly of the English market to British planters, and what had been large exports of sugar from England to continental Europe before 1713 declined materially. It was driven out by French sugar and attracted to England by a monopoly market. The conspicuous failure of an act of 1739 which empowered the British West Indies to export directly to Europe was an indication of the effectiveness of monopoly in England to raise sugar prices on an expanding domestic market.[71] Directly and indirectly a policy favorable to the British West Indies increased the incentive to smuggling in the northern colonies; and French sugar forced the English product from both the colonial and the European market.[72]

The expansion of trade in the northern colonies aided and was aided by the increase in smuggling and the accessibility of the products of the French West Indies. By 1741 Newport, Rhode Island, had 120 ships which were trading in such commodities as bread, flour, Indian corn,

70. L. F. Stock, *Proceedings and Debates of the British Parliament respecting North America* (Washington, 1924), IV, 162–163.

71. Pares, *op. cit.*, p. 475; C. M. Andrews, *The Colonial Period of American History* (New Haven, 1938), IV, 88–9.

72. It was claimed that French "sugar-island" supplies cost 30 to 40 per cent more than those of the British West Indies; but in spite of this, English sugar was forced out of the European market. See Otis Little, *The State of Trade in the Northern Colonies Considered* (London, 1748).

sugar, molasses, salt, rum, tar, and pipe staves. Provisions and tar, for example, might be exchanged in Newfoundland for refuse fish. The fish were traded to the West Indies for molasses, which was brought back to Newport to be converted into rum.[73] This was sold to those in the slave trade, to the Indians in the fur trade, to the fishing, shipbuilding, and lumbering industries, or exported to Newfoundland, the southern colonies, and Guinea. It was estimated in 1741 that the total trade in salt, sugar, cotton, rum, molasses, lime juice, and cordage imported from the West Indies to New England, and of cod, mackerel, herring, beef tallow, oil, oats, and horses from New England to the West Indies equaled the trade from New England to England itself. New England imported agricultural products from Ireland and the southern colonies. To the latter, small fishing vessels, idle during the winter, traded salt, rum, sugar, molasses, and dry goods in exchange for corn, pork, pitch, and tar. The growing market for slaves in the British West Indies led to an increasing demand for French West Indian molasses for the manufacture of rum to be traded in Africa for slaves. The general rise in sugar and molasses, especially from 1740 to 1750, under monopoly conditions, accelerated sugar growing, particularly considering the relatively stationary character of flour prices; and it also stimulated expansion in the fishing industry. But after 1750 the price of flour did tend to rise, and the price of sugar and molasses fluctuated and tended to decline, with serious effects for both sugar and fishing interests.[74]

Increasing demands for poorer grades of fish in the West Indies hastened the growth of the offshore fishery of Nova Scotia. Shipbuilding after 1713 and the evolution of the schooner enabled New England vessels to occupy, with greater certainty, the Banks[75] and the Nova Scotian fishing grounds, following the retreat of the French to Cape Breton. As a result of the development of the offshore fishery by the schooner, Canso became an important center. Sack ships from England brought goods for the Canso fishery and the New England trade, and returned with fish for Europe. The small New England schooner and

73. Weeden, *op. cit.*, II, 585, 641. Also W. E. B. Dubois, *The Suppression of the African Slave Trade from Africa to the United States of America 1638–1870* (New York, 1896), chap. iv.

74. See Bezanson, Gray, and Hussey, *Prices in Colonial Pennsylvania* (Philadelphia, 1935), *passim.*

75. Schooners fishing on the Banks procured five fares annually, beginning in March with a trip to Sable Island Bank, "as the fish taken there exceed any in the world," returning for a second trip to Brown's Bank and other banks near Cape Sable, then making a third and fourth trip to Georges Bank, and the last trip to Sable Island for winter cod. This deep-sea fishery was conducted by 50-ton schooners employing seven men and averaging 600 quintals. The fish brought to the shore were dried and exported.

the larger sack ship from the mother country became parts of a closely knit organization. As the larger English vessels carried an outbound cargo of fishing stores for the colonies, a return cargo of fish to Spain, Portugal, and the Mediterranean, and a return of freight to England, they were able to maintain an effective control of the fishery in the early part of its history. Canso[76] was at once a part of the fishery frontier and of the trading frontier between England and New England. Fishing stores brought via Canso both contributed to the expansion of the New England fishery and encouraged the development of shipbuilding and trade. New England's trade with Spain and Europe became increasingly important, as was exemplified in the development of a three-cornered trade between Boston, Canso, and Spain. The production of the best grades of fish suited to the European market, however, meant the development of a boat fishery rather than one by schooner, and, in turn, of the further growth of settlements. Small New England vessels built up the Banks as a dry fishery, and Canso tended to weaken the position of New and Old England in the European market by the inferior character of its product. But it strengthened the position of New England in the West Indian trade; and this expansion increased smuggling directly to the French West Indies or through Cape Breton.

The aggressiveness of New England commercialism already evident in the seventeenth century became still keener in the eighteenth. In spite of attempts on the part of France to consolidate her position, her inability to achieve an integrated empire became more apparent; and New England intensified and thrived on such French limitations. Later the frontier position of Cape Breton was also apparent, as shown by the decline of Canso, the prominent part played by New England in the capture of Louisburg in 1745, and by Louisburg's return to France by England in 1748. If Canso had been a frontier, so, too, had Cape Breton. Both England and New England knew the dangers of the French competition which Louisburg represented. And it was significant that when, in 1745, the frontier fortress was taken, it was the forces of New England that played the most important role. More than this, not only was New England striking its roots deep into the French Empire and sapping its strength; it was becoming always more restless under the restrictions of the British Empire. New England's influence became increasingly evident in Newfoundland and in the West Indies. The collapse of the French Empire, due in part to her, was quickly followed by New England's secession from the mother country.

76. See H. A. Innis, "Cape Breton and the French Régime," *Transactions of the Royal Society of Canada,* sec. 2, 1935; see also, for statistics on the Canso fishery, *Documents Relating to Currency, Exchange and Finance in Nova Scotia 1675–1758* (Ottawa, 1933), pp. 156, 174–180, 190–191, 194–195, 205.

FRANCE

The French by having the island of Cape Breton are enabled to be at market with their fish to Spain, Portugal etc. at least six weeks sooner than we can from Newfoundland.

WILLIAM WOOD, *A Survey of Trade* (1722)

HAVING failed during the period ending with the Treaty of Utrecht in their attempt to maintain an empire extending from Hudson Bay to Placentia, Nova Scotia, and the West Indies, the French withdrew from Hudson Bay, Placentia, and Nova Scotia. They consolidated their position on a narrower front, with the selection and fortification of Louisburg in Cape Breton.[77] Isle St. Jean (Prince Edward Island) was selected as a base for this centralized development.[78] But these plans were handicapped by a fishery conducted from a wide range of ports in France,[79] each with more or less distinct and widely separated fishing regions in the New World.

77. For an extended discussion see Innis, *op. cit.;* also A. M. Wilson, *French Foreign Policy during the Administration of Cardinal Fleury 1726–1743* (Cambridge, 1936), chap. x; and J. S. McLennan, *Louisbourg from its Foundation to its Fall 1713–1758* (London, 1918); C. P. Gould, "Trade between the Windward Islands and the Continental Colonies of the French Empire 1683–1763," *Mississippi Valley Historical Review,* March, 1939, pp. 473–491.

78. See D. C. Harvey, *The French Régime in Prince Edward Island* (New Haven, 1926), *passim,* but especially chap. iv, on the difficulties of enforcing the monopoly in the fishery. At St. Peters in 1728, 4,874 quintals were taken, and in 1730 four schooners and 23 boats were operating. A trading vessel of 100 tons came out from Granville to engage in trade and fishing in 1730, and a vessel from Bordeaux in 1731.

79. The relative importance of the various areas, and of the various types of fishery, even though the total is exaggerated, is indicated in statistics given out in 1745 by Governor Shirley, entitled *A Computation of the French Fishery as it was Managed before the Present War.* Six ships came to Gaspé, Quadre(?) and Port aux Basques, and each employed approximately 60 men, who caught 18,000 quintals of fish, or, for the 18 ships and 1,080 men, a total of 54,000 quintals. On the same rough reckoning, Les Trois Isles in Newfoundland had its 3 ships and 180 men, and took 9,000 quintals. From St. Malo and Granville 300 ships were sent out to fish at Petit Nord, in the Straits of Belle Isle, along the North Shore of the Gulf of St. Lawrence, and south to Gaspé. Cape Breton had about 500 shallops with 5 men each and they produced 300 quintals of fish, or a total of 150,000; also 60 brigs, schooners, etc., with 15 men each, which took about 600 quintals apiece, that is, a total of 900 men and 36,000 quintals. To carry home the grand total of 186,000 quintals, 93 ships, taking some 2,000 quintals each, were necessary and, with average crews of 20, they employed 1,860 men. The dry fishery, including that of Cape Breton, employed 414 ships, 24,520 men, and produced 1,149,000 quintals of fish. The bank fishery employed, from the River Leudre, 40 sail; from Olonne and Porteux, 60; Havre de Grace, 10; St. Malo, 20; other ports, 20, or a total of 150. Their ships were manned by from 16 to 24 men apiece, and caught from 22,000 to 30,000 fish; or, with an average of 20 men and 26,000 fish, a total of 3,000 men and 3,900,000 fish. The catch—dry and green—was sold in northern and southern France and to Spain and Italy. In addition, every 100 quintals of fish averaged, in oil, one hogshead of 60 gallons, or a total of 11,490 hogsheads. The green fish, estimating 4,000 fish to weigh 100 quintals, produced 975 hogsheads, a grand total of 12,465 hogsheads, or in weight, 3,116 tons. The dry fish, at 10 shillings a quintal, was valued at £574,500; the oil, averaging £18 per ton, at £56,092, the

But the "greatest part of the French ships employed in the cod-trade do not take in their lading at Louisbourg."[80] In 1719 it was said that 500 ships left Rouen, Dieppe, Fécamp, Le Havre, Honfleur, Granville, St. Malo, Nantes, La Rochelle, Les Sables, Bordeaux, and Bayonne.[81] It was estimated that the French had 200 vessels on the Banks which made voyages "without entering a port in America." The crews of their largest ships, also some 200, engaged in the dry fishery, "have no settled habitations but having erected houses and cleared small places for gardens they raise roots and herbage sufficient to serve them yearly for soup and salad with their return to France."[82] The latter fishery was conducted by "the ships company in fishing in the inhabited bays."[83]

About the Gaspé peninsula, large vessels continued to fish at Caraquet and Port Daniel and small boats at Pabos in Chaleur Bay, Renard River, Grand and Petit Estang, the Magdalene River, and at Mount Louis, St. Anne, and Matane. Lack of a harbor at the latter point restricted the industry. The number of vessels and boats fishing at various places along the North Shore of the Gulf is shown in the table below:

*The North Shore Fishery**

	Number of Vessels	*Men*	*Fish (quintals)*	*Oil (barrels)*
1726	15	1,008	35,910	...
1729	18	1,275	33,000	...
1731	18	1,406	34,900	335
1732	15	1,530	41,300	481
1733	15	1,243	46,000	861
1735	16	1,465	50,600	900
1736	13	1,071	52,510	582
1739	17	1,173	48,500	608
1742	17	1,231	55,700	491
1743	14	1,000	53,600	549

* This fishery remained practically stationary and the vessels came chiefly from St. Malo and Granville. In 1729, 14 were from St. Malo. They visited Isle au Bois, Blanc Sablon, Ile des Marmettes, Forteau, St. Modest, Lance au Loup. In 1717 Petite Rivière had 18 fishing ships, Forteau 8, Lance au Loup only 2 because of the lack of drying space, St. Modest 3, Baie Rouge 2. Twenty-seven vessels came in 1719, the largest drying 2,000 quintals. Ile au Bois declined slightly and Lance au Loup and Forteau increased in importance. See E. T. D. Chambers, *The Fisheries of the Province of Quebec* (Quebec, 1912), pp. 82 ff.

green fish, or "mud" fish, at 9 pence each, giving an added £146,250. Putting freight charges on 1,114,000 quintals at 3 shillings a quintal, or £172,350, the total intake was valued at £949,192. McLennan, *op. cit.*, pp. 380–381; also *Documents Relating to Currency*, etc., pp. 230–232.

80. De Ulloa, *op. cit.*, II, 405.
81. M. Bronkhorst, *La Pêche à la morue* (Paris, 1927), p. 26.
82. Little, *op. cit.*
83. De Ulloa, *op. cit.*, II, 405.

Indications of the character of the fishery are given in the following table:

Isle au Bois		*Blanc Sablon*		*Forteau*		*Lance au Loup*		*St. Modest*	
Size of Vessels (milliers*)	*Boats*	*Size of Vessels* (milliers*)	*Boats*	*Size of Vessels* (milliers*)	*Boats*	*Size of Vessels* (milliers*)	*Boats*	*Size of Vessels* (milliers*)	*Boats*
200	18	160	14	240	12	270	22	160	17
185	16	130	11	130	12	130	11	140	14
145	12	70	7	265	20			100	11
160	14	150	15	74	8			70	7
170	15	135	11	50	8			45	4
150	13	67	7	110	10			60	7
270	22	60	6	133	12				
260	23								
207	21								
218	20								

* One millier = about 1,000 pounds.

Fishing vessels went from Quebec as well as from France to participate along the North Shore in fishing and sealing under concessions.[84]

As a result of difficulties with the English, the French in the Petit Nord had moved north of Notre Dame Bay and west of Fleur de Lys. "The smaller whiter and better salted" fish of the Petit Nord obtained a premium of a dollar a quintal in such Mediterranean markets as Marseilles,[85] Genoa, Leghorn, Civitavecchia, Sardinia, Corsica, and Sicily.

84. See E. T. D. Chambers, *The Fisheries of the Province of Quebec* (Quebec, 1912), pp. 92 ff.; also J. N. Fautcux, *Essai sur l'industrie au Canada sous le régime Français* (Quebec, 1927), II, 531 ff.; *P.C.*, VII, *passim.* In 1734, 13 fishing boats went from Quebec to the lower St. Lawrence. Fishermen from France and from Quebec apparently secured permits from the owners of the posts. Chambers, *op. cit.*, pp. 89 ff. See a wage agreement of Jean Gatin St. Jean with Gabriel Amiot to engage in the cod fishery at 60 livres a month beginning May, 1717, and ending with the return of the expedition to Quebec. H. A. Innis, *Select Documents in Canadian Economic History 1497–1783* (Toronto, 1929), p. 393.

85. In the 1750's St. Malo sent out 80 vessels a year, for the most part in the Marseilles trade. For interesting letters—The Mégon Papers—exchanged by members of a firm engaged in the St. Malo fishery, see Henri Sée, *Le Commerce maritime de la Bretagne au XVIII siècle, mémoires et documents* (Paris, 1925). A member of the firm wrote in January, 1724, to a friend at Rennes: "I am supporting a venture to the cod fishery in Newfoundland in charge of a trustworthy individual. I have a high opinion of its prospects both because specie will be high and because provisions will be low as compared to money." A letter to Marseilles dated October of the same year commented on the poor returns of 2,000 quintals "in Provence weight." The captain, however, was an excellent trader and it was expected that "the vessel would return pound for pound of the capital if the local catch of fish failed at Marseilles and cod was worth from 20 to 24 *livres.*" In September, 1738, the writer advised his correspondent at Marseilles to send vessels with cod from Newfoundland to Alicante,

After the Treaty of Utrecht, boats were not left over the winter but 600 or 700 men were sent two months ahead to prepare cargo. A 200-ton ship would have some 70 men, and their work would be divided up in this way: fourteen small boats with 3 men each would do the fishing; two large boats, with 5 men in each, would carry the fish ashore; five experienced men would split fish, and a salter with 12 men would work with them. The salt required would average 10 hogsheads for every 100 quintals of fish. The cost of outfitting and provisioning the ship would be about £700. And she would take 4,000 quintals of cod to the Mediterranean and sell them at from 30 to 40 shillings a quintal. One third of what the fish would bring would pay the wages up to the time of selling the fish in the Levant. After that, and when the return cargo for France had been taken aboard, payment would be by the month. The rights of the admirals, or first arrivals, remained in force, some ships arriving so early that they found the harbors frozen.

The success or failure of this fishery depends indeed in a great measure on the station of the ship, and the conveniences for curing the fish. Besides, as the wages paid by the owners to the master, petty officers and men, is always one third of the sound fish brought to Europe, the shorter the time, the greater is the advantage to each man on board. . . . Such are the motives for which the masters of vessels in this trade are so eager to be among the first, that they may chuse their several conveniences in order to finish their fishery with the greater dispatch; and returning early to Europe may turn their cargo to a better account.[86]

The ice caused serious damage in 1750, and in the following year fewer vessels were sent out and wages were limited. "The equippers had made a rule that allowed of their limiting, in advance, what the sailors would receive."

In this region suitable drying beaches were more frequent and expensive flakes less necessary. The timber was cheaper, for it had not

Cartagena, or Cadiz, and to get a return cargo for Le Havre or St. Malo. "The freight charges are the only thing which will prevent a loss." He suggested that a cargo of soap should be purchased for sale in St. Malo and lower Brittany. The significance of the shipping industry to fishing and trade was important. "Many ships have come back from Marseilles empty, while foreign ships have loaded for Bordeaux, La Rochelle, Nantes, or Le Havre. The freight that one can pick up at Marseilles for one of these ports means profit or loss for the Newfoundland trip. . . . Foreign freight charges are lower. . . . Furthermore, French ships fraudulently sell English cod at lower prices, and pass them off as fish that they took themselves." *Considérations sur le commerce de Bretagne, 1756,* pp. 40–41. For illustrations offered by the records of British trade, see the protests against lowering the duties on Smyrna raisins because it would keep the raisins of Denia, Belvidera, and Lipari from being brought back by fishing ships. Stock, *op. cit.,* IV, 519–520.

86. De Ulloa, *op. cit.,* II, 405–409.

been cut as was the case farther south. According to Joseph Banks, "in the neatness of their stages and the manner of working they [the French] are much our superiors." The French "have the properest kind of salt for the purpose, of their own, which renders their voyages much shorter than ours; for we are obliged to go from hence to Rochelle, Oleron, St. Martins etc. to fetch that commodity which they have at their own doors, and thereby we most frequently spend a month or six weeks more on our voyages than they do."[87] Twenty vessels were engaged in this fishery in 1733. The Basques fished at Port au Choix on the west coast of Newfoundland and after the removal from Placentia the fishery was conducted on a smaller scale on the south coast. St. Jean de Luz, Bordeaux, La Rochelle, and Nantes took over 60 cargoes annually. Complaints were made that a Guernsey ship manned in St. Malo had engaged in the fishery at St. Pierre in 1714 and 1715, and at other ports. French deserters from Cape Breton at Port aux Basques (1724) were supplied by ships from Bayonne and St. Jean de Luz.

"Followed more by the French than any other nation, is . . . the mud fishery." The cod caught on the Grand Banks or near Sable Island lay salted in the hold of the ship "till it has sufficiently purged; then they shift its place and having salted it a second time, stow it for the voyage." "That caught in summer—after June or July—is inferior to that caught at the end of winter." From the beginning of February to the end of April was the best season, "the fish, which in winter retire to the deepest water, coming then on the banks and fattening extremely. . . . Sometimes they are known to make two voyages in a year. For 'tis the south part of the Bank that this fish chiefly haunts, and these likewise are accounted better than those taken on the north."[88] Green cod were taken chiefly in the Gulf of St. Lawrence, the Grand Banks, St. Pierre, and Sable Island by double-decked vessels of from 100 to

87. Malachy Postlethwayt, *The Universal Dictionary of Trade and Commerce* (London, 1751). The "sounds" were salted in barrels holding six or seven hundred pounds, and the tongues in barrels holding four or five hundred; and they were sold chiefly in Burgundy and Champagne. The roes were salted and used on the Brittany coast to attract pilchards. The oil, in barrels that also held four or five hundred pounds, was sent to Geneva and used by tanners in France. "The standard cod is that which is two feet in length with the head off. The second is smaller called the middling; the third is the least. The dealers in this commodity however subdivide it into seven or eight kinds; one of these is a fish in the opening of which, or in the severing the head, some fault has been committed." De Ulloa, *op. cit.,* II, 409. In 1733 vessels from Nantes to the Newfoundland fishery had declined from 20 to 7 or 8. From St. Malo, some 20 to 25 vessels took fish to Bordeaux and Bilbao and 40 or 50 from the Petit Nord to the Mediterranean. Henri Sée, "L'Industrie et le commerce de la Bretagne dans la première moitié du XVIII siècle," *Annales de Bretagne,* XXXV, 192, 202, 448–449.

88. De Ulloa, *op. cit.,* II, 409.

150 tons; and they brought home "30 to 35,000 at most for fear of spoiling before brought to France, especially those first caught unless salted with great care." It was important "to have a master who knows how to cut up the cod, one who is skilled to take the head off properly, and above all a good salter on which the preserving of them and consequently the success of the voyage depends. . . . Merchants of the Sands of Olone in Lower Poictou interest themselves, most of all the French, in this fishery and with the most success though this city be small and haven bad, having had some years 100 vessels." The Channel ports "trade little in time of war because of the risque there is in going out and in, the channel being commonly full of privateers." The wages of the master and crew were one third of the fish. The best caught on the south part of the Banks "are therefore chiefly reserved for Paris, where there is a great consumption of them. Those caught on the north side are commonly small and sell for much less." Fishermen could take 350 or 400 a day. They were split and stowed in the hold "in beds a fathom or two square, laying layers of salt and fish alternately but never mixing fish caught on different days." After being left until, in three or four days, the water had drained from them, "they are replaced in another part of the ship and salted again after which they are no more meddled with. They were sold in different places—Nantes, Rochelle and Bordeaux—and sorted according to size, *1°*, great cod of 100 to 90 pounds; *2°*, middling, from 100 to 60 pounds, *3°*, small, and *4°*, refuse."[89] But the last were not handled at La Rochelle and Bordeaux. "The greatest quantity comes from Nantes, the river Loire most conveniently transporting them to other cities and they are very cheap there, except in war time." At Nantes they were counted at 124 the hundred, at Orleans and in Normandy 132, and at Paris 108.

During the war with Spain from 1739 to 1744 the French displaced the English in that market. "Since the commencement of the war with Spain the French have found the sweets of supplying the Spanish markets." The capture of Louisburg in 1745 weakened their position in the fishery, and the Seven Years' War had serious consequences. The continued support given by New England to the extension of the settlements in Newfoundland and the pressure of surplus population on shore space involved a steady movement northward both of settlers and of the fishing ships displaced from the settled regions, and consequently there was a continuous encroachment on territory in which the French fishery prevailed and competition with the French in the markets of the Mediterranean.

89. Malachy Postlethwayt, *op. cit.*

ATTEMPTS on the part of France to consolidate the fishery emphasized the limitations of the French Empire more conspicuously. France's success in the fishery and the assistance it gave to her naval strength[90] were offset by the expansion of the carrying trade of England and of the fishing industry and carrying trade of New England. Her failure to link up her tropical, maritime, and north-temperate regions contributed to the successful growth of the British Empire. Expansion in the newly opened French West Indies and competition with the relatively exhausted soils of the British West Indies contributed to the expansion of the British slave trade and, in turn, of the export of British goods to Africa. Attempts to create a monopoly in Great Britain and the colonies for British West Indian sugar meant support given to the British slave trade and to British industry, and an increasing market for colonial products such as lumber, fish, and agricultural produce. On the other hand, with an increased market for these exports there went an increase in demand for the cheaper products of the French West Indies. Colonial trade,[91] and particularly that of New England, intensified the contradictions of British policy as expressed in the Navigation Acts and in the Molasses Act. Cheap supplies of molasses as a by-product of sugar and lack of a market for rum and molasses in France were valuable aids to English colonial trade. As for New England, she gained from the colonial policies of France and Great Britain both.

The expansion of the market for fish and other products called for an increasingly integrated industrial community concerned with production and trade in agricultural produce, lumber, rum and molasses, ships and fish. The fishing industry continued to capitalize the advantages of shipbuilding, shipping, and year-round navigation, and to join the resources of the land in its production of lumber and provisions to the contributions of the sea. In production and marketing it depended on individual initiative; and the unit of production—the small ship—

90. "The history both of France and England will show you that it is since their procuring leave to fish at Newfoundland that they have grown so formidable at sea; that their navy royal has augmented in proportion to the numbers of ships employed in that fishery." *The British Merchant* (2d ed., London, 1743), II, 257. "They are now so much our rivals in this trade and are increased to such a prodigious degree that they employ yearly from St. Malo, Granville, Rochelle, St. Marten's, Isle of Rea, Bayonne, St. Jean de Luz, Sibour [Sibiburo], etc., to carry on their fishery on the Great Banks of Newfoundland and on the coasts of that island, that is, in their wet and dry fish, upwards of four hundred sail of ships; they do not only now supply themselves with the fish they formerly had from us but furnish many parts of Spain and Italy therewith, and rival us there to our prodigious loss." *Idem*, pp. 255–256.

91. On the increasing disproportion between North American production and West Indian consumption, and the increasing trade between the English colonies and the French islands, see Pares, *op. cit.*, pp. 396 ff.

provided advantages of flexibility in marketing in the numerous islands of the West Indies, and also a flexibility in the handling of a diversity of commodities. For such cargoes large numbers of aggressive small owners were ceaselessly in search, either to buy and sell them for profit, or to use them to fill empty holds and cover the costs of outgoing or return voyages. The problem represented by ships that lacked either outgoing or return cargoes—bane of the fur trade of New France—was largely solved. Expanding industry and New England trade with the West Indies were aids to further expansion in the fishing industry and trade with the Mediterranean, Europe, and England. The aggressiveness of New England traders in small ships lowered freight rates and contributed to the development of a trading organization. Molasses was smuggled not only directly from the French West Indies but also indircctly from Cape Breton. Provisions and schooners were sold by New England to Cape Breton and a growing trade at Louisburg coincided with a decline of the New England fishing industry at Canso. Similarly, provisions and West Indian products were exchanged in Newfoundland for Mediterranean goods, English manufactures, and bills with which English ships "purchase their cargoes . . . at two months date which are very seldom protested." The extension of trade from New England to Cape Breton and Newfoundland was in part a result of the tightening of the British colonial system as covering the West Indies and the Mediterranean.

The support given to Cape Breton[92] enabled the French to improve their position in the European market. New England fishing interests, as contrasted with carrying interests, were linked to the British West Indies and the European market; English fishing interests in Newfoundland were also linked to the European market. The carrying interests profited since the French Empire was maintained in Cape Breton by provisions and supplies from New England, and by English colonial policy, because of the difficulty of developing trade between the St. Lawrence, the fishing regions, and the French West Indies. Restrictions in the French market on West Indian rum, designed to further the consumption of French brandy, increased the quantities of rum available

92. French encroachment on the market for Newfoundland fish in Italy was regarded as a result of France's possession of the fishery in Cape Breton. Gee, *op. cit.* "If we recover the island of Breton again we not only secure our own Newfoundland and New England fisheries but shall deprive the French of theirs." "They will have no port for their ships to lie in on the continent to secure them from us in time of war, nor to send out their men of war or privateers from, to endanger our trade." *Considerations on the State of the British Fisheries in America and their Consequence to Great Britain with Proposals for their Security by the Reduction of Cape Breton which were Humbly Offer'd by a Gentleman of a Large Trade of the City of London to His Majesty's Ministers in January 1744–45* (London, 1745).

for the English colonies. Having French rum, English competition intensified the difficulties of the French fur trade. The extension of the fur trade by way of the St. Lawrence into the interior and increasing competition with the English led to military measures restricting New England to the coast and in turn to its consequent extension to Nova Scotia. The fertility of the land occupied by the Acadians in the Bay of Fundy regions and, on the other hand, the smuggling across the narrow isthmus between Nova Scotia and Cape Breton linked the interests of New England's farmers and traders to the fisheries and led to the capture of Louisburg and the expulsion of the Acadians.

The French fishery at the end of the period was divided between the important branch conducted by fishing ships at Gaspé, Petit Nord, the Labrador, and the bank fishery, and the complex resident fishery at Cape Breton in which the fishing ships were conspicuous. In contrast with the importance of the fishing ship to the French was its decline in the English fishery in Newfoundland. The spread of settlements as an outgrowth of the English fishery, in spite of determined efforts to check them, contrasted strikingly with the obstacles in the way of establishing settlements in connection with the French fishery, in the face of determined efforts to encourage them. The fishing ship was equipped with provisions and supplies from the home port, and the fishery was an extension of French activity and not a basis for isolated settlements. Not only did a continental country such as France provide supplies and a market for the French fishery, but its long coast line meant a variety of markets from Paris to Marseilles. It also meant an attitude of independence on the part of fishing ships from the various ports. The fishery was prosecuted in a large number of widely separated fishing harbors and by many separate interests. The extent of the area over which the French conducted the fishery contrasted sharply with the restricted territory of the English. The concentration of the ports in the relatively infertile area of the western counties meant the need of depending on other sources for provisions and supplies to meet the demands of an expanding fishery. Cheap solar salt was purchased from tropical regions. Limitations in matters of supply accompanied limitations in the size of the market. The expansion of the English fishery depended fundamentally on foreign trade. Protestant England depended on markets in Catholic Europe. The concentration of the fishing-ship ports in the western counties of England favored a concentration of ships in harbors in Newfoundland between Bonavista and Cape Race. New England developed as an area from which the resources of the Gulf of Maine could be utilized and sold to the foreign markets of Spain, the Mediterranean, and the West Indies; and New England became a source of provisions and supplies for the Newfoundland fishery, contributing to the rise of settlements and the decline of the fishing ship. English dependence on trade and on the dry fishery favored the increasing specialization incidental to that fishery. Concentration facilitated the development of depots where the product could be graded for a variety of markets, whereas the dependence of the French on the bank fishery and on the participation by separate ports in isolated regions ultimately limited their effectiveness in the foreign dry-fish market. France developed along lines of self-sufficiency, England along lines of trade expansion.

A flexible economic organization of the fishery made easier a shifting to new areas, new technique, and new markets, and was enhanced by the rapid evolution of technique adapted to the varied demands of the fishery. In New England the small schooner became a most efficient type of vessel for the bank fishery and, in turn, for trade with the West Indies and Newfoundland. Whereas the French continued to rely chiefly on the fishing ship, the English fishery became divided into the boat fishery of Newfoundland and the schooner fishery of New England. New England was able to take immediate advantage of shifts in the economic development of other regions such as the opening of the gold mines in Brazil,[93] the increasing importance of sugar production in the foreign West Indies, and the increase of settlements in Newfoundland. The carrying trade based on the fishery capitalized the advantages of the sea, and the production deficiencies of England as an island called for an ever-growing expansion and a constant search for new resources.

The human labor required by the relatively exhausted soil of the British West Indies gave substantial support to Great Britain's African slave trade, so that the interests of the sugar industry, with the

93. The decline of sugar production in Brazil and the rise of gold mining increased the price of slaves and the cost of sugar production in the British West Indies in contrast with the islands owned by the French. Under the Methuen Treaty, wine was exported in increasing quantities from Portugal to England along with Brazilian gold. From 1730 to 1740 the exports from England to Portugal were valued at more than £1,000,000, and from Portugal to England at more than £300,000. From 1740 to 1760 imports to England declined from £428,857, the average for 1740–45, to £256,-600, the average for 1755–60, while average exports to Portugal increased from £1,115,100 to £1,300,681 for the same periods. A rise in the price of slaves and the advantages of the Asiento Treaty precipitated difficulties with Spain and the outbreak of war in 1739. In 1748 the Asiento Treaty was renewed for four years in the Treaty of Aix-la-Chapelle but it was terminated in 1750. "The Spanish War which began in 1739, was principally a colony quarrel. Its principal object was to prevent the search of the colony ships which carried on a contraband trade with the Spanish main. This whole expense is, in reality a bounty which has been given in order to support a monopoly. The pretended purpose of it was to encourage the manufactures, and to increase the commerce of Great Britain." Adam Smith, *op. cit.*, p. 581.

cost of slaves rapidly rising as a result of competition from the gold fields of Brazil, were linked to the slave-trade interests in their common desire to obtain a monopoly of the sugar market of Great Britain. Larger numbers of slaves meant an expanding market for colonial products, and the profitable character of the slave trade encouraged participation by the colonies. Participation was hastened by cheaper supplies of rum from the French West Indies and the aggressiveness of colonial traders. Newly opened areas in the French West Indies forced British West Indian sugar from the European market and enabled New England not only to avoid but to capitalize the weakness of monopoly control in Great Britain.[94] The expansion of colonial trade was an aid to the growth of the settlements in Newfoundland, but the demands of the West Indies for New England fish and for poorer grades of the Newfoundland cod made easier a growing competition from the French in the European market.

Economic flexibility demanded political flexibility. The gradual decline of the influence of the fishing ships of the western counties upon English policy, the growing up of settlements in Newfoundland, and the increasing independence of trade in New England and its opposition to the colonial policy were indications of the effect of the fishing industry on political development. It was significant that Faneuil Hall, the cradle of liberty, was named in honor of a man who prospered in the fishery. The expansion of the British Empire through the fishing industry and sea trade was dependent on flexibility; and, later, the breakup of the old empire was partly the result of maladjustment between an economic structure powerfully influenced by the fishing industry and a relatively inelastic political structure supported by vested interests such as the sugar plantations. But the efficient flexibility which characterized the fishery of Newfoundland and New England was sufficient to displace the French. The British Empire, in its connection with the fishing industry, had elements of strength as contrasted with the French Empire, but it was inevitable that the competitive elements incidental to this flexibility should persist and that the capture of Louisburg and Quebec should be followed by the revolt of the colonies farther south. France was eliminated because of her continental background by a people who lived in terms of islands, trade, settlement, and the sea. The English fishery became a more valuable nursery for the British navy in so far as it was more closely related to development of

94. Pares, *op. cit.,* pp. 475–476. See C. M. Andrews, "Anglo-French Commercial Rivalry, 1700–1750," *American Historical Review,* XX, 539–557, 761–781, particularly for a discussion of the West Indian problem with special reference to mercantilistic writings.

areas accessible to cheap water transport and to the growth of the carrying trade. The Treaty of Utrecht and the Treaty of Paris were milestones in the evolution of economic forces in which the geographical characteristics of an island, with its relative dependence on world trade, and the geographical characteristics of a continent, with its relative dependence on self-sufficiency, had become increasingly important.

Negotiations leading to the Treaty of Paris were designed to exclude the French from the fishery as a means of reducing enemy naval strength and of weakening a competitor. A petition of the merchants of Bristol in 1762 stated that the French monopoly,

> though not yet admitted by the English to extend so far as to exclude us from it, yet in fact by force and management they had engrossed to themselves this which is by much the most valuable part of the fishery of the island, and is now much more so than ever before as the fish of late years in a great measure left the eastern shores and shifted their course to the north parts of the island.[95]

> Whether the French fished on the northern or southern coasts of Newfoundland it made no difference to Great Britain, for in both cases they have always had the substance of those fisheries and we but little more than the shadow of them.[96]

The French insisted, as in 1713, that "the Newfoundland fishery is absolutely necessary for the support of the Kingdom in general and more particularly for the maritime provinces of western France; where thousands of families would be reduced to beggary in case that fishery be taken from them."[97] But French ports concerned in the fishery were unable to present a united front against English encroachment.

The Treaty of Paris[98] in 1763 reaffirmed the rights of the French to the shore fishery between Bonavista and Cape Race ceded in the Treaty of Utrecht; but in the Gulf of St. Lawrence they were not allowed to fish within three leagues of the shores of the islands and the continent, nor within fifteen leagues of the coast of Cape Breton.[99] They were granted St. Pierre and Miquelon, to quote the treaty, "in full right . . .

95. *C.O.* 194:15. In that year the French captured St. John's and destroyed a large part of the fishery.

96. Joseph Massie, *Historical Account of the Naval Power of France* (London, 1762), p. 18.

97. Paterson, *op. cit.;* see "Les Chambres de Commerce de France et la Cession du Canada," *Rapport de L'Archiviste de la Province de Québec pour 1924–1925,* pp. 201–228.

98. For an account of the diplomatic struggle and the part played by the fisheries see Pares, *op. cit.,* pp. 577 ff.

99. Prowse, *op. cit.,* pp. 311 ff. The French claimed, until 1778, that Point Riche was Cape Ray.

to serve as a shelter to the French fishermen; and his said most Christian Majesty engages not to fortify the said Islands, to erect no buildings upon them, but merely for the convenience of the fishery and to keep upon them a guard of fifty men only, for the police." The enforcement of the treaty was carried out under a proclamation of 1764 stating "that there should be no distinction or interruption given to the subjects of France in enjoyment of the fishery," and it included provisions for guidance in the interpretation of this clause.[100]

Appendix A

In 1742 returns of vessels entering and clearing St. John's (not allowing for smuggling) indicate that provisions came largely from England and Ireland. Cork and Waterford, particularly the former, supplied 883 firkins of butter of a total of 958. England supplied most of the beef (1,332 barrels), pork (1,742 barrels), and peas (307 hogsheads), beer (101 hogsheads), and slops (54 hogsheads). Of 9,940 quintals of bread imported, 5,750 came from England and the remainder as well as the flour (845 barrels) came from the colonies, chiefly New York and Philadelphia. Bacon (2,200 hundredweight) came almost entirely from Philadelphia. Boston supplied a small quantity of rum, 147 hogsheads (Barbados 644), sugar, 19 barrels, and most of the pitch, tar, and turpentine (123 barrels), the lumber, the livestock and poultry (210 head of sheep and 100 geese). The carrying trade was dominated by England and all imports of salt (15,348 hogsheads) were brought either from Lisbon direct (7,928 hogsheads) or from Calary (4,240 hogsheads), or indirectly via England (1,150 hogsheads), being reëxported to the colonies. Most of the fish was carried by English ships to Europe, and, of a total of 121,365 quintals (including 27,500 carried over from 1741), 19,339 were exported to Oporto, 36,230 to Lisbon, 20,346 to Leghorn, 6,600 to Naples, 1,097 to Gibraltar, 1,000 to Figuera, and 1,500 to Madeira. Ships clearing for Dartmouth carried 17,480 quintals, for Poole 2,150, for Teignmouth 2,800, and for Barbados 7,472. Only 100 quintals were sent to New York. Boston took 7,000 sealskins. Ships of from 100 to 150 tons brought salt from Lisbon or passengers and provisions from England, and left with fish for Lisbon, Oporto, or other Continental ports, or an English port with passengers and fish. With a carry-over of fish in Newfoundland it was possible for the *John and Jane* to arrive at St. John's on May 22 with 300 hogsheads of salt from Lisbon, to leave on June 10 with 1,000 quintals of fish for Lisbon, to return on August 23 with 342 hogsheads of salt, and to leave for England on October 2 with 30 passengers and 30 tons of oil. *C.O.* 194: 11.

100. *Idem,* pp. 318–319. These provisions were rigorously carried out both as regards the French and the English. See, for example, cases in which a French vessel was barred from trading and in which the French were forbidden to build boats. Prowse, *op. cit.,* pp. 320–321, especially p. 321 n.; also cases in which the French were forbidden to trade with the Eskimos and Indians. *Idem,* pp. 324, 333, 337–338.

Appendix B

The Charge of fitting out & Mentaining a fishing Boat for the Season*

Item	£	s	d
To a Boat	£20		
To 1 New Road	2		
To 1 Sute of Sails	4	10	
To Rigging & Blocks	1	1	
To Ropes for Sean Lines 3 hund^{d}. weight	3		
To 1 Small Anchor of 40 lb. & 1 Cillick	1	10	
To 3 dozen fishing Lines @ 6 ^{ce} each	0	18	
To 1 D^{o}. Sand. Ditto @ 10^{ce}	0	10	
To fishing Leads 56 pounds	0	10	6
To Sheet Lead 12 pounds	0	4	
To 6 Dozen Small Quarter hooks	0	1	9
To 1 Grose Middle Ditto	0	6	0
To 3 Dozen Bank hooks for Giggers	0	8	0
To 1 Boatmaster @ 23 £	23		
To 1 Midshipman @ 18	18		
To 1 Foreshipman @ 12	12		
To 1 Captain @ 7	7		
To 1 Splitter @ 20	20		
To 1 Salter @ 16	16		
To 2 Greenmen @ 5	10		
To 4 Barrels Pork @ 50/ pq	10		
To 2 Barrels Beef @ 40/ pq	4		
To 1 m of Bread @ 12/ pq	6		
To 3 Gallons Sweet Oyl @ 5/ pq	0	15	
To 1 ferkin Butter	2	0	
To 2 Bushels of Pease	0	10	
To 2 Gallons Rum @ 3/ pq	3		
To 11 Gallons Molasses	1	10	
To 1 Caplin Sain of 30 foot deep & 4 fathom Long	18		
To 3 Netts	2	10	0
To 40 Hhds Salt @ 8 pq	16	0	
	£205.	4	3

allowed to a Single Boat
- Stage room 16 feet wide & 70 feet Long.
- Flake room 50 Yards long & 40 yards wide
- N.B. 10 Hhds Salt allowed to Cure one hund^{d}. Quintals of Fish.

J. W. Webb

* Verbatim from *C.O.* 194:15.

Appendix C

A CATALOGUE OF THE NEWF. LAND NORTHARD FISHERY*

Item	£	s	d	Total
To 10 Boats at £20 p. boat	£200			
To 20 Roades for Ditto	40			
To 10 Sutes of Sails	45			
To Riggin & Blocks	5			
To ropes for Sean Lines etc. a 5 hund^{d} weight	5			
To 15 Small Anchors of 40 $^{lb.}$ Each	15			
To 30 dozen Fishing Lines @ 6 ^{ce} each	9			
To 10 Ditto Suad Ditto @ 10 ^{ce}	5			
To Fishing Leads 3 hund^{d} weight	3	3		
To Sheet Lead 56 pounds		10	6	
To 5 Grose small Quarter hooks	1	7	6	
To 10 Grose middle Quarter Ditto	3			
To 2 Grose Bank Ditto for G.e.gen	1	4		
To a Cooper with 3000 Staves	35			
				£ 368. 5. 0
To a Ship of 150 Tons	700			
To fitting out Ditto	250			
To 10 Boats Masters @ 23$^{£}$ each	230			
To 10 Midshipmen @ 18 each	180			
To 10 Foreshipmen @ 15 each	150			
To 10 Captains @ 12 each	120			
To 5 Splitters @ 20 each	100			
To 3 Salters @ 16 each	48			
To 13 Greenmen @ 10 each	130			
To 60 Barrels Pork @ 50/ p q	150			
To 20 Barrels Beef @ 20/ p q	40			£2098
To 10 m. weight bread @ 12/pq	60			
To 64 Gallons Sweet Oyl @ 5/	16			
To 10 ferkins Butter	20			
To 5 Hhds Pease	10			
To 10 Barrels Flouer	10			
To 500 Gallons Rum	75			
To 110 Gallons Malasses	15			
To 1 Lance Seain 22 foot deep and 75 fathom Long	15			
To 1 Ceaplin Sean 30 foot deep and 40 fathom Long	18			
To 12 Netts	10			
To 700 Hhds Salt @ 8/ p q	280			£ 529
				£2995. 5

N.B. A Ship of 100 Tons, with 10 Boats & 50 Men, will not mentain herself; as the People do not now, go by the Share.

* Verbatim from *C.O.* 194:15.

CHAPTER VII

THE COLLAPSE OF THE FIRST BRITISH EMPIRE 1763–1783

To increase the shipping and naval power of Great Britain by the extension of the fisheries of our colonies is an object which the legislature seems to have had almost constantly in view. Those fisheries upon this account have had all the encouragement which freedom can give them and they have flourished accordingly. The New England fishery in particular was before the late disturbances one of the most important perhaps in the world.

ADAM SMITH, *The Wealth of Nations*

THE encouragement of the fishing industry described by Adam Smith hastened the growth of commercialism in New England and sharpened the conflict with monopolistic aspects of mercantilism to the point which brought about its collapse. The losses sustained by the French, and the occupation by New England of the territory vacated, accelerated the disturbance.

After the conquest of New France, England attempted to expand her empire into wide new territories. The expulsion of the French from Cape Breton and the Gulf of St. Lawrence involved numerous major readjustments. Their losses during the Seven Years' War and their confinement, under the Treaty of Paris, to St. Pierre and Miquelon and the Newfoundland shore from Bonavista to Point Riche kept their fishery from making any extensive recovery, in spite of the encouragement given it in the form of bounties.[1] The French ships in the bank fishery numbered at least 130. They averaged 100 tons, with crews of 25 or more. They took some 2,000 fish per man; that is, with 65 fish to a quintal, a total in quintals of from 100,000 to 130,000. To this could be added from 1,000 to 2,600 hogsheads of oil. At St. Pierre and Miquelon schooners and small vessels of about 50 tons which engaged in fishing on adjacent banks, or went early in the spring to the Orphan Banks or to Chaleur Bay, grew in number from 38 in 1764 to 70 in 1768, and

1. In 1767 a bounty of 500 livres was paid to ships going to the region between Bonavista and Cape St. John. In 1768 the bounty was 500 livres for ships of 40 men or less, 750 livres for those of 40 to 60 men, and 1,000 livres for crews of more than 60. In 1767 a bounty of 25 sols a quintal was paid for cod exported to the West Indies. F. Louis-Legasse, *Evolution économique des Iles Saint-Pierre et Miquelon* (Paris, 1935), pp. 53–54; Henry Schlacther, *La Grande pêche maritime* (Paris, 1902), pp. 44–45.

the men they carried from 570 to 1,300. The catch increased from 22,800 quintals to 42,500. Stages increased from 20 to 70, and oil vats to the same extent. The resident fishery fluctuated.[2] English fishermen along Newfoundland's south shore were forbidden to sell fish to the French[3] and were removed to enable English fishing ships to occupy their places. New England bank vessels were likewise forbidden to participate in trade. Trading vessels from the West Indies declined from 20 in 1764 to 4 in 1768 as a result of the difficulty of disposing of rum and molasses in Newfoundland. In the region from Bonavista to Point Riche the number of ships increased rapidly.[4] They reached the coast as early as May 5 and left for their market ports from August 1 to the end of September. In 1765, of a total of 14,932 men engaged in the French fishery, 13,362 returned to France at the end of the season. Every fifth man was a green man, or one being trained in the industry.

2. Eight vessels and 30 to 40 boats were at St. Pierre in 1763. From 1765 to 1777 the average catch by residents at St. Pierre was estimated at 6,000 quintals. Some Acadians expelled from Nova Scotia took refuge in St. Pierre and Miquelon in 1764 but were compelled to leave in 1767. J. B. Brebner, *The Neutral Yankees in Nova Scotia* (New York, 1937), p. 107.

3. In 1764, 17 vessels were seized in Newfoundland for selling fish to the French on the Banks. *Quebec Gazette,* September 6, 1764: "The governor is very severe on this clandestine traffic so injurious to the fair trade." According to the *Quebec Gazette* of July 11, 1765, 230 vessels left St. Malo to engage in the fishery, from which it was inferred that smuggling was profitable. For an account of difficulties with the French smuggling at St. Pierre, because of its attempt to replace Louisburg in the trade of the English colonies with the French West Indies, and of difficulties in the concurrent fishery of the Petit Nord, due to the increasing importance of private property in the south and the migration of fishing to the north, particularly during the Seven Years' War, see G. O. Rothney, "The History of Newfoundland and Labrador, 1754–1783," master's thesis, unpublished, University of London, 1934, chaps. iv, v. In 1768, 2 French vessels from St. Pierre and 14 other vessels were seized for fishing beyond their limits south of Point Riche.

4. In 1763, 5 ships carried 13,600 quintals to market from the east coast, and on the west coast 5 St. Jean de Luz ships took 9,500 quintals from Codroy. In 1764 the number greatly increased, to 91 ships of 14,830 tons, with 5,315 men and 960 boats, and the yield totaled 146,270 quintals and 1,329 hogsheads of oil. The following year, 117 ships of 18,495 tons, with 7,862 men and 1,405 boats, took 292,790 quintals. There was a slight decline in 1768, but the number of stages had increased to 121. The smaller fish, it was estimated, produced one hogshead of oil per 100 quintals in contrast with two hogsheads from the fish on the south shore. The number of stations reported in the fishery increased from 20 in 1764 to 28 in 1768. Quirpon, as the largest, had 16 ships taking 17,900 quintals in 1764, and 10 ships taking 33,100 quintals in 1768; Old and New Port au Choix, 7 ships taking 12,000 quintals in 1764, and 8 ships taking 7,600 in 1768. The figures for Croc declined from 9 ships and 9,600 quintals in 1764 to 4 and 6,800 in 1768; the figures for Cape Rouge increased from 4 and 8,600 to 8 and 20,150. Sansfond, Paquet, Fleur de Lys, and La Scie had 5 ships which took 6,700 quintals in 1764, and 13 ships which took 14,400 in 1768. The larger stations with more than 5 ships in 1764 included Fichot, Great Goose Cove and St. Julien, and Canada Bay; and, in 1768, St. Anthony, Conche, Engele, and Canada Bay.

The French fishery fell off after 1770,[5] and in 1778 after the outbreak of war with England it suffered a bad blow by the destruction of St. Pierre and Miquelon.[6]

The retreat of the French from Cape Breton and the Gulf of St. Lawrence resulted in their rapid occupation by the English. The scattered and diversified character of the fishery in both areas meant that it was taken over by interests as various as those of the Channel Islands, Halifax, Quebec, Newfoundland, and New England. Halifax, established in 1749, served alike as an offset to Louisburg and as a base from which fishing operations could be satisfactorily prosecuted on the adjacent banks;[7] and the fishermen of New England rapidly extended the fishery, especially after the conquest of New France. From Long Island to La Have not less than 300 New England vessels went forth "to catch their summer fares." In 1760 Liverpool[8] was founded, and by 1762 contained some 90 families, or more than 500 people in all. At Port Senior 70 houses had been built. In 1761, 17 schooners were employed in the

5. See *C.O.* 194:16, 17, 18, 21; also 49.

THE FRENCH FISHERY*

Years	Number of Ships	Tonnage	Number of Boats	Number of Men	Cured Fish in Quintals	Oil in Hogsheads
1769	431	44,727	1,455	12,367	215,030†	3,153
1770	437	45,541	1,470	12,855	435,340	3,511
1771	419	42,369	1,327	12,640	239,864†	4,259
1772	330	37,257	1,468	15,248	388,800	4,687
1773	284	33,332	1,452	14,476	336,250	3,358
1774	273	31,530	1,614	15,137	386,215	3,377

† By tale i.e. count

* *Second Report from the Committee Appointed to Enquire into the State of Trade to Newfoundland 1792,* p. 57.

6. As to the position of Newfoundland in the Seven Years' War and the American Revolution see G. O. Rothney, *op. cit.*, chaps. i, ii, viii.

7. See Lorenzo Sabine, *Report on the Principal Fisheries of the American Seas* (Washington, 1853), pp. 62 ff.; also William Douglass, *A Summary Historical and Political . . . of the British Settlements in North America.* It was the expenditure involved that led to Burke's outburst: "Good God! What sums the nursing of that ill-thriven, hard-visaged and ill-favoured brat has cost this wittol nation! Sir, this colony has stood us in a sum not less than £700,000." Edmund Burke, *Works* (Oxford, n.d.), II, 370.

In 1750 Halifax produced 20,000 quintals. In 1751 a bounty to run for three years was granted amounting to sixpence a quintal for dry fish and one shilling a barrel for pickled fish. In 1757 further bounties were added: one shilling a quintal for merchantable fish and one shilling a barrel for pickled fish. A prize of £20 was offered for the largest catch of merchantable fish. *Documents Relating to Currency, Exchange and Finance in Nova Scotia 1675–1758* (Ottawa, 1933), pp. 340–341, 423–428, 459–460.

8. Brebner, *op. cit.*, pp. 55–56; see also *Report of the Board of Trustees of the Public Archives of Nova Scotia, 1933* (Halifax, 1934), pp. 21 ff. *Idem,* 1934 (Halifax, 1935), pp. 27 ff.; and D. Allison, "Notes on a General Return of the Several Townships in the Province of Nova Scotia for the First Day of January, 1767," *Collections of the Nova Scotia Historical Society,* VII, 45–71.

fishery. Cape Sable Harbor, which had formerly been occupied by 12 French families interested in the fishery and the fur trade, was granted in 1760 to 200 proprietors, chiefly fishermen from Cape Cod, Plymouth, and Nantucket. In 1761, 20 families arrived, and in 1762 several more came with vessels to establish a fishery. The area from Cape Sable to Long Island was granted to fishermen from Marblehead. The proprietors were largely part owners of fishing schooners. The township of Yarmouth[9] was granted to proprietors, farmers, and fishermen, and it was expected that a boat fishery would be developed.

Simeon Perkins' operations at Liverpool were typical of New England activities in Nova Scotia. Perkins came from Norwich, Connecticut, in 1762 and, except for two years from 1767 to 1769, he was involved in the business activities of Nova Scotia. In June, 1766, fishing vessels from Cape Ann and other New England ports visited Liverpool on the home voyage from the Banks. In the same month and in July others sent from Liverpool began to arrive with "fares" (cargoes) of 60 to 300 quintals. They left with fresh supplies of salt for the second fare, returned in September and October, and in some cases sailed for a third.[10] The fish were dried and sent in small bundles to Halifax and New England ports, especially Boston. Early in the new year fishing schooners carried fish to Dominica in the West Indies, returning late in March with salt, sugar, and molasses. In April vessels began to arrive from New England for the Banks; and schooners, having returned from the West Indies,[11] left for the Banks, Cape Breton, Chaleur Bay, and Prince Edward Island. On April 7, 1767, the *Jolly Fisherman* left with stores provided by Perkins for which he was to receive half the fish and oil.[12] "I pay for my part of the shoremen, the rest of the crew [are] on the common lay of this place.[13] They provide all small generals and pay their part of great generals and shoremen and draw three quarters of the remainder." Supplies were distributed to ports along the shore

9. See also an account of John Barnard, who had a brother at Yarmouth, and his plans for the fishery in Nova Scotia and trade with the West Indies in 1772, K. W. Porter, *The Jacksons and the Lees* (Cambridge, 1937), I, 77, 242–244.

10. A schooner arrived from Banquereau with 200 quintals on May 17, 1772, "the earliest fare of bank fish ever landed in this place." Simeon Perkins' diary. The references are from a copy in the Canadian Archives. Other extracts and references follow.

11. The *Sally* sailed on January 15 for Dominica and returned on May 8 to report that markets were poor there and that she had visited St. Kitts, Nevis, and St. Eustatia, and finally disposed of her cargo at St. Croix.

12. The *Jolly Fisherman* left for a second fare on July 9, 1767, went to Newfoundland, Labrador, the Straits of Belle Isle, and the Grand Banks, caught her fare of 150 quintals at Scatari, and returned September 20.

13. That is, "according to the customary arrangement here." What "generals" and "great generals" may mean is obscure.

in exchange for fish. Provisions and goods, in addition to those from the West Indies, were obtained from Halifax, New England, and other ports.[14] Expeditions to the Mediterranean were, on the whole, not successful.[15] A similar routine was followed in later years. Schooners to the West Indies added herrings and lumber to their cargoes[16] and brought back quantities of salt, particularly from St. Martins. On his return in 1769, Perkins resolved on June 19 "to sell principally on short credit to people I think punctual and will pay in fish." On January 11, 1773, he wrote: "I think Liverpool is going to decay, and it may be many years ere it is more than a fishing village." And on February 8: "People in poor circumstances. Everything needed is very high, their pay uncertain, the land hard and rocky, very few cattle of any kind and they kept mostly on salt hay. It costs 40s. to keep a cow through the winter and £5 to keep an ox fit for work." On March 20, schooners carried salt to people in exchange for potatoes. "We do not raise half a supply of potatoes and roots and very little corn. . . . Three quarters of the inhabitants out of bread and meat." Perkins turned to an increasing extent to lumber operations.

The advantages of New England became less conspicuous in the regions east of Halifax, and were offset by the effectiveness of competition from Quebec and Newfoundland, and also from the Channel Islanders[17] who had the advantages of bilingualism and were able to capitalize the position formerly occupied by Biscayan and other French ports in the European markets. After 1765 various Jerseymen fished at La Poile, Fortune Bay, Jersey Bay, Sablon, Burin, Placentia, and Ile

14. On July 22, 1767, a cargo from Maryland was purchased. It included 400 bushels of corn at 3 shillings; 42 casks of bread at 16 shillings; 10 barrels of flour at 16 shillings; 14 barrels of seconds at 15*s*. 6*d*.; and 14 barrels of thirds at 14 shillings.

15. On July 13, 1766, the schooner *Nabby* arrived from Bilbao, but reported poor markets there and at Lisbon. She went to Falmouth for a pass to the Mediterranean, and took a cargo of iron to Figuera, returning with salt to Liverpool (N.S.). She went to Newfoundland, returning on October 15 with 30,000 fish (400 quintals). She returned from Poole on May 12, 1767, with 1,000 bushels of salt and reported high expenses. "She makes a poor voyage. The cargo sold well, only the ship carried so little." She left for the fishery on May 29, calling at Halifax for boats and barrels, and returned on July 29 with 22,000 fish (250 quintals). She returned from her second voyage on September 27 with 140 quintals and was chartered on September 29 at £19 10*s*. per month for Fayal. The *Olive,* Captain Godfrey, was lost in 1773 on a return voyage from the Mediterranean.

16. A schooner arriving on February 28, 1773, reported the sale of a cargo of lumber at £4 and fish at 16 shillings; and another, arriving on April 6, 1773, reported a sale of lumber at £5 per 1,000 and of fish in hogsheads at $2 a quintal and in bulk at 6 shillings currency. In 1773, vessels went in the spring to Newfoundland for salmon, and in 1774 large quantities of alewives were taken at Liverpool. Both salmon and alewives were exported.

17. In Newfoundland the Jersey fishermen had begun to frequent Ferryland, particularly after Sir Walter Raleigh was appointed governor of Jersey in 1600.

Verte; but apparently large numbers left for territory vacated by the French. Joshua Mauger, a Jersey captain, became a merchant in Louisburg in 1745 and played a dominant role in the later economic and political history of Nova Scotia. As a result of his influence the production of rum and sugar was encouraged by a tax laid on these products by the colony; and by 1768 the distillery and sugarhouse at Halifax were regarded as "of great consequence to the trade of the province, as the molasses and raw sugars are purchased in the West Indies with boards and other lumber and with mackerell and other barrell fish as also with the inferior kind of dry cod fish the greatest part of which articles are unfit for any other market."[18] Two distilleries in Halifax had a capacity of 60,000 gallons of rum. The burden of this protected industry rested uneasily on the highly competitive fishing industry, and duties collected on West Indian products threatened to force a return of fishermen from Cape Breton to Newfoundland.

In eastern Nova Scotia, firms from the Channel Islands were quick to take advantage of the withdrawal of the French. The Robin firm[19] established a post at Arichat in 1765, and later an outpost at Cheticamp from which fishermen every year returned in the winter season. It possessed 13 vessels, employing about 300 men; and in 1776 it employed some 60 families who used 50 shallops in their fishery. Acadians who went to St. Pierre and Miquelon in 1764 returned to Madame Island in 1768. In that year, according to Holland, 6 decked vessels from New England were engaged in fishing on the Banks and in drying their fish at Dartmouth Harbor in Cape Breton. Complaints were made at this point and at Petit Lorembec that the shallow fishery suffered as a result of offal thrown overboard from these vessels. Jersey Islanders had 2 brigs and 18 shallops at Darnly Island, and a brig of 150 tons at Petit Degrat. A boat fishery was carried on at Little Bras d'Or which had 6 shallops. Main-à-Dieu had 15; Baleine, 6; Petit Lorembec, 2; Louisburg, 5 decked vessels and 2 shallops; Ardoise, 5 shallops; and Petit Degrat, 14. The boat fishery was sensitive to burdens imposed on the fishery. Main-à-Dieu fishermen were "daily deserting [to Newfoundland] to the great loss of their employers." At Louisburg, which was dependent on Nova Scotia, and

to whose distribution of justice at such a distance it must have recourse and

18. *Nova Scotia Archives* (hereafter *N.S.A.*), LXXXIII, 13. See an account of the various preferences given the domestic production of rum. Brebner, *op. cit.*, pp. 69, 253, and Appendix A; also, for the significance of the industry to the neutrality of Nova Scotia, *passim.*

19. Richard Brown, *A History of the Island of Cape Breton* (London, 1869), chap. xxii.

whose impositions and taxes, too little calculated for encouraging this branch of the business, must be obeyed, are such grand objections to the prosperity of this place that so long as Newfoundland affords more immunities it must always be what it is at present, without any person of substance to support any public improvement, without any trade to the mother country. . . . Those that supply the fishery with anything, have all their necessaries thro' two or three different channels which obliges them to charge so much additional price that the wages of the fishermen are inhansed thereby to so great a degree as makes it impossible any quantity of fish that can be caught should be able to leave any remarkable profit to the adventurer.[20]

The impost "of fifteen pence per gallon on all spirits [which] . . . more than doubles the price of that commodity to the consumer . . . hath such influence from the vast quantities used that the fishery must be entirely given over." Mauger's gains from his control over the distilling industry at Halifax clashed with those of his fellow countrymen in Cape Breton.[21] Nevertheless the fishery had expanded by 1774, as the following table indicates.

		Number of Vessels			
	Population	*With Topsails*	*Schooners*	*Shallops*	*Exports of fish, in Quintals*
Louisburg	144	..	5	4	920
Le Baleine	39	..	..	5	1,090
Main-à-Dieu	131	..	1	15	3,370
Meray	29	..	..	..	...
Little Bras d'Or	30	..	..	2	290
Chapeau Rouge	47	..	..	6	820
St. Peters Bay	186	..	..	19	3,620
Petit Degrat	168	5	..	36	8,520
Arichat	238	1	3	34	7,380
	1,012			121	26,010

In 1783, due in part to the effects of the war and in part to its people's complaints that they were being taxed, and unfairly taxed, without representation, Cape Breton was separated from Nova Scotia.

The effects of the general development of settlements and the fishery in Nova Scotia after 1763 were shown in the quantities of fish exported

20. D. C. Harvey, *Holland's Description of Cape Breton Island and Other Documents* (Halifax, 1935), p. 80.

21. Lawrence Kavanagh moved to St. Peters in 1774 after he had complained in 1773 of the impressment of men on his vessels. Lord Dartmouth wrote to Governor Legge: "I am informed that Mr. Lawrence Kavanagh is very largely concerned in the fishery carried on from Louisburg." (February 24, 1775.) Harvey, *op. cit.*, p. 30.

and the growth of trade generally. In 1764 Nova Scotia exported 66,-400 quintals of dried cod valued at £39,840[22]—of which 22,000 quintals or about one third were caught and "made" by the people of New England during the summer—and 7,200 barrels of "pickled fish of different sorts," valued at £7,770. In 1772 it was noted that, although only a few small vessels were owned in Nova Scotia, "there is a considerable and increasing boat fishery, which is a nursery for seamen, and the great advantages arising from the situation of the Banks so near the whole coast of Nova Scotia must in time draw thither the entire cod fishery of North America." In this year, 1772, 19 vessels totaling 2,175 tons brought from Great Britain and the Channel Islands imports to the value of £30,000. They included salt, fishing supplies, clothing, iron, sugar, beer, and other European and East Indian products. In return 14 vessels of 1,890 tons carried exports to England and the Channel Islands valued at £3,750; and they included fish, oil, lumber, furs, etc. Imports from southern Europe, Africa, the Azores, and the West Indies, carried in 9 vessels of 485 tons, totaled £2,000. Such imports included salt, rum, molasses, and brown sugar. Exports to the same areas, taken in 17 vessels of 1,025 tons, totaled £13,615 and included fish, oil, and lumber. Imports to Nova Scotia from the British colonies, in value £31,000, about equaled the imports from Great Britain. They were brought in 110 vessels of 3,996 tons, and included naval stores, provisions, salt, rum, molasses, sugar, and wine. Exports to the colonies, carried in 134 vessels of 4,807 tons, totaled £26,000 and included oil, fish, furs, grindstones, and rum.[23] In exports as in imports almost one half of the trade was with the colonies, especially New England. The remainder of the imports were from Great Britain; and, of the exports, the remainder were to the West Indies, southern Europe, the Azores, and Africa. Smuggling was carried on between the people of St. Pierre and Miquelon and those along the coast from Canso to Chaleur Bay.[24]

In the Gulf of St. Lawrence, firms from the Channel Islands occupied

22. *N.S.A.*, LXXV, 185. Other estimates of exports from the fisheries: 1763, £25,-500; 1764, £47,600; 1766, £35,700. Brebner, *op. cit.*, p. 130. The census of 1766 gives 367 boats, 119 schooners, 3 square-rigged ships, *idem*, p. 126. "The trade of this port [Halifax] does not seem very extensive; there are not above thirty vessels here at present [1774] and most of these are fishing schooners. They carry on a little trade to the West Indies, Philadelphia and New York where they send their fish and oil, and some fur and lumber; they have also a vessel or two that trades constantly to London. They have several breweries and distilleries and are famous for tanning the best leather in America." Patrick McRobert, *A Tour through Part of the North Provinces of America* (Edinburgh, 1776), republished April, 1935, p. 18, also *passim.*

23. *N.S.A.*, XC, 22–24; H. A. Innis, *Select Documents in Canadian Economic History* (Toronto, 1929), pp. 266–277; *Report of the Board of Trustees of the Public Archives of Nova Scotia*, 1933, pp. 30–31.

24. *N.S.A.*, LXXXIX, 36; *idem*, pp. 223–224; Brebner, *op. cit.*, p. 110.

a competitive position. The whale fishery was forwarded and prosecuted with great success by New England vessels.[25] The walrus fishery of the Magdalen Islands was carried on by a New Englander named Gridley. And a small number of Acadians began a sedentary fishery in these islands about 1773. Around Chaleur Bay, with its advantages arising from the relative scarcity of fogs and the early arrival of the cod, thanks to which the cured fish could be dispatched to market "six weeks sooner than in any part of America," competition was keen, following the disorganization which accompanied the withdrawal of the French. Fishermen of Rhode Island and Cape Cod came in sloops and schooners which they laid up in Gaspé while they carried on the fishery in whale boats that they bought at the different posts from Cape d'Espoir to Point St. Peter.[26] Settlers complained that fishermen stole "their master's boats and vessells in which they must necessarily be intrusted, full of their fish, to remote places in Newfoundland." "Others sell great part of their master's fish on the very banks to the New England schooners for spirituous liquors; who come to fish on the same banks or rendezvous in some of the harbours along the coast, and as long as the liquor lasts, neglect the remainder of their work, often to the total loss of the whole season to their masters." These schooners were also charged with running into the infant settlements along the coast and carrying off "the fish drying on the flakes under guard of women and children."[27] Complaints against the schooners for throwing refuse overboard and preventing the fish from going to shore resulted in regulations to end that abuse.

The numerous advantages possessed by fishing interests centering in the Channel Islands led to the establishment of a sedentary fishery. The withdrawal of the French fishermen from the Gulf of St. Lawrence induced Jacques Robin to petition in 1763 for a seigniorial grant at the mouth of the Miramichi.[28] The Jersey firm of Robin, Pipon and Company[29] in 1766 sent out the Jersey brig *Sea Flower*, 41 tons, with

25. In 1761 New England vessels were successful in this fishery. Brebner, *op. cit.*, pp. 51–52. An act was passed in 1764 for the encouragement of the whale fishery in the Gulf of St. Lawrence and on the coasts of His Majesty's dominions in America. In this act duties on whale fins were reduced from that date to 1770. The whale fishery was prosecuted along with the cod fishery and employed about 100 sloops and schooners of from 50 to 100 tons.

26. A. C. Saunders, *Jersey in the Eighteenth and Nineteenth Centuries* (Jersey, 1930), p. 213; and Brebner, *op. cit.*, p. 130.

27. Innis, *op. cit.*, pp. 165–166. An ordinance to deal with this came into effect on May 1, 1765.

28. Brebner, *op. cit.*, pp. 103–104.

29. Saunders, *op. cit.*, pp. 197 ff.; also J. M. Clarke, *The Heart of Gaspé* (New York, 1913), pp. 180–181. According to Clarke, letter books were in existence at Paspebiac as early as June 5, 1777.

Charles Robin[30] as supercargo. In the following year he was agent on the *Recovery*, 118 tons, arriving June 2. He sailed in a shallop along the coast to supply salt to planters at Bonaventure. He also engaged in the fur trade on the Restigouche. The *Endeavour*, 122 tons, was seized by the British government in 1767, and the *Sea Flower* and *Recovery* were seized in 1768 because of alleged illicit trade and failure to comply with an act passed in 1764 which required Jersey vessels to clear from English ports. The *Seville Trader* was chartered on September 8, 1769, by Robin, Pipon and Company to go to Seville, and the *Hope*, of 101 tons, was sent to Bilbao with 70 tierces of salmon. Recruits were brought out from Jersey to learn the business, and the practice of hiring men to fish was introduced. "I keep four shallops fishing and the Percé gang." Furs and whale and cod oil were sent to England in 1777. But in 1778 the *Hope*, carrying 1,400 quintals of cod and ready to sail for Lisbon about the middle of June, was captured by the Americans. Another vessel, the *Neptune*, on the way to Miscou to collect fish, was captured in July and 1,050 quintals were taken. In 1778 Charles Robin had a station at Carleton, but returned to Jersey in that year; and he came back to Paspebiac in 1783 as the chief partner in Charles Robin and Company.

In 1777 it was estimated that the Gaspé fishery employed on an average 12 vessels a year, and exported 16,000 quintals of fish. Fishermen who left New England on the outbreak of the war settled at Point St. Peters[31] and Mal Bay. Percé continued to be the most important post. Boats averaged, with two hands, a take of 350 quintals in nine weeks; and one man in seven weeks had been known to take 22,000 cod, or 208 quintals. At Paspebiac "thirty people from Europe" (chiefly from Jersey) and 10 resident families took between 12,000 and 14,000 quintals in one summer. The fishing began about May 25 and lasted for six weeks, at the end of which time boats went to Percé. From Bonaventure 10 ships were sent annually to the West Indies and Europe. The salmon fishery had been developed at Restigouche. The fishery was conducted with boats costing less than £8 in contrast with Newfoundland boats costing at least £60.

Halifax interests concerned themselves with the Bonaventure fishery, but failed after three years; and Quebec interests were scarcely more successful. Moore and Finlay and Alexander Mackinsay started

30. J. Robin was agent at Arichat, Madame Island, and Paspebiac. Charles Robin wintered at Arichat in 1769, but returned home in 1769–70. He returned to Arichat in 1770 and stayed in Canada in 1770–71. In 1772 he found that his house at Paspebiac had been burned.

31. At Point St. Peters fish were taken within a mile or a mile and a half from shore.

fishing establishments in Chaleur Bay. In 1767 Charles Robin entered a partnership with William Smith, formerly of Moore and Finlay, each with a vessel at Bonaventure and Paspebiac. But in two years other Quebec interests were in bankruptcy.

Quebec and New England interests in the Canadian Labrador came into conflict with those of Newfoundland. Grants along the Canadian Labrador given by the French were continued and extended by the English at Quebec in 1760.[32] In 1763 Labrador, the Magdalen Islands, and Anticosti were annexed to Newfoundland and regulations were introduced in 1765 providing for the prosecution of the Labrador fishery by fishing ships. English ships left Petit Nord territory which was reoccupied by the French after the war and went to Labrador.[33] Difficulties arose with New England whaling vessels on the Labrador, with the walrus fishery of the Magdalen Islands, and with Quebec merchants.[34] Governor Palliser wrote:

> It was therefore time and my indispensable duty to annul those monopolizing pernicious grants from Govr. Murray so injurious to the rest of the King's subjects, so prejudicial to the shipping trade and manufacture of this

32. See *P.C.* (*Privy Council*), VII, 3638 and *passim* and *P.C.*, III, *passim*.

33. D. W. Prowse, *A History of Newfoundland* (London, 1896), pp. 324, 327, 336. Innis, *op. cit.*, pp. 264–265. Regulations dated August 10, 1766, admitted the colonies to the Labrador fishery, but obvious handicaps of distance and regulations establishing the British ship fishery (August 10, 1767) hampered New England. *P.C.*, III, 986, 1016.

34. A petition signed by John Lymburner, John Gray, Hugh Finlay, John Isbister, and eight others dated November 1, 1765, stated that "we the subscribers merchants of Quebec, being largely in advance for the different seal fishing posts upon the Labrador coast, the settlement of which we have been accomplishing ever since the year 1761 with much trouble and at a great expense, can hardly express the consternation we are thrown into by His Excellency the Governor of Newfoundland's order dated the 28th August last . . . the purport of which seems entirely to deprive us and the great number of people we employ of the fruits which we hoped would accrue from the labour, industry and expense we have bestowed on these settlements." Fishermen had returned to Quebec as a result of threats of corporal punishment for violation of the regulations. "Thus having engaged our people, victualled them for eight months, furnished with all the expensive apparatus of a seal fishery, dispatched them in vessels built and purchased for the business, we have the mortification to see not only our labour and expense in this last outfit totally lost and rendered ineffectual, but also many of our settlements left deserted, our buildings, fixtures, fishing materials, provisions and merchandize totally exposed to destruction." Lower Canada Sundries, Canadian Archives. Regulations for the fishing-ship fishery were not adequate, especially to the seal fishery in Labrador. See G. O. Rothney, "The Case of Bayne and Brymer: An Incident in the Early History of Labrador," *Canadian Historical Review*, September, 1934, pp. 264–275; also Rothney, master's thesis, *op. cit.*, chaps. vi, vii. An extract from a letter from a London merchant dated July 2, 1766, to his Quebec correspondent and published in the *Quebec Gazette* expressed the hope that a suspension of Palliser's regulations would imply no interference in 1766 and 1767. "We are determined to leave nothing undone that can contribute to giving their Resolutions on the Fishery a Turn favourable to Canada, which, if effected, will be

Kingdom and to lay the fisheries open to all His Majesty's subjects from Britain and not suffer it to turn into a colony fishery but keep it a free British fishery agreeable to the laws and to what has ever been the policy of the nation respecting the fisheries.[35]

A system built up on the fishing industry of Newfoundland clashed with a Quebec system based upon the fur trade, first under the French, and later under the English, which depended on sealing during winter and on the salmon fishery in summer, and made settlements a necessity. The Quebec Act[36] restored Labrador to Canada and reëstablished the equilibrium in which Labrador, as developed from Quebec in the French regime, was returned to English control by Quebec. Just as a system based on the fur trade depended on continental types of structure in centralization and monopoly, a system founded on the fishing industry depended on a maritime and competitive type. The height of land between the two systems became evident in the Canadian Labrador.

The dried-fish industry was carried on particularly with a view to meeting the demands of the European market for the high-grade product. The Channel Islands inherited the region, especially Gaspé and Cape Breton where fish were taken suitable to European demands. Quebec was chiefly interested in the Labrador. New England became less successful in more distant areas, although the demand of the West Indies for poorer-grade fish enabled her to extend the fishery to the Gulf of St. Lawrence in competition with Nova Scotia. The inherently divisive character of the fishing industry showed itself in the emergence of a divided control of the fishing region.

The expansion of the Newfoundland fishery[37] and its conflict with

of such Consequence to it, that we cannot doubt those Gentlemen concerned will frankly contribute to the Expences of the Sollicitations which will be considerable, not less than from One to Two Hundred Pounds." Quebec merchants concerned included J. Gray, W. Brymer, D. Bayne, C. Grant, B. Price, J. Johnston, G. Fulton, W. Mackenzie, Duncan and Beller, and J. Philibot.

35. *C.O.* 194:16.

36. Adam Lymburner claimed in a petition of 1774, following the abrogation of the rights of fishing ships under the Quebec Act, that he owned posts at St. Augustine and Shikataika which were taken possession of by a fishing ship in 1767 under an order from Palliser. He was threatened with dispossession in 1775. Bradore and L'Ance St. Clair were summer posts, which had been occupied by his brother since 1761, and the latter had been a winter post since 1772. St. Modest had been a post since 1768 and Pied Noir a post since 1772. At Cape Charles, he had been dispossessed in 1771 by a British fisherman. In all they represented a total investment along the coast of £6,000. See *P.C.*, III, 1095 ff. See the instructions that private rights should not be disturbed and should not be extended, in 1775, *Royal Instructions to British Colonial Governors 1670–1776,* ed. L. W. Labaree (New York, n.d.), II, 693.

37. Edmund Burke wrote in 1766: "The most valuable trade we have in the world is that with Newfoundland."

New England fishing interests continued to receive support from the New England trade. That fishermen were attracted from Cape Breton to Newfoundland was an indication that the Newfoundland fishery was becoming more profitable. The withdrawal of the French during the war had facilitated an expansion in both regions. Trading ships from New England to Newfoundland increased from 83 in 1766 to 175 in 1774. In 1765 trade from New England to Newfoundland was estimated at "£102,304 sterling carried in 142 vessels, nine tenths of which is paid for in bills of exchange on England."[38] It was estimated that in 1764 smuggling, in spite of regulations, would increase the total trade to double that amount, and that by 1774 it had reached £300,000 or £400,000.

As a result of support from New England and of the decline of the French and other[39] fisheries, the Newfoundland fishery expanded and production increased.[40] Fishing ships were forced to the more distant

38. Prowse, *op. cit.*, p. 329, also pp. 323–324.

Goods Imported from the Plantations*

Great and Little Placentia	£ 7,011. 16. 2
Ferryland	5,918. 1. 2
St. John's	38,035. 0. 11
Harbor Grace	5,751. 14. 9
Other ports in Placentia Bay	7,011. 16. 2
Fortune Bay, Port aux Basques to Codroy	3,505. 18. 1
Trepassey and St. Mary's	8,000. 0. 0
Renewse and Fermeuse	5,918. 2.
Bay Bulls	4,900
Carbonear	4,750
Trinity Bay	10,501. 14. 9
Bonavista and Greenspond	1,000

* (*C.O.* 194:17.) A customhouse was established in 1762, presumably to check the importation of coarse Irish woolens, shoes, candles, and soap; and fees were regulated on the basis of Halifax as the nearest port. Merchants had combined to resist the payment of fees and there had been misinterpretation as to the difference between "bankers" and trading ships, that is, ships with cocketable goods.

39. The Spanish were definitely forbidden to engage in the fishery. See Geronimo Uztariz, *The Theory and Practice of Commerce and Maritime Affairs*, trans. by J. Kippax (London, 1751). It was estimated that the annual consumption of cod was 487,500 quintals. "The Mercantilism of Geronimo Uztariz; a Reëxamination," *Economics, Sociology and the Modern World* (Cambridge, 1935), pp. 111–129; Richard Pares, *War and Trade in the West Indies 1739–1763* (Oxford, 1936), pp. 564 ff.; Vera Lee Brown, "Spanish Claims to a Share in the Newfoundland Fisheries in the Eighteenth Century," *Canadian Historical Association Report*, 1935.

40. See Charles Pedley, *The History of Newfoundland* (London, 1863), chaps. vi, vii. The number of residents apparently dropped from 15,981 in 1764 to 10,949 in 1774; but the number of boats belonging to residents increased from 1,236 in 1764 and 1,117 in 1766 to a high point of 1,446 in 1774. They produced 352,690 quintals of cod in 1764, but did not reach the 300,000 mark again until 1773 when 366,446 quintals were produced and then 312,426 in 1774. The number of byeboatmen increased from 281 in 1764 to 643 in 1771, but declined to 555 in 1774; and servants of byeboatmen

outports and to Labrador. Poole interests extended the fishery north from Twillingate. In 1764, 17 English vessels with 794 men and 113 boats were engaged in fishing north of Fleur de Lys.[41] The migration to the north was accompanied by the development of other industries. The number of tierces of salmon carried to the foreign market fluctuated from 2,320 in 1764 to 649 in 1770, 3,543 in 1773, and 725 in 1784. The value of the seal oil produced fluctuated from £3,304 in 1764 to £26,388 in 1773 and £17,605 in 1774.

As a result of the regulations in 1765 favoring the fishing ships, small vessels went to Chateau on the Labrador at the end of the fishing season in Newfoundland. The Labrador fishery was conducted chiefly by English fishing ships, but the number of residents increased with the salmon, seal, and whale fishery and the fur trade. The expansion of the fishery to the north had been responsible for a decline of the fur trade, in so far as it had been carried on with the Beothicks—the natives of Newfoundland—near Bonavista. The fishing industry led to increased demands for bait and provisions. It also led to the destruction of bird life on Funk Island. Larger numbers of fishermen became interested in the salmon fishery and, in turn, in the fur trade. The beaver declined owing to the increase of trapping. The supply of furs diminished and, as a result, open warfare developed with the Indians. The fishing industry contrasted strikingly with the fur trade in the character of its relations with the Indians. Whereas the fur trade depended upon the welfare of

increased from 2,903 to 6,909 in 1772, and declined to 5,161 in 1774. Their boats also increased irregularly. In 1764 they numbered 366; in 1772, 605, but declined to 518 in 1774. Their catch fluctuated from 92,050 in 1764 to 155,847 in 1772 and 145,800 in 1774. The number of passengers brought out in British ships increased from 4,090 in 1764 to 7,695 in 1772, but declined to 4,925 in 1774. Ireland supplied the largest number, with England second and Jersey a poor third. The number of fishing ships from England increased from 141 in 1764 to a high point of 369 in 1771, but declined to 254 in 1774. The total for 1771 included 244 bankers, and declined to 100 in 1774. See George Chalmers, *Opinions on Interesting Subjects of Public Law and Commercial Policy Arising from American Independence* (London, 1784). Boats kept by British fishing ships increased from 210 in 1764 to 536 in 1766, declined to 430 in 1769, increased to 556 in 1771, and declined to 451 in 1774. Fishing-ship production increased from 116,570 quintals in 1764 to 305,391 quintals in 1772, but declined to 237,640 in 1774. The number of stages increased from 994 in 1764 to 1,208 in 1768, but declined and only recovered to 1,219 in 1774. Sack ships, "the greatest part of which arrive from foreign ports with salt or in ballast," declined from 116 in 1763 to 92 in 1767, but increased to 146 in 1772 and to 149 in 1774. Fish carried to foreign markets increased from 470,188 quintals in 1764 to 610,910 in 1771 and declined to 516,358 in 1774. Prices, on the whole, increased slightly from 11 shillings to 12*s.* 6*d.* a quintal in 1765, and to between 11 and 14 shillings in 1773.

41. Griguet had 2 ships with 64 men and 10 boats; Cremailliere, 1 with 24 men and 4 boats; White Arm, 1 with 30 men and 4 boats; Great Goose Cove and St. Julien, 2 with 115 men and 17 boats; Conche, 3 with 164 men and 23 boats; Engele, 7 with 240 men and 31 boats; and Canada, 1 with 77 men and 10 boats. *C.O.* 194:16.

the Indian, the fishery, with the large numbers which it involved, brought Indians and whites into direct competition. The disappearance of the Beothicks is related to the fishing industry as closely as the survival of the North American Indian is related to the fur trade.[42] Hostility and warfare with the Eskimos on the Labrador were a handicap to its expansion.

The problems of the Labrador in the development of both the fur trade and the salmon fishery, and the inadequacy of fishing-ship regulations to cope with these activities, became plain in the experience of Captain George Cartwright,[43] who was engaged in extending trade in this region.[44] Nicholas Derby of Bristol had established a post at Cape Charles in 1765 with 150 men, but in 1767 he had been forced by the Eskimos to abandon it, and English traders had been restricted to the territory south of Chateau Bay where, in 1765, the government had erected a blockhouse.[45] The work of the Moravian missions,[46] encouraged by Palliser, paved the way for the development of trade with the Eskimos. After two visits in 1766 and 1768 to Newfoundland and the north, Cartwright had become interested in the possibilities of sport and of profit; and in 1770, at Bristol, he made one in the partnership of Perkins, Coghlan, Cartwright and Lucas. An 80-ton schooner was taken over from Perkins and Coghlan, who were chiefly interested in the trade between Poole and Fogo Island. Cartwright bought a 50-ton vessel at Bristol, and after reaching Fogo he proceeded north and established the post formerly abandoned by Derby. The winter of 1770–71 was spent in trapping and seal hunting, and in 1771 boats were sent north to Point Spear and Cape St. Francis to fish for salmon. Communication was established with the base at Fogo from which a cargo of dry fish was sent to Oporto. But trouble arose upon the seizure of Charles River

42. *P.C.*, I, 17–24; also evidence of George Cartwright, *First Report of the House of Commons Committee on Newfoundland Trade 1793*, pp. 37–42; also *Second Report*, p. 25. "The fishermen of all countries as far as I have been able to ascertain, wherever their numbers predominate, conduct themselves towards the weaker party in the most overbearing and wanton manner." John McGregor, *British America* (London, 1828), I, 219. The Beothicks disappeared with the great auk.

43. *Captain Cartwright and his Labrador Journal*, ed. C. W. Townsend (Boston, 1911); also J. D. Rogers, *Newfoundland* (Oxford, 1911), pp. 143–144; W. G. Gosling, *Labrador, Its Discovery, Exploration, and Development* (Toronto, n.d.), chap. xii; S. C. Richardson, "Journal of William Richardson, who visited Labrador in 1771," *Canadian Historical Review*, March, 1935, pp. 55–61.

44. At the same time, in England, Edmund Cartwright, his brother, was working on the invention of the loom.

45. *P.C.*, I, 14; also III, 1059 ff.

46. J. E. Hutton, *A History of the Moravian Missions* (London, 1923), pp. 130 ff. Okkak was established in 1775, and Hopedale in 1782. Rogers, *op. cit.*, p. 146; also *P.C.*, III, 1311 ff.

in 1772 by the partnership of Noble and Pinson of Dartmouth, with the result that Cartwright was compelled to conduct his business operations —his salmon fishing, sealing, and his trading in furs and whalebone with the Eskimos—from Fogo. In 1772 fire destroyed the Fogo establishment with a loss of £500. But in spite of his loss, which was half of that, the termination of the partnership, and the difficulties with Noble and Pinson, he decided in 1773 to embark on his own capital. He was persuaded that "commerce will in progress of time have the same effect on these people that it ever has had on other nations; it will introduce luxury, which will increase their wants and urge them to much more industry than they at present possess."[47] He bought a brig of 80 tons and, after acquiring goods from London, Weymouth, and Waterford, nets from Bridport, and salt from Lymmington, he arrived at Charles Harbor on August 27, to find that only 12 tons of seal oil had been extracted and 50 tierces of salmon put down. Lack of capital was a deciding factor in limiting the expansion of the business, and in December, 1773, a new partnership was arranged in which Cartwright invested £2,000, plus the value of his vessels and stock, and the other partners advanced an equal sum in cash. The partnership purchased an additional 230-ton American-built vessel and dispatched it to Cadiz for a cargo of wine for Adam Lymburner of Quebec,[48] there to reload for Charles Harbor with plank, boards, hoops, hogshead and tierce packs, bread, flour, and other things to be had more cheaply in Quebec than in England. The 80-ton brig was wrecked, but was replaced by another ship which brought goods and provisions directly from England and Ireland. During the season men were engaged in salmon fishing, sealing, trapping, and making hoops for salmon tierces. Furs were taken in the winter, salmon at the opening of the season, and cod at a later date. Cod were caught in large numbers near the Dismal Islands, and apparently cod seines were used effectively. In 1775 the wrecked brig, after having undergone repairs, in its turn went to Barcelona, took on a

47. *Captain Cartwright and his Labrador Journal,* p. 90. Compare this with the comment of Sir George Simpson of the Hudson's Bay Company: "I believe [the conversion of the Indians] would be highly beneficial . . . as they would in time imbibe our manners and customs and imitate us in dress; our supplies would thus become necessary to them which would increase the consumption of European produce and manufactures, and in like measure increase and benefit our trade, as they would find it requisite to become more industrious and to turn their attention more seriously to the chase in order to be enabled to provide themselves with such supplies; we should moreover be enabled to pass through their lands in greater safety which would lighten the expense of transport, and supplies of provisions would be found at every village and among every tribe." Frederick Merk, *Fur Trade and Empire* (Cambridge, 1931), p. 108.

48. *Captain Cartwright and his Labrador Journal,* p. 144.

cargo of wine for Lymburner at Quebec, and also returned to Labrador with bread, flour, and other goods.

The enterprise was apparently not a success, as Cartwright's partners sold their shares to him for £1,200, or at a loss of nearly one half. The Quebec triangular trade*required constant attention and was also a failure although the heavy outgoing cargo always proved tempting to east- and southbound vessels. As a result of the outbreak of war, vessels for Newfoundland left under a convoy and most of their provisions were obtained from Waterford in Ireland. In 1778 an American prize—a ship from England of 80 tons—called at an Irish port and brought provisions, stores, and men; and a second vessel took corn to Leghorn and from Lisbon brought back salt for Labrador. Again, the cod fishery was successful and boats took fish at various points along the coast. As estimated, this new enterprise promised a profit of £1,500, but an attack by Boston privateers changed it to a loss of £14,000. Cartwright, however, carried on his venture until 1786 when further losses forced him to give up temporarily. He claimed that he cleared "above one hundred per cent for the last three years," or from 1790 to 1792, inclusive.

THE aggressive commercialism of New England which contributed to the breakup of the French Empire pushed relentlessly on to the breakup of the first empire established by Great Britain. Restrictions were now placed on New England's fishing industry. It had to endure competition from areas more advantageously located to supply the markets of Europe. In the case of colonial trade with the French West Indies there were other restrictions. All this culminated in the revolt of the colonies. The British Empire, with its lack of flexibility which allowed of its being dominated by vested interests from the West Indies, proved unequal to the strain imposed by the expanding commercialism of New England. The results were evident in the emergence of lines more sharply drawn between areas in which New England had special advantages in the fishing regions, particularly as regarded the West Indies, and those in which England had special advantages in the European market.

In spite of the Treaty of Paris and the withdrawal of the French, New England, in competition with Nova Scotia, the Channel Islands, and Newfoundland, failed to reap the full advantage from the ceded territory. She was concerned most largely with the West Indian market for her lower-grade product. The Channel Island interests and those of Newfoundland were more solidly entrenched in European trade and occupied more effectively the territory vacated by France. The debatable territory which centered about Canso in the earlier period was widely

*See Notes to Revised Edition, p. 509.

extended. New England became more interested in trading with Newfoundland and Nova Scotia, and the problem of the West Indian market became more acute.

In 1763 the Massachusetts fishery brought estimated annual returns of £164,000, employed vessels worth £100,000, and consumed provisions and supplies worth £22,000.[49] In the decade from 1765 to 1775 it was estimated that there was an average of 665 ships, of 25,630 tons, and 4,405 men in the fishery.[50] Marblehead and Gloucester had 150 and 146 vessels respectively, followed by Plymouth, Salem, Chatham, Ipswich, and Manchester. The vessels averaged about 40 tons and had crews of six or eight men, some of whom were employed on shore to dry the fish. About 350 vessels of from 70 to 180 tons, with an average of eight men, were employed in carrying the fish to market. New England continued to have "profitable and constant employment for their fishing vessels during the winter, whilst our ships were laid up for four or five months in that season in the ports of Dartmouth, Poole, etc."[51] In 1763 exports representing 64 per cent of the value of the cod taken that year by the New England fisheries went to the West Indies and the trade employed 150 vessels. During the three years ending in 1773, the exports of cod from the colonies were as follows:[52]

	Great Britain and Ireland	*Southern Europe*	*British and Foreign West Indies*	*Total*
Dry (quintals)	706	102,601	241,987	345,294
Green (barrels)	7	300	36,136	36,453

Of these totals 60,620 quintals and 6,280 barrels were purchased from Newfoundland, Canada, and Nova Scotia. The British West Indies imported 161,000 quintals of dried fish and 16,178 barrels of pickled.

The increasing importance of the Newfoundland fishery, with the support of New England trade, involved competition with New England fish in European markets. It accentuated the dependence of New England on the British West Indies for the sale of fish and other products,[53]

49. See Raymond McFarland, *A History of the New England Fisheries* (New York, 1911), pp. 104 ff.

50. *Idem*, pp. 111–112, 116 ff.; also W. B. Weeden, *Economic and Social History of New England, 1620–1789* (Boston, 1890), II, 750 ff.

51. John Lord Sheffield, *Observations on the Commerce of the American States* (London, 1784).

52. *Idem.*

53. From 1771 to 1773 the thirteen colonies sold to the British West Indies 76,767,695 feet of boards and timber (Nova Scotia and Canada, 232,040 feet), 59,586,194 shingles (Nova Scotia and Canada, 185,000), 57,998,661 staves (Nova Scotia and Canada, 27,350), 1,204,389 bushels of corn, 396,329 barrels of flour, 51,344 hogsheads of

and upon the French West Indies for supplies of molasses to meet the demands of expanding trade. The British West Indies could neither consume all the produce of New England nor provide supplies of molasses adequate to the demands of the northern colonies.[54] The total molasses production of the British West Indies would not equal two thirds of the molasses imports of Rhode Island. In 1763 Massachusetts imported 15,000 hogsheads from the French West Indies and 500 hogsheads from the British islands. With Rhode Island, the two colonies imported 29,000 hogsheads, of which 3,000 came from the British islands. In 1763 thirty distilleries in Rhode Island and sixty in Massachusetts made molasses into rum to be sent to Africa in exchange for slaves and for gold dust and other products. Rhode Island employed 184 vessels in foreign trade and 352 in colonial. The production and trade in rum in the British West Indies meant competition with rum produced in the northern colonies from French West Indian molasses. The market for rum expanded in the fishery, the slave trade, and also in the fur trade on the withdrawal of the French and the penetration of English traders into the interior.[55] The expansion of colonial and West Indian trade involved competition and an increasing dependence of the colonies on the French West Indies. Cape Breton had disappeared as a spillway for trade between the British and the French empires.

The strength of the monopoly exercised by British planters had been

fish (Nova Scotia and Canada, 449, and Newfoundland, 2,307), 44,782 barrels of beef and pork, 7,130 horses, and 3,189 barrels of oil. See H. C. Bell, "The West India Trade before the American Revolution," *American Historical Review*, January, 1917, pp. 272–287. In 1771 New England purchased 67,000 quintals from Newfoundland. For an account of the trade see Porter, *op. cit.*, I, 12–13, 78–79.

54. EXPORTS OF RUM IN 1773*

To	Gallons British West Indies	Gallons New England
Great Britain	10,963	961
Ireland	23,250	1,240
Southern Europe	6,688	68,412
Africa	530	419,366
West Indies	2,078	12,057
Nova Scotia, Canada and Newfoundland	50,716	608,025
	94,225	1,110,061

* Chalmers, *op. cit.*

55. For an account of the increasing consumption of rum in New England after 1713, its displacement of foodstuffs in New England trade after 1750, and its significance in the American Revolution, see C. W. Taussig, *Rum, Romance and Rebellion* (New York, 1928), chaps. i–iv. The importance of rum is overemphasized and that of the fishery neglected. Brebner, in *The Neutral Yankees*, gives an excellent account of the significance of rum in the attitude of Nova Scotia during the revolutionary period. See especially pp. 21–22, 70, 149.

evident in their success in preventing England from increasing her possessions in the West Indies by the Treaty of Paris. France was allowed to retain Guadeloupe in response to their demands, but it was opposed by the northern colonies. England obtained Canada in response to *their* demands and of those of the sugar plantations as, possibly, an additional market.[56] Smuggling in the French West Indies increased during the Seven Years' War and as a result of loss of French colonies in the Treaty of Paris. In August, 1763, the governor general of Guadeloupe encouraged smuggling by an order which authorized the importation of certain specified foreign goods, i.e., lumber, provisions, horses, and colonial products, provided that sugar and rum alone were taken in exchange;[57] and the effect was a competition with the British West Indies for supplies and a narrowing of the market for their products. A more rigid enforcement of the Navigation Acts, the reinforcing and modification of the regulations of the Molasses Act (6 Geo. II, c. 13) of 1733, and its amendments in the Sugar Act of 1764 (4 Geo. III, c. 15) were designed to counteract the measures taken by the French.[58]

The increasing interrelationship between the New England fishery and the trade of the French West Indies, and the dependence of the expansion of the fishery on increasing trade, were in direct conflict with a policy of restriction. "The publication of orders for the strict execution of the Molasses Act has caused a greater alarm in this country than the taking of Fort William Henry did in 1757. . . . The merchants say there is an end of the trade in this province; that it is sacrificed to the West Indian planters" (January 7, 1764).[59] Assuming an

56. See W. L. Grant, "Canada versus Guadeloupe, an Episode of the Seven Years' War"; *American Historical Review,* July, 1912, pp. 735–743; G. S. Graham, *British Policy and Canada 1774–1791* (London, 1930), pp. 1–10; and Pares, *op. cit.,* p. 216.

57. Arthur Girault, *The Colonial Tariff Policy of France* (Oxford, 1916), pp. 24–25.

58. The Molasses Act imposed a duty of 5 shillings a hundredweight on foreign sugar entering the colonies, 9 pence a gallon on foreign rum, and 6 pence a gallon on foreign molasses. The Sugar Act reduced the duty on foreign molasses and syrups to British colonies from 6 pence to 3 pence a gallon, with the difference, however, that attempts were to be made to check smuggling. Duties on foreign sugars were increased £1 2*s.* a hundredweight, and foreign rum or spirits were prohibited. See F. W. Pitman, *The Development of the British West Indies 1700–1763* (New Haven, 1917), especially chap. xiv; and G. L. Beer, *British Colonial Policy 1754–1765* (New York, 1922), chap. xiii. Wines imported directly from the Azores and Madeira were required to pay a high duty, but if imported by Great Britain the duty was low.

59. *Select Letters on the Trade and Government of America and the Principles of Law and Polity Applied to the American Colonies Written by Governor Bernard at Boston in the Years 1763–68* (London, 1774). See A. M. Schlesinger, *The Colonial Merchants and the American Revolution 1763–1776* (New York, 1917), pp. 42–43, 48–49, 59–60. It was claimed that heavy duties on foreign sugars would destroy navigation and the fishery, allow only the finest sugars to be imported into America, and give the French the advantage of manufacturing it. All sugar from the Continent

equal balance of trade between New England and the British West Indies at the date of the passing of the Molasses Act, "since that time North America has increased to above double; the British West Indies remain as they were. What is to become of half the produce of North America if it is not suffered to be carried to foreign markets upon practicable terms of trade?"[60] In Massachusetts in 1763 the restrictions upon the importation of roughly £100,000 worth of molasses, "to purchase which fish and lumber of near the same value must be sent from hence," meant the restriction of trade and the fishery.

Our pickled fish wholly, and a great part of our codfish, are only fit for the West India market. The British islands cannot take off one-third of the quantity caught; the other two-thirds must be lost or sent to foreign plantations, where molasses is given in exchange. The duty on this article will greatly diminish the importation hither; and being the only article allowed to be given in exchange for our fish, a less quantity of the latter will of course be exported . . . the obvious effect of which must be a diminution of the fish trade, not only to the West Indies but to Europe, fish suitable for both these markets being the produce of the same voyage. If, therefore, one of these markets be shut, the other cannot be supplied. The loss of one is the loss of both, as the fishery must fail with the loss of either.[61]

The limitations on the fishery on the Labrador[62] and in the Gulf of St. Lawrence aggravated the grievances; "and the fish trade of New England is of too great consequence to run any risque of checking it."[63]

being treated as French sugar, New England was kept from exchanging a valuable commodity for English manufactures. The Navigation Acts became more burdensome in requiring American vessels to call and unload in Great Britain, particularly those engaged in trade with Spain and Portugal, since it meant longer voyages and the loss of perishable commodities.

60. *Idem.*

61. Statement of the Council and House of Representatives of Massachusetts, 1764, cited by Sabine, *op. cit.*, pp. 136 ff.

62. "But the grand matter of complaint is the restraint laid on their fishery, no American being suffered to take cod in the Straits of Belle Isle or on Labrador shore; and thereby rendering our new watery acquisitions entirely useless and the restraint itself be attended with a very large expence, and instead of endeavouring to make the most of that extensive fishery, it is become a scene of violence between the Europeans and Americans; the interruption of the fishery is weakening our naval power and depriving the Americans of the most valuable source for taking of and paying for the manufactures of Great Britain." "Restraints . . . will not be fully removed but by an act of parliament to explain that of William 3rd and give free liberty to all the British subjects to improve the fishery to the utmost which greatly strengthens our naval power." Statement by Dennis de Bredt, agent of the House of Representatives of Massachusetts Bay, 1767, printed in A. B. Hart, *American History Told by Contemporaries* (New York, 1898), Vol. II. See E. T. D. Chambers, *The Fisheries of the Province of Quebec* (Quebec, 1912), pp. 98–103.

63. *Select Letters on the Trade and Government etc.*

The prosperity of interests which had thrived when possessing a monopoly of the British market was reflected in the strength of the West Indies' influence in Great Britain. It was exposed, on the other hand, to the weakness of that concentration on sugar production which characterized the plantation system, with its increasing demand for slaves, absentee control, and inefficient, conservative, agricultural technique. Adam Smith commented on the rapid increase in the market for sugar in Great Britain "within these twenty years," and on the effectiveness of the duty in limiting the manufacture of white and refined sugars to Great Britain; and he argued[64] that the expanding market was a support to the monopoly position of the British West Indian planters.

> Their whole produce falls short of the effectual demand of Europe and can be disposed of to those who are willing to give more than what is sufficient to pay the whole rent, profit and wages necessary for preparing and bringing it to market. . . . We see frequently societies of merchants in London and other trading towns purchase waste lands in our sugar colonies, which they expect to improve and cultivate with profit by means of factors and agents, notwithstanding the great distance and the uncertain returns from the defective administration of justice in those countries.[65]

Islands ceded by France after the Seven Years' War—Dominica, St. Vincent, Tobago, and Grenada—possessing virgin soil became magnets for small-scale white planters who had been displaced by the large plantations of the British West Indies. The competition from the sugar of these islands was sharpened by the lower taxes made possible by the decision of *Campbell* v. *Hall* in 1774.[66] Exports of muscovado sugar from Grenada increased from 65,699 hundredweight in 1764 to 198,159 in 1773; from St. Vincent, to 58,691 in 1773; and from Tobago, to 50,385 in 1775.

Monopoly profits in Great Britain led not only to increased production in the new islands of the British West Indies but also to increased exports from the French islands to the European market. The three-cornered trade from England to Africa, to the West Indies, and thence back again to England was displaced to an increasing extent by two trade routes: first, a direct trade from England to the West Indies; second, a three-cornered trade with New England. Exports, drawing largely on Rhode Island for rum, were sent to Africa and exchanged for slaves. The slaves were carried back to the British West Indies

64. Adam Smith, *Wealth of Nations* (New York, 1937), pp. 545, 548.
65. *Idem,* pp. 156–157.
66. Lord Mansfield's decision that the Crown made an irrevocable grant of its legislative power when an elected assembly was set up in a colony under instructions from the King.

where there was an increasing demand for them. With specie the New England ships purchased sugar and molasses from the French West Indies, and returned home.

The New England fishery had become the basis of commercial expansion and maritime activity on an extensive scale.[67] "Their earnest application to fisheries and the carrying trade," says David MacPherson, "together with their unremitting attention to the most minute article which could be made to yield a profit, obtained them the appellation of the Dutchmen of America." Attempts to check the fishery and the trade with the West Indies after the collapse of war prosperity evoked determined protests. As a result of them the Stamp Act of 1765 was repealed[68] and the tariff on foreign molasses was lowered to one penny, which was imposed on all molasses in 1766. Duties on foreign sugar remained at a high level, but export duties on British West Indian sugar were lowered. A tax on *all* molasses, on the other hand, was a *revenue* tax, and, followed by the Townshend Acts of 1767, led to the nonimportation measures of 1769.[69] It was claimed that restraints on trade to the foreign West Indies, Africa, Madeira, and the Mediterranean involved losses to 400 vessels in the fisheries, 180 vessels in the lumber and provisions trade with the West Indies, and a sharp decrease in shipbuilding.[70] The struggle against restrictions on trade in European wine and West Indian molasses was accompanied by protests against the East India Company's monopoly in tea. The Continental Association of October 20 declared against the importation of goods

67. See Schlesinger, *op. cit.*, pp. 22 ff. on the commercial provinces.

68. "In her present condition, Great Britain resembles one of those unwholesome bodies in which some of the vital parts are overgrown, and which, upon that account, are liable to many dangerous disorders scarce incident to those in which all the parts are more properly proportioned. A small stop in that great blood-vessel, which has been artificially swelled beyond its natural dimensions, and through which an unnatural proportion of the industry and commerce of the country has been forced to circulate, it is very likely to bring on the most dangerous disorders upon the whole body politic. The expectation of a rupture with the colonies, accordingly, has struck the people of Great Britain with more terror than they ever felt for a Spanish armada or a French invasion. It was this terror, whether well or ill grounded, which rendered the repeal of the stamp act, among the merchants at least, a popular measure. In the total exclusion from the colony market, was it to last only for a few years, the greater part of our merchants used to fancy that they foresaw an entire stop to their trade; the greater part of our master manufacturers, the entire ruin of their business; and the greater part of our workmen, an end of their employment." Adam Smith, *op. cit.*, p. 571.

69. Schlesinger, *op. cit., passim;* also V. D. Harrington, *The New York Merchant on the Eve of the Revolution* (New York, 1935); C. M. Andrews, *The Colonial Period of American History* (New Haven, 1938), IV, 106–107.

70. Schlesinger, *op. cit.*, pp. 133–134. At the outbreak of the Revolution nearly one hundred ships were carrying one fourth of the dried and pickled fish and one sixth of the wheat, flour, and rice to the Mediterranean.

on December 1, 1774. The answer was the Restraining Act of 15 Geo. III, c. 10,[71] effective July 12. Passed by Great Britain, it restricted New England trade to English ports and provided that no English vessel should be permitted, except by special license, to engage in any fishery on any part of the coast of North America. In May an embargo was placed on trade with Nova Scotia, Newfoundland, and the West Indies and hostilities followed quickly.[72] A nonexportation regulation was adopted on September 10, 1775, New England schooners became privateers, and fishermen active seamen.[73]

The outbreak of the American Revolution brought about important changes in the Nova Scotia fishery. The Restraining Act and the embargo stimulated industry and direct trade with the West Indies. From 1763 to 1774 "New England colonists engrossed almost the whole of the fisheries both great and small." For Nova Scotia war conditions, the exclusion of New England from the fisheries and the West Indian trade, and the expenditures upon the army and navy meant marked prosperity.

Settlements turned their trade from New England ports to Halifax.[74] "The poor people, . . . in a manner of anticipation, mortgaged their *catch* in the spring of the year to those merchants and shopkeepers in Halifax who advanced supplies to them for that purpose." Country traders disposed of goods "to those fishermen who did not leave their own harbours in the autumn and spring, but still the fish taken centred in Halifax. . . . By passing, however, through one other channel it became somewhat enhanced in price, yet the trade to the West Indies and the fisheries continued to increase."[75] Liverpool developed as a port suitable to larger vessels which brought goods to, and took them from,

71. See Sabine, *op. cit.*, pp. 139 ff.; also V. G. Setser, *The Commercial Reciprocity Policy of the United States 1774–1829* (Philadelphia, 1937), pp. 6 ff.

72. The West Indies were the object of hostility as a result of their influence upon Great Britain in encouraging restrictive legislation. See Schlesinger, *op. cit.*, pp. 403–404, also pp. 416, 420–421, 425–426, 489, 568–570, 586–588; for the effect of the Revolution on the fisheries, see pp. 531–533, 538, 565.

73. McFarland, *op. cit.*, chap. vii, and C. B. Elliott, *The United States and the Northeastern Fisheries* (Minneapolis, 1887), pp. 23–24.

74. For a list of sea-going craft entering the ports of Nova Scotia between July 4, 1778, and November 15, 1781, see the *Report of the Board of Trustees of the Public Archives of Nova Scotia, 1936* (Halifax, 1937), Appendix C. For a discussion of the significance of the merchant class see Brebner, *The Neutral Yankees, passim,* and V. F. Barnes, "Francis Legge, Governor of Loyalist Nova Scotia 1773–1776," *New England Quarterly,* April, 1931, pp. 420–447. For a criticism see W. B. Kerr, "Merchants of Nova Scotia and the American Revolution," *Canadian Historical Review,* March, 1932, and "Nova Scotia in the Critical Years 1775–1776," *Dalhousie Review,* April, 1932, pp. 97–107.

75. "A Petition of Merchants of Halifax, January 23, 1818," *Acadian Recorder,* January 24, 1818.

distant markets to be distributed and collected at smaller ports along the coast. Trade expanded on flexible New England lines; for instance, small amounts of capital were invested in shares of ships and cargoes.[76] Family relationships, as in the case of that of Perkins, where the major part of the family remained in Connecticut, made this easier even during the Revolution. The firm of Cochrans at Halifax took the place of Russell of Boston in Perkins' business activities. Cargoes of fish purchased chiefly alongshore were sold to Bermuda and to Halifax. Rum,[77] molasses, and salt were brought from Bermuda in exchange for spars, boards and lumber, and fish.[78] Flour was difficult to purchase, particularly when embargoes were imposed at Quebec. On January 15, 1778, no pork was to be had at Halifax, flour was very dear, rum 6*s.* 6*d.* a gallon, and cod reached a low point of 10 shillings a quintal.

The bank fishery ceased because of the dangers from New England privateers. On October 16, 1776, Perkins wrote regarding their depredations: "This is the fourth loss I have met with by my countrymen." In 1779–80 a privateer, the *Lucy*, was fitted out at Liverpool and succeeded in taking several prizes carrying a variety of goods, one with a cargo worth £2,000. The collapse of the bank and Newfoundland fisheries led to trade in pickled fish of other varieties—alewives, salmon, and herring. Herrings were taken in large numbers, but salt was scarce in September, both in 1778 and 1779.

In Newfoundland the embargo on exports of food from New England had serious consequences.[79] Palliser's Act (15 Geo. III, c. 31) 1775, had been intended, together with the Restraining Act, to over-ride the effects of the embargo by reëstablishing the bank fishery, limiting the resident fishery, and prohibiting the trading and fishing of Americans.[80]

76. Perkins commonly took one eighth of the cargo; and of thirty-two shares of the privateer he took four valued at £77 9*s.* 8*d.*

77. The import tax of 5 pence on imported rum which protected local distillers and hampered trade to the West Indies was offset in November, 1774, by legislation admitting goods from the West Indies to Nova Scotia duty-free when at least two thirds of the cargo was paid for in goods of the province. Imports from elsewhere than the British West Indies were subject to a duty of 10 pence per gallon on rum, 5 pence on molasses, and 5 shillings a hundredweight on brown sugar. There were, however, re-export drawbacks of a half penny on molasses and 6 pence on brown sugar.

78. In August, 1779, a vessel from Bermuda was offered 5 shillings a gallon for rum, 3*s.* 6*d.* for molasses, 70 shillings a hundredweight for brown sugar, 80 shillings for "clayed sugar," 2 shillings for lime, and 2*s.* 6*d.* for salt, the purchaser to pay provincial duties in return for fish at 14 shillings a quintal. But these offers were declined. See W. B. Kerr, *Bermuda and the American Revolution 1760–1783* (Princeton, 1936).

79. See Ougier's evidence, *First Report of the House of Commons Committee on Newfoundland Trade 1793,* p. 45.

80. Prowse, *op. cit.,* pp. 344–345.

Provisions from England and Ireland were admitted free of duty. The preamble read:

> Whereas the fisheries carried on by His Majesty's subjects of Great Britain and of the British dominions in Europe have been found to be the best nurseries for able and experienced seamen, always ready to man the Royal Navy when occasions require; and it is therefore of the highest national importance to give all due encouragement to the said fisheries, and to endeavour to secure the annual return of the fishermen, sailors, and others employed therein to the ports of Great Britain . . . at the end of every fishing season:

For a period of eleven years bounties were to be paid on British-built and British-owned vessels of 50 tons and over with not less than 15 men, three fourths of them being British subjects, that left an English port on January 1, each year, caught not less than 10,000 fish, and landed them on the east coast of Newfoundland before July 15. The first 25 vessels making two trips to the Banks were given a bounty of £40 each and the next 100 vessels £20 each. This act and the Restraining Act were counterparts of the Quebec Act and served to exclude New England from the fishing areas belonging to England and obtained from France, as the Quebec Act attempted to prevent expansion from the coastal colonies to the interior.

The stimulus given to the bank fishery by the bounties, by the outbreak of the Revolution, and by the problems of the Revolution brought about a renewal of the struggle between fishing ships and settlers.[81] It was claimed that the "design of this act was to favour, and keep alive, the principle of a ship-fishery carried on from England—and they [the merchants] have, many of them, no scruple to say, that since Sir Hugh Palliser's Act, it is with the greatest difficulty that merchants can carry on the fishery with profit to themselves." The petition[82] from the merchants, boatkeepers, and principal inhabitants of St. John's, Petty Harbor, and Tor Bay (1775) asked for amendments to Palliser's Act which would improve their position; and, in particular, they asked for amendments to this effect: that bounties be given to vessels employed in the bank fishery with ten men each; that property in the island should be left subject to attachment; that servants should be given less freedom in making complaints against their masters; that fishing admirals be allowed to appoint deputies to "determine matters relative to the

81. Rogers, *op. cit.*, pp. 147–148. The decline of the trade of Bideford, Barnstaple, Giverts, St. Loo, Mevagissey, Fowey, and Topsham meant an increasing importance for Dartmouth and Poole. See Rogers, *op. cit.*, 138; also W. A. Miles, *Remarks on an Act of Parliament passed in the 15th Year of His Majesty's Regime on the Credit of Vice-Admiral Sir Hugh Palliser's Information* (London, 1779).

82. Prowse, *op. cit.*, pp. 341–343.

fishery"; that wanton destruction of trees by "rinding" or cutting should be prohibited; that the waste of small fish consequent on the use of cod seines should be checked; that during the breeding season of birds in the northern part of the island their destruction for the feather trade and for purposes other than the providing of food and bait should be checked; that oil[83] and cork might be imported duty-free; and that the price of bread and flour should not be allowed to rise above 12 shillings a hundredweight. They asked also that the practice of landing passengers without an adequate supply of provisions should be checked and that vessels should not be allowed to carry away more provisions than were necessary for their own needs. Further, masters of vessels should be required to provide for the return of unemployed passengers. The number of licensed houses had increased to more than eighty. The petitioners asked that they should be reduced to twelve

> and that each person so authorized to vend liquors should be obliged to keep a fishing shallop and cure all the fish said shallop may catch. . . . The number of shop keepers and retailers of goods have increased lately in St. John's to the great detriment of the fish catchers, as formerly every employer had the supplying [of] his own servants, which we apprehend in equity they are entitled to. From the very great wages given to them for the short season of prosecuting the fishery, the profits arising from such a supply was a small emolument to reduce the enormous wages given; but at present the masters are deprived of this, by their servants being supplied at those retail shops before alluded to, who in the fall of the year collect their bills, in consequence of which the servants are often reduced to great distress during the winter to prevent which we pray that each shop-keeper of goods may in future be obliged to keep a shallop on the fishery, otherwise to have six months liberty to sell off his goods and leave this island, as we deem every person not immediately concerned in the fishery (except his Majesty's servants) is a burthen to the island.

In spite of favorable circumstances, such as the closing of Portuguese ports to New England fish[84] and the expansion of the market that accompanied the withdrawal of New England during the Revolution, the severing of trade with New England restricted production. On March 21, 1778, Dartmouth had complained that the fishery was "in a most alarming and distressed state from the loss of a great number of fishermen and seamen already impressed into His Majesty's service and

83. Olive oil was an essential item in a diet of salt fish and was expensive because of its shipment via England rather than directly from Lisbon. Griffith Williams, *An Account of the Island of Newfoundland.*

84. English trade with Portugal declined, however, with the establishment of Portuguese companies and the increasing importance of direct trade between England and Brazil. Cottons had also begun to displace gold as an export from Brazil.

from the impossibility of getting those who have hitherto escaped, to work on board the vessels in order to fit them out for the present intended fishery, owing to the strictness of the officers employed in the impress service."[85] During the Revolution the population of Newfoundland dropped slightly, to 10,701 in 1784. The number of boats fell to 1,068, and the catch to 212,616 quintals. Byeboatmen decreased to 289, their servants to 2,317, their boats to 344, and their catch to 93,050 quintals. In 1767 the number of men returning to England was "double what it ever has been for sixty years past." By 1779 the byeboatkeeper was unable to continue in the fishery because of the increased cost of provisions. In 1784, English fishing ships had decreased slightly to 236, passengers numbered 3,187, and boats 572.[86] But the catch was only 131,650 quintals. The number of stages declined to 942. Exports of fish totaled 497,884 quintals. The number of sack ships fell off to 60.

By the Treaty of Versailles, in 1783, the nations concerned in the fishery reached a new equilibrium. New England fishermen were given the right to fish on the Grand Bank and other banks of Newfoundland and in the Gulf of St. Lawrence and to take fish on the British portion of the Newfoundland coast, although they were not allowed to dry them. They were also given "liberty to dry and cure fish in any of the unsettled bays, harbours, and creeks of Nova Scotia, Magdalen Islands, and Labrador, so long as the same shall remain unsettled."[87] The French continued in possession of St. Pierre and Miquelon[88] but abandoned rights on the coast from Cape Bonavista to Cape St. John in return for rights granted along the coast from Point Riche to Cape Ray.

The withdrawal of the French from Cape Breton and the mainland made it still harder for New England to obtain her needed supplies of West Indian products and for the French West Indies to obtain continental products. It also demanded, particularly with the expansion of the fishery, an increase of trade directly with the French West Indies. In the twenty years between the Treaty of Paris and the Treaty of Versailles the trend of the preceding periods was more marked and reached its ultimate conclusion. British colonial policy encouraged the British West Indies by restrictive legislation, widened the market for New Eng-

85. *C.O.* 194:19.

86. For a comparison of the Newfoundland fishery from 1764 to 1784 see George Chalmers, *An Estimate of the Comparative Strength of Great Britain* (London, 1804).

87. McFarland, *op. cit.*, pp. 127–128; also Sabine, *op. cit.*, pp. 149 ff.

88. Prowse, *op. cit.*, p. 353; see D. D. Irvine, "The Newfoundland Fishery, a French Objective in the War of American Independence," *Canadian Historical Review*, September, 1932, pp. 268–284; also E. S. Corwin, *French Policy and the American Alliance of 1778* (Princeton, 1916); "Peace at the Newfoundland Fisheries," *The Writings of Thomas Paine* (New York, 1906), II, pp. 1–25.

land exports, and increased the necessity of depending on the French West Indies for imports. Encouragement to New England, indirectly by assistance to the West Indies and directly by the encouragement of the fishing industry, increased the disproportion between temperate and tropical development within the empire. The lack of political elasticity which made possible the control by the sugar planters conflicted with the demands of an expanding economic structure. Rigidity of control based on sugar production clashed with the divisive commercialism of the fishery. As Pitman has pointed out, the temperate zone of the northern colonies was too large for the small tropical area, whereas with France the tropical area was too large for the small temperate area. The inevitable tendency toward equilibrium produced a lack of political balance which finally broke the control of both first empires.

The inherent characteristics of the fishing industry continued to assert themselves. Nova Scotia became a new base of operations, and Halifax a new center, linked to London and in competition with Boston. Channel Islands interests replaced the French in Cape Breton and the St. Lawrence. The Newfoundland fishery continued to expand, particularly toward the north, and to occupy the regions formerly controlled by those French areas which produced grades of fish more suited to the Mediterranean markets. New England continued to give support to the growth of Newfoundland, the African slave trade, the fur trade in the interior, and the West Indian trade in lumber, agricultural products, and the lower grades of fish. Demands for molasses and rum also brought about an increase of trade with the French West Indies. The ultimate necessity of carrying on the fishery on the grounds nearest the coast, and the slowness with which these advantages made themselves felt, were behind the essentially divisive character of the fishery. On the outbreak of the Revolution this was accentuated and Nova Scotia and Newfoundland broke off relations with New England.

The industry's intensely competitive nature increased the importance of recognizing and taking advantage of geographical conditions. The burden of government in Nova Scotia, with its revenue system based on rum and molasses, forced fishermen to migrate from Cape Breton to Newfoundland. As the merchants and fishing ships of the West Country broke the control of the companies in Newfoundland, the merchants and fishermen of New England had an outstanding part in breaking the control of Great Britain in North America. So, too, Nova Scotia became independent of New England;[89] and Newfoundland began to develop along more independent lines and also to develop a trading or-

89. Brebner, *op. cit.*, pp. 252, 303–304.

ganization. The influence of the commercialism of the colonies and the Atlantic made itself felt in the breakup of the old empire and in the appearance in 1776 of Adam Smith's *Wealth of Nations.*

John Adams, thinking of what New Englanders used molasses for and how they resented the Molasses Act, once said he did not know why they should blush to confess that molasses was an essential ingredient in American independence. Of Nova Scotia it might be said that the rum made in Halifax from molasses was just the opposite, an essential ingredient in Nova Scotian loyalism. The rum industry provides the clue for unravelling the close-knit fabric of trade and finance in London and Nova Scotia, which goes far to explain the policy of the little commercial group which dominated Nova Scotian behavior far more effectively than the merchants of Boston, Philadelphia and New York were able to do once they had loosed the revolutionary spirit of their populace.[90]

The decline of France made easy a rapid expansion from the Channel Islands,[91] as exemplified in the activities of Joshua Mauger. The trade of Halifax and, in turn, its commercial policy were dictated from London rather than from Boston. Nova Scotia was converted from an outpost of New England into an outpost of Old England, and, behind this, mercantile interests were able to profit by the disadvantages of New England in the West Indies, particularly after the outbreak of hostilities, and later, the Treaty of Versailles. The British Empire retreated to more solid ground and began the task of consolidating its position by substituting Nova Scotia for New England. The vigor of commercialism based on the fishing industry broke the control of companies in the West Country, in Newfoundland, and in New England. In turn it broke the control of the navigation system which had been elaborated to succeed them.

The problem of the British and French empires was in part a result of the inability to control an aggressive commercialism based on the fishing industry. Its impact on company control in the West Country, Newfoundland, and New England, and in turn on the navigation system, was in contrast with the restraining effect of commercialism centering on diverse ports in France, and on the economic development of New France. Company control became inadequate in the fur trade of the St. Lawrence and led to active government intervention rather than commercialism. Such intervention was evident in wars with the Iroquois, expenditures on fortifications, valorization schemes, and major monetary disturbances through inflation in the years preceding 1713 and

90. *Idem,* p. 149; also pp. 21–22, 70.

91. The Channel Islands engaged in illicit trade in brandy, canvas, and cordage from France.

1760. The effectiveness of short-term credit in commercialism based on the fishing industry was in striking contrast to the limitations of long-term credit in the fur trade and in the plantation colonies.[92] Staples demanding long-term credit were dependent on capital control in relation to the metropolitan development of Great Britain. Under these conditions the effectiveness of staple interests was evident in political influence and legislation.* With dependence on commercialism as in New England the essentially close relationship between the economic and the political institutions of the British Empire disappeared.

92. See Adam Smith, *op. cit.;* also W. R. Scott, *The Constitution and Finance of English, Scottish and Irish Joint Stock Companies to 1720* (Cambridge, 1912), Vol. I, chap. xxii.

CHAPTER VIII

THE EFFECTS OF WAR ON FRANCE AND NEW ENGLAND, 1783–1833

FRANCE

Agreeably to the policy acted on at all times by the French, bounties were, immediately after the treaty, granted to encourage and support the French Newfoundland fisheries. These bounties, if the fish be exported to meet us in foreign markets, are about equal to the expense of catching and curing, and which, if imported into France, is sufficient to protect against loss. No encouragement, however, is given but with the proviso of training seamen.

JOHN MCGREGOR, *British America* (1833)

FRANCE suffered from the effects of the American Revolution, the French Revolution, and war with England lasting almost continuously from 1793 to 1815. The narrowing of the frontiers of the French Empire under English aggression had also contributed to the breakdown of the Old Regime in France. New England recovered after the American Revolution and gained from the difficulties of the French until, from 1807 to 1813, she was involved in difficulties herself. Newfoundland and Nova Scotia were favored by the handicaps of their rivals, but after 1815 they suffered from the competition which followed American and French recovery.

In 1785 France set herself to make good the injuries inflicted upon her Newfoundland fishery by the struggle in America. Her measures took the form of bounties. The first was one of 10 livres[1] a quintal on dried fish carried in French vessels to French settlements in the West Indies or continental America. This, coupled with a duty of 5 livres a quintal on foreign fish, practically gave her a monopoly of the market. The second was one of 5 livres a quintal on fish carried to European markets.[2] But in spite of this encouragement the French fishery still declined, as indicated in the table below.[3]

1. See Appendix, p. 27.
2. Emile Hervé, *Le French-Shore et l'arrangement du 8 Avril 1904* (Rennes, 1905), chaps. iv and v; also Ougier's evidence, *Second Report of the House of Commons Committee on Newfoundland Trade 1793*, p. 26; also *First Report*, p. 33; and J. M. Grossetête, *La Grande pêche de Terre-Neuve et d'Islande* (Rennes, 1921), p. 385; Henry Schlacther, *La Grande pêche maritime* (Paris, 1902), pp. 45–46.
3. *See opposite page for note.*

The outbreak of war with England in 1793 brought about a collapse. The chief establishments[4] at St. Pierre and Miquelon were destroyed and the people migrated to the Magdalen Islands.[5] The islands were returned to France in 1802 and in that year 8 brigs and 5 schooners with 50 men from Havre de Grace, St. Malo, Bordeaux, and Bayonne were engaged in this fishery. But St. Pierre and Miquelon were again in English possession from 1803 to 1814.

In the fishery along the French Shore of Newfoundland,[6] interests

3. THE FRENCH FISHERY, 1786–1792

	Number of Ships	Tonnage*	Number of Boats	Number of Men Employed	Fish Cured (quintals)	Oil (tons)
1786	86	22,640	1,532	7,859	426,400	1,059
1787	73	15,690	1,342	6,402	128,590	323
1788	86	20,130	1,560	7,433	241,262	603
1789	58	15,900	1,035	7,314	239,000	603
1791	42	10,417	628	5,895	40,580	121
1792	46	9,180	689	3,397	94,000	174

* In the period from 1769 to 1774 the average tonnage per ship had been 108, with 35 men to every 100 tons, in contrast with 238 tons with 40 men to every 100 tons in the period from 1786 to 1792. *Second Report from the Committee Appointed to Enquire into the State of Trade to Newfoundland 1793*, p. 57.

4. In 1792 an estimate was as follows:

	St. Pierre	*Miquelon*
Forty- to seventy-ton vessels	33	9
Shallops	200	76
Flat fishing boats	400	300

John Waldron of Fortune Bay criticized these figures and stated that in August, 1792, there were 40 brigs and ships of an average of 150 tons; and their crews, with the residents, employed 640 "flats" handled by two men, 110 to 120 fishing shallops, with three men apiece, and 100 bankers with eight men each. See also *Second Report, op. cit.*, Appendix No. 6. In 1787 a small schooner of 50 tons and 12 men caught 700 quintals of cod on St. Pierre Bank and cured them at Codroy. In 1788 St. Pierre and Miquelon had an extremely successful fishery, and this led to the sending of vessels to Bay of Islands, Quirpon, Croc, and other stations for spare supplies of "bay" or rock salt.

5. Ferdinand Louis-Legasse, *Evolution économique des Iles Saint-Pierre et Miquelon* (Paris, 1935), p. 23.

6. English fishermen were barred from areas held under French fishing rights. Five shallops arriving in 1786 from the station of Noble, Kingsworth and Company on the coast of Labrador to fish at Quirpon were ordered to leave. Thomas Spelt, for twelve years a resident at Noddy Bay, possessing houses and anxious to remain during the winters "for the convenience of furring, and killing seals," was advised to move to some part of the coast beyond the French boundaries. British subjects in White Bay with salmon brooks and fishing rooms were warned to remove their fixed establishments, as occurred in the case of two Englishmen at Sops Arm. They had 2 boats and 5 hands who took 290 tierces in 1786. Pinson and Noble, from Labrador, were reported as having 2 boats and 5 men who took 220 tierces; and in 1786 they were still in possession of Southwest Brook at Hare Bay. An Englishman carried on a salmon fishery on the Humber River in 1787, and brought 76 tierces of salmon and £265 worth of furs to St. John's. But this establishment was apparently abandoned the following year.

centering in St. Malo and Granville were chiefly concerned, although vessels also came from Dunkirk, St. Brieuc, Binic, and Paimpol. In 1786, after the granting of the bounties, a company of merchants had 15 vessels fishing from Quirpon Bay south to St. Lunair. The ice kept ships from arriving at Quirpon before June 6, but a cargo of fish was sent to the Mediterranean as early as July 18. The fishing was done within less than two miles of the shore. Of 14 stages in 5 ports, 10 were at Quirpon. These were 100 feet in length by 50 feet in breadth, and were neatly thatched with wood and brush and covered with canvas. Between St. Anthony and St. Julien, 29 vessels from Granville arrived about the end of May, each vessel as usual having a stage and an oil vat. Croc, because of its easy access and convenient location, was the chief rendezvous for warships and convoys, but in many cases boats were compelled to go two or three leagues for fish, with the result that no fishing rooms or stages were built in the harbor. Seven vessels totaling 1,120 tons, with 560 men, were engaged at this port. At Carouge and Conche Harbor the fish were found near the shore and were taken with nets. Twenty-two vessels with 2,040 men from Granville were stationed at these points. St. Brieuc had a ship of 300 tons and 90 men at Orange Bay. White Bay was not a successful fishing ground. Paimpol sent vessels to Fleur de Lys and Bay of Pine; and St. Brieuc sent them to Fleur de Lys and Pacquet. Complaints were made of Indian disturbances at Cap Rouge, Conche Harbor, and Pacquet. In 1785 a special bounty of 75 livres a man was granted to all crews of ships fishing in the Bay of Islands, "on account of its being a rocky dangerous coast."

The French were described as having a large home market and a specialized market at Nice for fish which were not dried as much as the English product. The French were also able to make their voyages in two thirds of the time required by the English, and were not burdened with heavy fees. But they had numerous disadvantages. Without permission to winter, they were unable to keep their flakes, stages, and equipment in good condition;[7] and, being unable to get covering from the woods for their stages, they were forced to use canvas, which was more expensive. It was held that their boats were smaller and less seaworthy, and their seamen were so much less expert that they did not take more than half as many fish. After splitting and heading, the fish were piled in salt for fifteen to twenty days, then washed and cured in

7. Every year the French brought out a fresh supply of boats in frames and sections stowed in the salt in the ships' holds. They were quickly put together and used with the boats which had been left over the winter. Salt remaining from the fishery was buried and used the following season. Bay or rock salt—darkish in color—was brought out in bulk and used at the rate of 10 hogsheads to a hundred quintals of fish.

the sun. "They use no flakes and dry their fish upon rows of shrubbery spread on the rocks." Wages were lower,[8] provisions were poorer, and the French found their shore not as well adapted to carrying on the fishery as the English. The cod were as large,[9] but the weather, especially near the Straits of Belle Isle, was less favorable to drying; navigation was more dangerous and losses heavier; the bait fish were less plentiful and arrived later. In many ways, indeed, the fishery was apparently less dependable.[10] Following the outbreak of war in 1793 the French fishery declined sharply.[11] Bounties were suspended, and their revival in 1802 was of slight consequence.

8. The boats, with three men each, were out early and late and in all kinds of weather. The men from Granville were given an allowance of a fifth of the fish, and those from St. Malo £16 for the voyage. They were allowed seven ounces of pork a week and as much bread as they wished. The expense of the fishery was put at 100 écus the boat or *batteau.*

9. From Fleur de Lys to St. Anthony the fish were said to be smaller, but to the north and at Quirpon after the end of July they were as large, it was claimed, as those of the Banks. To the south, 50 or 60 quintals produced a barrel of oil; but to the north it took 90 or 120. The small size were taken with capelin to the end of July, and the large with herring during the remainder of the season.

10. In 1787 north of Hare Bay to Quirpon the fishery was tolerably successful, but elsewhere poor. St. Lunair was deserted. Some vessels were engaged at Port Saunders but 37 boats were drawn up on shore at Bay of Islands. A French ship and a snow came early in the spring but left without participating in the fishery. From Bay of Islands around the north coast to Croc the fishery, in 1788, in contrast with the previous year, was a failure; but south of Croc, in Canada and Orange bays and East of White Bay to Cape St. John it was a success. Five vessels from Bayonne were at Port au Choix and Ferrole on the west coast, and took 8,400 quintals. In 1791 the fishery was described as very bad and, for the third year in succession, it was a failure in 1792, probably averaging less than 100 quintals to a boat (a saving voyage). At Quirpon the number of vessels declined from 10 of 3,500 tons in 1786 to 8 of 2,350 tons in 1792; at Fichot, from 7 of 1,900 tons to 3 of 240 tons; at Croc, from 7 of 1,120 to 3 of 690; at Cap Rouge, from 12 of 3,300 to 9 of 1,540; and, at Conche, from 10 of 2,500 to 3 of 470. The total number of men employed in boats dropped from 4,627 in 1786 to 2,007 in 1792. The number working on shore dropped from 3,232 to 1,390, and the grand total from 7,859 to 3,397.

11. THE FRENCH FISHERIES IN 1802*

Port	Ships	Tons	Home Ports	Boats	Men in Boats	Men Ashore	Cod (Quintals)	Oil
Pacquet	2	144	St. Brieuc	8	24	18	2,400	40 bbls.
La Scie	2	262	St. Malo	13	39	47	3,000	14½ tons
Fleur de Lys	1	260	Binic	16	50	24	3,330	66 bbls.
Cap Rouge	6	1,032	Granville	59	62	236	11,000	218 bbls.
Conche	2	250	"	9	23	21	2,400 dry 900 green	49 bbls.
St. Julien	4		St. Malo	36			11,000	200 hhds.
Fichot	3		Granville	26			6,000	90
Zealot	3		"	36			11,050	200
Goose Cove	3		"	27			7,000	100
St. Anthony	2		"	17			3,600	60

Five vessels visited Quirpon, and two each, Canada Bay and Fourchette.

* *C.O.,* 194:43.

To assist in the recovery after the Napoleonic wars, in 1816 and to run for three years, bounties were granted of 50 francs a man on vessels in the fishery of St. Pierre and Miquelon and the coast of Newfoundland, and 15 francs a man on vessels in the North Sea and on the Grand Banks. Exports to French colonies in French vessels were given 24 francs a quintal; from French ports to the Mediterranean in French vessels or Spain, Portugal, Italy, and the Levant, 12 francs; and from fishing grounds to Italy, Spain, and Portugal, 10 francs. In 1818 bounties were increased to 40 francs on every quintal and shipped direct to colonies. Modifications were introduced in 1822 and 1832. The total bounty payments in premiums and drawbacks increased from 365,000 francs in 1817 to 4,400,000 in 1829, and were estimated to equal the cost of catching and curing the fish, assuming that green men formed from one fourth to one third of the total. A tariff of 44 francs a hundred kilos was imposed on cod in 1791 and of 40 francs in 1814.

As a result of this energetic support, the annual French fishery was estimated in 1830 to consist of from 300 to 400 vessels, or approximately 50,000 tons and 12,000 men.[12] Of an average catch for five years, 245,000 quintals, 27,000 quintals went to the French West Indies, 17,000 to Spain, Portugal, and Italy, 160,000 to France, 29,000 being reëxported from France. It was claimed that 296 vessels of from 100 to 350 tons were employed in the shore fishery;[13] from Granville there were 116; from St. Malo, 110; Paimpol and Binic, 30 each; Havre, 4; and Nantes, 6—each ship having about 50 men and 5 boys. Each establishment was equipped with cod seines and capelin seines. In the banks fishery

some of the French ships make two voyages . . . carrying the fish back to France to be cured. Others make one voyage to the banks, and when they

12. See R. M. Martin, *History of the British Colonies* (London, 1834), III, 470–474; also John McGregor, *British America* (Edinburgh, 1828), pp. 214–217. The estimates vary but all agree that there was rapid expansion. McGregor gives a total for 1832 of 325 vessels running from 100 to 400 tons, and 14,000 men. These ships were fitted out at St. Malo, Bordeaux, Brest, Marseilles, Dieppe, and Granville. The average from 1823 to 1827 was 319 vessels and 6,413 men taking 20,700,000 kilos, and from 1829 to 1832, 362 vessels, 7,823 men, taking 26,800,000 kilos. Henry Schlacther, *op. cit.*, p. 50. In 1822 bounties on the Banks were increased from 15 francs a man to 50 francs provided the product was dried in Newfoundland and in 1829 to 30 francs if not dried in Newfoundland. The bounties on exports direct to the colonies were reduced to 30 francs and increased to 40 francs indirect shipment in 1822 but the arrangement was reversed in 1832 when direct exports were paid 30 francs and indirect 24 francs with effective results. *Idem*, pp. 48 ff.

13. In 1815 detailed regulations divided ships into three classes: 142 tons and over with 30 men, 90 to 142 tons with 25 men, and 90 tons or less with 20 men. Fishing stations were divided into those with 15 boats, those with 10 to 14 boats, and those with 9 boats or fewer.

complete a cargo proceed with it to St. Pierre . . . where they cure the fish. The principal part of the crews are, in the meantime, employed fishing along the shores in boats; and the fish caught by them makes up the deficiency in weight and bulk occasioned by drying the cargo caught on the banks. Sometimes these ships, if their cargoes are not complete, stop, on their return from the coast, to catch fish on the Banks, which they carry in a wet or green state to France.[14]

The costs of drying were heavy as St. Pierre and Miquelon had no supplies of wood for stages. The population of the islands increased from 800 in 1820 to 1,100 in 1831. In addition to the advantages of bounties "they obtain all their articles of outfit cheaper; the wages of labour are, with them, lower and . . . as well as having the markets of the world open to them, [they have] a great home market." Trawl fishing had been introduced and probably contributed to the expansion.[15]

NEW ENGLAND

You cannot but be aware that the 3rd article of the treaty of Peace of 1783 contained two distinct stipulations, the one recognizing the rights which the United States had to take fish upon the high seas, the other granting to the United States the privilege of fishing within the British jurisdiction and of using, under certain conditions, the Shores and Territory of His Majesty for purposes connected with the Fishery; of these, the former being considered permanent, cannot be altered or affected by any change of the relative situation of the two countries, but the other being a privilege of fishing within the British jurisdiction derived from the Treaty of 1783 alone, was as to its duration necessarily limited to the duration of the Treaty itself. On the Declaration of War by the American Government and the consequent abrogation of the then existing treaties, the United States forfeited, with respect to the Fisheries, those privileges which are purely conventional, (and have not been renewed by a stipulation in the present Treaty). The subjects of the United States can have no pretence to any right to fish within the British jurisdiction or to use the British territory for the purposes connected with the fishery.

LORD BATHURST TO SIR RICHARD KEATS, JUNE 17, 1815

14. McGregor, *op. cit.*, p. 215.

15. The trawl was apparently devised by fishermen from Dieppe. In 1832 Dieppe had from 15 to 20 vessels averaging from 60 to 190 tons. Every man fished with from 15 to 20 lines, each having 120 to 130 hooks, and they were set out every evening and taken up in the morning. The vessels left in March or April and returned in August or September. M. L. Vitet, *Histoire des anciennes villes de France:* Dieppe (Paris, 1833), II, 256–257.

THE difficulties of the French fishery rendered easier the recovery of New England, which had suffered, on the outbreak of the Revolution, both from the effects of the struggle and from the cessation of trade between New England, Newfoundland, and the West Indies. Markets in Europe increased with the decline of the French fishery and the disappearance of the restrictions of the colonial system. On the other hand, new British regulations followed almost a decade of warfare.[16] They were introduced by an act passed in 1783[17] which supported an order in council of July 2, 1783, and restricted trade between the United States and the British colonies to British ships and prohibited American trade in fish with the British West Indies. The market of New England was narrowed and competition from Nova Scotian fisheries encouraged.*

The difficulty of securing united action in measures of retaliation against British policy contributed to the movement for the adoption of the American federal constitution;[18] and, to begin with, it led to the passing of an act which discriminated in tonnage duties against foreign-built and foreign-owned ships. This was followed by attempts to aid in the recovery of the fishery by the payment, in 1789, of a bounty of 5 cents a quintal on dried and 5 cents a barrel on pickled fish when exported, and the placing of a duty of 50 cents a quintal and 75 cents a

16. Raymond McFarland, *A History of the New England Fisheries* (New York, 1911), chap. viii; S. E. Morison, *Maritime History of Massachusetts 1783–1860* (Boston and New York, 1921), chap. x; E. R. Johnson, *History of Domestic and Foreign Commerce of the United States* (Washington, 1915), pp. 157 ff. "Their fisheries being almost at a stand, money very scarce, no market for their lumber or fish. . . ." Joseph Hadfield, *An Englishman in America, 1785* (Toronto, 1933), p. 185. The Cabots at Salem had about 100 vessels "principally employed in coasting the West Indies trade where they exchange fish and lumber for the produce of the islands. They have two vessels from this port engaged in the African trade. Upon the whole Salem may be considered as the second place of importance in the state." Beverly had a standing much like that of Salem. *Idem,* pp. 195–196. Another estimate reckoned that Marblehead possessed one fourth of the tonnage, and Plymouth, Salem, and Beverly controlled one fifth of the fishery. The average earnings of Marblehead vessels declined from $483 in 1787 to $456 in 1788 and to $273 in 1789.

17. G. S. Graham, *British Policy and Canada 1774–1791* (London, 1930), p. 64. The act, modified by orders in council and renewed from year to year, was made permanent in 1788. It permitted trade in British bottoms between the colonies and West Indies, but forbade exports of American meat, dairy produce, and fish.

18. On the significance of Massachusetts to the federalist movement see Morison, *op. cit.,* chap. xii. In 1791 Jefferson issued a report pointing out the decline of, and the importance of assistance to, the fishery. He cited the natural advantages of skilled labor, family employment, low insurance, winter fisheries and year-round occupation, small vessels and a small amount of capital, cheap ships, provisions, and casks. Tonnage duties and the tariff on salt and other supplies involved an estimated burden of $5.25 a man in the season or $57.75 for a vessel of 65 tons. The loss of the Mediterranean markets, duties in foreign markets, and bounties by foreign producers were further handicaps. *The Writings of Thomas Jefferson* (New York, 1854), VII, 538 ff.

*See Notes to Revised Edition, p. 509.

barrel on foreign imports of fish.[19] In 1792 these bounties were abolished and specific allowances were paid on vessels to the extent of $1.00 a ton annually on ships of from 5 to 20 tons, and $2.50 on those of 20 to 30 tons, with the provision that no vessel was to receive more than $170 in all. These rates were increased by one fifth later in the year. In 1799 they were changed to $1.60 a ton on vessels of less than 20 tons, $2.50 on ships above 20, and a maximum of $272. In 1793, vessels were given permits to obtain salt and fishing equipment at foreign ports without the payment of duties.

This support was singularly effective upon the outbreak of war between England and France in 1793 and the issue of proclamations in the various islands of the British West Indies permitting trade in American vessels.[20] Although attempts to evade the unfortunate results of restrictions upon such trade—attempts made both by the West Indies and the United States through proposed negotiations in the Jay Treaty—had ended in failure, the desired result was brought about in part by the exigencies of war. From 1786 to 1790 it was estimated that an annual average of 539 ships totaling 19,185 tons and 3,292 men exported 250,650 quintals of fish, 108,600 valued at $325,800 to Europe, and 142,050 valued at $284,100 to the West Indies. The tonnage of American ships entering British West Indian ports had increased from a total of 4,461 in the three years ending September 30, 1792, to 58,989 in the year ending October 1, 1794. Exports from the United States to the British West Indies had grown from $2,144,638 for the year ending September 30, 1792, to $9,699,722 for the year ending September 30, 1801. The tonnage engaged in the fishery suffered and was put at 50,163 in 1793, at 28,671 in 1794, at 42,746 in 1798, and at 29,000 in 1799. Of a total of 392,726 quintals exported in 1800, 62 per cent or 244,353 quintals went to the West Indies and 144,493 to Europe, of which Spain purchased 76 per cent.[21] The renewal of war between England and France after the Peace of Amiens revived

19. See Lorenzo Sabine, *Report on the Principal Fisheries of the American Seas* (Washington, 1853), pp. 159 ff., and C. B. Elliott, *The United States and the Northeastern Fisheries* (Minneapolis, 1887).

20. See L. J. Ragatz, *The Fall of the Planter Class in the British Caribbean 1763–1833* (New York, 1928), pp. 231–236, for an account of trade between the United States and the British West Indies during the war with France.

21. According to another estimate, of the total New England fishery production about 65 per cent of the value went to Europe and the remainder to the British and foreign West Indies. More than three fourths of the trade was carried on with the money of British merchants, for which credit it was claimed they had not paid more than 12 shillings on the pound. For statistics of the fishery from 1789 to 1833 see *Fishery Interests of the United States and Trade with Canada* (Washington, 1887), I, 311–312; also Sabine, *op. cit.*, pp. 176–177.

earlier difficulties. In spite of numerous efforts, Great Britain refused to grant concessions, and on June 27, 1805, she opened the British West Indies to the products of all colonies or countries in America "belonging to or under the dominion of any foreign European sovereign or state," in any foreign single-decked vessel owned and navigated by persons inhabiting any of those said colonies or countries in America.

The Revolution had been followed by the expansion of other fisheries, e.g., mackerel, herring, sea bass, and by the taking of clams and lobsters. With that had come a growing dependence on local markets. Fishing increased in smaller centers which were inadequately equipped with either capital or experience for engaging in distant trade and using larger vessels; and settlements spread along the coast of New Hampshire and Maine. In 1792 Cape Ann possessed 133 Chebacco boats averaging 11 tons, and some 200 such boats in 1804. Piscataqua employed 27 schooners of 630 tons and 20 boats, with 250 men. In 1791, New Hampshire, including the Isle of Shoals, produced 5,170 quintals of merchantable fish, 14,217 quintals of Jamaica fish, and 6,463 of scalefish, or a total of 25,850 quintals. Plymouth Bay gradually increased its fisheries of cod, herring, and mackerel. In 1802 Wellfleet was employing 25 vessels in these diverse fisheries. Yarmouth and Chatham employed 35 vessels. Whale fishing had been important for New Bedford and Nantucket, and again became active after 1800.

The recovery of the New England fishery and its extension to the Banks, the Labrador, and the Gulf of St. Lawrence were necessarily slow because vessels of larger size were needed. It was expedited by bounties and by markets in the West Indies and the Mediterranean. In 1792 American vessels were described as taking "unwarrantable liberties on the [Labrador] coast," and as driving British fishermen from the fishery. In 1797, 35 American vessels were engaged in drying fish on the Magdalen Islands, each with two crews, one for taking and the other for drying the fish; and they were said to have recently discovered the St. George Bay bank in Newfoundland. In 1804 American vessels were reported on the Labrador,[22] mostly schooners of 45 tons from Cape Cod, Plymouth, and Boston. It was estimated that New England sent 300 ships and 10,600 men to the Gulf of St. Lawrence. In 1805 "there were not less than 900 sail of American vessels engaged in

22. They arrived early in June and left about September 10. The fish were caught from ninety to one hundred miles out and brought in to be dried on shore, the better grades being sent directly to Alicante, Leghorn, Naples, Marseilles, and the Mediterranean generally, and the poorer grades to the West Indies. The owners usually subscribed one third; and the share system—in which the crew owned half the fish—brought from $280 to $300 to each fisherman in a successful season, or wages of from $16 to $20 a month.

trading and fishing upon our shores from Davis Straights thro' the straights of Belleisle and up as far as the isle of Anticosti."[23] Provincetown, in 1790, had 20 ships; in 1802, 33, of 1,722 tons; and in 1807, 62, of from 38 to 162 tons. They fished on the Banks, on the Labrador coast, and in Chaleur Bay, taking a yearly average of 33,000 quintals. The fish were cured either at the place where they were taken or in Provincetown itself, and thence exported, chiefly to Spain and Portugal. Newburyport sent its first vessel to Labrador in 1794, and by 1806 its fleet had increased to 45. By 1808 fishing towns between New London on the Thames and Schoodic on the Maine coast were sending large numbers of ships of various sizes through the Strait of Canso to the Gulf of St. Lawrence. The smaller fish of Gaspé and the Labrador, which had been of importance to France in the Mediterranean market, and the large fish for domestic consumption and the Spanish market were being caught by New England. From the low ebb of 1789 Marblehead gradually improved and, after 1800, in the winters it exported large quantities of cod to France, Spain, and the West Indies in ships which had fished during the summer. Salem, which had been largely interested in trade, sent to the Banks about 20 fishing vessels aggregating some 1,300 tons and 160 men. Though sending from 45 to 60 ships to the Banks in 1788–89, Gloucester's fleet, on the other hand, declined by 1804 to 8 of over 30 tons. It was estimated that between 1790 and 1810 about 1,232 vessels went annually, an average of 584 to the Banks and 648 to Chaleur Bay and Labrador. The bankers, with an average total tonnage of 36,540, and employing 4,627 men and boys, made three "fares" or trips a year, used 81,170 hogsheads of salt, took 510,700 quintals of fish of a value of $6 a quintal in foreign markets, and extracted 1,752 barrels of oil worth $10 a barrel. Excluding salt, outfitting involved a cost of some $900 which, with the vessel, made a total of about $2,900. Ships that went to Chaleur Bay and Labrador in those years averaged a total of 41,600 tons and carried 5,832 men and boys. They made one fare a year. They brought home an average of 648,000 quintals of fish worth $5 a quintal and 20,000 barrels of oil. This involved an outlay for 97,200 hogsheads of salt, and in addition one of their ships would cost some $1,600 and its equipment $150.

The fishery[24] suffered a serious blow from the Embargo Act of 1807, and in 1808 and 1809 American vessels on the Labrador and Newfoundland coasts declined to small numbers. The total tonnage dropped from 69,306 in 1807, of which Marblehead had 21,068, to 34,486 in 1809;

23. D. C. Harvey, "Uniacke's Memorandum to Windham, 1806," *Canadian Historical Review,* March, 1936, p. 52.
24. Morison, *op. cit.,* chap. xii.

and by 1814, as a result of the war, to 17,855. The average tonnage increased from 34,024 in the decade 1789–98 to 48,208 in 1799–1808, but declined to 40,071 in 1809–18. Tonnage increased to 64,807 in 1817, and to 69,107 in 1818. Exports of dried cod rose from an average of 394,198 quintals between 1790 and 1798 to an average of 438,453 from 1799 to 1808, and declined to 200,437 from 1809 to 1818. In 1807, of a total of 473,924 quintals, 56 per cent or 268,332 went to the West Indies, and 192,981 to Europe, France and Spain sharing almost equally. In 1816, of a total of 219,991 quintals, the West Indies again took 56 per cent; and, of a total of 89,192 quintals sent to Europe, France purchased 45 per cent. Exports to France rose to an important position from 1807 to 1816 as a result of the difficulties of the French fishery in wartime. It was estimated that about $9,000,000 worth of pickled fish and more than $49,000,000 worth of dried fish were exported between 1789 and 1818, and that, between 1791 and 1818, 73,928,614 bushels of salt were imported from Maia, Lisbon, and Turks Island. In 1800 there were 136 saltworks in the vicinity of Cape Cod.

The Treaty of Ghent, in 1814, omitted all reference to the fisheries because of the opposition of Clay, representing the West, to British navigation of the Mississippi, and because of the insistence of Adams on the retention by the United States of former fishing privileges. The increasing importance of the interior split the delegation and weakened its fisheries stand accordingly.[25] The Convention of 1818 granted fishing rights to American fishermen on the southern coast of Newfoundland between Cape Ray and the Ramea Islands, on the western and northern coasts between Cape Ray and the Quirpon Islands, in the waters surrounding the Magdalen Islands, and on the coast of Labrador from Mount Joly east to the Straits of Belle Isle and north on the Labrador coast. The United States renounced "any liberty heretofore enjoyed to take, dry and cure fish on or within three marine miles of any of the coasts, bays, creeks or harbours . . . Provided however that the American fishermen shall be admitted to enter such bays or harbours for the purpose of shelter and of repairing damages therein, of purchasing wood, and of obtaining water, and for no other purpose whatever."[26] Moreover, they were forbidden to fish, after settlements had become established, on those portions of the coast line on which fishing

25. Elliott, *op. cit.*, pp. 50–57; also Sabine, *op. cit.*, pp. 161 ff.; and John Quincy Adams, *The Duplicate Letters, the Fisheries and the Mississippi* (Washington, 1822).

26. *Proceedings of the North Atlantic Coast Fisheries Arbitration* (Washington, 1912), IV, 52–55; C. E. Cayley, *The North Atlantic Fisheries in United States Canadian Relations,* doctor's thesis, University of Chicago, 1931.

rights were granted. Various rights were left undetermined, but the shore fishery had been definitely forbidden to Americans.

The Convention of 1818 was offset by renewed encouragement. A rapid postwar recovery[27] was assisted by a tariff, imposed in 1816, of $1.00 a quintal on smoked and dried fish, $2.00 a barrel on salmon, $1.50 on màckerel, and $1.00 a barrel on other varieties of pickled fish. Exports of pickled fish cured with foreign salt were allowed 25 cents a barrel rebate. Bounties on the cod fishery were increased in 1819 to $3.50 per ton for vessels of from 5 to 30 tons employed four months in the fishery; for ships of more than 30 tons, $4.00 a ton; and, for such ships, if employing not less than ten persons for three and a half months, $3.50 per ton.

The recovery of the French fishery after the Napoleonic wars, the disappearance of the French market for New England fish, and increasing competition from Norwegian fish in Spain reduced the importance of European trade. Cod was exported chiefly to the West Indies and Surinam. The growing value of a protected home market, especially in the southern states, led to increasing attention to other species of fish. In 1804 an inspection act was passed by Massachusetts to cover the marketing of pickled fish, and in that year 10,000 barrels of mackerel had been taken. With the introduction of the "mackerel jig,"[28] hand-line fishing meant an increase to 297,986 barrels by 1827, and to a record catch of 450,000 barrels in 1831. In 1830 nearly 900 vessels were engaged in this fishery.[29] Large-scale methods of operation were introduced and adapted to fishing for other varieties. In 1831 Americans were reported to be fishing for herring off the Magdalen Islands, and, for the first time, with nets. Gloucester vessels began to fish for cod and halibut on Georges Bank in 1821, and had a substantial fleet by 1833.

27. See McFarland, *op. cit.,* pp. 134 ff.

28. "This method of taking Mackarel was invented by the fishermen of Massachusetts, about twenty years ago, and has since become a great source of wealth to that country, it employs nearly two thousand vessels, and a proportionate number of men. The Mackarel is often seen in great plenty when it will take the hook; the voyage is similar to that of a whaling voyage, and requires a great deal of cool rigid perseverance. Vessels are often four and five weeks cruising among large shoals of this fish without taking any only with the gaff, when of a sudden it will bite so eagerly, that fleets of five and six hundred sail of vessels got their loads in a few days, and it is equally uncertain where and when to find it in the humour for the hook. But [it] has hitherto been confined to different parts of the entrance of the Bay of Fundy, between Cape Sable and Nantucket and in the Bay Chaleur." *Journals of the Assembly, Nova Scotia,* 1834, Appendix 31, p. 40. Bait-cutting machines were introduced in 1824.

29. Sabine, *op. cit.,* pp. 178 ff.; also O. E. Sette and A. W. H. Needler, *Statistics of the Mackerel Fishery off the East Coast of North America 1804–1930,* Investigational Report No. 19, Bureau of Fisheries (Washington, 1934); W. G. Pierce, *Goin' Fishin,' the Story of the Deep Sea Fishermen of New England* (Salem, 1934).

An expanding domestic market enabled the New England fishermen to return to Chaleur Bay and the Labrador. In 1824 large numbers of fishing vessels[30] from the United States were there reported. In 1829, on the Labrador, of a total of 2,108 vessels and 24,110 men who brought back 1,773,000 hundredweight of fish and 17,730 hogsheads of oil, it was estimated that the United States had 1,500 vessels and 15,000 men, their returns in fish being 1,100,000 hundredweight, and in oil, 11,000 hogsheads. The estimate was probably high but the expansion had been great.

30. Schooners reported at Beverly with fish from Chaleur Bay included the *Romp,* with 75,000 fish; the *Hope,* with 61,000; the *Girl,* 55,000; the *Angler,* 48,000; the *Active,* 52,000; the *Elizabeth and Rebecca,* 60,000; and the *Pelican,* 62,000. *Acadian Recorder,* October 20, 1824. At Marblehead, the *Osprey* was reported with 20,000; at Newburyport, the *Mary Ann* from Labrador with 80,000; and several others. *Montreal Gazette,* September 22, 1824. In 1827 it was claimed that more than a thousand American vessels went to the Labrador and Newfoundland. See W. G. Gosling, *Labrador* (Toronto, n.d.), pp. 373–374.

CHAPTER IX

THE RISE OF NOVA SCOTIA, 1783–1833

TO THE CONVENTION OF 1818

You will further perceive, by the Commercial arrangements which have since taken place between the two Countries, that our Parent State evinces a determination to prevent all foreign interference with the welfare of her Colonies. The British-North-American Provinces will, consequently, be enabled to supply our West-India Islands with fish and lumber, without the dread of any competition from their American neighbours in these branches of commerce. Prospects so encouraging will, I doubt not, be taken advantage of by the industries and enterprising inhabitants of Nova Scotia.

Address of the Lieutenant-Governor, Nova Scotia, 1816

With the disappearance of New England from the British Empire, Nova Scotia secured the advantages of the British commercial system. The fishing industry became the source from which commercial interests emerged and insisted on realignments. It continued to stimulate trade between old and new civilizations and between temperate and tropical zones by the cheapest type of navigation and in ships which meant the growth of industry. The struggle with the West Indies carried on by New England in the old empire was continued by Nova Scotia in the new. Her aggressiveness showed itself in a realignment of political structure, in the support she gave to Newfoundland's increasing population, and in the disappearance of the West Country fishing ships. She occupied a frontier position in the correction of the maladjustments which had wrecked the old empire and paved the way for the solidarity of the new.

New England had the advantage of possessing markets in Europe and the foreign West Indies in consequence of the difficulties of the French fishery. In spite of attempts to restrict the trade of the British West Indies to Nova Scotia, New England had access to them at various intervals and, with the recovery of France and the increasing importance of Norway after the Napoleonic wars, she succeeded in gaining complete access. The increasingly active commercialism of Nova Scotia had opposed the admittance of American fish to the British West Indies. It attempted to restrict the New England fishery, to develop an entrepôt trade of American products to that market, and to develop the Nova Scotian fishery and trade to the St. Lawrence. Compelled to meet competition from a region no longer shackled by the colonial system, it pressed for a revision to meet new demands. While New England had failed to offset the West Indian planters' influence in Great Britain, Nova Scotia, with the stimulus of New England experience and aggressiveness, succeeded and thereby contributed to the evolution of the second British Empire.

Restrictions on trade from the United States had intensified the difficulties of the British West Indian planters.[1] Cheap, bulky commodities such as lumber, commodities perishable in a warm climate such as flour, and fresh provisions and livestock could not be handled save in short voyages and by frequent and regular shipments; and attempts to reduce such imports from the United States had serious consequences.*A decline in the prices commanded by tropical products in Great Britain, with increased imports after the American Revolution, and a continuance of heavy customs duties were added to difficulties arising from the increased cost of supplies and of slaves.[2] Later, revolution in San Domingo[3] and war with France resulted in a scarcity of tropical products and raised prices of sugar in Great Britain in the period from 1792 to 1796. These conditions stimulated imports of East Indian sugar which, from 3,839 hundredweight in 1792, rose to 220,836 hundredweight in 1800; and, upon the adoption of Bourbon and Tahitian varieties of cane, there followed an expansion of sugar production in the British West Indies. Tobago, lost to the French in 1783, was recaptured in 1793; and Martinique, Guadeloupe, and St. Lucia were reduced in 1794. Then, under the stimulus of high prices, foreign areas such as Brazil and the neutral West Indies exported larger quantities of sugar to Europe, carried by American shipping, and without the burden of British duties. A drop in prices from 1800 to 1802 was followed by the Peace of Amiens (1802), the restoration of acquired territories except Trinidad, and an improvement in the market for sugar in Great Britain. But exports to Europe from Cuba and foreign territory, with the assistance of the American carrying trade, increasing competition from East Indies sugar, the exporting of foreign sugar to Great Britain in British bottoms after the opening of free ports in the British West

1. See L. J. Ragatz, *The Fall of the Planter Class in the British Caribbean 1763–1833* (New York, 1928), chap. vi *et seq.*

2. See H. C. Bell, "British Commercial Policy in the West Indies 1783–93," *English Historical Review,* July, 1916, pp. 429–441, for material suggesting that the burden was not a heavy one for the West Indies. Efforts to become less dependent on the United States were evident in the expedition to obtain breadfruit in Tahiti, which was marred by the incidents leading to the mutiny of the *Bounty.*

3. For an account of the connection between the French Revolution and the revolt in San Domingo, led by Toussaint L'Ouverture, and the movement toward the abolition of slavery see W. E. B. Dubois, *The Suppression of the African Slave Trade from Africa to the United States of America 1638–1870* (New York, 1896), chap. vii.

*See Notes to Revised Edition, p. 509.

Indies, and, finally, the abolition of the slave trade in 1807 meant debt and despair for the planters. The renewal of the war with France, the decline of the European market upon the introduction of the continental system, and increasing duties in Great Britain to support war finance made further difficulties for the British West Indies; and protests led to the relaxing of restrictions on American shipping to the islands, which had been imposed under the pressure of the agitation of British shipping interests in 1804. The effects of the American embargo in 1807, the War of 1812, and the growing cost of supplies were offset in part by the collapse of the continental system, growing exports to Europe, and increasing prices up to 1815. But, again, the addition of St. Lucia, Tobago, Demerara, Essequibo, and Mauritius to the British Empire in 1814 caused fresh competition. The continuance[4] of the war duties, competition from the slave-produced sugar of Brazil and Cuba, the labor problems resulting from the cessation of the slave trade, reprisals, and the difficulty of obtaining supplies from the United States forced the fortunes of the British West Indies to the low point which preceded the Colonial Trade Act of 1822.

Production in Brazil expanded with the extension of the New England trade that had been excluded from the islands. Direct trade from England to Brazil had increased along with the expansion of Brazilian cotton production after the American Revolution. The flight of the King of Portugal to Brazil in 1807 had its part in bringing on the collapse of Portuguese commercial monopoly, which came to an end with the Brazilian Declaration of Independence in 1822. The Monroe Doctrine, coinciding with this development, was offset by the colonial trade acts by which Huskisson opened South American trade to Great Britain and British North America.[5]

The political influence of the West Indian planters ceased to be effective.[6] It declined because of the troubles of the planter class, the trend toward free trade, the rise of humanitarianism, and the increasing power of the East Indies and British North America. Mauritius was placed on a·basis of equality with the West Indies in 1825, and the

4. Ragatz, *op. cit.,* chaps. x, xii.

5. C. R. Fay, "South American and Imperial Problems," *University of Toronto Quarterly,* January, 1932, pp. 183–196; also Herbert Heaton, "When a Whole Royal Family Came to America," *Canadian Historical Association Report, 1939.*

6. See Lilian M. Penson, *The Colonial Agents of the British West Indies* (London, 1924), chap. xii, for the part played by the abolition of slavery in the downfall of West Indian influence; also L. M. Penson, "The London West India Interest in the Eighteenth Century," *English Historical Review,* July, 1921, pp. 373–392; also Helen T. Manning, *British Colonial Government after the American Revolution, 1782–1820* (New Haven, 1933), pp. 54–56.

preference, introduced in 1816, which favored the West Indies as against the East Indies was lowered in 1830. The struggle for the emancipation of the slaves and its success in 1834 were made easier by the demands of the East Indian planters for equality of treatment with those of the West Indies. As a result of the American Revolution, United Empire Loyalists and others had migrated from New England to Nova Scotia,* and attempts to develop the provinces of British North America as a source of exports to the British West Indies to replace the exports of the United States meant longer and fewer voyages, greater charges, higher labor costs, and, for the British West Indies, paying more for supplies.

In British North America more fish and lumber were produced and they were sent to the British West Indies under the stimulus of higher prices and the restrictions on American trade.[7] "If the New England traders could find a profit in sending their vessels to this coast for fish, those who inhabit its borders can carry on the business to much greater advantage." The exclusion of the United States from the British West Indies contributed to the expansion of shipbuilding, lumbering, and fishing. It was claimed that New Brunswick built 93 square-rigged vessels and 71 sloops and schooners in the decade following 1783. With the migration of the Loyalists and others, Shelburne had a population of 12,000 and several fishing vessels;[8] and whalers and fishermen were settling at ports along the Atlantic coast.[9] In 1787 Canso[10] was of "more real value and consequence . . . to Great Britain as a nursery for seamen and fishing than all the remaining coast of Nova Scotia"; and, in

7. See Ragatz, *op. cit.*, pp. 184–188, for the trade statistics.

8. Shelburne had 4 sail in the cod fishery in 1784, and 10 in 1785 as well as several boats, but its success was of limited duration.

9. G. S. Graham, "The Nantucket Whale Fishery," *New England Quarterly*, June, 1935, pp. 179–202; Margaret Ells, "The Dartmouth Whalers," *Dalhousie Review*, April, 1935, pp. 85–95. [See also Notes to Revised Edition, p. 509.]

10. From April 9 to July 9, 1789, over fifteen small vessels of tonnage up to 60 from Halifax; five from Saint John, New Brunswick; and one each from Liverpool, Shelburne, and two or three other ports, arrived at Canso with provisions, British merchandise, shipbuilding material, and salt. Two brigs belonging to Guysborough of 84 and 122 tons returned, the first with 1,952 bushels of salt, and 25 puncheons of rum, 6 hogsheads of sugar from Turks Island, the second with 4,000 bushels of salt from Sicily. The mackerel fishery was of first importance, only a small vessel going to the Banks. Vessels carried furs and skins and small quantities of timber, wheat, and oats to Halifax. During the same period vessels passed through the Gut of Canso for Prince Edward Island from Halifax, New London, Boston, and Barnstable; for Miramichi from Halifax; for Chaleur Bay from Halifax; for Bonaventure Island from Halifax, New London, Salem, and Rhode Island; for Cape Breton from Cape Cod, and Stonington, Connecticut; for the Magdalen Islands from the Jersey Islands; and for Quebec from Bermuda with flour and Indian corn. These vessels carried chiefly salt and provisions. Vessels not examined included 124 British and 125 American entering and 80 British and 39 American leaving. Canadian Archives, *N.S.A.*, 115.

1790, 37 vessels arrived and 27 cleared. The fishery flourished at Liverpool, Cape Sable, and Poboncour. In the years following 1783 Simeon Perkins sent lumber from Liverpool to Port Roseway, Port Mouton, and the newly settled areas. Vessels left for the West Indies with lumber and fish. The fish were shipped at Liverpool or purchased at points along the Atlantic coast, and the lumber came from Frenchman's Bay and kindred points; later from the Bay of Fundy and New England.[11] Vessels returned either directly with salt, molasses, and rum, or, after calling at American ports, with flour, corn, and provisions. Schooners proceeded in the spring to Newfoundland and the Labrador for salmon, in spite of difficulties with the French and Quebec authorities. In the summer they went for mackerel to Crow Harbor, Margaret's Bay, Deep Cove, and the South-west Harbor, and to Prospect for herrings, chiefly for export to the United States.[12] The flexibility which characterized the New England fishery became more evident in Nova Scotia. Ships were built on shares, chartered for short periods, were available for freight or for the owner's cargo, and were loaded by the owner, either individually or in partnership. The divisibility of both vessels and cargoes and the relatively small size of the ships made it easy for merchants and others along the coast to take a very extensive part in trade.

The character of the commercial organization meant having to depend on a large central port such as Halifax[13] for supplies of European goods that were brought in large vessels, though she was subject to competition from other central ports such as Boston. "Very little has been done in the fisheries this year [1784] and the fisheries must be the source of wealth to this place [Halifax]. Unless they prosper this must decline. . . . Three or four vessels are out upon the whaling business. . . . The cod fishery has not been very productive."[14] The total exports of the province in 1789 were estimated at 20,000 quintals of cod at, say, 12 shillings; 10,000 barrels of mackerel, salmon, and herring at, roughly, 20 shillings; 1,500 barrels of whale and other fish oil at about 30 shillings; and 10,000 pounds of whalebone at about 2*s.* 6*d.* In 1792

11. For a discussion of trade regulation, see Manning, *op. cit.*, chap. ix.

12. The act prohibiting trade with the United States except under a license from the governor was regarded by Perkins as serious for Liverpool, "particularly those that carry on the salmon and mackerell fishery as Boston and other parts of the U. S. is the best market for them articles at present." May 10, 1788. The references are from Perkins' diary. See page 186.

13. "The township of Barrington . . . was settled by people from Cape Cod and Nantucket in Massachusetts, who emigrated there with a view to carrying on the fishery. Previous to the American Revolution, Boston was to them what Halifax now is to the present generation, there they got their supplies—and that was the home market for their fish." *Acadian Recorder,* March 24, 1827.

14. Walter to Shelburne, November 20, 1784, Canadian Archives, *N.S.A.*, 106.

Halifax cleared outward for the West Indies 6,489 tons and inward 6,571 tons.

The position of Nova Scotia during the American Revolution and the increasing importance of Halifax intensified the difficulties of administration in the maritime region, so characterized by diversified development. Military and colonial policy[15] strengthened the trend toward independent development, as the Maritimes were now divided into Nova Scotia, New Brunswick, Cape Breton, and Prince Edward Island. Commercial interests[16] which had migrated from Great Britain to Newfoundland and Nova Scotia sought to increase the trade of Halifax and to restrict the trade and fishery of the United States. In 1786 bounties[17] were paid by Nova Scotia for one year on ships of 40 tons and over. They were renewed in 1787, and paid on vessels of more than 75 tons in 1788. To control American encroachments on the fishery and on trade, George Leonard was appointed superintendent of trade and fisheries in May, 1786, and four deputies were added in 1789.[18] Parr, writing to Evan Nepean in July, 1787, said, "I wish to shut the rascals quite out of our coast but am afraid to get into a scrape owing to that part of the treaty which says they may fish with consent of the inhabitants etc.; that consent they will ever get in some parts of this extensive coast and to prevent smuggling is impossible."[19] Hundreds of American vessels were reported to be taking advantage of this interpretation of the treaty. Many paid fees to enter and clear and, under the pretense of landing salt and provisions for their own use, they sold large quantities of American products, especially rum, and bought Nova Scotia fish. In 1787 Leonard gave orders that fish caught by United States vessels should be brought to shore in vessels belonging to the King's subjects and cured with the assistance of the crews of those vessels. Permits to fish were granted on payment of two dollars to the masters of shallops "built in this province," if the master took the "oaths of fidelity and allegiance to His Majesty King George the Third." A petition submitted on August 8, 1788, asked for a ship and small shallops to check the encroachments by American fishermen and prevent the sale of fish and rum in the creeks where "most part of the

15. See Marion Gilroy, "The Partition of Nova Scotia, 1784," *Canadian Historical Review,* December, 1933, pp. 375–392. Government, according to Lord Durham, was facilitated "by means of division to break them down as much as possible into petty isolated communities, incapable of combination, and possessing no sufficient strength for individual resistance to the Empire." Also Manning, *op. cit.,* pp. 35–36.

16. The Forsyth interests centered in Greenock, and Foreman Grassie and Company were supported by Brook Watson, William Goodall, and John Turner of London.

17. A. T. Smith, "Transportation and Communication in Nova Scotia 1749–1815," master's thesis, Dalhousie University.

18. Manning, *op. cit.,* p. 261.

19. Canadian Archives, *N.S.A.,* 109.

improper traffic is carried on." At Liverpool, in the following year, the drying of fish in the harbor by American fishermen was prohibited, and they were allowed in only under special consideration. A complaint in March, 1789, claimed that Americans secured their fish at a lower rate than the fishermen of Cape Breton, "and by their illicit practices forestal us even in our own markets. Their outfit in provisions (the chief expense) is obtained at least thirty per cent cheaper than ours, and they barter their fish at the neutral islands in the West Indies which [*sic*] are immediately smuggled into our islands to the great disadvantage of the British traders." In 1792 new complaints were made that American fishermen came in for bait and shelter and engaged in trade along the western shore and throughout the province. Twelve sail of Americans sent men to dig clams for bait in the port of Le Bear. In the same year the lack of a road to Halifax was held responsible for the purchase by Americans of fish and provisions at Pictou. In 1793, 40 or 50 vessels near Annapolis were charged with throwing offal overboard, "which is not allowed by our own vessels."

The desire to restrict the American fishery was one expression of the desire to restrict American trade. In 1784 merchants and farmers petitioned against the importation of New England provisions, but the petition was ignored because of the possible injury to fishing and lumbering. In 1785 imports of lumber and provisions from the United States were prohibited, but in 1786 livestock, provisions, and lumber were admitted in British vessels manned by British subjects. To secure revenue with which to construct lighthouses, check smuggling, and restrict trade,[20] an act was passed in 1793 imposing duties on foreign ships that entered the harbors. To secure money to meet the demands of the public debt and further to restrict trade, a provincial act imposing duties on goods imported into Nova Scotia (32 Geo. III, c. 13) was passed, and it was not disallowed.

The advantages possessed by New England in the Pacific and European trade and arising from the recovery of her fishery became effective, and efforts by the Assembly to check smuggling were not conspicuously successful. In 1797 it was stated in the House

20. A Petition of William Kidston and others of Halifax, Merchants and Traders, was presented by Mr. Cochran, stating "that great injury is occasioned to the Commerce and Revenue of this County by Strangers and transient Persons bringing with them into Halifax, and other Parts of the Province, Articles of Merchandize for Sale, and interfering with the established Traders of the Town and Province; and praying the House would pass an Act for laying a Duty on all Articles of Merchandize imported for Sale by Strangers and transient Persons into this Province, or grant them such other Relief in the Premises as to them may seem meet." *Journals of the Assembly, Nova Scotia,* June 13, 1792, p. 136; see Manning, *op. cit.,* p. 251; also *A Calendar of Official Correspondence and Legislative Papers, Nova Scotia, 1802–15, Public Archives of Nova Scotia, Publications No. 3* (Halifax, 1936), p. 303.

that the greatest Part of the Settlements on the whole Coast of the Province are, with very little Interruption from the Officers of his Majesty's Customs, continually and almost wholly supplied with India Goods, Articles liable to Duty by the Laws of this Province, and with all other Articles of Merchandize from the United States; either by small Vessels owned in this Province, or by Fishing Vessels belonging to the said States to the Ruin of the Trade with the parent State and her Colonies, as our Merchants are unable to form any probable Estimate of the Quantity of Goods they may safely import from Great Britain, or the British Plantations. They are also deprived of the Productions they have contracted to receive for any Supplies by them entrusted to the Planters on the Coast, as those Planters generally barter them for those Articles so clandestinely and illegally imported.[21]

The Council, in opposition, argued that "the high Provincial Duty of Ten per Cent. and the other Restrictions under which the American Trade labours, [were] the principal Cause of [the] smuggling complained of; and therefore they esteem it a preferable Measure to lessen the Duties, thereby to render the Gains of the contraband Trader less, and to place the fair Trader on a better footing in the Market."[22] At the same time the House of Assembly introduced legislation restricting the size of vessels engaged in American trade to 60 tons and upward, but this was rejected by the Council on the grounds that it would

greatly reduce in value the vessels under sixty tons burthen owned in this province, and employed in the trade with the States and would render it im-

21. *Journals of the Assembly,* July 6, 1797. A petition signed by Lawrence Hartshorne and Company, William Forsyth and Company, and a number of others, "the principal merchants of the town of Halifax," stated:

"That the Petitioners have been accustomed to make annually from Great-Britain, large Importations of British Manufactures and other Merchandize for the Consumption of the Inhabitants of this Province, upon which they have paid a Provincial Duty of two and a half per Cent. besides a particular Duty upon Teas and other specific Articles.

"That the Merchants of the United States of America, trading directly with the different Foreign Ports in Europe and in the East-Indies, avoid many Charges and expences, which the Petitioners trading wholly with Great-Britain are subject to; and are therefore able to undersell the Petitioners in all Articles of the Produce or Manufactures of those Countries.

"That from these Causes, as well as by evading the Provincial Duty above mentioned, there have been for several Years past great Quantities of East-India and other Goods fraudulently smuggled into the Province from the United States, more especially into the Western Ports and the Harbours in the Bay of Fundy; whereby the Demand of those Articles from the British Importer is diminished; the Trade with Great-Britain decreased; the British and provincial Merchant injured; and the Revenue of the Province greatly defrauded.

"That the Petitioners therefore confiding in the Wisdom of the House and their Disposition to encourage the fair Trade of the Province, and to discountenance and detect Smugglers, respectfully pray that the house would take into consideration so serious an evil and devise such remedy for it, as to the House shall seem fit." *Idem.*

22. *Idem,* June 30, 1797.

possible for the settlers in many of the harbours of this province (which from the shallowness of the water will not admit of vessels of sixty tons) to carry as their practice is, their fish to the States in small vessels and to bring from thence the necessary supply of bread corn for the winter.[23]

It was also felt that such legislation was beyond the jurisdiction of the province. In 1800 a proposed reduction[24] of three pence a gallon on the duties on spirits and wines was favored by the Council but opposed by country members of the Assembly. The Council in turn opposed expenditures on roads. Pleas were made for additional duties on molasses to check indirect trade through the United States.[25] The *Earl of Moira* was sent to the Gulf of St. Lawrence to protect the fishery and to restrict smuggling.[26] In 1800 two craft from the United States were seized. The effect of opposition to New England was slight.[27]

23. *Idem.* Later, the Council opposed and the Assembly favored the schooner interest.

24. From a speech of the lieutenant governor in 1800: "The losses which the Merchants have sustained by the capture of Vessels employed in the Foreign Trade, and the Embarrassments to which our Commerce and Fisheries have become liable from the deficiency of circulation Money, may make it expedient to lessen the Duties on Spirits and Wines, and the state of our Public Funds perhaps renders the present time the most proper for such Reduction." *Idem.* Andrew Belcher and other importers of British and West Indian goods, pleaded that they "felt most sensibly difficulties arising from the Capture of their Vessels, the general Suspension of Trade, and the great scarcity of Specie, and many of them having large Stocks of European and West India Goods on hand, which they cannot dispose of, beg leave to represent to this House that it is with extreme Difficulty they can raise the Sums of Money necessary for the Payment of the Duties on these Articles." *Idem,* March 12, 1800.

25. Forsyth, Smith and Company and other merchants and traders in the town of Halifax, in a petition to the House of Assembly: "Large Quantities of Molasses have been lately imported from the United States of America, in Order to be sold and consumed in this Province, which naturally Effects the Interests of the Petitioners as they are concerned extensively in the Export and Import Trade from this Country to the West India Islands, and praying, that an additional Duty of Impost and Excise may be levied on the Importation of Molasses from the United States of America." *Idem,* March 3, 1800. See also a petition by James Fraser and other merchants of Halifax complaining of "transient persons" and "illicit trade." *Idem,* March 6, 1802. The speech of the lieutenant governor in 1800 referred to the great disadvantages to Nova Scotia "arising from the present intercourse between the American states and his Majesty's West India islands which has almost annihilated the fisheries."

26. "Greatly to the benefit of fair traders, who are now exceedingly disappointed in their remittances of fish oil, peltry and furs, which are clandestinely obtained by these adventurers for present supplys to the people on the shores, who have previously taken credits from our merchants for their support in the preceding winter and outfits in the spring and by these means are unable to pay their debts; of course the merchants there are so far prevented in making their remittances to Great Britain." Lieutenant Governor Wentworth to George Portland, July 23, 1800, Canadian Archives, *N.S.A.,* 132.

27. From a letter written December 18, 1800: "I am informed that numbers of American vessels come from the States and carry away a great part of the fish caught by the inhabitants and in return for it, leave contraband articles of various sorts, to the detriment of His Majesty's revenue and of the fair trader." At Passa-

The difficulties of the fishery could be followed in the fortunes of Simeon Perkins at Liverpool. From that port schooners had extended trade to Quebec, Newfoundland, Fayal, and Great Britain. As for the coasting trade in provisions, lumber, plaster of Paris, and fish, it had been extended to the Bay of Fundy, Lunenburg, Halifax, Arichat, Miramichi. Chaleur Bay, Gaspé, the Magdalen Islands, and Labrador. A friendly society of merchants was formed on March 10, 1795, which apparently was able to set the price of fish on August 4, 1801. Many salmon-fishing voyages were made to Newfoundland and particularly to the Labrador where rivers, for example, the Natashkwan and St. Paul's River, were leased from Quebec interests.[28] Mackerel were taken at Crow Harbor and Fox Island, and herring at Harbor Bouché. An establishment for the taking of fish and the purchase of fish and lumber was acquired by Perkins through a partnership at Port Medway. Quantities of pickled fish were sent to Boston, New York, Philadelphia, and North Carolina in exchange for provisions and marine supplies such as pitch and tar. Other vessels carried dried fish and lumber to the West Indies to be exchanged for sugar, molasses, and rum; and in particular they went to Turks Island for salt. Lumber provided a balanced cargo for the West Indies when shipped with exports of fish in the fall, at the end of the season, and was likewise a good cargo for the slack season in the spring when sent to Halifax, Newfoundland, and the West Indies. Coal, and cattle and sheep from Nova Scotia or New England were carried to Newfoundland. On the outbreak of war with France and Spain, fish were sent to the United States, even by the Channel Islands merchants at Arichat. There was competition from Newfoundland fish, which had been excluded from the European market.[29] From 1798 to

maquoddy, Americans were smuggling tea, textiles, rum, brandy, and gin to small seagoing vessels carrying plaster of Paris, lumber, grindstones, and other products to be reshipped in larger vessels for American ports. "Their illegally purchasing fish taken by British subjects in small boats or from the drying grounds and thereby becoming the exporters of that article to foreign markets . . . must evidently tend to the disadvantage of Great Britain by preventing the fisheries being conducted on so extensive a scale as they would otherwise in that quarter and in [due] course in some measure check her nursery for seamen in these colonies." Leonard to Portland, November 10, 1800, Canadian Archives, *N.S.A.*, 132.

28. On August 1, 1800, Perkins complained of interference from Lymburner and Crawford on St. Paul's River. In that year he secured a lease of the river for three years for £500 from Lloyds of Quebec. Other statements from Simeon Perkins' diary follow.

29. On August 4, 1801, a schooner from Newfoundland reported that the price of Madeira fish had fallen from 18 shillings to 13 shillings and that 100,000 quintals had been shipped to the West Indies and the United States. On December 9 a schooner from Boston reported that news of peace "made a stagnation of business and dumped the sales of fish, and in some measure the produce of the country; flour and corn, had fallen a little." *Idem.*

1801, fish brought low prices in the United States and the cost of supplies was high.[30] Perkins could write on August 7, 1801: "The prospect of privateering is very gloomy and every other branch of business has the same aspect. Fish of every kind I fear will decline in price. Codfish have already fallen in all the markets, salt and provisions remain very high, the fishermen on this shore have made a poor hand of it." On August 17 a schooner from Labrador reported "very poor traiding by reason of a number of vessels being on the coast on the same traiding voyage and a vessel load of bread and flour from Quebec. Add to that, the principal gentleman, viz. Mr. William Smith, is dead."

With the temporary end of the war with France and the publication of the preliminary articles of peace on October 1, 1801, the outlook for markets in the West Indies led to "a great spirit in the people for building vessels," and the result was the construction of many new ships. On November 11, 1802, an insurance broker began to transact business in Liverpool. The fishery expanded and trade increased. But the renewal of the war with France brought the boom to a sudden end. Perkins wrote, on July 4, 1803:

> It happens very unfortunate for the country that we are involved in a war with France, the settlement in particular will suffer much as our traders have invested most of their trading stock in vessels for merchants business and the fisheries which will probably be mostly unemployed, and will not sell for half the cost of them, and probable the price of fish will fall one quarter in their value. We have now on the fishery business[31] 4 brigs, 15 schooners, 1,365 tons, 171 men, with 1,605 hhds. salt. We had last year 4 schooners, 598 tons, fish entered at the custom house 1,429 bbls. salmon, 2,710 qtls. codfish.

On September 5, "The price of fish stumbles me so that I am at a loss how to manage." Early in 1804 losses were reported on vessels as a result of high wages and the high prices of salt and provisions. High insurance rates, difficulties of preparing fish[32] for markets other than

30. "The various fisherys of our coasts are considerably diminished this year [1800] by the high price of wages, being from 5 to 7 guineas per month and salt and provisions, cordage, sail cloth and other necessarys for navigation greatly enhanced." Wentworth to King, Canadian Archives, *N.S.A.*, 132.

31. The port of Liverpool had one ship of 200 tons, 14 brigs totaling 1,807 tons, 25 schooners totaling 1,394 tons, and one sloop of 42 tons. Simeon Perkins' diary.

32. August 22, 1803: "Fish in general very salt and a great deal of salt is dry among them which is a waste of salt and damage to the fish. . . . They prove very salt indeed so that they break to pieces many of them tho they have had only 2 days son." November 12, 1811: "It is rare to get them dry at this season. When they are brought in boats and when we git them early they are seldom made thoroughly, all which is much against the small trade with the boat fishermen." December 31, 1811: "The shore fish are so dry and horney that we get only 8 ct. [cwt.?] into a common

the West Indies, even salmon for the United States,[33] and the decline of the salmon fishery[34] led to serious losses.

On June 25, 1804, Perkins had this to say:

> Our West India trade is so embarrassed by the great risk and high premium on our vessels and the dull and low markets for fish and lumber occasioned by the United States having, in a manner, free trade to the West Indies that we cannot carry on the business without great loss and there does not appear to be any other opening for trade. What will be the event, time will discover, but at present the prospect is very gloomy as we have no other dependence but the fishery and trade, having no farms to resort to when trade fails, for these reasons the war opperates more particular against this town, but is felt more or less by every traiding town in the province.

In this he reflected the demands of the merchants of Nova Scotia for the exclusion of the United States from the British West Indian market. Temporary adjustment involved an increasing dependence on Newfoundland for supplies of fish and a growing exchange of lumber for salt from Liverpool, England, rather than from Portugal or the West Indies. The needs of Newfoundland made for a greater development of shipping in Nova Scotia, and increased the demands for provisions from the United States. In 1805 the exclusion of Americans from the trade in either fish or salt provisions, following representations from Nova Scotia and Ireland, lowered the price of fish in Boston; but the outbreak of war between England and Spain strengthened the position of the United States in the sale of merchantable fish. The high prices of flour in the United States were a handicap to Nova Scotia and Newfoundland. In 1805, fish prices were high in the West Indies, but lumber was low in Newfoundland. As ever, Quebec was a high market for provisions. Privateering was unprofitable. On August 9, 1805, Perkins wrote: "As the cruize [of the privateer] is likely to turn out it will not be a very lucrative business but in these hard times I am glad to undertake any lawfull business to support my family and pay my debts." People were reported to be leaving Port Mouton for the United States and absconding without paying their debts.

hhd. 6½ into the second size and 6." Labrador fish were "very limber and [we] get 10 ct. in hhd." *Idem.*

33. September 3, 1803, when loading salmon for Boston: "Barrels very bad, we have to start some of them, this is principally owing to the staves being very thin and the crows not cut deep enough and the heads small." *Idem.*

34. August 22, 1804: "It seems the salmon have nearly done on that coast, and the Americans engross the codfishing. So that all prospect of making a hand of fishing on that coast is over and we shall make a very poor hand of it in our vessels this year. My business fails where ever I undertake." *Idem.*

Various efforts were made to offset the disastrous effects of war.[35] In 1802 a bounty of one shilling per quintal was granted on codfish caught and cured by citizens of Nova Scotia, but a year later this was regarded as unsatisfactory. In 1803, the sum of £1,350 was appropriated to provide for

> a bounty, to be distributed in equal proportions per Ton on each vessel, [to] be paid to the owners of all vessels of twenty tons and upwards, owned and registered in this Province, and which shall be wholly employed in the Fisheries for any term, not less than three months, between the first day of April, and the thirtieth day of November, in the year 1804, provided that one half at least of each respective Crew of said Vessels shall be persons employed on shares and not on wages.[36]

The proposal was, however, rejected by the Council.

With the renewal of war between England and France, the relaxing of the restrictions on trade between the United States and the British West Indies, and the encouragement of the New England fishery by bounties, Nova Scotia was handicapped in the West Indian market still further. It was held that duties on imports to the West Indies by British subjects were higher than those on imports by Americans admitted under a proclamation. The neutrality of the United States enabled American vessels to obtain from 10 to 12½ per cent lower insurance. "From those circumstances so unable are the petitioners to contend with the Americans in the West India markets that they derive greater advantage by selling their fish at an inferior price in the United States; whence the Americans re-export them to the West India Islands under the above-mentioned advantages, so as to make a profit even on their

35. A petition was presented on April 2, 1800, to the House of Assembly by William Smith, John Powell, and John Martin of Ketch Harbor, "in behalf of themselves, and others, Fishermen on the Coast of the Province, That from the high price of Salt and all other Supplies necessary for the fisheries and the low price of the Fish when cured and fit for the Market the Petitioners and others employed in the Fishery are reduced to great Poverty and Distress, and with difficulty enabled to keep their Families from Want, and praying the House will take their Case into Consideration, and grant a Bounty on Fish and Oil when cured for Market, or such other relief as to them may seem fit." A committee recommended on April 5 that a bounty be allowed "to revive the almost annihilated Fisheries of this County, it being the principal Staple from whence springs the Revenue of the Province . . . of One Shilling and Six Pence, on each Quintal of merchantable and one Shilling on each Quintal of West India Fish, And One Shilling and Six Pence on each Barrel of Pickled Fish caught by Vessels owned and manned by Inhabitants of this Province." However, the House was unable "to do anything for the accomplishment of objects of such great publick utility." *Journals of the Assembly, Nova Scotia,* May 20, 1800.

36. *Idem,* July 8, 1803. A committee of the Assembly recommended a bounty of 7*s.* 6*d.* a ton on such vessels and that a sum should be set aside to pay a bounty on about 3,660 tons. *Journals of the Assembly, Nova Scotia,* June 23, 1803.

outward voyage."[37] Low prices in the West Indies and the loss of vessels by capture in 1804 and 1805 aroused strenuous resistance[38] which found its expression in the organization of a committee of merchants in 1804, and in the preparation of petitions and memorials against a proposal to extend by treaty the rights of the United States to West Indian trade. The merchants proclaimed their ability to supply the West Indies with dried and pickled fish and "at no very distant date" with all other articles "except perhaps flour, Indian meal, corn, and oak staves." As a result, the governors of the West Indies were warned in 1804 not to admit American goods except in cases of very urgent necessity,[39] and this was in spite of protests from the colonial agent of Barbados. Following a change of administration in England in 1806 the restrictions were not enforced, but difficulties in reaching an agreement with England led the United States to pass an Embargo Act, which was proclaimed on December 22, 1807.[40]

Efforts to check trade between the United States and the West Indies were accompanied by a renewed demand for bounties to build up the Nova Scotia fishery and to offset the effects of other bounties and the

37. See the petition of the merchants and other inhabitants of Halifax, March 23, 1804; also "The memorial and petition of the merchants and other inhabitants of New Brunswick, May 11, 1804," Hugh Gray, *Letters from Canada* (London, 1809), p. 393.

38. From 1793 to 1804 Nova Scotia fish "found very little vent except in the United States; where strange as it may appear it was sold for the purpose of reshipment to the *British West India Islands;* for the colonial merchants except under circumstances by no means of a general nature could not, unprotected as they were, stand a competition with the Americans, who exclusive of other advantages navigated their vessels as neutrals, under charges greatly below the British shippers of these colonies, who independent of extravagant outfits, laboured under the pressure of a quadruple insurance, high wages, impressments, delays, demurrage, numerous uncertainties, loss of convoys and in that event the almost certainty of capture. Under the state of depression the colonial trade, even of their own produce passed in a great measure into the hands of the Americans." *Acadian Recorder,* January 23, 1818. See also Gray, *op. cit.,* pp. 385–399. For a general argument to support the British North American colonies see *idem,* pp. 231–242. On the other hand the overwhelming importance of trade from the United States to the British West Indies is shown in statistics for 1804, 1805, and 1806, *idem,* Appendix No. VII.

39. *A Calendar of Official Correspondence, op. cit.,* p. 67. Illegal trade with the United States was the object of further protest in 1805. *Idem,* p. 82. To encourage trade between Nova Scotia and the island of St. Vincent, Lieutenant Governor Wentworth wrote (March 3, 1803) to inform Governor Bentwick of St. Vincent that all St. Vincent's exporters to Nova Scotia would be granted a drawback "of all the provincial import duties upon your produce imported here, equally to the non-resident as to the inhabitant—provided those commoditys were purchased in the British islands, or paid for here, with the produce of this province." *Idem,* pp. 31–32. Complaint was made against a 3 per cent transient duty to which Nova Scotian goods were liable and not those of the United States. *Idem,* pp. 48, 51–53. In 1807 Jamaica paid a bounty on imports of British North American fish. See page 207, note 77.

40. L. M. Sears, *Jefferson and the Embargo* (Durham, 1927), especially chap. vi; and E. F. Heckscher, *The Continental System* (Oxford, 1922), pp. 127 ff.

migration of labor to New England.[41] They were paid on imports of salt from 1806 to 1808; and in 1806 they were also granted on the basis of the tonnage of vessels[42] to encourage the Labrador and bank fishery. The measure operated "as a great stimulus to exertion, [and] was productive of very beneficial consequences." But it was superseded by an arrangement that set aside £2,000 to pay a bounty of 1*s.* 6*d.* a quintal on exports of fish to the West Indies in 1807. On such exports from June 1, 1806, to June 1, 1807, the Imperial government also paid a bounty of 2 shillings a quintal on Newfoundland and British American salt fish; 1*s.* 6*d.* a 32-gallon barrel on shad; 2*s.* 6*d.* a barrel on herring; 3 shillings a barrel on mackerel; and 4 shillings a barrel on salmon. In 1808 complaint was made by the Assembly that the payment of the bounty on fish "does not go direct to those who are actually engaged in the fisheries and the small proportion of it which may eventually reach them, is not considered as affording any encouragement to their exertions." The Council, on the other hand, would not agree to the payment of bounties on ships in spite of the Assembly's insistence that they were "a direct encouragement to the banking branch of this business which, in the opinion of this house is most conducive to the health, morality and emolument of those who are engaged in it and best promotes the joint interest of this province and the parent state."[43] The issue was joined between the Council, representing Halifax interests favorable to the bounty on fish, and the Assembly, representing the outports and favoring the bounty on ships. The Council was opposed to the "schooner interest" on the ground that fish would be sent to the United States and smuggled goods brought back in return.

It was in fact giving a bounty to His Majesty's subjects to furnish Americans with an essential staple article of export. For, the contiguity of the American shipping ports and the high prices which their neutral situation and superior advantages in carrying on navigation enabled them to give for the article, naturally induced the catchers of fish . . . to sell them . . . in

41. "From various causes," merchants were obliged to state, "so great has been the emigration of Fishermen, and others, from this Province to the American States, that the customary offers of the Merchants, which is all they can possibly afford, have hitherto proved insufficient to draw them back again to this Province; on the contrary, during the last season, even a great many industrious families have gone to that country. This has been, in a great measure, occasioned by the encouragement by bounties held out by the Legislatures of those States, and, partly, by the burthens, expences, inconveniences and depressions, to which this Trade is peculiarly subject in time of war; and which, during this last summer, have, in one instance, at least, increased beyond any former precedent." *Journals of the Assembly, Nova Scotia,* December 21, 1805.

42. They were 10 shillings a ton on vessels of 30 tons, 15 shillings on those of 30 to 40 tons, 20 shillings on those of 40 to 50 tons, and 25 shillings on all above 50 tons. *Journals of the Assembly, Nova Scotia,* December 28, 1805. See page 261, note 85.

43. *Journals of the Assembly, Nova Scotia,* February 3, 1808.

small craft for the American market, [and] bring back in return foreign articles to the great injury of the fair trader.[44]

With measures to stimulate trade to the West Indies and with encouragement by bounties, exports of fish from Halifax shifted as follows:

Exports of Fish from Halifax

	COD		HERRING	
	Dried (*quintals*)	*Pickled* (*barrels*)	*Smoked* (*barrels*)	*Packed* (*boxes*)
1806				
West Indies	38,896	18,779	242	1,228
United States	19,769	16,681	106	191
1807				
West Indies	54,155	27,117	48	5,248
United States	11,009	14,445	20	195

The Embargo Act increased the trend.[45] An armed schooner was employed in 1807 and 1808 to restrict trade with the United States. In the words of the lieutenant governor:

We can now fully and fairly estimate the effects of the Embargo, so long and so rigorously imposed on the commerce of the United States, by the Government of that country. The manner in which their general restriction of trade has been carried into execution, leaves no doubt as to the real object intended to be accomplished by it. The project has totally failed; and the British Nation has derived sufficient experience from the measure, to be convinced that her Colonies and Commerce can be as little affected by the Embargo of America, as by the Blockading Decrees of France. New sources have been resorted to with success, to supply the deficiencies produced by so sudden an interruption of commerce; and the vast increase of Imports and Exports of this Province proves that the Embargo is a measure well adapted to promote the true interest of his Majesty's North American Colonies.[46]

Immediately preceding the embargo large numbers of vessels cleared from American ports with cargoes to be exchanged at sea or at Passamaquoddy, and inland trade between Vermont and Quebec was extremely active. After the declaration of the embargo, coasting vessels seemed too often to be driven into Nova Scotia ports by stress of weather or broken masts. British goods were smuggled to the United States by ships which cleared coastwise in ballast with foreign commodities on their manifest, received their goods at sea from a foreign vessel, and arrived at the port for which they had cleared. British naval au-

44. *A Calendar of Official Correspondence*, pp. 140, 143, 152.
45. *Idem*, pp. 112, 128, 142, 158.
46. *Journals of the Assembly, Nova Scotia*, November 24, 1808.

thorities were ordered to encourage American ships sailing from the United States to the West Indies.

An entrepôt trade was built up.[47] On May 4, 1808, a proclamation admitted specified American goods for reëxport to the West Indies[48] and on June 23 it was revised to admit American vessels with a long list of products to Halifax, Liverpool, Shelburne, and Yarmouth,[49] but later limited by the British government to Halifax and Shelburne in an order in council of October 26, 1808. Americans were permitted to sell cargoes and to purchase export cargoes, but after August 15, 1809, the trade was confined to British subjects until Halifax was opened to neutral shipping in December, 1811. In 1809 a duty of one shilling a quintal was placed on fish imported into the West Indies from the United States and of 5 shillings a ton on American vessels trading there. The embargo was lifted on March 3, 1809, but on May 20 it was followed by a nonintercourse act against France and England. Colonial produce reëxported from the United States declined from a total of $59,643,558

47. As early as 1790 a request was made "that a free port may be established in the province for the reception of American and other produce; that by this means our vessels would be furnished with cargoes for the West Indies nearly as cheap as from the United States; the trade of the province would be greatly increased and the mother country ultimately benefitted by the sale of large quantities of British goods which the trade would take off; and the money thence acruing would at last centre in Great Britain." Beamish Murdoch, *A History of Nova Scotia* (Halifax, 1867), III, 85–86. A long memorandum dated February 18, 1806, advocated this step. D. C. Harvey, "Uniacke's Memorandum to Windham, 1806," *Canadian Historical Review,* March, 1936, pp. 41–58.

48. The colonial merchants in Halifax attempted to extend their control over West Indian trade under the embargo, and on November 12, 1808, petitioned for the total exclusion of all foreigners from this trade, claiming that the northern provinces were able to supply fish and lumber. They revived their pleas for the opening of a free port to receive American produce. The carrying trade would be transferred from the United States to Great Britain, the West Indian planter would be able to obtain American produce, smuggling would decline, the colonies would obtain control over the fisheries, especially if restrictions were removed on Mediterranean trade, and emigration would be deflected from the United States. On November 28 in a memorial to Lord Castlereagh they asked that foreigners "may at least be prohibited from carrying there [to the West Indies] the articles of dried and pickled fish, and that a countervailing and protecting duty on all the different articles of lumber be paid by foreigners so admitted, to be given in bounties to British subjects importing the same article." A committee of the House of Commons was not convinced and pointed out that while they could supply some lumber and fish they had little flour and "that the trade now carried on between the British West Indies and the United States of America is very convenient and advantageous to the inhabitants of our colonies and one which they could not relinquish without essential detriment unless it were compensated by other advantages." *A Calendar of Official Correspondence,* pp. 186, 192.

49. The proclamation of June 23, 1808, was declared illegal, but its general effectiveness in causing discontent in Massachusetts apparently warranted its continuation. *A Calendar of Official Correspondence,* pp. 148, 154–155, 158, 203. See page 250, note 65.

in 1807 to $16,022,790 in 1811. While American ships were idle, those of Quebec and maritime ports were active. Trade between New Brunswick and Nova Scotia expanded rapidly. Five vessels arrived in Nova Scotia in 1807, 38 in 1808, 58 in 1809, 47 in 1810, and 80 in 1811. Of 64 vessels from the United States entering Halifax in 1808, 14 were American; of 176 in 1809, 108 were American; and in 1810 all of the 60 vessels entered were British.

From December, 1811, on, restrictions were placed on American vessels trading to Halifax; and on August 1, 1812, following the outbreak of war in June, the port was closed to American ships. In 1812, duties were increased on wines, rum, and spirits, and in 1813 a duty of 10 per cent was imposed on imports from the United States with the exception of grain, flour, and naval stores. Revenue receipts trebled[50] in two years, smuggling increased, and great quantities of goods were exchanged at sea. Governors' licenses were issued in 1812 to permit trade with Americans, which was facilitated by the hostility of New England to the war. Vessels left Halifax with cargoes of dry goods to meet coasting vessels from American ports in ballast, but with manifests for goods similar to those shipped from Halifax to another coast port. Late in 1813 a blockade of the American coast caused great inconvenience.

Commercial interests became increasingly alert in pressing for legislation such as bounties on salt, fish and fishing vessels, and cleared land that would be favorable to trade and the fishery. The formal organization of 1804 was renewed annually in Nova Scotia, and after 1811 it extended its activities to Great Britain. A petition to the Assembly stated

> that in consequence of the late indulgencies granted to foreigners in supplying the West-India Islands, and the depression thereby occasioned to the trade of this Province, and others his Majesty's North American Colonies, the petitioners were appointed, at a public meeting held in Halifax, a committee to solicit his Majesty's Ministers on that subject, and to watch over, and endeavor to promote, the general commercial interests of the community.[51]

Another petition, of February 21, 1811, complained that they were having "very great difficulties in promoting their applications in England, the same being frequently neglected or misrepresented by persons interested in opposing them." They were "informed that the West India Planters were accustomed, on particular occasions, to employ Special

50. For an account of the extent of trade during the War of 1812 see W. R. Copp, "Nova Scotian Trade during the War of 1812," *Canadian Historical Review*, June, 1937, pp. 141–146. On June 23, 1811, a schooner "sails for a voyage towards the states expecting to have a cargo brought out to him." July 2: "No money stirring. I suppose it is mostly gone to Eastport and to the states for provisions." Simeon Perkins' diary.

51. *Journals of the Assembly, Nova Scotia,* December 31, 1808.

Agents in London to solicit their interests—being Gentlemen intimately versant in the object of their immediate pursuit," and were advised to do the same thing. The merchants reported in 1812 that they had

> adopted the plan recommended to them, and have the satisfaction to say that their success has, in consequence, been answerable to their expectations. That, having thus employed Nathaniel Atcheson, Esquire, of London, in this service, and thereby received from his exertions most essential benefits towards promoting the general and particular commercial interests of this Province at large, and being in expectation, from the completion of his pending applications to His Majesty's Ministers, to receive further benefits, the Petitioners respectfully solicit this Honorable Assembly, that they will be pleased to grant such a sum of Money as may be deemed adequate to the remuneration of the said Nathaniel Atcheson, Esquire, for his past services.

A committee of the Assembly[52] reported favorably on this petition on March 3, 1812. Atcheson's position as secretary of a committee of merchants in England interested in the trade of the North American colonies did not entitle him, however, to any special privileges.[53]

In 1813 Atcheson succeeded in having foreign fish excluded from the West Indies, and in 1814 in having the preference renewed on colonial timber. Legislation in 1813 also permitted the British colonies "to extend the whole of their importations to and from each other, an indulgence, as is also that of trading to the Mediterranean without touching at an English port, often solicited by, but never granted to the old colonies."[54] The merchants presented a memorial on October 8, 1813.[55]

52. *A Calendar of Official Correspondence*, p. 247.

53. "This House entertains a high sense of the zeal and ability with which Nathaniel Atcheson, Esq. has solicited many important commercial privileges for this Province; but at the same time it is the determination of this House not to acknowledge itself a party to any measure that shall be solicited otherwise than through the medium of the Speaker, or of a Joint Committee of this House and His Majesty's Council, or through the regular channel of the Legislature, and signified by the King's Representative." *Journals of the Assembly, Nova Scotia,* March 23, 1816.

54. As early as 1790 merchants asked that salt, wines, and fruit might be imported directly from the Mediterranean. Nova Scotia protested that vessels carrying fish to Europe, "being restrained to an additional passage to England, landing and expenses there ruins the whole voyage, and of course compels the Nova Scotia Merchants to send their best fish to the United States, who derive all the benefit of the freight to, and profit at, the Foreign Market." Lieutenant Governor Wentworth to the Secretary of State, January 15, 1803. Landing European goods in England "operates such a loss in the perishable nature of the goods, and such an additional expence upon others, that it is found more profitable to export our best fish to the United States . . . who freight it in their own bottoms to all the ports in Europe, and our consumption of most of the returned articles are smuggled in upon our extensive coasts . . . thus rendering this people a fishing colony to the United States instead of to Great Britain." May 26, 1803, Lieutenant Governor Wentworth to the Duke of Clarence, *A Calendar of Official Correspondence,* pp. 31, 35, 49.

55. *Idem,* pp. 299–300.

It urged the disadvantages of American admission to the West Indian trade and argued that from 1783 to 1793, when American produce was imported under a heavy duty through the French West Indies, prices were low; and that from 1793 to 1806, when Americans were admitted during the war with France, prices rose but moderately; and that from 1806 to 1813, when Americans were excluded,

> prices have in no instance risen higher than the risks and expenses incurred and in one solitary instance in the present year beyond the medium price of eight dollars a quintal of cod-fish. But if the Americans on restoration of peace are admitted to the British West Indies these colonies . . . will remain in perpetual infancy and not have the power to supply the islands when a sudden fit of displeasure may seize hold of the government of the United States. Their conduct . . . will prove that no indulgence or forbearance of Great Britain, consistently with the safety and dignity of government, can avail.

An address of the Council and Assembly on March 5, 1814, supported the memorial.

The commercial interests were active not only in attempting to exclude the United States from the West Indian markets but also in seeking to have American fishing vessels excluded from British waters. Atcheson submitted a memorial on December 14, 1812, on the necessity of protection to Nova Scotia in the negotiations for peace. The Assembly,[56] with the support of the merchants, protested vigorously against the possible continuance of Article 3 of the Treaty of 1783 in the proposed treaty with the United States. A committee of the Assembly stated on February 15, 1814, that it introduced

> into the harbours of this Province, and on the Labradore Coast, such numerous Foreigners, that the Fishermen of these Colonies have been deprived of the chief means which Providence has assigned to them, of procuring a livelihood. This third article of the Treaty, has during the period, from the conclusion of the last American Peace, to the commencement of the present War with that Nation, impeded the fisheries of these Colonies, to the very great improvement of those of the United States, which, by reason of the cheaper outfits of their Vessels, are able to undersell our Merchants in the British Islands.
>
> The committee must also observe to the House, that the American Fishermen are a People over whom there is little restraint from any sense of propriety . . . and that they resort to these Coasts and the Labradore in such great numbers (to the amount lately before the War, as it is said, of fifteen

56. *Acadian Recorder,* July 31, 1813. *A Calendar of Official Correspondence,* pp. 279–280.

hundred Vessels in a season,) that by throwing their Gurry or Offal overboard, contrary to the Provincial Act of this Province, (10 Geo. III, c. 10, page 162,) [they] have greatly injured the fishery to their own immediate detriment as well as that of the Inhabitants; a decided proof of which is, that, during the last Season, the British Fishermen have experienced a very great increase of Fish, on the Banks and on the Labradore: The intercourse is also very injurious to the political morality of the lower classes of People of these Provinces, whose attachment to the Mother Country will be best secured by being debarred from such contagious principles.

The London merchants petitioned Lord Bathurst in June, 1815, "that Americans should be excluded from fishing on the shores of British Colonies and that in future all intercourse with the British West India islands in American vessels be prohibited." In the same year H.M.S. *Jasseur* began seizing American vessels on the ground that the treaty had been abrogated by the outbreak of war in 1812. Exclusion was not vigorously pressed in 1815;[57] but in the following year there were indications of a change.[58] H.M.S. *Menai*[59] was engaged to warn American vessels from British American ports, and in 1817 several[60] were seized

57. See *A Calendar of Official Correspondence,* pp. 346, 348. "The Americans fish in our very harbours and will continue to do so; they know we are afraid of them, and however humiliating we must admit the fact. Was it an emanation of the pusillanimity of our government at home that dictated the release of eight fishing vessels without trial?" *Acadian Recorder,* July 15, 1815.

58. "You will prevent them . . . [save as hereafter excepted] using the British territory for purposes connected with the fishing vessels from bays, harbours, rivers, creeks, and inlets of all his majesty's possessions. In case, however, it should have happened that the fishermen of the United States, through ignorance of the circumstances which affect this question, should, previous to your arrival, have already commenced a fishery similar to that carried on by them previous to the late war, and should have occupied the British territory, which could not be suddenly abandoned without very considerable loss, his royal highness the Prince Regent, willing to give every indulgence to the citizens of the United States, which is compatible with His Majesty's rights, has commanded me to instruct you to abstain from molesting such fishermen, or impeding the progress of their fishing during the present year, unless they should, by attempts to carry on a contraband trade, render themselves unworthy of protection or indulgence; you will however not fail to communicate to them the tenor of the instructions which you have received, and the view which His Majesty's government takes of the question of the fishery, and you will, above all, be careful to explain to them that they are not in any future season to expect a continuance of the same indulgence. Signed, BATHURST." Printed in the *Quebec Gazette* (1816).

59. The following endorsement was made May 15, by the commander of His Britannic Majesty's ship *Menai,* of 64 guns in the Bay of Fundy, on the back of the license of the schooner *Clarissa,* Lear, master, belonging to Newcastle. "Warned from fishing in the ports, harbours, creeks, or bays, within jurisdiction of His Britannic Majesty's North American Colonies, or using any port thereof for any purposes connected with the Fishery." *Montreal Gazette,* July 1, 1816. See a letter of protest against American encroachment. *Acadian Recorder,* May 3, 1816.

60. "I have it in command from His Excellency the Lieutenant Governor to ap-

near Cape Sable early in the season. They were later released on the payment of costs.[61] But the Convention of 1818 finally excluded American fishermen from British inshore waters.[62]

prize you, that American Fishermen are not permitted to frequent the Harbours, Bays or Creeks, of this Province, unless driven into them by actual distress; and I have to desire that you, on no account, ask or receive any Light Money, Anchorage, or any other Fees whatsoever from Vessels belonging to American Subjects." Circular dated Halifax, June 24, 1817, signed by Rupert D. George, Secretary.

61. See a letter of protest printed in the *London Sun* and reported in the *Montreal Gazette,* February 11, 1818, against the release of these vessels. "The recent decision of the Vice Admiralty Court of Halifax, by which the American fishing vessels have been restored, notwithstanding they were seized on the shores of the King's Colonies, has excited great attention in British North America, and at Poole, Dartmouth, and various other places interested in the Fisheries at Newfoundland, and in the Gulf of St. Lawrence; more especially as the unequivocal and explicit declaration of Lord Castlereagh, in an answer to a question from Sir John Newport in the Session before the last on this subject, led the public to believe the Americans were to be considered in the same light as *other foreigners,* and that they would not be allowed to fish *nearer* the coasts of any of his Majesty's dependencies than the latter, namely, than three leagues; which are the limits laid down by all the writers on the Law of Nations. If the Americans are thus to be encouraged, the Fisheries of the King's Colonies must be abandoned, and the subjects of other Nations will probably claim the same right to fish on these coasts, and how their vessels can be seized by his Majesty's cruisers or revenue officers, and condemned, whilst the Americans are allowed to fish within the same limits with impunity, is beyond our comprehension; unless in order to conciliate the Americans, the Nation is prepared to quarrel with the other Maritime Powers, whose subjects may be disposed to fish on these coasts and waters.

"We really did hope a more generous, firm and decisive policy would have been adopted towards British North America. It should be recollected, that its inhabitants are the remainder of the Loyalists who quitted the other parts of that Continent on the breaking out of the first American war—abandoning their property and adhering with fidelity to their Sovereign and to the Country of their ancestors.—Surely these are not the people who ought thus to be treated! For whilst the Americans are permitted to interfere with our Colonial Fishermen, the latter have no chance whatever of competition in markets with the former." Vessels were also seized in 1818, notably the *Nabby,* for the illegal importation and exportation of goods, and for taking and curing fish. *Acadian Recorder,* September 5, 1818.

62. In spite of its stipulations, a report of the joint committee of the Council and Assembly in 1819 complained "that human ingenuity would scarcely have devised a more destructive measure for British America than this convention. . . . The convention is far more ruinous to the colonies in North America than the treaty of 1783; but at the same time, it is but justice in your committee to observe that excluding the Americans from any intercourse with the West Indies was a point of the utmost importance to the best interests of Great Britain and her colonies." As a compensation for the loss, the report asked assistance from the British government for a wide range of projects. See D. C. Harvey, "Nova Scotia and the Convention of 1818," *Royal Society of Canada,* 3d Series, sec. 2, 1933. A joint address of the Council and Assembly of March 26, 1818, protested "against the right of foreigners to use the straight of Canso and to pass into the Gulf of St. Lawrence or into the Bay of Fundy for the purpose of fishing therein, or within the line which separates the territory of His Majesty in the last mentioned bay from the territory of the United States." The address also protested the granting of "destructive monopolies claimed on the Labrador shore and the improvident grant of the Magdalen Ids.," which gave support to American fishing vessels and constituted a serious detriment to the Nova Scotia fisheries.

THE REVISION OF THE COLONIAL SYSTEM

At this moment one of the greatest sources of trade belonging to Nova Scotia was their fishery. Their rivals were in the neighbouring provinces belonging to the United States. Supposing they took their cargoes to New York they then were upon equal terms. But neither at the Brazils nor up the Mediterranean were they as the law now stood at all upon an equality, and if the Nova Scotian went to the Baltic he could not exchange his cargo for the produce of those countries. This gave an effectual premium and bounty to the fishery of the United States and by this we lost the benefit of the trade of Nova Scotia.

WILLIAM HUSKISSON

THE commercial interests of Nova Scotia had emerged as an effectively organized group and, with the outbreak of difficulties between Great Britain and the United States, they pressed steadily for a position as intermediaries between the United States and the British West Indies. Whereas Nova Scotia had traded through the United States prior to 1807, she attempted to force the United States to trade through her during and after the War of 1812. She had succeeded under the exigencies of the war, and she resisted vigorously efforts of the United States to penetrate the British West Indies market on its cessation. After the war, as a result of continued pressure Great Britain restricted trade between the United States and the West Indies. In spite of petitions and protests West Indian governors were warned not to admit American vessels in 1815. American vessels were not admitted to North American ports except at St. Georges and Hamilton in Bermuda, and were subject to port charges and duties and restricted to enumerated goods. British ships were allowed to carry on a triangular trade with the United States and the West Indies, and British tonnage increased accordingly.[63] British ships carried cargoes to American ports, thence to the West Indies, and thence to England, whereas United States vessels were restricted to a direct trade with Great Britain. By 1816 three fourths of the tonnage in the West Indian trade was monopolized by the British.

The United States attempted to retaliate by a discriminating tonnage duty of 44 cents—later increased to 94—and a 10 per cent duty on merchandise; and, finally, early in 1817,[64] passed an act imposing an additional duty of two dollars a ton on foreign vessels from ports ordi-

63. F. L. Benns, "The American Struggle for the British West India Carrying Trade, 1815–1830," *Indiana University Studies*, X, 1923, pp. 94–95.

64. V. G. Setser, *The Commercial Reciprocity Policy of the United States 1774–1829* (Philadelphia, 1937), and J. M. Callahan, *American Foreign Policy in Canadian Relations* (New York, 1937), chap. vi.

narily closed to American ships. British vessels sailing from Great Britain to the United States and then to the West Indies were not affected. An American Navigation Act, passed on March 1, 1817, went farther and prohibited the importation into the United States of non-British produce in British vessels. Failure to secure concessions led the Americans to pass another Navigation Act on April 18, 1818, which, after September 30, 1818, closed American ports to British vessels coming from ports ordinarily closed to American vessels. British vessels sailing from American ports were put under bond not to land their cargoes in any port closed to American vessels. The act was effective in reducing British shipping in American ports but was weakened by increased trade through the free ports in Bermuda, partly in American and partly in British vessels, the longest part of the voyage being reserved to the British ships. A free-port act[65] on May 8, 1818, authorized the designation of ports in New Brunswick and Nova Scotia as free ports for three years, through which, first, there was permitted the importation of food, lumber, and naval supplies either in British or foreign ships—but in the latter case, this applied only to goods which were the growth or product of their own country; second, through the same free ports the exporting of certain goods in either British or foreign ships was permitted—but in the latter case this applied only to exports to the country to which the ships belonged. Halifax was declared a free port on August 16, 1818, to date from July 16; Saint John was so declared in the same year; and St. Andrews was added in 1821. The United States declared these ports were not open "by the ordinary laws of navigation and trade," and refused to admit British vessels from them to the United States; but American vessels were free to resort to them with their cargoes. British vessels consequently loaded American goods at Halifax for the British West Indies. In addition to the Free Port Act a further measure permitted importation in British vessels to the British West Indies of tobacco, rice, grain, peas, beans, and flour from any colony or possession in the West Indies or on the continent of North America under any foreign *European* state, and consequently made easy the smuggling of American produce through the foreign

65. This act followed the earlier legislation which, in 1808 (48 Geo. III, c. 125) permitted imports of enumerated articles from the United States to the British North American colonies, and the export of other enumerated articles from the British North American colonies to the United States, up to March 25, 1809, and was continued in the following year by an act allowing any goods imported or exported to Nova Scotia and New Brunswick in any ship up to March 25, 1812, and in 1811 (51 Geo. III, c. 97) allowing sundry articles to be imported and exported from certain ports in New Brunswick and Nova Scotia to certain foreign ports.

West Indies. In 1818 and 1819 American exports arrived in the British West Indies by devious routes thus provided, as follows:

	1818	*1819*
British West Indies, directly	$3,488,653	$ 843,312
British North American Provinces	2,355,700	3,038,995
Swedish Islands	278,846	345,793
Danish Islands	983,583	1,120,857

Governors in the British West Indies attempted to overcome the difficulties by admitting to their ports vessels with provisions and lumber from the United States.

American restrictions were made more drastic[66] in the Navigation Act of 1820, which closed American ports after September 30, 1820, to British vessels from ports in Lower Canada, New Brunswick, Nova Scotia, Newfoundland, Prince Edward Island, Cape Breton, the Bermudas, the Bahamas, the Caicos Islands, or any British possession in the West Indies or in America south of the boundary of the United States, and prohibited imports unless wholly the growth, product, or manufacture of the colony where laden and whence directly imported. This legislation precipitated demands for the revision of the Navigation Acts. The British West Indies suffered from a scarcity of provisions and a drain of money, since the British North American provinces with an inadequate market for rum and molasses also demanded specie.[67] The legislature of Antigua complained of a marked increase in the prices of provisions and supplies. Halifax merchants protested against the de-

66. Benns, *op. cit., passim.* See also *Acadian Recorder,* March 13, 1822; also G. F. Butler, *Nova Scotia in the Struggle for the British West Indies Trade, 1783–1830,* master's thesis, Dalhousie University, *passim;* and Setser, *op. cit.,* pp. 223 ff.

67. A table giving returns on twelve representative voyages from Halifax to the West Indies by ships of three of the principal houses showed small profits, and also showed that profits on outbound voyages were offset by losses on homebound voyages. The largest cargo sold on the north side of Jamaica had an outbound invoice of £5,955 19*s.* 8*d.* The net sales, less freight charges, amounted to £6,071 5*s.* 1*d.* The homebound invoice was £6,508 8*s.* 5*d.* The net sales, less freight charges, in this case amounted to £5,373 2*s.* 11*d.*, that is, the gain on the outbound voyage was £115 5*s.* 5*d.*, and the loss on the homebound voyage, £1,135 5*s.* 5*d.* Thus the net loss was £1,020. Only five ships carried homebound cargoes. All of these meant losses; and only two showed gains on both the outbound and homebound voyages. *Journals of the Assembly, Nova Scotia,* 1822, p. 207. A joint address of the Council and Assembly, March 19, 1822, complained of piracies in West Indian waters as a result of disturbances in the South American states, and asked the home government to suppress them. "Premiums of insurance have risen to an extravagant height and the personal danger to which mariners are exposed makes it difficult to man vessels in that trade: Convoys being of little use as the delay in such a climate proves destructive to the commodities of this colony which are generally of a perishable nature." *Idem,* p. 172.

mands from the West Indies "which," they said, "if conceded to the extent desired by the memorialists must necessarily effect the destruction of our limited trade." A public meeting in Halifax on January 28, 1822, resulted in the revival of "the commercial society." In reply to petitions that trade should be permitted in American ships[68] and to statements that the British North American colonies "neither afford a market of such importance nor are capable of supplying beyond a very limited extent those articles which are wanted for the West Indies, and are in fact themselves in a very great measure dependent on the States for flour and provisions for their own support," they presented a memorial pointing out that prices had declined in the West Indies as a result of the ending of the war and that the British North American colonies had supplied them abundantly. The admission of the Americans, the memorial asserted, "would prove a measure pregnant with mischief and the application of the petitioners is founded upon the most narrow and partial views of local and perhaps individual interest, and is utterly subversive of that system which has so long been the settled constitution of the colonies and has preserved in the mother country the pre-eminence she now enjoys." It was claimed that they could supply fish, oil, white-pine lumber, and hoops of better quality and at lower prices than the United States, and that increasing quantities of flour, bread, white- and red-oak staves, salted provisions, and butter could be available.

> When these various exports and the daily increasing population of these provinces are kept in view and their value as a nursery for seamen and ample stores of timber, masts, spars and other articles for naval purposes, especially in the case of differences with the northern powers, is dispassionately considered your memorialists will not shun a comparison with any of the sugar colonies as to their respective importance to the nation at large.

The proposals of Nova Scotia were outlined in detail[69] and were influential as the basis of a compromise with the West Indies in the adop-

68. Shipowners in Great Britain supported Nova Scotia. See *Acadian Recorder*, June 8, 1822. Later, certain interests in Nova Scotia became much more sympathetic toward the West Indies. See the *Acadian Recorder*, August 5, 1826, which points out the dangers that lay in the abolition of slavery and its possible repercussions on Nova Scotia. An editorial of June 25, 1831, on the other hand, was anxious to dissociate Nova Scotia from the activities of the West Indian planters in opposing the abolition of slavery.

69. The relief proposed by Nova Scotia was designed first to extend 52 Geo. III, c. 98, 55 Geo. III, c. 29, and 57 Geo. III, c. 4, permitting direct exportation of West Indian produce to Malta and Gibraltar and to other ports south of Cape Finisterre; second, to allow free exports of West Indian produce in British ships to any part of Europe and to abolish licenses, bonds, etc.; third, rather than restrict exports from the above regions to the "direct" return of identical ships which had first proceeded from the West Indies, to allow any British vessel to go from any port in Europe to the British colonies with lumber, staves, etc., that would not compete with exports

tion in 1823 of legislation designed to improve the position of both groups of colonies (3 Geo. IV, cc. 44 and 45). The first act[70] permitted the importation of certain enumerated articles—lumber, naval stores, cotton, livestock, provisions, wool, and tobacco, but not fish and salted provisions—into specified ports in the British North American colonies and the West Indies from any foreign country in North or South America or from any foreign island of the West Indies, in British ships or ships of the country of which the articles were the growth, produce, or manufacture, provided the said imports were brought directly from such country. It imposed duties on imports to British North America and the British West Indies of 5 shillings a barrel on flour, £10 per hundred-pound value on cattle, and £1 1*s.* per thousand staves from non-British North America and the West Indies. It permitted the export from these ports of any articles, except arms and naval stores, in British or in foreign ships, provided they were sent directly to the country to which the ships belonged. The same import duties were charged on foreign goods carried in British as in foreign vessels, and British vessels from foreign ports were subject to the same restrictions as foreign vessels from those ports. These privileges were to be confined to ships of countries which gave similar privileges to British ships in their ports. Halifax, Saint John, and St. Andrews were opened to trade and several British West Indian ports were made free ports. The second act (3 Geo. IV, c. 45)[71] permitted the exporting of commodities, the growth, produce, or manufacture of the British colonies, to certain ports in Europe in British ships.

These concessions were made partly because of the distress following the American Navigation Act of 1820, and partly in recognition of the importance of American shipping. The United States pressed her ad-

from Great Britain; fourth, to extend to British North America the privileges given to Bermuda in 52 Geo. III, c. 79, 53 Geo. III, c. 50, 55 Geo. III, c. 29, and 57 Geo. III, c. 28, and to make such ports bonding ports; fifth, to permit imports of British West Indian produce to the Channel Islands directly from the British colonies or from Great Britain; and, sixth, to reduce fees collected from ships in the West Indies. *Acadian Recorder,* June 8, 1822.

70. Benns, *op. cit.,* pp. 82–85, and Alexander Brady, *William Huskisson and Liberal Reform* (London, 1928), pp. 90–94.

71. Other legislation in the Wallace Robinson code of 1823 (3 Geo. IV, cc. 41, 42, 43) included the repeal of clause 3 of 12 Car. II, c. 18 (1660) which required goods and merchandise the growth, production, or manufacture of Asia, Africa, or America to be imported into Great Britain or Ireland from *any place whatever* in British-built ships. Masts, timbers, boards, potash, salt, pitch, tar, tallow, rosin, hemp, flax, currants, raisins, figs, prunes, olive oil, corn, grain, wine, sugar, vinegar, brandy, and tobacco could be imported into Great Britain in British-built ships or in ships belonging to the country or place in Europe of which such goods were the growth, produce, or manufacture, or in ships belonging to any port in Europe into which such goods were brought or imported, and in which they were loaded. See *Acadian Recorder,* August 10, 1822.

vantage further and the President, under the authority of an act passed in 1822, issued a proclamation on August 24, 1822, giving notice that American ports should be open to British vessels from enumerated British North American ports as well as from the British West Indies, the imports to be restricted to West Indian goods if from the islands, or to the produce of the British North American colonies if from their ports, consequently treating these areas as separate regions. A discriminating tonnage duty of one dollar a ton on British vessels and of 10 per cent on their cargoes was still to remain in force. Great Britain protested but Congress passed an act on March 1, 1823, opening ports to any British vessel coming directly from enumerated British colonial ports with articles from the colonies, provided that the importation of articles of a like nature "from elsewhere" was not prohibited; and that goods might be exported from enumerated ports to the United States on equal terms in vessels of either state. Further, on proof that vessels of the United States were admitted to enumerated British colonial ports with no higher duties or charges than on British vessels, or on like goods or merchandise imported by the colonial ports "from elsewhere"—i.e., other British colonies as well as foreign nations—the tonnage duties might be remitted. Goods could be imported from British colonial ports only in vessels coming directly from those ports and when said goods were the growth, produce, or manufacture of those colonies. Goods might be exported to enumerated colonial ports in British vessels provided they had come directly from one of those ports.[72] But Great Britain refused to concede the assumption that the United States had the right to dictate conditions under which goods were exported from British colonial ports to British colonial ports, and an order in council was issued on July 17, 1823, imposing duties similar to those levied by the United States.

With a continuous loss of carrying trade to the United States[73] and

72. Benns, *op. cit.*, pp. 94–95. "The United States insisted that as long as Great Britain reserved the right to favor British shipping by manipulating customs duties so as to draw American supplies through adjoining colonies, the United States would penalize British shipping in the direct trade with the colonies by continuing to collect the discriminating duties on tonnage and merchandise." Setser, *op. cit.*, pp. 230–232.

73.

	Tonnage Entering American Ports from the British West Indies		Tonnage Clearing	
	American	British	American	British
1822	33,719	715	28,720	101
1823	71,346	9,520	68,350	8,654
1824	93,933	6,501	91,637	7,567
1825	101,604	6,207	93,967	6,742

Figures from Benns, *op. cit.*, p. 104.

their dominance of the West Indian trade, and in recognition of the independence of the South American colonies, new legislation was introduced in 1825. Halifax merchants submitted and forwarded to England an elaborate statement which read in part as follows:

> The advantages of a well assorted cargo for the continent of Europe or a South American market need not be enumerated and such cannot be obtained in this province without the privilege of making our shipments of foreign articles free of duties. . . . The right of warehousing, free of duty, merchandise of all kinds from foreign countries would be essentially important to the trade of this province in giving increasing employment to shipping which, under the existing regulations of colonial trade, is often materially affected in the most injurious manner—our vessels being frequently obliged to return from foreign places in ballast, when cargoes could be obtained which would meet a market in other foreign countries; for in the mother country had we the right to warehouse and then to export goods so imported and warehoused to any country whatever open to British commerce, such a privilege could not, if bestowed on this colony, prejudice the interest of either the British manufacturer or ship-owner; as an extension of the trade of this province would add to the consumption of British manufactures in the same ratio that it would increase the wealth of its inhabitants and the building of ships; and by opening a wider field for the enterprise of our merchants, would bring large sums into circulation which are now unemployed, and also afford employment for many of our respectable young men in the mercantile line who have now little inducement to venture abroad. The rights thus obtained would only militate with [*sic;* against?] the interests of foreigners, as our merchants would be enabled to assort their cargoes from this province to meet markets in South America and elsewhere in the same manner as is practised by their neighbours in the United States. Indeed a market could often be had in the United States for foreign sugars, etc., which if shipped from a foreign port direct to the United States in an American vessel or a vessel owned where the article is produced—one of which must be the case by the laws of the United States—the freight is lost to our ship and the funds produced by the outward cargo from this province are thus deviated from their natural channel and employed for the benefit of foreign vessels. Now, as the ships of the United States are permitted to take from hence foreign goods we would take advantage of the freight in our ships to this province and merely give the American ships the short freight from hence to the United States which would always be extremely low, as they frequently return home from this port in ballast. The privilege now asked would often prevent the sacrifice of a cargo abroad when payment can only be had in some inadmissible articles at a ruinous price.[74]

They asked that foreign articles consumed in the province should be

74. *Acadian Recorder,* February 12, 1825.

made free of duty if this was not to the disadvantage of the mother country. The effect of the statement was evident in Huskisson's speech of March 21, 1825, when he uttered the words that have been put at the head of this section.[75] As a result of the arrival and the reading of his speech in Nova Scotia "the generous flame of enthusiastic and patriotic joy," it was reported, "burst forth into the most ardent and heartfelt expressions of congratulations among our delighted citizens. . . . Such a day of cheerfulness has not been witnessed for ten years."[76]

The new legislation (6 Geo. IV, cc. 109 and 114), effective January 5, 1826, opened free ports in the British West Indies and North America, except Newfoundland, to vessels from foreign countries, and to all goods with a few exceptions, which included dried or salted fish, salted beef and pork, with a 15 per cent duty on all foreign goods, one tenth to be remitted when the goods were imported through the United Kingdom into these British colonies. Goods could be exported from the free ports, except those of Newfoundland, directly to the country from which the ships came; but trade could be prohibited by an order in council with any country in Europe having possessions in America or the West Indies if similar privileges were not granted to British ships in the possessions of such a country in America.[77] The intercourse of foreigners was restricted to "the ships of those countries which having colonial possessions shall grant the like privileges of trading with these possessions to British ships or which not having colonial possessions shall place the commerce and navigation of this country and of its possessions abroad upon the footing of the most favoured nation." Ships from European countries could not carry produce of their own countries to other European ports and secure a cargo for British ports. The warehousing system was extended and five free warehousing ports were opened: Bridgetown, Kingston (Jamaica), Halifax, Quebec, and Saint John. Foreign countries were admitted in competition with the United States, and a number of American products—wheat, barley, oats, potatoes, Indian corn meal, cotton, wool, hides, staves, hay, tobacco, tallow, pitch, tar, and turpentine—were subject to a duty. All fees and sinecure offices were abolished.[78]

75. *The Speeches of the Right Honourable William Huskisson* (London, 1831), pp. 304 ff.

76. *Acadian Recorder,* April 30, 1825.

77. French trade from the French West Indies to the British West Indies in articles the growth and produce of Great Britain and Ireland "occasioned the decline in exports to the colonies from Great Britain by about one million last year." *Acadian Recorder,* August 18, 1827.

78. These gains were largely offset in Jamaica by an act imposing tonnage duties

Reciprocal advantages were demanded from the United States, but without success. The port of Halifax was closed to American vessels on January 5, but reopened by a provincial order in council of January 23, 1826, which admitted American vessels subject to duties and tonnage dues. On July 28, 1826, a 10 per cent ad valorem duty levied on produce in British vessels in American ports was imposed on American vessels entering Nova Scotian ports. An order in council of July 27, 1826—effective after December 1—excluded American vessels from all British colonies except the East Indies and British North America and in their case continued the discriminating tonnage and import duties. As a result, on March 17, 1827, the President issued a proclamation under the act of 1823 which implied that, since all British colonial ports, with the exception of those of the East India Company, open by treaty, and ports of the British North American provinces were closed to American vessels, therefore, all American ports were closed against British vessels from any British colony in the Western Hemisphere. The West Indies were probably not seriously affected as American produce continued to be imported through the British provinces and the non-British islands. An act of 1827 admitted certain articles essential to the West Indies duty-free to Canada,[79] and made Kingston and Montreal free ports for goods brought by sea or inland navigation. Staves, timber, wood, hoops, and shingles were considered the produce of Canada

of 2*s.* 6*d.*, Halifax currency, to provide salaries for the customs officer, in lieu of fees. It was estimated that Nova Scotia sent 5,000 tons of shipping annually to Jamaica, and with the continued restriction on United States exports, the total would increase. Since the chief exports were fish, which it was necessary to sell shortly after their arrival in Jamaica, the tax tended to fall on Nova Scotia rather than on the consumer. "It is impossible to doubt that the levying this tonnage is in pointed opposition to the objects and spirit of the new navigation system—which took effect on the Fifth of January, 1826." See for a full statement of objections, *Journals of the Assembly, Nova Scotia,* February 14, 1827.

79. It also provided "that masts, timber, staves, shingles, lathwood, cordwood for fuel, raw hides, tallow, ashes, fresh meat, fresh fish, and horses, carriages and equipages of travellers, being brought by land or inland navigation into British possessions in America, shall be brought duty free." "All articles the produce of any of the British North American provinces are to be suffered to be removed from province to province for exportation, the same as if they had been exported direct from any one of the provinces, and all articles from the West India colonies and wine in casks from Gibraltar and Malta, are to be suffered to be imported into and removed from province to province in British North America without paying a higher or an additional duty than if imported in any one of these provinces direct. These and other regulations will be a severe blow to brother Jonathan, be of incalculable advantage to our whole colonial trade and lay a foundation for the improvement and prosperity of our valuable North American possessions, to an amount and to an extent which exceeds calculation. Already the accounts from Newfoundland state . . . that shutting the Americans out of our colonial trade has doubled the number of vessels employed in the fishery of that settlement." *Acadian Recorder,* August 18, 1827.

when imported from any British province. Customhouse officials in the British West Indies admitted American products from the foreign islands in British vessels subject to the same duties as if brought directly from the United States, provided that the goods had actually been landed in a foreign port.[80]

The election of Andrew Jackson as President in 1828 was followed by a more conciliatory attitude on the part of the United States, and, in spite of the most energetic opposition from the British North American colonies to the reopening of the colonial trade in the West Indies to the United States and the efforts of a newly appointed provincial agent,[81] the President's proclamation on October 5, 1830, removing discriminating duties and opening American ports to vessels from the British colonies was followed by a British order in council on November 5, 1830, repealing the order in council of July 27, 1826. In the West Indian reciprocity agreement of 1830 Great Britain allowed a trade in all products in the vessels of either country, and the United States waived its demands that American produce should be admitted on the same terms as like produce from British possessions, and no longer insisted that trade should be restricted to vessels coming directly from the islands. Another act was passed by Great Britain bestowing upon the British North American provinces the power to impose protective duties. This injured American shipping interests but encouraged growers of American produce. There grew up a triangular trade between the United States, the West Indies, and British North America which was

80. More than half of the most valuable kinds of timber which had previously gone from the United States to the West Indies passed through the foreign islands. Flour imported from the foreign islands to the British West Indies increased from 21,090 barrels in 1825 to 142,090 in 1828 and corn and grain from 9,249 barrels to 126,221 barrels. Benns, *op. cit.*, p. 156.

81. Nova Scotia had ceased to have an agent in 1826. "Those Islands now receive through these Colonies a regular supply of the Articles which they require from the Continent of America, for the greater part of which they pay with their own produce. This not only creates a most beneficial Barter Trade between the Northern Colonies, and the British West Indies, but increases the Intercourse between the Northern Colonies themselves. Nova Scotia and New Brunswick, in consequence of their situation on the Atlantic, become the Carriers between the Canadas and the West Indies. The Fish caught on the Coast of British America, is carried in the Vessels of the Atlantic Colonies to the Islands, and there disposed of for Rum, Sugar and Molasses, with which those Vessels return to supply the wants not only of the Atlantic Colonies, but of the Canadas also. The Canadians pay for these supplies in Flour, Pork, and other Articles of Agricultural Produce which are required for the Fishermen of the Seaboard, and all the Colonies are thus made to feel how beneficial they are to each other." *Journals of Assembly, Nova Scotia,* February 23, 1830. See also "The Joint Address of the House of Assembly and the Council," of which it could be said, "The words, 'loyal and faithful subjects' made up the greater part of it"; also the "Report of the Commercial Society," *Acadian Recorder,* February 6, 1830.

shared by American and British shipping, and the trade to England continued to be monopolized by British ships.

While Nova Scotia was, in the end, unsuccessful in prohibiting trade between the United States and the British West Indies, she had contributed powerfully to the extension of trade through her influence in the revision of British commercial policy. The effects were summarized as follows:

The Committee first refer to the new Colonial system, which commenced in 1826 . . . and, after examining it in its general operation, agree in opinion, that it was both intended to confer, and has actually procured to this Province, very great and essential advantages. Until that time our Shipping had, with Foreign Countries, an intercourse only of the most limited kind. . . . Now it is general, and without restriction. Formerly the circuitous voyages, now very advantageously pursued, could not be attempted . . . and the produce of our Fisheries was principally exported to the West Indies. Under the new System, this Market is relieved by the demand for that produce in more distant Countries, heretofore seldom visited by Colonial Vessels; and to Foreign Ships, against which they had, with few exceptions been always closed, our ports are now freely opened. The convenience of Warehousing Merchandise, as practised in England, is extended hither; and, for the first time in the History of British America, its navigation participates in the conveyance of Foreign Produce wholly or partially, to its ultimate destination.

These privileges present a very gratifying contrast, not only to the ancient system of restriction and prohibition, by which a barrier was interposed against all intercourse between British and Foreign Possessions, but also to the later and changeable policy, which periodically, and in a limited degree, afforded openings for Colonial enterprise. And whatever may have been the immediate objects of the Imperial Parliament in conferring these privileges, whether for the promotion of the Manufactures, Navigation or Commerce, of the Mother Country; or for the more liberal and generous purpose of restoring to the Colonists their equal rights with the Native British subjects, it cannot be questioned that the new system must become fruitful in benefits, wherever it can freely operate. But it was accompanied by these further advantages: The intricate, confused and indigested code of Plantation Laws was succeeded by simple and perspicuous enactments . . . and the abolition of all Fees, with the vexatious and illegal exactions in too many Ports attending them, gave general relief to all engaged in Navigation, and effected an immense saving to the Shipping interest. In all these points of view the new System of Inter-Colonial and Foreign Intercourse came recommended by powerful claims on the approval and gratitude of the Colonists; and entitled His Majesty's Government to their sincerest acknowledgments for the benefits thus conferred.[82]

82. *Journals of the Assembly, Nova Scotia,* March 23, 1829.

The commercial interests attempted not only to exclude American fishermen from Canadian waters and to exclude American trade with the West Indies, but also to build up the fishery and the trade with Canada. The bounty system was extended and modified in 1810 and 1811.[83] A scarcity of salt[84] in 1814 as a result of the war involving France and Spain resulted in a petition which stated

that the small quantity of that article at market was so very high in price during the last Spring and Summer that not only many of the Fishermen were deprived of their accustomed occupations, but eventually the West India supply became much less than it would otherwise have been, and thereby, in consequence of a limited exportation, the Merchants fell short, also in the amount of their imports, from whence the Provincial Revenue principally arises.

That it is almost superfluous to observe to this Honourable House, that when the Fisheries and the attendant Commerce flourishes, the effects circulate beneficially through every branch of industry, and that the farmer, the woodsman, and the mechanic, all there-from receive an immediate and general impulse; but the remote consequences to this, and the other British Provinces, will be in all probability, either adversely or prosperously incalculably great: for should the West India Islands suffer from a deficient supply, when the whole trade is confessedly in possession of British Subjects the consequence would prove as injurious to these Provinces, as a plentiful supply at this period would become beneficial to them, by offering to those who have the controul of British Commerce an undeniable proof of the incompetency or the competency of these Colonies in time of peace, to furnish what may be required without the intervention of Foreign vessels and traders.

That the article of Salt is of a bulky nature, compared with its price, and can only be imported from loss [*sic*] under circumstances which are liable to great fluctuations. That to apply a remedy to this inconveniency, the wisdom of the Legislature of this Province gave a proportionate bounty on Salt, when that article was below a certain fixed price, by which policy the importation became at any rate a certainty, and induced the Merchants in England and Scotland without hazard to employ their Capitals partly in this branch of trade, which not only supplied the Fisheries directly, but the Timber trade also partook of the other part of their Capitals, which furnished shipping for a return voyage, and thus the freight out and home was put upon two articles, instead of becoming an over burthen upon one; which at times, by reason of

83. 50 Geo. III, c. 10, and 51 Geo. III, c. 18. On September 18, 1811, Perkins sent his crew out for a short fishing trip "to make up four months to entitle us to the bounty." In 1812, it was recommended that £25 should be paid on a schooner of 36 tons taking 76 quintals, and that £74 should be paid on a schooner of 76 tons taking 360 quintals on the Labrador shore. A schooner of 70 tons with 650 quintals, lost in 1811, was recommended for a bounty of £100 10*s*.

84. A bounty on salt brought imports from Setubal via Newfoundland. *A Calendar of Official Correspondence*, p. 244.

low prices of Timber at home ceased altogether, to the ruin of the woodsman; the fruits of the whole winter labour often found no customers in the spring and summer, and

That certainty being confessedly the best friend of Commerce, the providence of the wisest Legislators has, at all times, afforded to it their aid and support, where favorable opportunities presented themselves, nor can a more unquestionable means of nourishing the Fisheries of this Province occur, than that of renewing a Bounty on Salt, so extended in its duration as to place the distant Merchant in security.

The Council had defeated a bill in 1814 but had granted a bounty in 1815[85] which kept the average price of salt at 10 shillings a hogshead (55 Geo. III, c. 20). With the postwar depression, the price of fish fell from 27*s.* 6*d.* a quintal in June, 1814, to 12 shillings in June, 1817; and while the price of white flour declined from 100 shillings per barrel in January, 1815, to 45 shillings in October it rose to 85 shillings in May, 1817. Salt reached 15 shillings and even 20 shillings a hogshead and, on March 3, 1817, there was a petition for a renewal of the salt bounty. The bounties granted by France and New England to their respective fisheries led to demands for the same support for Nova Scotia fishing; and the following year brought a petition[86] for a bounty on the tonnage of Labrador and bank vessels "above a fixed burthen, fishing upon shares or otherwise." This tonnage demand was disregarded. But, by 58 Geo. III, c. 20 (1818), a bounty was paid on salt, another of 1*s.* 3*d.* a quintal on cod, and a third of 10 pence on scalefish, with the proviso that the total expenditure should not exceed £3,500 (58 Geo. III, c. 21 and 59 Geo. III, c. 21).

In 1821 the lumber trade of New Brunswick, through the return of a great number of her lumber ships "ballasted with salt," gave that province "a cheap and abundant supply." The result was that many fisher-

85. *Journals of the Assembly, Nova Scotia,* March 1, 1815. A committee recommended that the tonnage bounty should be revived and the former bounty on salt—that of 1806—of 15 shillings a hogshead should be increased to 17*s.* 6*d.* *Idem,* March 14, 1814.

86. "The Petitioners are now experiencing a state of great depression arising from an increased population, without a means of affording it employment; a rivalship at every Market to which they resort; themselves left destitute, whilst those against whom they have now the misfortune to contend are supported by the most liberal policy of their respective Governments, being assisted by ample Bounties upon those Fisheries which a few years since were in a state of complete ruin but are now reviving with a vigour proportioned to the encouragement received; under such a contrasted state existing between Foreigners and the Petitioners, their utmost efforts must prove unavailing, unless supported by other aid than is within their individual power to afford." *Journals of the Assembly, Nova Scotia,* February 10, 1818. See a letter by "Probus" arguing against bounties and for expenditures on roads and bridges. *Acadian Recorder,* February 5, 1818.

men in western Nova Scotia went to Saint John for their supplies, and the intercourse thus commenced was "followed up by the shipment of their fish to that port in payment of such supplies, whereby the trade of Nova Scotia with the European and West India markets is greatly diminished."[87] A bounty on salt was recommended to induce fishermen to obtain their supplies from Halifax. In 1822 complaints were made of the irregularity of the supplies of salt obtained from transient vessels in the timber trade, of the low prices of fish[88] which followed the exclusion from Spain of the Newfoundland product by high duties, and of competition in the West Indies. It was proposed that a bounty should be granted on merchantable cod. The object was to increase exports to the European and South American markets, "to lead the resident merchant into regular importations of salt from Europe, enable him to extend his remittance to the mother country and at the same time increase the catch and improve the fish for the West India market."[89] At the same time he would be able to take advantage of the opening of South American and Mediterranean markets under the colonial-trade acts. Careful regulations were drawn up in 1823 to enable merchants and traders to supply fishermen with salt so that they might "be enabled at least to purchase a hogshead of salt for a quintal of fish."[90] In 1824 bounties of 1*s.* 6*d.* a quintal (4 Geo. IV, c. 2) were paid on exports of merchantable fish "cured in the province and exported to Europe, the Cape of Good Hope, Brazil, River of Plata, round Cape Horn, Mexico, and Columbia in ships or vessels registered and solely owned in the province." In 1824 and 1825 New Brunswick paid a bounty of 20 shillings a ton on vessels.[91] The advantage she possessed because of being in

87. *Journals of the Assembly, Nova Scotia,* February 24, 1821.

88. See prices 1820 to 1822. *Idem,* 1822, pp. 167, 206–207.

89. *Idem,* March 16, 1822.

90. *Idem,* February 26, 1823. See a letter protesting against bounties which neglected the boat fishery comprising three fourths of the fishery, and favoring Liverpool, Shelburne, and Yarmouth. *Acadian Recorder,* May 1, 1824. Adam Smith of the "land of porrich and political economy" was constantly cited. On the other hand it was argued that since agriculture was secondary to fishing in various parts of the province a stimulus to the fishery was a stimulus to agriculture.

91. The Halifax Chamber of Commerce declined to support a petition of the St. John's Chamber of Commerce encouraging "a tendency to limit the privileges of British subjects in their trade with foreign countries or be at variance with the spirit of those laws so recently made for the purposes of establishing and regulating such trade." *Acadian Recorder,* February 9, 1828. "Nova Scotia has but a very limited interest [in timber duties]." *Idem,* April 20, 1833. On the other hand, it was stated "that not only the export of Timber from the Colonies, and the consumption of British Goods, would be affected by the adoption of such a measure [the lowering of the preference on colonial timber], but the Fisheries also would become greatly injured thereby, as the supply of Salt and other Articles for the Fisheries is principally obtained from Great Britain, in return for the Timber exported thither, and is brought

the timber trade had lowered the price of salt to 9*s.* 6*d.*—at Halifax it cost from 13 to 14 shillings—and, as said, it attracted fishermen from Nova Scotia. In 1826 fish sold in Halifax at from 11 shillings to 12*s.* 6*d.* a quintal, and in Saint John at from 8 shillings to 9*s.* 1*d.* Larger vessels from the western coast of Nova Scotia, by going to Saint John, compelled Nova Scotia to purchase fish from Newfoundland or to submit to a loss of markets. Imports from Newfoundland were followed by complaints of consequent "injury of the fishermen, the farmers and the merchants of Nova Scotia." Vessels from the Labrador complained of a poor season in 1827 and of competition from Newfoundland, which had "the very great advantage of consuming all foreign articles free of duty besides that of receiving their salt on much lower terms." The depressed state of the fisheries[92] led to a discussion of methods to extend the industry, particularly by a bounty on tonnage.[93] To take advantage of the colonial-trade acts effective in 1826 the merchantable fish was obtained from Newfoundland and exported to South America. A bounty of 6 pence a quintal on cod exported from the province was recom-

to British North America in Ships which would otherwise be unemployed; that the very existence of Trade in these Northern Colonies depends upon the prosperity of the fisheries, which are the principal support of the Trade to the West Indies: We could not supply the Islands with Timber, and numerous other Articles, if our Fisheries failed, as that staple article affects directly or indirectly every other branch of Commerce from these Atlantic Colonies." *Journals of the Assembly, Nova Scotia,* January 8, 1831. The abundance of salmon in New Brunswick rivers supported an export trade to the West Indies. "The great risque would be, and which has already hurt the lower settlements on this river, that the vast abundance of fish might induce the settlers to apply more to the fishing than to the cultivation of their lands." Patrick Campbell, *Travels in the Interior Inhabited Parts of North America in the Years 1791 and 1792* (Toronto, 1937), p. 75, also pp. 27, 63, 66.

92. See Appendix B.

93. "From the Ports of the Bay of Fundy, the Fishery for Mackarel is becoming extensive, and the Province is surrounded by Banks abounding with Cod Fish; which as well as the Fisheries on the Coast of Labrador are almost wholly engrossed by the subjects of a Foreign State: to enable the Inhabitants of the Province to avail themselves of advantages they possess, in their favourable situation, for prosecuting these important Branches of Industry it appears necessary to offer that degree of encouragement which will assist them in the change necessary for pursuing the Sea Fishery, which requires greater preparation, a more extensive outfit, and more numerous crews than the present Boat Fishery can employ. The Committee also believe that a more extensive prosecution of the Bank Fisheries, as well as those on the Labrador Coast, will improve the habits of that class of Men, preserve them from many accidents, to which they are exposed in their open Boats; and above all will render them Good and Valuable Seamen. . . . At least the Committee apprehend these results to have been obtained by the Fishermen of our neighbours. For these reasons the Committee recommend that a bounty of ten shillings per Ton on the registered burthen of all Vessels owned and fitted out in this Province, and employed in the Bank, Sea, or Labrador Fisheries, during a certain limited time, or returning with a specific quantity of Fish, for each Ton of the Vessel's burthen." *Journals of the Assembly, Nova Scotia,* March 13, 1827.

mended as a means of improving methods of curing in both the boat and the vessel fishery. In 1828 it was decided to appropriate £5,000 a year for three years,[94] of which £2,000 provided a bounty of one shilling a quintal on merchantable fish for South America and Europe (9 Geo. IV, cc. 7, 20); and the regulations were altered in 1829 to make possible the extension of the bounty to fish shipped in "vessels of all nations and markets."[95] In 1833 further proposals were made to assist the fishery by payment of bounties,[96] but the opposition of agricultural interests was added to the opposition of the Council and they prevailed.

Similar attempts were made to give aid to the mackerel and other fisheries.[97] The catch of mackerel at Digby increased from 630 barrels in 1824 to 3,011 in 1825, and to 5,629 in 1826. A small mackerel trade between Pictou and the United Kingdom had begun in 1824. In 1825 complaints were made of the unfortunate influence on the markets of badly cured mackerel and herring. As a partial remedy, in 1827 an act was passed regulating the manufacture of barrels. At least seven pickled-fish laws and amendments were also placed on the statute books, but with little effect.[98] An inspection act was passed in 1828 and an

94. A society for the encouragement of fisheries was formed in 1828 to offer premiums both for fish taken on the Banks, the Labrador, and in the Gulf of St. Lawrence, and also for vessels landing the largest quantity of "merchantable" fish in Halifax.

95. *Idem,* February 14, 1828, and February 24, 1829.

96. To support the cod and mackerel fishery it was recommended that £4,000 be granted annually for three years. *Idem,* March 30 and April 6, 1833. The decline of interest in bounties for the fishery was evident in a facetious report of Mr. Homer of Barrington on February 4, 1832: "*This* is *my* own, my native land. . . . (At the emphatic word the hon. gentleman put his hand rather lower than the region of his heart.)"

97. Whaling ships, supported by a bounty, were sent to the Brazil Bank. In 1826 "the whaling association" launched the *Pacific.* See John McGregor, *British America* (Edinburgh, 1833), I, 340–341, 400–401. The *Susan and Sarah* arrived first on June 29, 1829, and she was followed by the *Trusty,* the *Rose,* and the *Pacific.* In 1832 the *Susan and Sarah* returned from the Pacific on June 2 with a full cargo of oil. Whale oil was admitted to Great Britain under nominal duties, but the industry was not profitable.

In 1827 a Nova Scotia vessel took 1,275 seals on the south shore of Newfoundland and four vessels with crews totaling 84 sailed in 1828. A seal fishery was prosecuted from Cape Breton, particularly from Cheticamp and Margaree, by about 20 small sail in the Gulf of St. Lawrence. In 1829 a seal fishery carried on in vessels from Halifax, Lunenburg, and Liverpool was prosecuted on the east coast of Newfoundland, but met with disappointing results. Bounties of £750 annually were recommended, to be paid at the rate of 15 shillings a ton on vessels above 50 tons, and 10 shillings on all vessels under 50 tons engaged in the seal fishery and "fitted out and owned in this province." *Acadian Recorder.*

98. Mr. Homer claimed in a speech to the Assembly of February 15, 1832, that pickled fish was sent to the West Indies in barrels which lacked uniformity and were in bad condition. Vessels for Demerara, Barbados, and other islands (excluding Jamaica which purchased on consignment) forced to "run for a market," were in a dif-

amendment in 1829.[99] Since, too, the problems of appointment and administration proved difficult, in 1833 a bill was passed which provided for the appointment of inspectors in all districts.

In the attempt to expand her fisheries, Nova Scotia had paid bounties to offset the effects of competition from the bounty-fed fisheries of the United States and France, to consolidate her position in the West Indies market, to increase her trade in the Mediterranean, and to take

ficult position when they sought to make collections. Nova Scotia alewives had "almost entirely lost their repute" and were displaced by Scotch herring. "Alewives taken going down the river after spawning" were "poison fish," and were "the very worst food that can be given to slaves, as it both disheartens them, keeps them continually murmuring and brings on those scorbutic diseases, so common among the negroes in that climate." Merchants supplied the planters and received returns at crop time but "hucksters or small dealers who buy their fish from the merchants when in the casks or barrels and retail it out to the town negroes by the pound or single fish" were much more careful in their purchases. Slaves who were hired out to masters at a stipulated payment, and who found their own food, were very particular in their choice. The abolition of slavery led to demands for better grades of fish. "These slaveholders talk of high feeding and low feeding their slaves. . . . Those who low feed them will purchase for them any cheap article of food no matter how stale or unpalatable it may be so that it will support nature and prolong their miserable existence." "It is the common practice to sell this damaged fish at public auction to the highest bidder and that there are always some of those hardened slave-holders who for the sake of getting cheap provisions will purchase it, when it in the end becomes the nauseous food of the forlorn African slaves." An inspection law "would be an act of humanity." Support of the mackerel trade in the West Indies would enable vessels to obtain the satisfactory coarse salt at Turks Island, Long Island, and Exuma. Cape Breton, Prince Edward Island, and Pictou also sent fish to Montreal in unsatisfactory condition. Very little "No. 1" mackerel was available for the Boston market. This grade was sold for domestic consumption; "No. 2" for slaves in the Carolinas, Georgia, and New Orleans; and "No. 3" to Cuba and the Spanish and French West Indies. Merchants in Halifax sending a consignment to Jamaica acted as their own inspectors with very beneficial results. *Acadian Recorder,* March 3, 1832.

99. "It appears that the Seine Mackarel Fishery at Canso has heretofore been a great injury to the character of our Mackarel, as at that place, for these several years past, they have been in the habit of drawing on Shore larger quantities of this Fish than they could well save, and much of it has been allowed to get tainted before it was split, and was then put up into Barrels, and imposed on this and other Markets, to the manifest injury of the Merchant, Consumer, and all concerned. We therefore recommend to all the Inspectors of Pickled Fish, to give particular attention to the Mackarel from Canso. Besides the evil before mentioned, it is supposed that by hauling large quantities of Fish in Seines, and allowing that Fish to get putrid in the Land Wash, and sometimes tripping Seines with a great number of half-dead mackarel therein, it has had the effect of driving that valuable Fish from our Coast, we therefore hope that by rigidly enforcing this Law, and by condemning all bad Fish it will cause the Fisherman to be more careful in Curing his Mackarel, which will probably enhance the value, and increase the demand for that Article, when it will become an object to send our small Vessels on Mackarel Voyages from our shores, and in the Bay of Fundy, and generally to enter more extensively into that business." *Journals of the Assembly, Nova Scotia,* March 6, 1829. See the same for an extensive account of the problems of administration.

advantage of the Convention of 1818. These measures were accompanied by various others. To increase the production of merchantable fish for the Mediterranean market, demands were made by the Halifax Chamber of Commerce for lower duties on wines in 1823, and they were reduced from 2*s.* 6*d.* in 1825. To encourage shipping and manufactures, duties were lowered on hemp, wet or dry hides, tallow, tobacco, flaxseed, cocoa, and raw turpentine; also on mahogany or other materials used by cabinetmakers.[100] Numerous articles formerly carrying high duties were admitted duty-free in 1828 "if brought from a warehouse in England." The long struggle against customs fees was ended by their abolition and the substitution of salaries for customs officials.[101] The position of naval officer was abolished in 1824. Halifax became a free warehousing port and depot for the East India Company in 1825.[102] Pictou and Sydney were made free warehousing ports in 1828, and Liverpool and Yarmouth warehousing ports in 1833.[103]

100. *Idem,* March 10 and April 3, 1826.

101. See Marion Gilroy, "Customs Fees in Nova Scotia," *Canadian Historical Review,* March, 1936, pp. 9–22.

"Few, if any, of the Articles which are liable to pay a Duty to His Majesty are carried in Coasters between the Ports of Nova Scotia and New Brunswick, but . . . the principal Trade consists in the carrying of Plaster of Paris, or Fish, Lumber, or Articles to be consumed in the Fisheries, which Trade is confined to small Vessels, upon which the payment of these fees is very injurious and oppressive." *Journals of the Assembly, Nova Scotia,* January 29, 1821.

"The House of Assembly are duly sensible that the Shipping Interest of this Province will derive very great advantage from the total reduction of the Fees formerly payable to the Officers of the Customs within the British Possessions; but they at the same time beg to represent, that those advantages will not be so extensive when it is considered that our Colonial Vessels still remain liable to heavy charges in Foreign Ports; while British Ships and Ships of Forcigncrs, which formerly paid towards the support of the Customs House, are now admitted free from any fees or impositions whatsoever." *Idem,* April 4, 1826.

102. The East India Company was allowed to export directly from China to the British North American colonies (5 Geo. IV, c. 88). Samuel Cunard was appointed agent. The *Trusty* sailed in 1825 and returned July 1, 1826.

103. By 3 and 4 Wm. IV, c. 59. In 1790 and later, outports demanded the establishment of customhouses to end the inconvenience of going to Halifax to enter vessels. The House of Assembly argued that "every port in the Province in which there was stationed an Officer of the Customs should be declared a free port, and that the House of Assembly would willingly grant money to effect these desirable objects, did they not conceive their constituents do already pay sufficient to support such an Establishment of the Customs as the Concession of these privileges would require. . . . The House of Assembly would humbly represent that confining the Foreign Trade of this Province to any specified number of its Ports is injurious to the others, without any corresponding advantage to the Colony or to the Empire at large; to compel the Ship-owners of the Distant Ports to resort to Halifax, Pictou or Sydney, to enter their Foreign commodities, before they are permitted to dispose of them, is to subject them to restrictions which are destructive to their spirit of enterprise, and which have a manifest tendency to induce and encourage illicit Trade." *Journals of the Assembly, Nova Scotia,* April 11, 1832.

In the development of trade to the West Indies, Nova Scotia attempted to build up an entrepôt trade not only for goods from the United States but also for goods from Canada. In 1818 it was held that imports of flour from Canada and restrictions on imports from the United States would assist British shipowners and injure foreign merchants.

A very large portion of the proceeds of the fish and lumber exported from hence to the West Indies returns to this country in flour, grain, etc., in the same vessels from the United States, the proceeds of the fish, etc., being vested in rum, sugar and other produce in the West Indies, and the vessel proceeding to a port in the United States whence the cargo is exchangeable for flour. . . . This trade is not confined to the port of Halifax but is carried on from the different outports in the province to a very considerable extent and has employed a large portion of our shipping since the peace with the United States.[104]

It was urged that steamship connections would enable Nova Scotia to become a center for imports of West Indies products for Canada and for Canadian products for the West Indies. Coal could be sent from Pictou and Cape Breton to Canada, and wheat, pork, butter, and lard could there be obtained more cheaply than in Boston or Ireland. Moreover, by so doing the drain of specie to the United States would be checked. A committee of the Assembly recommended "that the Trade with the Canadas should be fostered and encouraged as an outlet for the surplus West India Produce imported into the Province, and to furnish in return the necessary supply of Flour for the consumption of the Province, in place of the ruinous Flour Trade now carried on with the Americans."[105] The Assembly committee further recommended that a duty of 2*s*. 6*d*. a barrel should be imposed on flour from the United States, that there should be a bounty of 2*s*. 6*d*. a barrel on "flour imported from the Canadas" and of 2*s*. 6*d*. a hundred on flour made of wheat grown in Nova Scotia, and that the surplus of revenue from the duty, over and above the bounties, should be used to encourage the fisheries.

In 1825 the Assembly of Lower Canada pointed out that about 225 vessels were employed in trade between Quebec and the lower ports. They took from twenty-one to twenty-three days to go from Quebec to Halifax, and the return voyage upstream even with westerly winds was very tedious, while a steamship would make either run in six days. The advantage of possessing steamship service was obvious, and was recognized by the government. A bill was passed which provided assistance to

104. *Acadian Recorder,* March 21, 1818; also January 6, 1821.
105. See *Journals of the Assembly, Nova Scotia,* March 16, 18, 19, 1822.

the amount of £1,500, and a company was finally organized. In 1831 the S.S. *Royal William*[106] was launched at Quebec; and she made one round trip in nineteen days. But the cholera epidemic of 1832 made it necessary to cancel her sailings and she was sold in 1833 for £5,000. After she had coaled at Pictou she was sent to England.[107]

Nova Scotia had also sought to develop agriculture as a means of lessening her dependence on the United States. The advantages she possessed in the production of livestock were a handicap in the production of wheat and flour;[108] but there was an increase in such exports from the Bay of Fundy region, from Prince Edward Island, and various pocket settlements. Beginning in 1818, John Young published in the *Acadian Recorder* his famous letters signed "Agricola."[109] One result was the formation of a provincial agricultural society in the following year, and it was supported by an annual grant until 1826. Duties imposed on American produce in 1820[110] were increased in 1821 and in 1826. A sharp reduction in prices in 1821,[111] because of a depression in Newfoundland, was followed by complaints against increases in Nova Scotia's cereal crops, which were attributed to the premiums of the agricultural society.[112] In 1822 Pictou was supplying flour to the fisheries,[113] and in 1823 it was claimed[114] that there were more than thirty

106. The ship cost £15,607. Of 569 shares, Quebec took 204, Montreal 135, Halifax 139, and Miramichi 91. See A. M. Payne, "The Life of Sir Samuel Cunard," *Collections of the Nova Scotia Historical Society,* XIX, 75–91; and F. L. Babcock, *Spanning the Atlantic* (New York, 1931).

107. It was claimed that she was the first to cross the Atlantic under steam. "If our memory serves us, about ten years ago a steamer went from New York to St. Petersburg and it is only five or six years ago since the *Munster Lass,* a steamer built in New Brunswick, safely crossed to Ireland." *Acadian Recorder,* August 24, 1833.

108. Cumberland sent butter to Halifax; and Annapolis, Digby, and Saint John sent potatoes, oats, apples, cider, etc. In 1815, 8,299 cattle and 9,047 sheep were driven to Halifax. *Acadian Recorder,* January 20, 1816.

109. *Letters from Agricola* (Halifax, 1922).

110. The rates of duty were as follows: Horses 40 shillings, oxen 25 shillings, cows 10 shillings, sheep 3 pence, hogs 5 shillings. They were increased in 1821 to: Horses £2 10*s*., oxen 35 shillings, cows 15 shillings, sheep 1*s*. 3*d*., and hogs 20 shillings. *Acadian Recorder,* January 6, 1821.

111. Prince Edward Island had exported cattle and sheep since the 1790's and was particularly affected. In 1821 she exported 27,000 bushels of potatoes to Halifax. *Acadian Recorder,* December 1, 1821.

112. See a letter from a Yarmouth farmer. *Idem.* In 1822 it was claimed that the province was approaching self-sufficiency in the matter of wheat, but this was extremely optimistic.

113. In 1803, 50 vessels sailed from Pictou with fish, oil, and cattle for the West Indies and Newfoundland and timber for Great Britain. From June 1 to December, 1818, 107 vessels totaling 23,681 tons left Pictou with cargoes of lumber products and fish. In 1831 Pictou merchants complained of serious inroads by American smuggling upon the Magdalen Islands trade.

114. John Young claimed that a given ratio existed between the price of flour, beef,

oatmeal mills in the province. Wheat production[115] was increasing in Antigonish. By 1833 Nova Scotia produced flour and pork, potatoes and oats; and the importance of agriculture made itself felt in the growing power of the Assembly, with its demands for expenditures on roads and bridges and for control over the revenue. Prince Edward Island also exported substantial quantities of foodstuffs.[116]

The initiative of Nova Scotia directed toward the revision of British commercial policy and the development of the fishery and trade also expressed itself in demands for control over the public funds. The Declaratory Act of 1778[117] and an act passed in 1822 provided that the net revenue of duties regulating colonial commerce should go to the local treasurers. When the fee system had been abolished the board of customs had issued instructions for the appropriation of part of the returns from duties. The response of the Assembly was this:

> It is a duty which we owe equally to the Government of the Mother Country and to His Majesty's Subjects in this Province, most explicitly to state to your Excellency, that we consider all the Duties imposed by, and payable under, the said Act of Parliament, except such Duties as are payable to His

and cattle in which the two latter advanced "in geometric progression" to the first. In Nova Scotia, however, the neglect of green crops and especially turnips forced farmers to sell their beef in the fall and to glut the market. The increase in oat production was linked to the low price of beef. *Acadian Recorder,* November 15 and December 6, 1823. Land was appropriated to the scythe rather than the sickle. Young was probably defeated in the election of 1823 by fishermen's votes. In that year he argued that a duty of 5 shillings on flour had no effect on price, but encouraged trade from Canada and the West Indies, and was therefore of advantage to the fishermen. *Idem,* August 30, 1823.

115. In 1825 Pictou, Colchester, and Stewiacke were reported as producing sufficient for their needs, but Annapolis imported large quantities of flour and Indian corn.

116. Prince Edward Island exports in 1831 were: 11,749 bushels of wheat, 17,754 bushels of barley, 116,703 bushels of oats, 214,056 bushels of potatoes, 2,693 bushels of turnips, 153 barrels of pearl barley, 1,192 barrels of flour, 175,289 barrels of oats, 78 barrels of beef, 330 barrels of pork. Her first cargo of wheat was sent to Great Britain in that year.

117. "Whereas the Parliament of Great-Britain, in and by an Act, made and passed in the 18th year of His late Majesty's Reign, entitled An Act for the removing of all doubts and apprehensions concerning taxation by the Parliament of Great-Britain, in any of the Colonies, Provinces and Plantations, in North America, have declared that they will not impose any duty, tax or assessment, whatever, payable in any of His Majesty's Colonies, Provinces or Plantations, in North America or the West Indies, except only such Duties as may be expedient to impose for the regulation of Commerce, the net produce of which Duties are to be paid and applied as therein directed: Resolved, That no duty, tax or assessment save and except such duties as are in the above in part recited Act excepted, can, since the passing of the said Statute, be imposed upon the Inhabitants of this Province, other than by the assent of their Representatives in General Assembly." *Journals of the Assembly, Nova Scotia,* January 13, 1821.

Majesty under Acts passed previous to the 18th year of His late Majesty's reign, to be entirely under the controul, and at the disposal, of the Colonial Legislatures; and that no other authority whatever can legally direct the Collector of His Majesty's Customs to pay over those Duties, or any part of them, to any Person but the Treasurer of the Province.—This House therefore most respectfully submit, that the Order of their Lordships is in direct opposition to the 13th Section of the said Act of the Imperial Parliament, which directs the manner in which all these Duties are to be paid and appropriated, and cannot be warranted by any Clause in the said Statute.[118]

In 1829 the Assembly assumed control over the disposal of the customs revenues[119] and in 1830 control over the salaries of customs officers.

From thus achieving control over monies formerly in the hands of the Imperial government, the Assembly proceeded to demand further powers from the executive. They complained that the act passed in 1825 (6 Geo. IV, c. 114) imposed duties which were too high and that

in its present shape, it may be asserted that this Act places at the disposal of the Executive a very large sum of duties, not annually but permanently granted; it diminishes in no moderate degree that indispensable and constitutional protection which the right of originating and applying the taxes raised on the People affords to their Representatives; and in times of excited feeling, happily as yet foreign to Us, may restrain within narrow limits that legitimate influence which this House, as the proper source whence the wants of the Provincial Government are supplied, ought ever to retain. Surely, if such results be possible under any circumstances, measures of prevention now become indispensable; and it nearly concerns the House to provide that, so long as this permanent and productive Impost exists, no part of its Revenue shall be applied without the concurrence of the Assembly. To the local Legislature the thirteenth section of the Statute confines the appropriation; and in considering the express and unequivocal terms there used, the Committee find it difficult to comprehend on what principle of construction the right of the Province to the whole produce of these duties has ever been questioned.[120]

The brandy dispute of 1830 precipitated a crisis.[121] A tax of 1*s.* 4*d.* had been imposed on brandy in 1826 but only one shilling of this had been

118. *Idem,* April 4, 1826.

119. In reply to a proposal that three fourths of the average duties for three years should be paid to the colonial treasurer, and the remainder used to pay customs salaries, the Assembly "asserted with all dependence and respect, but firmly and distinctly that the duties imposed by the Imperial parliament do of right belong to and are by the statutes placed at the sole disposal of the colonial legislature, and that their appropriation can only originate in this house." *Idem,* March 23, 1829.

120. *Idem.*

121. J. S. Martell, "The Origins of Self Government in Nova Scotia, 1815–1836," doctor's thesis. See I. W. Wilson, *A Geography and History of the County of Digby, Nova Scotia* (Halifax, 1900), chap. xix.

paid into the treasury of the province. The Assembly asserted the justice of its demand that the whole amount should be paid over, and insisted that it be met.

> To freely dispose of the produce of their industry, to grant to the Crown such aids as they deem proper, and to limit and regulate their application, are rights inherent in British Subjects. . . . When they cease to possess them, they cease to be free. As the Representatives of the free people of Nova Scotia, the House of Assembly hold it to be their undoubted right, of which nothing has deprived nor can divest them, in Bills of rates and impositions on Merchandize to fix the matter, the measure and the time, the terms, limitations, conditions and qualifications, without augmentation, alteration or diminution, by His Majesty's Council.[122]

As a result of this insistence[123] the Assembly acquired control over the tariff.

In the half century between 1783 and 1833 the fishing industry had increased rapidly as a result of the Revolutionary War, of the coming of the Loyalists, of the difficulties of France and the United States in later wars, and of the aggressive efforts of commercial organization. At the beginning it was confined to small craft capable of going out for only one or two days or a week and taking fish largely for local consumption.[124] Speaking in round numbers, 10,000 men were employed. They took 120,000 quintals of fish, of which 40,000 were exported. Yarmouth shipping increased from 26 vessels, totaling 544 tons, in 1790 to 41, totaling 1,880 tons, in 1808, and to 65, of some 3,000 tons, in 1828. About twenty voyages a year were made to the West Indies, the remaining activities being a matter of coasting and fishing. In 1828 Yarmouth had seven trading establishments, Milton three, and Chebogue four.[125]

122. *Idem,* April 8, 1830.

123. "Of the Constitution which secures and will perpetuate to us these advantages, no principle has ever been held more sacred than that by which your Majesty's Subjects are entitled to direct and control the expenditure of all Monies paid by them for the purposes of Government, and in no portion of the Empire is this principle more anxiously cherished than in Your loyal Province of Nova Scotia. Its Inhabitants feel that a Revenue derived from their labour, and expended without the control of their Representatives, by which a fund is secured which may hereafter be applied to create an influence that may endanger the independence of this, the popular, branch of the Legislature, is at variance with the existence of their undoubted rights; and calculated to weaken that affectionate attachment which now universally prevails towards Your Majesty's Government." *Journals of the Assembly, Nova Scotia,* March 27, 1833.

124. *An Account of the Present State of Nova Scotia* (Edinburgh, 1786); also S. Hollingsworth, *The Present State of Nova Scotia* (London, 1787).

125. See T. C. Haliburton, *An Historical and Statistical Account of Nova Scotia* (Halifax, 1829), Vol. II, *passim;* also McGregor, *op. cit.,* Vol. I, Book IV; and R. M. Martin, *History of the British Colonies* (London, 1834), Vol. III, chaps. iii, iv.

Tusket River produced about 2,000 barrels of alewives annually, and Pubnico was an important center for the eel fishery. Barrington had declined as a salmon and alewives fishery and was producing only about 250 barrels; but its cod fishery produced 22,000 quintals. The port had 69 vessels totaling 2,710 tons, of which 2 brigs and 4 schooners were employed in the West Indies trade, 15 in the coasting trade, 8 in the Labrador fishery, 41 in the shore fishery and, in addition, 62 boats. Shelburne had declined rapidly from a population of 12,000. Its boats were described[126] in 1823 as small, badly built, unsuited to deep water, and confined to the taking of small fish, that is, those averaging 140 to the quintal. Fishing villages such as Sable River were scattered along the coast to Port Hebert. Port Joli and Port Mouton were engaged in fishing and lumbering. Whereas Shelburne boats made 20 quintals a boat in a season, Liverpool boats made 60.[127] Ample drying space favored the Labrador fishery and the port had 56 sail, or 4,150 tons. "The new commercial regulations have augmented its commerce and have occasioned a vast increase in its coasting trade." Port Medway had an excellent fishery, taking salmon, mackerel, and alewives, as well as cod in the shore and Labrador fishery. La Have produced a wide variety. Lunenburg exported 20,000 quintals annually, one third from the shore fishery.[128] It had more than 100 vessels engaged in the coasting and foreign trade and in the fishery. Nineteen of Lunenburg's ships, totaling 1,500 tons, made voyages to the West Indies. It obtained mackerel from Canso and salmon from the Labrador. Cargoes of lumber, potatoes, and fish were also early sent from Lunenburg to the West Indies; and, arriving before the ships from England, they returned with substantial profits invested in rum, sugar, and molasses for Halifax, Quebec, and Newfoundland. Vegetables, fresh meat, and cattle were sent to Newfoundland and traded for fish. Chester had 14 schooners and sloops. The Guysborough region, centering in Canso, was the chief source of fisheries. At Tracadie and Harbor Bouché, Acadians were engaged in the fisheries and the coasting trade in summer, and in shipbuilding and the cutting of staves and hoops for the fishery in winter.

126. *Report of the Board of Trustees of the Public Archives of Nova Scotia* (Halifax, 1938), Appendix B.

127. A proposal to annex Liverpool to Shelburne rather than Halifax for the payment of customs was met by determined opposition in 1803, and later in 1809, on the ground that it would "damp the spirit of industry and enterprise . . . and depreciate the value of property." Simeon Perkins' diary.

128. In 1824, 16 vessels of a total of 788 tons, employing 91 men and 9 boys, arrived at Lunenburg from Labrador and Canso. They belonged to eight separate interests. The largest, with 6 vessels, brought 7,282 quintals of fish and 250 barrels of oil. In 1822 ships of a tonnage of 355, employing 47 men and 6 boys, took 4,030 quintals of fish and extracted 130 barrels oil. *Acadian Recorder,* March 26, 1825.

In the Gut of Canso, Loyalist descendants engaged alternately in fishing and farming. "To this unprofitable system not only they, but most of those who have subsequently settled have always adhered." Irish families from Newfoundland were not accustomed to agriculture, and the importance of the fisheries kept it in a very secondary place. In Chedabucto Bay, cod and pollock, or scalefish, were taken early in the season, herrings in the summer and early autumn, and mackerel in the spring and early autumn. Both seasons in 1824 and 1825 averaged 50,000 barrels of mackerel. The center of the fishery shifted from Crow Harbor to Fox Island and again to Crow Harbor. Difficulties over lands granted to settlers were finally settled in 1811 by a court decision against the fishermen, who were required to pay rental[129] in the form of a barrel of cured mackerel for a hut and an additional quantity in proportion to the land occupied. Pictou was a center for the Gulf of St. Lawrence fishery and sent large quantities of fish and oil to the West Indies. On the Bay of Fundy, up to 1819, Digby exported from 60,000 to 100,000 half-bushel boxes of herring annually, but the industry then began to decline.[130] In April the first run of Granville fish were taken in nets and averaged from 50 to 70 to the box. The later run, in late May, were taken in weirs and ran nearly 200 to the box. After 1824 small craft, from 20 to 55 tons, engaged in the mackerel fishery,[131] and sailed early in June on voyages of from four to six weeks. The men, receiving half the catch, earned from £5 to £9 a month, and continued to fish up to November 1, but never made more than three voyages. On the Clare shore small vessels engaged in fishing and carrying lumber, livestock, oats, and barley to Saint John. In 1830 Nova Scotia and New Brunswick, "but chiefly the former," sent to the Labrador between 100 and 200 vessels, totaling some 6,000 or 7,000 tons, and 1,200 men.

Cape Breton, because of its separation from Nova Scotia and also because of the immigration of Loyalists and others, in 1781 began to experience an expansion of its fishery. In 1785 Sydney, Main-à-Dieu, Louisburg, St. Peters, and Arichat exported 30,580 quintals and 174 barrels of fish and 403 barrels of oil, while, in addition, quantities were shipped from L'Indiene (Lingan), St. Anns, Port Hood, Gabarus, and

129. See a petition from the inhabitants of Arichat, Guysborough, Pictou, and Egerton asking that an armed vessel be stationed in Chedabucto Bay. *Journals of the Assembly, Nova Scotia,* April 12, 1832. See H. C. Hart, "History of Canso," *Collections of the Nova Scotia Historical Society,* XXI, 22 ff.

130. I. W. Wilson, *op. cit.,* pp. 104–106.

131. They followed the American practice of cutting up stale mackerel to attract the fish, "the great secret of the mackerel fishery." The fish were taken by hook and line, split and salted in barrels, and sold at wholesale at 17*s.* 6*d.,* or at retail at £1. Exported from Saint John, N.B., they brought up to $6 a barrel. Salting in the hold in kenches was less satisfactory than in barrels.

L'Ardoise. In 1787[132] petitions were made for grants of land at Conway Harbor and elsewhere for settlers, "the only support of the fishery on this island." The outbreak of war with France and the declaration of the Republic resulted, in 1793, in the migration of many skilled fishermen, Acadians from St. Pierre and Miquelon. From July 1, 1797, to July 1, 1798, 135 vessels entered at Sydney and 195 vessels cleared, while 54 entered at Arichat and 64 cleared. In 1801 Arichat and the northwest shore possessed 192 vessels and in 1804 cleared 86. These were mostly of 40 to 50 tons. Exports from Arichat in the latter year totaled 22,000 quintals and 1,533 barrels of fish. In 1816 exports from Cape Breton included 34,039 quintals and 4,408 barrels of fish, and 6,341 gallons of oil, of a value of £29,423; and Cape Breton's total exports amounted to £38,783.

The formation of a separate government in 1784 brought about revenue difficulties and uncertainty as to the legality of taxes,[133] and precipitated problems incident to the most suitable system of representation and the creation of an Assembly; but they were solved by a reannexation to Nova Scotia in 1820. Problems of competition between Cape Breton and Nova Scotia in the fishing regions near Canso and Arichat, the increasing prominence of Halifax as a trading center, and an increase in the number of traders[134] in Cape Breton weakened the control of Jersey merchants, and, in spite of protests, particularly from Sydney, intensified the need for union brought to the fore by the struggle to meet American competition.

In 1828 Cape Breton exported 41,000 quintals of dry fish, 18,000 barrels of pickled fish, and 2,209 barrels of oil. The island had 340 registered vessels averaging about 50 tons, and large numbers of small vessels, not registered, with probably a total tonnage of over 1,000. "About 300 boats are thus engaged [in the fisheries] and owned on the N.E. coast, by no means the most populous in resident fishermen, although great numbers repair thither in vessels from the southern shore and elsewhere." Traders other than Jersey merchants were established

132. In 1787 an ordinance prohibited the dumping of offal within three leagues from shore. Scattered settlements necessitated the alteration of an ordinance of May 13, 1790, appointing officers to cull and survey fish after August 24, 1792, and permitting the sale and export of fish so long as purchasers and vendors "mutually agree among themselves as to the quality of the said fish."

133. See Helen T. Manning, *British Colonial Government after the American Revolution, 1782–1820* (New Haven, 1933), pp. 57–58, 66.

134. Sir James Kempt wrote on October 19, 1821: "A very strong address [favorable to annexation] was presented to me at Arichat by the principal inhabitants of that district, by far the most populous and important in the island." Richard Brown, *A History of the Island of Cape Breton* (London, 1869), p. 445, also *passim*.

at Arichat,[135] Ship Harbor, St. Peters, L'Ardoise, Sydney, Main-à-Dieu, St. Anns, and Margaree,

> who supply the fishermen in those and the intermediate places, and in payment receive the fish, part of which are sold and consumed in the country and part exported. Traders also visit the coast and furnish the inhabitants with various articles taking fish and agricultural produce in return. . . . The agricultural exports consist principally of livestock, potatoes, oats, butter, cheese, salted beef and pork which find a market in Newfoundland; and wheat from the Gulf shore taken to Halifax.[136]

Imports included about 40,000 barrels of flour, 38,000 gallons of rum, and 32,000 gallons of molasses.

The results of expansion in Nova Scotia[137] and Cape Breton showed themselves in the growth of trade, especially at Halifax. In 1826, following the acts passed in 1825, Nova Scotia sent ships into the import trade which was formerly dominated by New Brunswick and British vessels. The number of ships inward to Nova Scotia increased from 1,427 in 1825 to 1,846 in 1826; imports[138] rose from £512,735 to £738,181, and exports from £390,371 to £454,621.

135. "The chief port is Arichat, long the seat of the trade carried on by merchants in the Island of Jersey . . . who employ the inhabitants and their vessels in taking the fish which are then exported in the Jersey ships to Spain, the Mediterranean, the West Indies and the Brazils. Arichat is indisputably the first commercial port in Cape Breton and exports much of the agricultural produce of the Island." Haliburton, *op. cit.,* II, 221; also D. C. Harvey, *Holland's Description of Cape Breton Island and Other Documents* (Halifax, 1935), Appendixes A and B.

136. Haliburton, *op. cit.,* II, 252–253.

137. POPULATION OF NOVA SCOTIA*

		1817	1827
Districts of	Halifax	16,487	24,876
	Colchester	4,972	7,703
	Pictou	8,737	13,949
Counties of	Annapolis	9,817	14,661
	Shelburne	8,440	12,018
	King's County	7,155	10,208
	Sydney	6,991	12,760
	Hants	6,685	8,627
	Lunenburg	6,628	9,405
	Queen's County	3,098	4,225
	Cumberland	3,043	5,416
		82,053	123,848

* *Journals of the Assembly, Nova Scotia,* March 31, 1828.

138. Imports from Great Britain increased £109,292; from foreign sources, £104,299; and from the West Indies declined £4,550. Exports to the West Indies increased £47,647; to Great Britain they decreased £18,799, and to foreign ports, £1,204. *Acadian Recorder,* March 17, 1827.

In 1828 Halifax had 6 ships, 67 brigs, and 77 schooners. Of these 70 were engaged in the West Indian trade, 6 in Brazil and foreign European trade, 4 in British trade, and the remainder in the coasting trade and the fisheries.

Its manufactures are still in an infant state, most of them dating only from about the year 1815. They consist of sugar refinery, distilleries of rum, gin, whiskey etc., breweries of porter, ale, etc., cabinet work, soap and candles, glue, leather, carriages, chocolates, linseed oil, combs, brushes, paper, snuff and other manufactured tobacco, flour, cordage, etc. Halifax in common with every part of British America experienced in its trade the embarrassments and difficulties incidental to a sudden transition from war to peace, but as the merchants of this place have always traded within the limits of their capital, the shock, though severe, was not such as to induce either ruin or distress. Business is conducted in a safe and honorable manner, and it is a fact highly creditable to the mercantile community that only one bankruptcy occurred among the respectable part of the merchants during the whole of the administration of his Excellency Sir James Kempt, a period of eight years.[139]

QUEBEC

What did it signify whether the fish they eat came from Gaspé or from Nova Scotia or New Brunswick?

Louis Joseph Papineau

In Quebec commercial interests were similarly engaged in competition with the United States; but their fishery was more widely scattered, and they were less effective than those of Nova Scotia. The Magdalen Islands had been exhausted as a center of the walrus fishery before the American Revolution. J. Janvrin of Jersey had an establishment on the islands in the years following 1782. As has been said, in 1793 Acadians had migrated from St. Pierre and Miquelon. The population increased, approximately, from 500 in 1797 to 1,000 in 1828[140] in spite of a grant of the island under letters patent to Admiral Isaac Coffin on August 24, 1798;[141] but the activities of such settlers were restricted, chiefly as a

139. "We have upwards of 100 licensed houses and perhaps as many more which retail spiritous liquors without license, so that the business of one half of the town is to sell rum and the other half to drink it." In 1825, of the total imports of brandy and gin, which amounted to 41,541 gallons, Halifax imported 32,361; of 563,708 gallons of rum, Halifax imported 386,248; and of 44,626 gallons of wine, 43,209. Haliburton, *op. cit.,* II, 13, 19.

140. Paul Hubert, *Les Iles de la Madeleine et les Madelinots* (Rimouski, 1926), Appendixes vii, viii.

141. See Lieutenant Baddeley, "Reports on the Magdalen Islands," *Transactions of the Literary and Historical Society of Quebec,* April, 1833, pp. 140–141.

result of the expansion of the New England fishery. The Convention of 1818 was less rigid in its application to the Magdalen Islands and to other relatively unsettled regions. American vessels arrived for the most part in April and paid the residents 10 per cent "for the privilege and trouble of drying their fish upon the beaches and flakes"; and French vessels from St. Pierre and Miquelon arrived in July and August. In 1831 it was claimed that 65 American ships were engaged in the herring fishery, and they took about 14,000 barrels, worth £7,000. Bay of Fundy vessels caught some 1,000 barrels, worth £500. About 12 schooners of from 30 to 60 tons, owned by residents of Pictou, Halifax, and Quebec, were engaged in trade. Some 27 were fitted out in the islands, 10 being in the Labrador fishery, and manned by from five to seven men on the share principle. In spite of smuggling, six stores and two extensive fishing establishments obtained seal oil and skins valued at £4,000, to which were added 12,000 quintals of dry cod, making a total value of about £12,000.

The recovery of the Gaspé fishery after the Treaty of Versailles was dependent on the Channel Islands. New England vessels were at a greater disadvantage than in the Magdalen Islands. The widely scattered character[142] of the Channel Islands interests enabled small groups of merchants to dominate. William Smith of Quebec had had three times the capital of Robin, and ships in proportion. John Sholbred, his representative in 1784, lost sums totaling several thousand pounds. Control by Quebec suffered a general decline.[143] After six years of activity, an establishment at Bonaventure, which succeeded that of William Smith, failed. Charles Robin went into business again in 1783 but lost enormous sums. Thomas Le Mesurier started a Guernsey establishment at Gaspé in 1784 but it failed with heavy losses as did its successor, a Jersey firm, Nicholas Fiott and Company, under the agency of George le Geyt at Percé. The same fate overtook the firm of Hamond Dumaresq and Company under their agent at Bonaventure, that of Johnson at Mal Bay, those of Edward Square and John Le Montais both at Point

142. In a petition asking to be allowed to sail without a convoy, in 1796, it was stated "that in particular the fishing trade carried on by different societies of merchants on the coasts of Newfoundland, Nova Scotia, and Labrador, where they have establishments at considerable distances from one another, obliges them to despatch single vessels at different times of the year to carry out the sundry articles wanted for the fishery; others during the winter season to carry to a market the produce of those fisheries and to fetch salt wanted for the ensuing year, renders it impossible to get those vessels together at any particular fixed time." A. C. Saunders, *Jersey in the Eighteenth and Nineteenth Centuries* (Jersey, 1930), pp. 90–91.

143. Three poor years in succession caused heavy losses to Canadians who came down each year to fish in shallops. In 1786 the district produced 50,000 quintals of cod and 1,000 tierces of salmon.

St. Peter, and another Jersey firm which, represented by Daniel le Geyt, failed after several years' struggle at Bonaventure. On Chaleur Bay, John Lee, supported by London interests, another firm, Math, Stewart and Company, and John Rimphoff, supported by similar connections, each lost about £12,000 sterling. Failure awaited a Jersey company with establishments at Port Daniel and Miscou under the agency of Philip LeCouteur. Daniel MacPherson, with fishery and supply establishments at Point St. Peter, Mal Bay, and Douglastown, finally succeeded. Guernsey men settled at Grand Grève in 1783 and at Indian Cove —Simonds—in 1798. The firm of Janvrin,[144] established at Grand Grève in 1770, moved to Percé in 1798. "Thus it is evident that if there is no competition at present it is because the place is poor."[145] The problems of the industry were eventually met by the evolution of efficient business organizations[146] and the monopolistic advantages possessed by Channel Islands merchants.

Established organizations demanded the revision of government regulations and protection from fishing ships. On February 12, 1787, Charles Robin[147] complained of the high duties on molasses and rum, of the difficulties inherent in Gaspé's greater distance from Europe, of the ice in spring and fall, and in having to compete with Newfoundland, which paid bounties to bankers (26 Geo. III, c. 26) and admitted rum and molasses free. It was claimed that Gaspé fishermen were compelled to stay with their fishing rooms while the Newfoundland planter could dispose of his rooms as he pleased. In Gaspé slight and temporary structures were built. They required heavy annual outlays for repairs, and this meant the loss of a valuable fortnight in the spring. Livestock damaged the equipment. Fishing vessels dumped offal overboard to the injury of the fishery. Salmon on the Restigouche were speared by the Indians and were consequently fit only for the West Indies, whereas the use of nets would make it easy to build up a European market.

Several large vessels were engaged in the fisheries of Chaleur Bay,

144. See the account books of this firm, 1796–99, in the Canadian Archives.

145. Saunders, *op. cit.,* pp. 213–214.

146. "Ceaseless industry, frugality, and caution and especially in the strict enforcement of the rule that no person shall be retained about the business who cannot be profitably employed, have long secured it the most solid prosperity." M. H. Perley, *Report on the Sea and River Fisheries of New Brunswick* (Fredericton, 1852). "Six *commis* [clerks] were appointed, each of whom returned to Jersey at the end of two years to report on the business. Minute regulations were introduced, and apprentices were brought out at the age of fourteen."

147. *Quebec Gazette,* April 30 and May 8, 1788. Provision was made for the inspection of pickled fish before they were put on the market, and for the adjustment of disputes between the seller and buyer of dried codfish. Hugh Munro was appointed inspector and culler of salmon and other wet fish.

Gaspé, Bonaventure Island,[148] and Percé in 1788; and regulations (28 Geo. III, c. 6) were put in force to encourage a free fishery between Cap Chat and the first rapids in the Restigouche. "The commander of every ship or vessel fitted out from Great Britain or the Dominions thereunto belonging and entering into any creek or harbour may reserve to himself so much beach or flakes, or both, as are needful for the number of boats he shall there use, provided they are unoccupied by any other person, or are not in this and the preceding cases private property by grant from His Majesty, or by grant"—that is, a grant made before 1760. As a result of the privileges given to fishing ships, Robin complained by letter from Paspebiac in January, 1790, that large numbers of American vessels secured foreign registers in Halifax and participated in the Chaleur Bay fishery.

Between 1808 and 1814 Gaspé gained from the high prices of foodstuffs in Europe due to the war, the disappearance of the United States from the fishery, and the consequent rise in the price of cod. The cost of supplies in part offset the rise in price. Charles Robin and Company and Philip and Francis Janvrin, in October, 1813, complained by petition that duties imposed on salt in an act (53 Geo. III, c. 1) from 1812 to March 25, 1815, without any arrangement for a drawback, were "very injurious" to the fisheries, for they had been forced to compete with other colonies which had no salt duties. After the Convention of 1818 it was said that the fishery declined rapidly in Chaleur Bay and in Gaspé because of the recovery of the New England fishery.

> It is generally supposed by persons who have practised the cod fishery in this bay, both in schooners and boats, that it receives its chief supply of fish from the southward on the Orphan Bank. It is beyond any manner of doubt ascertained that many hundred American craft (chiefly schooners) catch their load of fish in the Gulf of St. Lawrence and chiefly on the Orphan Bank, and many close to the islands of Miscou and Shippigan; as soon as the Gulf is free of ice, the American craft take their station so that before the 30th May there are generally several hundreds on the Orphan bank only and its vicinity.[149]

The organization of the Channel Islands merchants made it easier to grade for various markets and particularly for the Mediterranean, and they were supported by the colonial-trade acts in an extension to South America.

148. Lower Canada Sundries. Canadian Archives.

149. Letter of Charles Robin and Company to H. Bourchier, September 4, 1822. E. T. D. Chambers, *The Fisheries of the Province of Quebec* (Quebec, 1912), pp. 117–120.

A special committee of the Assembly,[150] appointed to investigate the fishing industry in 1823, indicated its increasing importance to the Province of Lower Canada. It gave special attention to the pickled-fish industry—that is, to salmon, herring, shad, and sturgeon—and concentrated upon the problem of grading. As a result of the recommendations of the committee, an act was passed in 1823 which required the appointment of inspectors at Montreal and Quebec. Another act in 1824 introduced regulations in the interest of conservation, particularly of salmon.[151] The evidence indicated an important internal market for pickled fish, and for cod from New Brunswick, Newfoundland, and Labrador; but it also brought out serious weaknesses in the West Indies market for cod, salmon, and particularly herring.

The attitude of the government of Lower Canada put serious burdens on the Gaspé fishery when competing with fisheries given support by Nova Scotia and New Brunswick. Proposals[152] were made to develop the Quebec fishery, and in 1827 there were protests against the imposition of duties on West Indian products entering Lower Canada; it was claimed that these duties were a handicap to the development of trade and the fishery. "If," said one objector, "you release the West Indies from taking the products of Canada and of its adjoining seas, release us as well from taking the rum and sugar of the West Indies. If you relax one part of the system, relax also the rest. Be consistent—do not tie up our arms and loosen those of our rivals."[153] In 1829 a bill was introduced which was designed to improve the markets in the West Indies, but failed to do so.[154]

150. *Journals of the Assembly, Lower Canada,* 1823, Appendix P.

151. Chambers, *op. cit.,* pp. 120–134.

152. In 1823 a bounty on cod and on vessels was proposed, to give "employment to a multitude of river craft rendered useless to the owners by the introduction of steamboats."

153. S. Atkinson, "The Effects of the New System of Free Trade upon our Shipping, Colonies and Commerce Exposed in a Letter to the Right Hon. W. Huskisson, President of the Board of Trade," *Montreal Gazette,* April 5, 1827.

154. See a proposal to provide a drawback of 2*s.* 6*d.* on every hundredweight of dry codfish shipped to the West Indies, to be deducted in part payment of the provincial duties paid on the West Indian produce purchased with the fish. This would in part enable the dried cod of the Canadian merchant to compete with those of Nova Scotia and Newfoundland in the West Indian market for that article. "The drawbacks, with the advantage the [Canadian] West India Merchant has over those of the sister Provinces of Nova Scotia, New Brunswick, and Newfoundland, in assorting his cargo with Beef, Pork, Flour, Butter, Lard, and other small articles in Quebec, would in part compensate him for the dangerous navigation of the Gulf of St. Lawrence the Canadian vessel has to encounter, thereby giving him a small prospect of gain on his outward cargo. A Bill was introduced into the House of Assembly last Session by a worthy Member, (I mean Mr. Christie) which, if it had passed into a law, would, as far as I had an opportunity of judging, have been of great service to the Fishermen and Fishdealer." *Montreal Gazette,* January 11, 1830.

Demands presented in the Assembly in 1830 in favor of the Gaspé fishery,[155] and in 1831 in favor of the Gulf and North Shore fishery, were opposed by agricultural interests. It was claimed that a bounty would place Lower Canada on a parity with Nova Scotia and New Brunswick and break up the monopoly of the Jersey merchants.

Many fishermen had in consequence left Gaspé and gone thither, and it being a condition to entitle them to the bounty, they must purchase every article used in the fishery within those Provinces, the Lower Canada merchants were deprived of so much of the market for their goods which the Gaspé fisheries afforded. . . . Although the fishery had been far more abundant this year than in former years, a less quantity, only 20,000 quintals, had been brought to the Quebec market. The premiums given in the adjacent Provinces did not alone affect the Gaspé fisheries, but had the same effect, though in rather a less degree, upon the very considerable fisheries of the North Shore. In consequence of the present state of things, the exportation of fish from Quebec to the West Indies had almost entirely fallen off, and had been diverted to Halifax. He [Mr. Thibaudeau] therefore moved that it is expedient to establish a bounty on the produce of the fisheries in this Province, and on such parts of that produce as are exported from Quebec.[156]

The opposition of agriculture was voiced in the protests of Papineau.

It was encouraging a species of industry the least proper for this country; for every fisherman they created they withdrew a cultivator from the soil, a pursuit that is infinitely more fit for Canada than any fishery. They had done quite as much as they need do, in not doing anything to injure the fishermen, and in taking off the duties on the salt and all materials they required; if they did not prosper, it was a sign that the pursuit was not profitable. And if it be not so, let them turn to cultivation . . . to the settlement of lands; they have enough around them fit for it. They must avoid the complicated system of commercial legislation, by means of bounties, drawbacks, and prohibitions, from which they had had the good fortune to escape hitherto. . . . If cultivation was carried on in Gaspé to the extent it was capable of, never mind the desertion of their fishermen; a more advantageous barter trade will be carried on with them, and the other Provinces; they will bring in their fish, and take in return the produce of the soil. What did it signify whether the fish they eat came from Gaspé or from Nova Scotia, or New Brunswick, and they would get it all the cheaper from them on account of the premiums.

155. For a list of exports see Chambers, *op. cit.*

156. *Montreal Gazette,* January 10, 1832. "Mr. Neilson argued it was a branch of industry not so much to be despised; it might furnish them an abundant supply of food—of a food particularly desirable in a Catholic country—for people would not always eat eggs and potatoes. It was advantageous in promoting the coasting trade, and forming seamen. . . . Every branch of industry was useful to the community, one more and the other less, but all ought to be fostered and encouraged."

A meeting of freeholders and citizens of Cape Cove and Ance à Beaufils on September 24, 1832, protested

> that the duty of 2½ per cent on importation into this District, of Dry Goods, is, in the total absence of all encouragement by Bounties or otherwise to the fisheries, peculiarly burdensome, and the more felt, as the trade carried on in connection with the fisheries on the south side of Bay Chaleur, belonging to New Brunswick, in our immediate neighbourhood, is exempt from any such duty, and as such ought in the meantime to be represented; and its repeal, (as well as the repeal of all duties upon Sugar and Molasses, and upon produce necessary to the fisheries, consumed in the District, Rum and Wines excepted, or at least an equalization with those of New Brunswick,) [ought] to be solicited from the Home Government . . . the above duty originally imposed by the Colonial Legislature, being continued in virtue of an Act of the British Parliament known as the Canada Trade Act . . . and that it be represented in support of this application that the motives of policy and justice towards the Sister Province of Upper Canada, which rendered necessary the continuance of those duties by Act of Parliament on importations to Quebec and Montreal . . . great proportion of which pass into, and are consumed in Upper Canada . . . do not apply to importations into this District, from whence they cannot without afterwards passing through those ports and payment of the duties there, reach that Province . . . and that it will essentially contribute to the relief of the "fisheries carried on in the District of Gaspé, if the said duty be taken off from all importations hither directly from Britain, or a drawback allowed, in case of importation from Quebec or Montreal of Merchandise, having there paid the said duty, and an equalization of all other duties with these of New Brunswick be made."[157]

In spite of the energetic support of a former Nova Scotian, Robert Christie, who represented the constituency of Gaspé, and who was successively elected and expelled five times between 1822 and 1834, the position of the Quebec fisheries remained subordinate to agriculture.

Here, too, the effects of monopoly control appeared clearly in the glaring evils of the truck system.[158] It was said that through this sys-

157. *Montreal Gazette,* March 5, 1833. See notices of meetings at Percé September 28, 1832; Sandy Beach, Bay of Gaspé, October 4, 1832; Gaspé, October 5, 1832; Douglastown, October 7, 1832; the north side of Bay of Gaspé, October 9, 1832; Point St. Peter and Mal Bay, October 12, 1832—all protesting against the expulsion of Robert Christie, member for Gaspé. *Idem.*

158. For his fish the fisherman was paid half in cash and half in goods, and of necessity the money received had to be spent in the company's store. "The people, to whom the company heads have made themselves necessary, live in a sort of serfdom and are wholly dependent upon them. For every thirty-three there is allotted an area of ground measuring, in arpents, thirty-three by ten, or only ten square arpents apiece; that is, not enough to live on. The owner's only resource is to fish. And as he is in no position to equip himself for that, he is always in debt to the merchant, always at his disposition; and he is even forced to take his chance of being put aboard

tem Robin was able to retire in 1802 after laying the foundations for his firm and bringing his nephews into the organization. Men trained in the firm of Charles Robin and Company also established the firm of Le Bouthillier* in 1830. In 1828 it could be recorded that

the establishment of the Robin Company at Paspebiac comprises eight dwelling houses, ten store houses with a salt loft, rigging loft, and mould loft for ship building and eleven sheds. The annual amount of outfits and supplies imported from Europe is upwards of £10,000 sterling; they export from the district 22 to 27,000 quintals of dry codfish, about 100 bbls. pickled fish, and 30 to 50 tons of cod liver oil. Besides the above they have an extensive fishing port at Percée, one at Grand River, and one at Newport, where the ship's crews and a number of servants from the parishes in the environs of Quebec—in all about, and sometimes above, three hundred and fifty men—are employed from the beginning of May to the latter end of August, and about half that number till the close of the navigation in the latter end of November. The trade they carry on in the district supports about eight hundred families which they supply with all necessaries for the fisheries, wearing apparel, etc.[159]

In 1807 Lymburner and others purchased the seigniory of Mingan, but the decline of the seal fishery led to failure in 1820. However, the expansion of the Labrador fishery from Newfoundland led to the reannexation of the area—save for the Magdalen Islands—which had been returned to Canada by the Quebec Act, in 1809 (49 Geo. III, c. 27). Protests from Quebec interests engaged in the salmon and seal fishery resulted in the final return of this region to Lower Canada in 1825

one of the company's ships and sent to Europe as a sailor when he owes so much that his fish cannot pay the debt. For that reason it is not unusual to find fishermen who have been to Cadiz, Messina and Palermo." This was written of conditions in 1811. Faucher de Saint-Maurice, *De tribord à babord* (Montreal, 1877), p. 361.

"When they make any motion to shake off their chains and do their buying elsewhere, the threat is made that they will be charged with their debts before the local courts, which they fear. Perforce, they bow the neck once more and in long penances have reason to regret their attempts to free themselves.

"Under the regulations governing the company agents they are forbidden to let the fishermen have anything before a given time; filled with supplies though the warehouses may be, not a single biscuit will be given out before the date assigned. Since the fishermen are paid only in goods they can put nothing aside for the future. When they have possessed themselves of such things as they have to have, they can only take out the rest in luxuries. That means that the ladies here are better dressed than the damsels of Quebec's faubourgs.

"Schools are ruled out. 'They have no need of education,' wrote M. Phillipe Robin to his clerks; 'If they were educated would they be any cleverer as fishermen?'" (1836.)

The above conditions remained in existence until 1882, and much longer in modified guises. De Saint-Maurice, *op. cit.,* pp. 360–361.

159. *Montreal Gazette,* November 20, 1828.

(6 Geo. IV, c. 59).[160] Eventually the territory was controlled by Jersey firms and by W. H. Whiteley from Boston, with an establishment at Bonne Esperance. As a result of the bounty-supported industry of the United States and the advantageous position of Nova Scotia and New Brunswick, large numbers of vessels of the latter were on the Labrador in 1829; and the Lower Canadian industry there had declined from over 20 vessels of from 40 to 80 tons in 1822 to 8. "Some of the fish caught by them is sent to Europe and the rest carried to Quebec; besides which they carry about £6,000 worth of furs, oil and salmon to Canada."

To conclude, the disturbance to the New England and French fisheries caused by wars in Europe and North America up to 1815 had contributed to the expansion of the fisheries of Nova Scotia and Newfoundland. The fishing industry and, in turn, shipping in Nova Scotia made for the organization of a merchant class which pressed for advantages that were to be had by the modification of the commercial system, as made in the colonial-trade acts. The independence of the United States and the rise of independent states in South America strengthened the trend toward flexibility within the British Empire. This involved a lessened control by the British West Indies and the exercise of an increasing influence by Nova Scotia in the interests of expanding shipping and trade. Nova Scotia inherited the traditions of New England and made still stronger the trend toward flexibility.[161] Increasing flexibility aided in the expansion of the Newfoundland fishery, which, with the recovery of the French and New England fisheries after 1815, pressed Nova Scotia along other lines of development, particularly agriculture. The defeat of Nova Scotia in the West Indies by the United States was offset by increasing demands from Newfoundland. The position of Nova Scotia, made plain by her external policy, was also evident in her internal development. Agriculture and other interests, working through the Assembly, pressed for expenditures on roads and bridges; and the rise of temperance societies was an indication of the limitations of public finance based chiefly on revenues from rum and spirits, and the increasing importance of banks. Demands from Great Britain for control over revenues were accompanied by demands for control by the Assembly. Divergence of interests necessitated political adjustment and, in turn, developed the political capacity which became conspicuous in the struggle for responsible government. Shipping and trade made for a

160. *P.C.* (*Privy Council*), I, 205 ff.

161. For a discussion of the significance of the development of the Assembly in Nova Scotia, especially in 1800, as contrasted with other British colonies, see Manning, *op. cit.*, pp. 130–131, 138–139. The disappearance of the debt in 1797 was a factor in hastening the Assembly's demands for control. *Idem.*, pp. 196–197.

Nova Scotian interest in the Imperial sphere. But, with increasing flexibility and the trend toward free trade, the province was forced to depend to an increasing extent on her own resources; and, in her persistent struggle with New England, she turned eventually from Empire to Confederation. Newfoundland entered upon the following period with a system of government destined to take lines of growth similar, but in opposition, to those of Nova Scotia. As in Nova Scotia, but at a later date, increase in population brought representative institutions and control over revenues. Newfoundland moved toward independence in opposition to a decentralized federation to which Nova Scotia agreed in the end.

Appendix A

PRICES IN THE AMERICAN MARKET OF COD FISH, DRY, PER QUINTAL*

	Low	*High*		*Low*	*High*		*Low*	*High*
1797	$4.00	$6.00	1810	3.00	4.50	1823	2.37	3.00
1798	2.50	4.75	1811	3.50	4.00	1824	2.37	3.50
1799	2.50	3.75	1812	3.50	3.75	1825	1.88	3.25
1800	2.75	4.50	1813	3.75	4.25	1826	1.88	2.50
1801	4.00	5.50	1814	3.75	6.00	1827	2.62	4.00
1802	3.25	5.00	1815	4.50	6.00	1828	2.37	3.00
1803	3.50	5.00	1816	2.50	5.00	1829	2.00	2.33
1804	3.50	5.00	1817	2.50	4.00	1830	2.00	2.50
1805	3.75	5.00	1818	2.75	3.75	1831	2.33	3.00
1806	4.00	5.00	1819	2.50	3.75	1832	2.50	3.25
1807	4.00	5.00	1820	2.50	2.87	1833	2.17	2.75
1808	4.00	4.75	1821	2.50	2.75			
1809	3.00	5.00	1822	2.50	3.62			

* From *New York Price Current,* 1797 to 1815; *New York Shipping and Commercial List,* 1816–1824; *Boston Daily Advertiser,* 1825–1841. I am indebted to Professor W. B. Smith of Williams College for these statistics.

Appendix B

"It is not expedient to renew the Bounty on Salt as the lately reduced price of that article, the increased resort of Vessels for Timber, etc., to the Ports of the Province, and the low price of Freight, for which Salt may be imported, promise a sufficient and regular supply of this Article. . . . The bounty on the general catch of Fish would encourage the Fisherman in a very direct and comprehensive manner; yet, that the amount necessary to be drawn for that purpose, from the Treasury, the extreme difficulty of guarding against misapplication, and the fact that one of the main inducements for granting bounties, viz., to encourage a new branch of Trade, does not apply to every portion of the Fishery, to which such a Bounty would attach, appear conclusive objections against this mode of encouragement." *Journals of the Assembly, Nova Scotia,* March 13, 1827. Mr. Homer, from Barrington, refused to sign a report of the committee of the Assembly and complained of small returns to the fishery, of migration to New Brunswick to take advantage of the bounty, and of desertion from the fishery. In 1816 Barrington took 23,700 quintals, the average earnings of the fishermen being £36; and in 1826, 16,600 quintals, the fishermen's average share being less than £12. Fishermen engaged in the Straits of Belle Isle and selling in New Brunswick earned £24, and in Halifax £18. He argued that by improving the quality of fish, exports of fish tal qual to the West Indies would be reduced and the risks of there easily becoming a glutted market for New Brunswick and Newfoundland fish minimized. The market was relatively inelastic as the slaves were given a weekly allowance of fish, and the annual requirements of the plantations were fairly accurately known. The Brazils provided a market for merchantable fish "and although the proceeds of the cargoes of merchantable fish shipped to the Brazils are not invested in homeward cargoes which would pay as great a proportion of immediate revenue as fish shipped to the West Indies would do, yet we have the returns in specie and it serves as an indirect remittance to Great Britain. The home cargoes that are bonded to be reshipped again serve doubly to extend our navigation. . . . The cotton, hides, tallow, etc., which are admissable to this province for a small duty facilitates the growth of our infant manufactures." "The abolition of the slave trade . . . has materially affected the fish trade of this country," as did the introduction of East Indian sugar to Great Britain and the heavy duties imposed by Spain and other countries. Homer recommended a bounty on merchantable fish in the province and a bounty of double that amount for merchantable fish caught on the Labrador, as it would facilitate competition with Newfoundland. "Whereas schooners proceeding to the Straits of Belle Isle and entering the western harbours take capelin for bait and split and salt the cod on board, then move eastward as the capelin moves in that direction (making possibly three or four moves), and finally either make their fish on the Labrador, or proceed to Cape Breton, or if from New Brunswick to Cape Sable, for that purpose, the former selling their fish in Halifax about the 10th or 15th of October and the latter in St. John about the 15th or 20th of December, they could afford with a bounty to have the fish made in Labrador and avoid keeping it for such a length of time in salt." At Barrington "early in the summer as soon as the first fish is made fit for market small vessels take it on board on freight to Halifax or St. John, etc., but mostly Halifax. On board one of those freighting vessels there will perhaps be fifty or sixty different shippers who send a few quintals each to purchase their own little necessaries. But when the fishing season is over then almost every vessel brings its own fish to market." Consequently, for boat fishermen and vessel fishermen, a system of inspection at Halifax would ensure the grade of merchantable fish. However, this would strengthen the monopoly position of Halifax and enable the merchants to gain advantage from the bounties, "the merchants here being few in number and the mercantile genius always capable of taking care of itself." "I have never yet known among my constituents one solitary instance of a man getting beforehand by fishing and fishing only; those who own vessels or parts of vessels did not earn them by fishing; they earned them in better times, by sailing coastwise, carrying plaster of paris, etc. . . . As a proof positive of fishing being an unprofitable business, is there a single schooner owned, fitted and manned from the port of Halifax? I doubt if there is a solitary one." The fishermen "are the main staff and support of the commerce of their country, they are the greatest source of revenue; from their labours originate the principal article of exportation; their hard earnings have helped to enrich many of those who are engaged in commercial pursuits, and have served to aggrandize their country. But they themselves, although they compose a large proportion of the population are literally in a state of bondage. . . . Their unprofitable callings have rendered them destitute of the means of improvement, and doomed them to perpetual servitude, their education and morals being almost totally neglected causes them to become a degenerate people in comparison with those who have these advantages."*

* *Acadian Recorder,* March 24, 1827.

CHAPTER X

THE AMERICAN REVOLUTION AND NEWFOUNDLAND, 1783–1833

[The American] Revolution has made an alteration in the value and importance of Newfoundland, which seems to me never to have been sufficiently considered. It appears to me, that since the peace [of] 1783, Newfoundland has been more completely our own, that it has been a more genuine British fishery, and of more value to the Mother country, than it ever was before. It is become a sort of cul de sac; what does not stay there must come to Great Britain and Ireland; there is no longer the competition and interloping trade of the New Englanders so much complained of heretofore by the merchants.

JUSTICE REEVES ON NEWFOUNDLAND, 1793

THE difficulties of New England and France during the long period of wars that had ended in 1815 had been advantageous to Newfoundland. The West Country had been handicapped by the war in Europe and the fishing ships had practically disappeared, but population had increased; and this brought about far-reaching changes which were reflected in the evolution of political institutions.

Newfoundland benefited from the demands of Europe for fish, while in the years following 1783 the French and New England fisheries were still recovering.[1] Increased trade from Great Britain and the increasing importance of the Maritime Provinces as a base of trade in supplies and provisions had largely offset the effects of the disappearance of the trade from New England. And, in the main, the drain of men and bills to New England had been checked.[2]

1. See the table of prices at St. John's, from 1782 to 1792, in Routh's evidence, *Third Report from the Committee Appointed to Enquire into the State of Trade to Newfoundland, 1793*, p. 49.

PRICES OF DRIED COD PER QUINTAL

Year	From	To	Year	From	To
1782	from 12*s*.	to 13*s*.	1788	from 13*s*. 6*d*.	to 10*s*. and 11*s*.
1783	15*s*. 6*d*.	to 16	1789	10	to 11
1784	11*s*. 6*d*.	to 12*s*. 6*d*.	1790	10	to 11*s*. 6*d*.
1785	15	to 16	1791	13	to 14
1786*	13	to 14	1792	15	to 16
1787	16	to 17			

* In 1786 prices quoted were: "Large merchantable" cod, 15 shillings to 16*s*. 6*d*. a quintal; "Small merchantable," 13 shillings to 14*s*. 6*d*.; "Madeira" (for the West Indies), 10 to 11 shillings; salmon, 40 to 55 shillings a tierce; and oil, £15 to £18 a ton.

2. Ougier's evidence, *First Report from the Committee Appointed to Enquire into the State of the Trade to Newfoundland, 1793*.

The effects of increased trade with Great Britain[3] were described by Chief Justice Reeves:

It is in the memory of several persons when the trade at St. John's was in the hands of five or six merchants; these persons brought out sufficient supplies for the people they employed either as servants or boatkeepers to catch fish for freighting their own ships. . . . At present the number of persons who can furnish supplies in the town of St. John's is so increased, that all monopoly is broken, and a very active competition is come in its place. All

3. In 1792 the largest ports engaged in the trade included London with 19 ships totaling 2,624 tons, Dartmouth with 85 ships of 6,954 tons, Exeter with 43 ships of 3,551 tons, Poole with 65 ships of 7,791 tons, Liverpool with 11 ships of 1,263 tons, and a total of 230 ships of 22,909 tons. "Biddeford and Barnstaple were once great towns in this trade and have long since ceased to employ any ship at all." Reeves's evidence, *Third Report,* p. 166. Ports in Scotland increased their trade. "Messrs. Andrew Thompson and Company, Crawford and Company, Stevensons and Company and Stuarts and Rennie are in the habit of transacting from Greenoch, Port Glasgow and other situations to Newfoundland business nearly equal to half of the amount of that from the port of Dartmouth to that district of St. John's." *Idem;* also Routh's evidence, *Third Report,* p. 32. In 1792 Glasgow had 5 ships totaling 747 tons, and Greenock, 8 ships of 1,423. *Second Report,* Appendix No. 6b, p. 56. The increasing interest of diverse ports in England in the Newfoundland trade coincided with the increasing importance of merchants in Newfoundland. A letter signed by forty-eight residents of St. John's was sent to "the Committee of merchants of Great Britain appointed by the merchants and employers carrying on fisheries and trading to Newfoundland" to thank them for preventing the passage of two bills in 1792. *Second Report,* pp. 21–22. An elaborate inquiry, taking two averages of six years, 1769 to 1774 and 1787 to 1792, showed a decrease in the number of ships from 580 to 509; an increase in tonnage from 41,448 to 53,771, or an increase in the average size of the ship from 71 tons to 105 tons; a decrease in seamen from 5,715 to 4,608; a decline in passengers from 6,924 to 4,681; an increase in boats from 2,306 to 2,349; an increase in fish caught from 684,746 to 698,365 or an increase in the average per ship from 1,181 to 1,372; an increase in the salmon pack from 1,767 tierces to 3,516; a decline in oil production from 718,848 gallons to 601,856; an increase in fishing stages from 1,157 to 1,592. With an average per ship of 22 men (10 seamen and 12 passengers), assuming that the passengers were boatmen and that the boats caught all the fish, each boat had 3 men who took 297 quintals of fish; and, assuming that seamen and passengers were fishermen, in the first period, each man took 54 quintals of fish and 57 gallons of oil. In the second, with 18 men (9 seamen and 9 passengers), assuming that the passengers were boatmen, each boat of 2 men took 297 quintals of fish and, assuming that passengers and seamen were fishermen, each man took 75 quintals of fish and 64 gallons of oil. This general conclusion indicated that a decline in the number of European seamen and passengers had been accompanied by a reduction of expenses, increasing efficiency, and, with a fairly stable price in foreign markets, a larger profit. *Third Report,* pp. 197 ff.; see also pp. 9–10; Appendix B and Appendix No. 7. This analysis showed, for London, an increase from 14 ships of 1,794 tons in the first period to 18 ships and 2,507 tons in the last; for Poole, an increase from 71 ships of 5,314 tons to 78 ships of 9,775 tons; for Dartmouth, an increase of from 64 ships of 4,295 tons to 97 ships of 8,897 tons; and a total increase from 235 ships of 16,840 tons to 263 ships of 27,611. "The trade and fishery of Newfoundland was in a much more flourishing state during the last six years than ever it was before." See Appendix No. 6a. See *Second Report,* Appendix No. 55, for a detailed statement of the ships and tonnage of individual ports for the years concerned.

the consequences of competition have followed; the prices of supplies are lowered, and boatkeepers less dependent, having more persons to take their fish and supply them with necessaries; hence the murmur of the western merchants against hucksters and adventurers, and hence the notion that the trade is ruined. It is true that some of the persons who sell supplies at St. John's do not carry on the fishery, but they sell their supplies to those who do; the produce of the fishery is still the object of the trade; fish and oil are still the staple commodities, and I do not see but that persons who make it their object to deal in these articles, must be reckoned among the encouragers of the fishery, although they do not themselves engage in keeping boats or ships. . . . As to this mode of carrying on the trade, whatever the West Countrymen may say against those who practice it, they certainly introduced it themselves. It is well known at Newfoundland that the most profitable way of carrying on the fishery is by supplying boat-keepers and taking in payment for the supplies the fish and oil they catch. It was this induced the Western merchants, as well as those of Poole to encourage the settlement of persons there many years ago; as these increased, the necessity of bringing men from England must diminish. The merchants found it to their interest to promote the former, and it was in vain to depend upon regulations to force them to another course. Residency and population have increased, because it is generally held the cheapest and most profitable way of carrying on the fishery by residents; when this was known, it was easily seen that any man who could land at Newfoundland with a cargo of supplies was as fitted for carrying on the fishery as a regular bred fisherman; from this observation arose the number of adventurers who have of late years come into the trade, and who are so much censured by the Western merchants for following the example they had set. These new comers have mostly resorted to St. John's and to Conception Bay, where there is more population, and where people are less united and more at liberty to engage with any new merchants that present themselves. In Trinity Bay and Placentia Bay I believe these new adventurers make very little impression. . . . It may happen, indeed, that through this the great gains of the trade may change hands, or it may even happen that the gains in the trade may be less to the individual merchants concerned; but the boatkeepers, who catch the fish and oil, and who thus create the property by which the merchant is to thrive, must certainly be gainers by this competition, for there are more bidders for their fish and oil, and they have more chances of getting their supplies cheap; at any rate if the sum total of fish caught and of ships and men employed, is the same, or if they are increased, as appears by what some gentlemen have said; and if the whole concern is in the hands of many merchants instead of a few (which in a commercial light is deemed always beneficial) what does it matter that this or that man or town is falling to decay, or this or that mode of supplying is practised? These modes all commence of themselves; they must of themselves change and die away; fashions of trade must be taken as they are, and cannot be controlled by regulations. . . . I cannot help saying that the grand means employed to prevent the increase of inhabitants has in my opinion contributed to their increase.[4]

Aaron Graham testified[5] that most of the merchants of St. John's

not only send out a sufficiency for themselves and such boatkeepers as it is proper they should supply for the purpose of filling their ships with fish and oil for market, but they also undertake to purchase, upon commission, fish and oil for ships not regularly engaged in the fishery and for the payment of which they send also provisions and other articles that may be wanted, both by Europeans and Island fishermen; nor is it an uncommon thing to purchase in the country whole cargoes of various articles from the colonies and West India islands, which they send to the different parts of the island for the purpose of collecting the cargoes of fish so purchased by them upon commission and such cannot be done but at a very great risk.

Graham is also quoted, in part, indirectly: "The trading merchants, however they may have increased the trade, have certainly at the same time occasioned great alteration in the fishery. . . . He does not think that the trading merchants have occasioned any decline of the fishery from Great Britain; he thinks that the trade has increased in a very great degree but the fishing not so much in proportion."[6]

It was generally conceded, however, that the growth of a trading organization in Newfoundland was accompanied by the decline of the fishing ships.[7] William Newman, a Dartmouth merchant formerly engaged

4. Reeves's evidence, *Third Report,* pp. 167–169, 172.
5. Graham's evidence, *Third Report,* p. 15.
6. *Idem,* p. 13.
7. In opposition, but really in support of this conclusion, Richard Routh, collector of the Island of Newfoundland, and familiar with the trade from 1782, pointed out that the firm of Messrs. Philip, Leigh and Company had lost £70,000 because of overexpansion: "Trading above their capital, ten times its amount, [they] had, or employed, upwards of twenty sail of ships and vessels and failed in consequence thereof." And Messrs. Thomas Trimlitt and Company had lost £50,000. These failures and the reduction in the trade of another firm through the death of the chief partner were held to account for the decline of Dartmouth. Another death at Plymouth accounted for the decline of that port. Weymouth had withdrawn from the trade about 1780, and Exeter had suffered from the bankruptcies of three firms caused by a scarcity of fish in the locality in which one firm was engaged, speculations, and unsound connections not concerned with the Newfoundland trade. *Third Report,* pp. 31–32. The fishing industry demanded detailed attention on the part of the members of the firms engaged. Overexpansion and the deaths of company heads had serious results. "The natural points on which the welfare of this trade depend [comprise] the purchase of goods in Great Britain with ready money, constant attention and labour, night and day, during the fishing season; faithful agents to see the duty done; the early or late catch of fish. If an early one, and good weather, the fish gets early to market, and remittances arrive soon to enable individuals to perform their payments; if a late fish, the fish gets late to market, and if the adventurer trades beyond his capital (which is more or less the case in all trades) he is of course pushed for ways

in the purchasing of fish on a commission basis,[8] stated that the Dartmouth fishery had declined by 31 ships, Plymouth by 14, Weymouth by 3, Wick in Cornwall by 3, and Exeter by 13. From 1784 to 1791 the number, he said, had fallen to an average of 480, with 4,475 men and 4,662 passengers. The number of men from Great Britain had decreased by 3,130, including decreases from Dartmouth of 1,400; Exeter, 800; Tinmouth, 700; and the decrease from Ireland had been about 1,500. The number of passengers going out from Dartmouth and Exeter in the spring and returning in the fall had dropped from about 1,500 in 1778 to a very small number in 1793. Since it was estimated that three fourths of the passengers were seamen, and that they included "one green man in every six and one other that has been only one voyage before," the effect of the decline on the position of Newfoundland as a "nursery for seamen" was obvious. Bankruptcies between 1771 and 1791 involved losses of £178,000, and merchants of Bristol, Dartmouth, Falmouth, Weymouth, Plymouth, Penzance, and Exeter lost considerable fortunes.

Ships became fewer particularly after 1788. The unusually heavy catch of that year precipitated difficulties. Prices declined in Spain and Portugal. Bad weather made curing difficult. The fish were "of a thin bony kind and of an inferior quality which could not be so well preserved by salt." Merchants refused to take fish in payment for supplies. From 1789 to 1792 harbors northward from Ferryland, as had happened thirty years before, suffered from a migration of fish because of an unusually large quantity of ice, which chilled the water. Poole reported an unprofitable fishery to the east and north and a successful fishery in St. Mary's Bay and Placentia. The number of fishing ships[9] increased from 236 totaling 22,535 tons and 2,603 men in 1784 to 389 of 34,846 tons and 4,306 men in 1788, dropped to 245 of 21,422 tons and 2,255 men in 1791, and recovered to 276 of 18,838 tons and 2,351 men in 1792. The number of boats owned by fishing ships dropped from 572 in 1784 to 273 in 1788, and, after a sharp rise to 413 in 1789, fell to 150 in 1792. Boats varied in size from 5 to 25 tons and employed

and means, and as in every other trade, falls a victim in time to the inconvenience of his situation." *Third Report,* p. 46. "In order to embarke in it a very great capital is necessary; the risque is increased by its being conducted at so great a distance from home; the plantations and utensils for the fishery which are purchased at a great expense are only valuable for the immediate purpose of carrying on the fishery." Jeffery's evidence, *First Report,* p. 11. Absentee ownership and the conduct of business through agents were important causes of serious losses. Routh's evidence, *Third Report,* p. 72. On the difficulties created by agents for the courts see Reeves's evidence, pp. 108 ff.; also Palliser's evidence, *Third Report,* p. 26.

8. *First Report,* pp. 1 ff.

9. *P.C.* (*Privy Council*), IV, 2006A.

crews of from three to seven men, with two or three men on shore. The number of quintals of fish made by fishing ships increased from 131,650 in 1784 to 412,550 in 1788, but declined to 97,815 in 1789 and totaled 160,910 in 1792.

Attempts to restrict colonization and increase the fishing ships by legislation failed. The bounty to build up the bank fishery provided in Palliser's Act (15 Geo. III, c. 31), 1775, was modified and continued for ten years in 1786 (26 Geo. III, c. 26).[10] William Knox held that the bounty given for 20,000 "tail" taken by July 15 had the effect of giving a bonus to the good fishery and penalizing a poor fishery.[11] To obtain the necessary quota, ships stayed on the Banks until July 15 and missed the capelin bait which "came on shore" early in the month. Largely restricted to the territory from Trinity south to Trepassey, the bank fishing was carried on by small vessels of from 40 to 120 tons and from 7 to 12 men. They kept three on shore to cure the fish, and made three or four trips a season. The ship lay at anchor three or four weeks, catching and salting fish, and then the full cargo was landed, cured, and dried. These vessels took from 700 to 1,400 quintals each. They increased from 141 in 1785 to 198 in 1789, and dropped to 157 in 1790. The St. John's bank fishery decreased from 140 sail in 1788 to 70 in 1792. Production by the bankers declined from 228,494 quintals in 1789 to 112,404 in 1791, but increased to 139,450 in 1792. The poorer grade of bank fish prevented it from making permanent inroads in the European market, in spite of the bounties.[12]

The difficulties encountered by the fishing ships in the Avalon Peninsula hastened the movement of West Country merchants to the north.

10. *Report of the Lords of the Committee of Privy Council for Trade* (March 17, 1786), p. 7, pointed out their ineffectiveness but recommended their renewal for ten years with changes, especially giving shares instead of wages. The act (26 Geo. III, c. 26) gave a drawback on salt and introduced regulations requiring the meshes of nets or seines to be increased from 3½ inches to 4 inches to avoid the destruction of small fish. Birds valuable for food or bait were not to be destroyed for feathers. See *P.C.*, I, 257 ff. The bounties were continued in various acts to 1803 (41 Geo. III, c. 97).

11. Newman argued that these bounties sustained the bank fishery and kept it from getting into the hands of the residents.

12. The increase in bank ships is probably deceptive as fees were reduced on ships engaged only in the bank fishery, and ships engaging in trade masqueraded as bank ships. Such deception was facilitated by the departure of bank ships at the end of the season with passengers and oil for England. Routh's evidence, *Third Report,* pp. 62–64. Routh claimed that "bankers" had increased from an average of 25 in 1760–70 to an average of 100 about 1790. *Idem,* p. 56. See also Newman's evidence, *First Report,* pp. 3–4. In 1787, of a total 204 vessels (13,177 tons and 2,086 men) about 90 were fitted out for the bounty granted by 26 Geo. III, c. 26, 1786. The bank fishery required substantial capital, and consequently was owned chiefly in England as the residents were too poor to acquire vessels. See Appendix B. Routh's evidence, *Third Report,* pp. 42, 57; Jeffery's evidence, *First Report,* p. 19.

Merchants of Poole and other ports left servants over the winter to engage in the fur trade and the seal fishery, to build ships which were sent with freight to Europe and sold in England, and to look after increasingly valuable property.[13] The salmon fishery was extended.[14]

The highly competitive character of the industry and the efficiency of the byeboatkeeper and of the residents contributed to the decline of large West Country merchants even when fish brought good prices. The byeboatkeeper became a boatkeeper, a merchant, a shipowner, and a possessor of a large fortune.[15]

13. Routh referred to the extensive profits made by merchants of Poole, Dartmouth, and other towns apparently engaged in this fishery. *Third Report,* p. 45. English byeboatkeepers and planters in White Bay generally sent their shallops to the north and to Labrador for fish, but took furs and seals in winter and, "with the advantage of the largest and finest timber on all Newfoundland," built vessels and shallops. In 1785, 8 fishing ships totaling 1,130 tons, with 93 men, 142 passengers, and 45 boats, took 12,300 quintals of fish on the Labrador. The coast supplied £1,572 worth of seal oil. In 1792, 4 fishing ships from Great Britain, totaling 468 tons, with 144 men and 30 boats, were engaged at Forteau Bay and took 138 tierces of salmon, 5,000 quintals of fish, and 95 barrels of oil; and 2 at Blanc Sablon of 320 tons, with 11 boats and 63 men, took 2,700 quintals and 26 barrels of oil. A plea for the extension of the power of government was made, and stated that the numerous "furriers" and planters "are entirely subject to the oppression of the merchant, who imposes whatever price he pleases; and upon any debt however small being incurred, and not being paid upon immediate demand, the boats and other effects of the debtor are seized without any authority for so doing, sold and purchased by the creditors for sometimes one-sixth of their value. The prices upon the coast are enormous and want great regulation, one hundred weight of course bisket charged to the planter at 30s. and other provisions equally dear in proportion; man of wars slops condemned at home by government are bought up by the merchants of Labrador and sold to the inhabitants at one guinea per packet." *C.O.* (*Colonial Office*), 194:21. See page 305, note 59.

14. In 1786 salmon fisheries in British limits were carried on by Payton and Miller at River Exploits; and Charles Brook, with 2 boats and 6 men, took 260 tierces. Charles Rousel at Halls Bay with one boat and 4 men took 40 tierces; John Crease at Loo Bay with one boat and one man, 25 tierces; John Slade, Jr., and Company at Indian Arm and Dog Bay with 2 boats and 4 men, 60 tierces; Mathew Ward and Company at New Bay with one boat and 2 men, 40 tierces; Lester and Company at Ragged Harbor, Dog Bay, Freshwater Bay, and Indian Bay with 4 boats and 14 men, 335 tierces; and Jeffery and Street at Gander Bay with one boat and 6 men, 400 tierces. Salmon weirs and nets of four-inch mesh were used at all locations except Ragged Harbor and Gander Bay, where only nets were used. *C.O.*, 194:21.

15. *Third Report,* p. 49; also Graham's evidence, pp. 12, 21–22. The resident merchants had "increased their capitals very considerably." One merchant had expanded his property from a small shop in 1779 until he had become "the owner of great fishing plantations in Quiddey Viddey, stores in St. John's and [was] also concerned in trading ships." *Idem,* p. 71. William Compton, who had been a merchant or byeboatkeeper at Petty Harbor for thirteen years, or from 1777 to 1790, hired 29 men and employed 5 boats in an agreement to run from the time of arrival to October 10 in return for fixed wages. He chartered the *Lord Longford* to arrive and depart by August 20 and to carry 1,000 quintals of fish to Leghorn or elsewhere. *Second Report,* Appendix No. 3, pp. 48–52. Ougier apparently lost money at Bay Bulls, but, after shifting his business to St. John's and there engaging in the business of purchasing fish on commission, he had one ship which took out goods for the fishery in

Formerly, say 20 or 30 years ago, a hedger or ditcher had only to go to Newfoundland and announce his intention to keep boats, and could find people to set him up. . . . At this time the trade was a sort of barter account. . . . Bread and flour was sold at 30s. per cwt., pork at £6 and other things in proportion; but if an independent man traded, and it was a bill account, one-third was deducted, or, to bring a bill account into barter, half was added; if the boat keeper was not an independent man he was sure in all probability to be brought into debt in one, two, or three years, which debt always claimed the preference of future dealings. The trade is now entirely altered in that particular; for the merchants found from experience, that although they apparently had the person's labour, yet there being a discontent on the side of the boatkeepers at not always having their wants supplied at the store, they had private dealings with others, and the fish went to pay for the same. A merchant therefore, in the present day, will only deal with independent boatkeepers and such only find their way to Newfoundland—men whose dealings are as secure and whose bills are paid as punctually as any merchant who trades to the island. Many men therefore who formerly kept boats are now servants for that very reason. . . . Every merchant is become a boatkeeper, and where one bye boat is lost many are kept by the merchants, so that on the whole the bye boat fishery is increased.[16] The merchants may be generally regarded as the principal parties concerned, as they supply the boat keepers with fishermen, with provisions, with cloathing and with implements for the fishery, in the same manner as if the concern was entirely their own and receive in return as payment the produce of the voyage.[17]

The number of byeboatmen in Newfoundland increased from 289 in 1784 to 583 in 1786, but declined to 290 in 1788, while the number of servants rose from 2,317 in 1784 to 4,743 in 1786 and fell to 2,107 in 1788. Boats owned by byeboatmen increased from 344 in 1784 to 540 in 1785, and fluctuated widely between a minimum of 317 in 1788 and a maximum of 584 in 1791. The catch by byeboatmen differed greatly from year to year, falling to 83,870 quintals, the lowest point, in 1790, and mounting to 123,023, the highest, in 1791. The number of passengers from England increased from 1,483 in 1785 to 2,024 in 1786, fell to 1,070 in 1790, and rose to 1,526 in 1792. The number of passengers from Ireland increased from 2,622 in 1785 to 3,630 in 1786, dropped to 1,551 in 1790, and rose to 2,455 in 1792. The total, including a small number from Jersey, increased from 3,187 in 1784 to 6,202 in

1790, took two cargoes of fish to Portugal, and on one trip carried passengers to Ireland and England. In 1791 a sloop of 60 tons, owned by him, went from London to Newfoundland, touched at two ports, and went to Spain. Moreover he was interested in sending out ships that first took goods to Newfoundland and later went to the Banks. Ougier's evidence, *First Report,* pp. 51 ff.

16. Routh's evidence, *Third Report,* p. 56. There were from 100 to 150 byeboatkeepers in St. John's.

17. Jeffery's evidence, *First Report,* p. 10; also Newman's evidence, *idem,* p. 25. It was estimated that an investment of £80 a boat was involved. See page 300, note 33.

1788, decreased to 3,122 in 1790, and rose to 4,256 in 1792. The large number of Irish passengers probably indicated that the resident fishery was important, and expanded at the expense of the byeboat fishery and the fishing ships. The total population rose rapidly from 10,244 in 1785 to 19,106 in 1789. Residents' boats maintained a preponderant position and increased from 1,068 in 1784 to 2,090 in 1788, but fell to 1,259 in 1791. The catch by residents rose from 212,616 quintals in 1784 to 457,105 in 1788, declined to 229,770 in 1791, and rose to 395,900 in 1792.

The growing importance of the resident fishery and of expanding trade could be seen in the cargoes of both fishing and sack ships.[18] The number of sack ships increased from 60 totaling 6,297 tons and 547 men in 1784 to 173 of 16,828 tons and 1,426 men in 1786, fell to 143 of 9,881 tons and 1,496 men in 1790, and rose to 161 of 21,275 tons and 1,319 men in 1792. Of a total catch of 3,841,483 quintals from 1787 to 1791, 3,492,303 were exported to Madeira, Spain, Portugal, and Italy; and of 2,704 ships totaling 295,679 tons, more than one half, or 1,600 with a tonnage of 194,425, were engaged in this trade.[19]

It is a branch of commerce highly important and most invaluable to Great Britain not only as a very considerable nursery for seamen but as it affords a consumpt for the growth, produce and manufactures of this country, to the amount of not less than half a million annually. . . . The fish caught in this trade is sent to foreign markets, namely, Spain, Portugal and Italy, and the returns I judge to be nearly 39/40ths in specie or in bills of exchange.[20]

18. The total number of quintals produced increased from 437,316 in 1784 to 948,970 in 1788, and declined to 552,260 in 1792. Another estimate puts the total at 692,554 in 1786–87; 894,587 in 1788–89; 738,976 in 1789–90; 751,296 in 1790–91. The number of quintals of fish carried to foreign markets rose from 497,884 in 1784 to 782,791 in 1789, but dropped to 452,402 in 1792. Salmon tierces increased from 725 in 1784 to 4,598 in 1792. The number of tons of train oil increased from 2,146 in 1784 to 2,847 in 1788 but fell to 2,091 in 1792. Seal oil mounted in value from £3,382 in 1784 to £11,920 in 1792; and furs from £540 to £2,280 in 1792. The total number of boats fluctuated widely, increasing from 1,984 in 1784 to 2,680 in 1788, and falling to 2,147 in 1792. After 1789 servants on ships became fewer, dropping from 4,799 to 2,438 in 1792; servants in boats from 7,323 in 1789 to 7,138 in 1792; and servants on shore from 6,152 in 1789 to 4,465 in 1792. The total number of servants dropped from 18,274 in 1789 to 14,041 in 1792. The number of stages rose from 942 in 1784 to 1,578 in 1788, and to 2,356 in 1792; and the number of oil vats from 673 in 1784 to 932 in 1789, decreasing to 654 in 1792.

19. In 1787, 277 vessels, totaling 34,405 tons, with 2,505 men, carried 640,725 quintals to Spain, Portugal, Italy, and Madeira; in 1789, 387 vessels took 807,927 quintals; and in 1790, 461,441 quintals were shipped. In 1787, 35 vessels (3,599 tons, 290 men) took 35,432 quintals to the West Indies; in 1789, 58 vessels carried 61,862, and in 1790, 51,287 quintals were shipped. Fifteen to twenty-five thousand quintals of dried fish and about 20,000 quintals of core fish went, in some 126 or 150 vessels, to Great Britain. *C.O.,* 194:21.

20. Jeffery's evidence, *First Report,* pp. 9–11; also Newman's evidence, *idem,* pp. 24 ff. "This trade is in its nature the most advantageous to the nation of any that she

As in the case of the cod, practically all the salmon (17,898 tierces) was exported to Madeira, Spain, Portugal, and Italy. The herring (4,559 barrels) went to the West Indies. The oil (14,799 tons), the sealskins (164,979), and the planks and boards (75,210 feet) went to Great Britain, Ireland, and the Channel Islands. About one half of the staves and shingles were taken by the West Indies.[21] The small vessels which were best calculated for that branch of trade could not afford "to pay the enormous port charges to which they are liable in the West Indies."

The value of exports from England grew from £88,056, to take a four years' average before the war, to £168,796 in 1784, to which must be added a marked increase in the export of salted provisions from Ireland.[22] The British West Indies' exports of rum into Newfoundland[23] increased to 178,870 gallons in 1785.[24] "The price of it in time of peace

can possess; its outfit is made wholly from the mother country (except a trifling amount of foreign salt) and the returns are made to the merchants exporting the fish, wholly in bills of exchange and bullion from foreign countries the amount of which would otherwise be included in the balance of trade against us, to be paid in specie by this country."

21. Appendix No. 2a. *Third Report,* p. 187.

22. In a total of 516 vessels in 1787 (56,884 tons and 5,015 men) bringing goods to Newfoundland, 355 (37,418 tons and 3,561 men) brought 77,156 hundredweight of bread and flour, 22,996 barrels of beef and pork, 9,432 hundredweight of butter and cheese, 227,168 bushels of salt, 87,422 pounds of tea, 1,114 hundredweight of refined sugar, 33 hundredweight of muscovado sugar, 31,075 gallons of molasses, 12,194 gallons of rum, 6,375 gallons of corn brandy and gin, 4,749 gallons of wine, 1,186 hogsheads of cider and beer, 1,013 pounds of coffee, 71,622 pounds of tobacco, 125,086 pounds of soap and candles, 1,132 chaldrons of coal, 1,464 barrels of pitch and tar, 5,300 feet of board and planks, 2,000 staves, 22 sheep and hogs, 25 dozen poultry. The proportion of pork to beef was at least four to one, and of butter to cheese nineteen to one. The salt was brought chiefly from Spain and Portugal, scarcely one fifteenth being British made. A large proportion of the vessels carried fish to Spain, Portugal, and Italy at the end of the season, returned to England and brought out provisions and supplies in large quantities. But 81 vessels (9,999 tons and 757 men) brought 337,692 bushels of salt directly from Spain, Portugal, Italy, France, and Madeira. These accounts did not include imports brought from Poole to Fogo, Twillingate, and Greenspond in 7 or 8 vessels. The total number of ships importing goods increased to 549 (60,863 tons and 5,263 men) in 1788 and to 594 (65,322 tons and 5,207 men) in 1789. In the latter year 383 vessels (39,395 tons and 3,483 men) brought, from Great Britain, Ireland, Jersey, and Guernsey, 62,421 hundredweight of bread and flour, 21,505 barrels of beef and pork, 9,216 hundredweight of butter and cheese, 406,896 bushels of salt, 58,519 pounds of tea, 795 hundredweight of sugar, 1,156 hundredweight of muscovado sugar, 6,711 gallons of molasses, 12,092 gallons of rum, 4,453 gallons of wine, 4,687 gallons of brandy and gin, 937 hogsheads of beer and cider, 29 pounds of coffee, 38,018 pounds of tobacco, 157,573 pounds of soap and candles, 325 chaldrons of coal, 706 barrels of pitch and tar; 116 vessels (16,489 tons and 1,054 men) brought 449,815 bushels of salt and 6,183 gallons of wine from Spain, Portugal, France, Italy, and Madeira. *C.O.,* 194:21.

23. *Report of Committee of Privy Council for Trade, etc.,* pp. 11–12.

24. In 1787, 31 vessels (3,276 tons and 258 men) brought 70 hundredweight of bread and flour, 400 bushels of salt, 2,059 hundredweight of muscovado sugar, 111,376 gallons of molasses, 195,451 gallons of rum, 672 gallons of wine, 17,256 pounds of coffee,

is so low as almost entirely to exclude all other spirituous liquors." The number of trading ships from the British American colonies fluctuated between a low point of 34 in 1786 and a high point of 76 in 1791.[25] In 1787 the United States sent to Newfoundland 11 ships (1,395 tons) with 10,450 hundredweight of flour and bread, 167 oxen and calves, 539 sheep and hogs, 136 dozen poultry, and 1,670 bushels of Indian corn.[26]

The channels of trade had changed materially after the Revolution, for it was necessary to restrict trade from New England and depend increasingly on Great Britain. Ships became engaged to a growing extent in trade rather than the fishery and assisted in the expansion of the resident fishery.[27] The effects of the Treaty of Versailles (1783) on the Newfoundland fishery were registered in the number of inquiries regarding its problems and especially in the investigation of 1793, and were admirably summed up by Justice Reeves:

2,410 pounds of tea. 17 barrels of pitch and tar, 5,000 staves, and 8 bushels of Indian corn from the West Indies; and in 1789, 59 vessels (5,470 tons and 420 men) brought 2,782 hundredweight of bread and flour, 4,064 hundredweight of muscovado sugar, 200,220 gallons of molasses, 226,602 gallons of rum, 1,683 gallons of wine, 16,256 pounds of coffee, 4,383 pounds of tobacco, 350 pounds of soap and candles, 102 barrels of pitch and tar, 1,500 staves, 154 bushels of Indian corn. Imports of rum from the West Indies were as follows: in 1790, 169,605 gallons; in 1791, 183,239; in 1792, 83,600; and in 1793, 120,937. Corn brandy was sent to ports having little connection with St. John's as that port handled most of the West Indian produce.

25. From Canada and Nova Scotia in 1786–87, 38 ships (4,796 tons and 333 men) brought 13,509 hundredweight of flour, 5½ hundredweight of butter and cheese, 60 hundredweight of muscovado sugar, 5,300 gallons of molasses, 396 gallons of wine, 12,096 pounds of tobacco, 354 chaldrons of coal, 12 barrels of pitch and tar, 645,860 feet of planks and boards, 11,090 staves, 312,500 shingles, 329 masts and spars, 76 oxen and calves, 222 sheep and hogs, and 2 dozen poultry. Prince Edward Island sent livestock and ships, especially to meet the demand of the Labrador and seal fishery after 1800. A. B. Warburton, *A History of Prince Edward Island* (St. John's, 1923), pp. 279–280, 287, 360.

26. In 1785 an act—25 Geo. III, c. 1—renewed in 26 Geo. III, c. 1, 1786, and in 28 Geo. III, c. 6, permitted flour and livestock imports from the United States to Newfoundland in specially licensed vessels; Helen T. Manning, *British Colonial Government after the American Revolution, 1782–1820* (New Haven, 1933), p. 263. The importance of provisions and other supplies from the United States was made apparent by the disappearance of the hostile attitude of Dartmouth and Poole and by petitions from the merchants of Poole, especially from 1788 to 1791, for permission to allow the importation of provisions. In 1789 imports from the United States included 11 hundredweight of muscovado sugar, 3,000 gallons of molasses, 3,942 gallons of rum, 302 chaldrons of coal, 9 barrels of pitch and tar, 784,000 feet of boards and planks, 460,000 of shingles, 1,300 staves, 503 masts and spars, 91 oxen and calves, 139 hogs and sheep, 567 bushels of Indian corn; also 3 ships (of 383 tons and 24 men) carrying 5,433 hundredweight of bread and flour. See G. S. Graham, *British Policy and Canada, 1774–1791* (London, 1930), chap. vii. For smuggling and the shift of trade from the United States to Ireland and the West Indies, see *Third Report,* pp. 33–34, 47, 54–55.

27. Small ships were built chiefly at Trinity and Harbor Grace. Eighteen of 1,743 tons were built in 1787, 34 in 1788, and 22 in 1789. *C.O.,* 194:21.

I cannot help thinking also, that since Newfoundland is so severed from New England, some of the topics respecting the population of the Island and the fears about colonization deserve less regard. Notwithstanding the increase of inhabitants, Newfoundland is still nothing but a great ship, dependent upon the mother country for every thing they eat, drink and wear or for the funds to procure them; the number of inhabitants seems to me rather to increase this dependence inasmuch as their necessities are thereby increased. They all look to the sea alone for support; nine-tenths of the people procure from the soil nothing but potatoes; and those who carry cultivation furthest reap no produce but what can be furnished by a garden. In some places hay is cut, but corn is never thought of; neither the soil or the climate having encouraged the few attempts that have been made to grow it. The population, though said to be great, is scattered as thinly as the products of the earth. Distant harbours and coves, not easily accessible by sea, are the places chosen for residence, the people of which have little knowledge or connection with one another to unite them.[28]

By the beginning of the 1790's the expansion of the fishery in other countries began to have its effect on Newfoundland. Norwegian stockfish taken in Iceland were being sold at lower prices, and Barcelona was almost wholly supplied from Norway.[29] Papal indulgences in the matter of diet were decreasing the consumption of fish in Spain. With the same result, in 1785 Spain and Portugal increased duties on dry cod, and in 1792 Spain raised the duty still higher—to 4*s.* 7½*d.* the English hundredweight. Although the French home market consumed 200,000 quintals,[30] 15,000 quintals were exported from France to Alicante in 1788 and 20,000 quintals in 1791, with the usual disturbing effect on prices produced by the dumping of small quantities. It was claimed that New England was beginning to dominate her old markets at Bilbao, San Sebastian, and Santander.[31] With cheaper ships, cheaper provisions, and lower wages, her competition became increasingly effective.[32] Newfound-

28. Reeves's evidence, *Third Report,* pp. 171–172.
29. Jeffery's evidence, *First Report,* p. 18.
30. *Idem;* also Routh's evidence, *Third Report,* p. 46.
31. Knox's evidence, *Second Report,* p. 21.
32. *Idem,* pp. 18–19. Routh argued, on the other hand, that the expenses of New England were too high to admit of her becoming a rival. *Idem,* p. 46. It was claimed that New England attempted to regain admission to the Newfoundland fishery through Bermuda in 1788. An expedition of 34 sloops of from 30 to 60 tons with crews of 8 to 12 men, of whom three quarters were blacks and, among the remainder, men skilled in the Newfoundland fishery, brought salt from Turks Island and after catching their fish on the Banks proceeded to land and cure them on stages rented from Newfoundland masters. One result was that measures were adopted prohibiting the landing and curing of fish. The privilege of drying fish in Newfoundland was restricted to vessels from Great Britain and Ireland by an act passed in 1789 (29 Geo. III, c. 53). D. W. Prowse, *A History of Newfoundland* (London, 1896), pp. 345–348. But it was possible that the Bermudians, having lost a position as carrier between the

land was hampered by restrictions on trade with New England, limited support from the Maritime Provinces, and the high cost of supplies from Great Britain.[33] Customs fees were an important source of complaint in the investigations of 1793,[34] and smuggling from both the United States and St. Pierre and Miquelon had increased.[35]

The outbreak of war with France meant new difficulties. New England occupied a strategic neutral position in the trade with European markets. In 1798 there was a smaller catch, though of better quality; but a carry-over of 109,050 quintals from 1797 and low prices in Oporto caused losses. A report made in 1799[36] stated that "Madeira" fish was

American colonies and the sugar colonies, had themselves attempted to open new fields. *Idem,* pp. 416–418.

33. The amount of cash was very limited, and money consisted chiefly of bills drawn on Great Britain and Ireland, two thirds of which were of denominations ranging from 30 shillings to £10. In 1797 the poverty in Newfoundland was said to be the result of the draining away of coin, and the consequent high wages and high prices of provisions. Copper coins were introduced in 1798. The cost of a common fishing boat and its operation for a year was about £213, made up as follows: four men's wages at £21, or £84; provisions for four men, £40; two shoremen at £19, £38; their provisions, £20; bait, £11; and boat, rod, lines, hooks, etc., £20. With a catch of 280 quintals bringing 9 shillings a quintal, earnings would be £126. "The necessaries of life are uncommonly scarce and dear. . . . Clothing also is exorbitantly high." Routh's evidence, *Third Report,* p. 41. "We know that in all new colonies the price of labour is greater than in an ancient country like this, and where there is so much active industry as in the great fishery of Newfoundland, the value of time and of labour is still higher. I am told that the wages to carpenters and masons is 4s., 5s., 6s., and 7s. per day in Newfoundland." A pair of shoes costing 2 shillings or 2*s.* 6*d.* in London cost 8 shillings in St. John's. The price of rum was double that of the prerevolutionary period. Reeves's evidence, *idem,* pp. 137–138.

34. Deputies were appointed to the outbays in 1782, including Trinity Bay, Conception Bay, Bay Bulls, Ferryland, Trepassey and St. Mary's, Great and Little Placentia, and Fortune Bay. *Second Report,* Appendix, p. 59. See, in particular the evidence of Newman and of Ougier, *First Report,* and Ougier's evidence, *Second Report,* pp. 5 ff. For a list of customs charges given by Newman see *First Report,* p. 23, and a refutation with a statement of correct charges, *Third Report,* pp. 39–40. The charges were estimated to be £20,000, or 4 per cent on £500,000 in merchandise, plus £2,000, a burden of 10 per cent on the capital invested in ships. *First Report,* p. 20. Routh's estimate was £1,600 at the most. *Third Report,* pp. 44–45.

35. Pork from the United States and salt, wine, oil, and foreign-made cordage were brought in illegally. Waldron's evidence, *Second Report,* p. 4. Street's evidence, *Third Report,* p. 24. See a list of goods seized between 1776 and 1792, which included olive oil, wine, cork, tobacco, tea, apples, onions, molasses, and salt. *Second Report,* p. 61.

36. Large merchantable fish, bringing an average price of 14 shillings a quintal in 1799, were caught on the Banks, and the small alongshore. The small sold in Italy at one shilling more than the large, but with the closing of Italian ports the large fish commanded a better price in Spain and Portugal. "This kind of fish is stow'd in bulk and keeps far better than the inferior sorts." It was cheaper to ship than "Madeira," and sold at 1*s.* 6*d.* to 2 shillings less because it had been slightly damaged in the cure. It was always "screw'd in casks in order for its better preservation." The West Indian or refuse fish was also "screwed in casks," and sold for 6 pence to one shilling less than the "Madeira." "Dumb [dun?] fish" or fish which had been kept too long on

being sent "of late years to the West Indies" and that "the greater quantity of the West Indian fish is now consumed by the inhabitants of Newfoundland as of late years the worst quality of the Madeira fish has been sent to the West Indies in lieu of the former." In the same year, 1799, an excellent fishery was marred by bad weather for curing, and only 46 vessels left for Portugal. In 1800, 336 ships (33,289 tons) carried to market 434,622 quintals of cod, 1,801 tons of oil, and 1,797 tierces of salmon.[37] In 1802, 266 vessels were entered inward, 247 outward, and 137 sailed from the outports. The catch was 600,000 quintals. In that year import duties in Spain were increased to 9*s.* 6½*d.* per hundredweight, and Newfoundland's dependence on the West Indies became greater.[38]

the flakes were seldom sent to market, being preferred by the Newfoundland people, and "in most use at the first tables." *C.O.*, 194:42.

37.

	Ships	Total Tonnage	Number of Men	Cod (quintals)	Oil (tons)	Salmon (tierces)
St. John's	240	24,012	1,838	286,076	1,005	1,797
Ferryland	23	1,801	113	31,110	29	
Placentia	31	2,703	187	45,535	194	
Conception Bay	42	4,773	326	71,901	573	

In the above year, imports in 342 ships (34,373 tons and 2,612 men) totaled 2,011 tons of bread and flour, of which 1,727 tons went to St. John's, 78 to Ferryland, 158 to Placentia, 48 to Conception Bay; 2,276 hundredweight of butter, of which 2,145 went to St. John's; 8,110 barrels of beef and pork, of which 6,124 went to St. John's and 1,235 to Conception Bay; 384,230 bushels of salt, of which 256,046 went to St. John's, 36,040 to Ferryland, 51,798 to Placentia, and 40,346 to Conception Bay; 157 pounds of refined sugar, 72,497 pounds of tea, 69 tons of wine, 497,783 gallons of rum and spirits, 147,833 gallons of molasses, 704 hogsheads of beer and cider, 925 chaldrons of coal, 1,030 barrels of pitch and tar, 119,137 pounds of soap and candles, 65,015 yards of sailcloth, 85 tons of iron, 140,955 pounds of tobacco, 101 tons of cordage, 599 dozen hats, 45,371 pounds of leatherware, 263,762 pounds of brown sugar, 613,261 feet of boards and planks, 14,610 pounds of coffee, 1,000 bushels of Indian corn, 470 sheep, and 363 oxen.

38. EXPORTS OF COD, 1804–1816
(*in quintals*)

	Spain, Portugal, and Italy	Great Britain	West Indies	British America	United States	The Brazils	TOTAL
1804	354,661	189,320	55,998	18,167	43,131		661,277
1805	377,293	65,979	81,488	22,776	77,983		625,519
1806	433,918	84,241	100,936	32,555	116,159		767,809
1807	262,366	130,400	103,418	23,541	155,085		674,810
1808	154,069	208,254	115,677	40,874	56,658		575,532
1809	326,781	292,068	133,359	41,894	16,117		810,219
1810							884,470
1811	611,960	139,561	152,184	18,621	1,214		923,540
1812	545,451	67,020	91,867	4,121		2,600	711,059
1813	706,939	50,678	119,354	14,389			891,360
1814	768,010	55,721	97,249	24,712		2,049	947,741
1815	952,116	46,116	159,233	24,608	588		1,182,661
1816	770,693	59,341	167,603	37,443	2,545		1,037,625

As a result of the war in Europe, and particularly after its renewal in 1803, trade was exposed to numerous shifts. In 1801 and in 1803 and from 1806 to 1811, bounties of 3 shillings per quintal[39] were paid on exports to Great Britain to be reëxported in neutral bottoms. Handicaps in the shape of the high grain prices in Great Britain during the war and weak salt imported from Liverpool instead of Portugal resulted, however, in exports to New England. The price of bread increased in 1805 to between 42 shillings and 50 shillings a hundredweight, and of flour to 70 shillings per barrel of 196 pounds.[40] Imports of bread and flour from New England were exchanged for fish for reëxport to the West Indies and Europe.[41] In the late war British North America and the United States exported 360,000 quintals to the West Indies, of which Newfoundland supplied 70,000 quintals and the United States 175,000. With Spain and Italy cut off from Newfoundland in 1806, 100,000 quintals of fish suited to those markets were exported to the United States. Exporters "disposed of great quantities of fish, in exchange for bread and flour, with the Americans who cannot cure it sufficiently well to suit European markets."[42] In the same year exports from Newfoundland to the West Indies increased markedly.[43] Activity in both Newfoundland and Nova Scotia followed the payment of Imperial bounties of 2 shillings a quintal from June 1, 1806, to June 1, 1807, on exports to the West Indies. But a glut was produced in the markets of Barbados, Grenada, and St. Vincent, the principal markets, which reduced the price of fish "so low as not to defray the first cost and expense of transportation." The Embargo Act, passed in 1807, and the consequent increased demand of the West Indies for bread and flour from the British American colonies raised the price of bread from 24 shillings a hundredweight to 38 shillings, but it provided a monopoly market in the West Indies for British American and Newfoundland fish. In 1808 Spanish duties on fish were reduced to 3*s*. 8*d*. a hundredweight; and in 1808 and 1809 convoys took cod-laden vessels to Gibraltar and Lisbon. But a scarcity of salt and a carry-over of 100,000 quintals in 1809 meant that the difficulties were not at an end.

The troubles of the fishery during the war period made for early changes in the industry. The number of fishing ships decreased. In 1806

39. 41 Geo. III, c. 77; 43 Geo. III, c. 154; 50 Geo. III, c. 80.

40. See lists of prices: R. M. Martin, *History of the British Colonies* (London, 1834), III, 496; Prowse, *op. cit.*, p. 379.

41. In 1804 licenses for imports of provisions from the United States by British subjects in British ships excluded salt pork and beef, which were plentiful in Ireland.

42. *C.O.*, 193:14.

43. Newfoundland exports were chiefly carried by Bermuda ships manned by slaves.

there were "but few ships—and those chiefly from Jersey—whose crews are at all employed in fishing, and even the principal part of *their* cargo is caught by resident fishermen."[44] "The bye-boat fishery . . . is wholly laid aside."[45] "The impress[46] in the port of Teignmouth, from whence the principal part of them fit out, appears greatly to have impeded it [the fishery] by increasing the expense, difficulty and uncertainty of procuring men."[47] And this was in spite of the fact that in 1805 the Admiralty allowed every vessel to take out four experienced men. Bankers, owned and fitted out in Newfoundland, were able to proceed earlier to the Banks and to make one or two more trips a season than bankers from England. They could be sailed more cheaply and were free from wartime molestations. The larger ones were sent home with core fish and oil. But the number of passengers brought out by fishing ships fell off, and they were forced to depend on relatively unskilled Irish laborers who were not entitled to a bounty because they were of no help to the navy. "The bank fishery is looked up to as a nursery for seamen. I think I could safely answer that there are but few that an officer of the navy would deem a seaman who are employed in the bankers. [They are] meer lubbards, and the Irish labourers on shore are not even used to the management of boats."[48] Bounties were withdrawn in 1803, and the bank fishery declined "as more expensive and less profitable than other modes of fishing and . . . exposed to many interruptions from war."[49] Bay Bulls, Cape Broyle, Ferryland, Capelin Bay, Renewse, Fermeuse, and Trepassey, which formerly supported more than 200 bankers, had almost none in 1807. The sack ships were "the most effectual nursery for seamen, as the vessels that sail for foreign Europe as well as those bound directly home close their voyage in Great Britain, excepting those few that return to Newfoundland with salt."[50]

The war years saw the beginnings of the seal fishery. The resident fishermen were drawn into it because the difficulties of the French during the war let them push northward; and the local bank fishery provided the necessary ships. In 1799 the ice carried the seals far south and the people of St. John's captured some 80,000 and those of Conception Bay about half as many. In 1800, 12,806 skins were exported from St. John's and 17,638 from Conception Bay. The following year the seal fishery was only partly successful because of the wages principle; but in 1802 it was reported as very successful. In 1804, 4,666 tons of shipping and 1,600 men engaged in it and took 156,000 seals. In 1805

44. *C.O.*, 194:45.
45. *Idem.*
46. The seizing of fishermen for the navy.
47. *Idem.*
48. *C.O.*, 194:43 (1801).
49. *C.O.*, 194:45.
50. *Idem.*

it employed 131 ships and 1,547 men, but 25 ships were lost in the ice. The fishery, formerly confined to Trinity, Bonavista, and northwards, "was carried on [in] the early part of the winter by setting netts for the seals. But it appears that the embarrassments under which the inhabitants laboured during the late war, in the cod fishery, obliged them to have recourse to a mode of fishing for seals in the spring of the year which had not been practiced before."[51] South of Bonavista,[52] where the seals did not so often come in with the ice, the fishermen "have of late undertaken to fit out schooners from thirty to sixty tons carrying ten to fifteen hands each and proceed to the northward to meet the ice."[53] The seal fishery was carried on from the middle of March to early in May, and by a share system which gave the owners of vessels one half of the proceeds and divided the remainder in equal portions among the crew, who sometimes cleared from £5 to £25 each. The owners' profits were generally "sufficient to defray the expense of fitting out the same vessel in the cod fishery which commences about the time of her return from sealing."[54]

In Conception Bay the expansion of the seal fishery was rapid.[55] This, and the fact that the merchants employed "so many craft on their own account, [which] would take up more of their attention than they

51. *Idem.*

52. Prowse, *op. cit.,* pp. 419–420; also Abraham Kean, *Old and Young Ahead* (London, 1935), pp. 127–128.

53. *C.O.,* 194:45.

54. *Idem.*

55. About the end of the century the capture of seals along the shore and the use of nets spread in narrow places, at the end of the fishing season, were supplemented by the use of vessels. Merchants of St. John's and Conception Bay sent out ships of from 30 to 50 tons with 12 or 14 men hired at a fixed wage. In 1800 about 50 vessels were employed. Schooners in the seal fishery included large decked boats of from 25 to 35 tons, and larger vessels of from 40 to 75 tons, with crews of 13 to 18 men. Poles were fastened alongside to prevent cutting from the ice. The gunners were provided with guns and free berths, and the remainder of the men with 40 shillings each for provisions. About the middle of March vessels proceeded north to the ice. On the location of seals the gunners shot the largest, and the other men killed with clubs. The seals were then dragged to the schooner and skinned, or "pelted." If successful, in about five weeks the vessels returned, unloaded, and started on a second trip. The seals were sold to the merchants at an agreed price, or graded into three sizes, or sold tal qual by auction, the average trip bringing from £9 to £12 a man. After the seals were landed the fat was cut into small pieces, put into puncheons, and left to melt in the sun. It was then put into square vats made of thick planks dovetailed and tarred, the corners being braced with iron clamps. They held from fifteen to twenty tons. Inside, a grating of rods slanted from the rim to the bottom. A faucet six inches from the bottom drew off the water and another two thirds of the way up drew off the "white" oil. After the oil was extracted, the blubber, or remainder, was boiled in copper caldrons and produced common seal oil. The use of blubber "was first introduced not many years ago at Harbour Grace," and was "productive of considerable profits." L. A. Anspach, *A History of the Island of Newfoundland* (London, 1819). See also John McGregor, *British America* (London, 1828), I, 197–199.

could properly bestow, encouraged the people to build craft for themselves." They advanced "all the necessary supplies on credit for three years, which was a good spur to industry."[56]

The seal fishery and the ownership of vessels in Newfoundland contributed to the decline of the West Country ship fishery. "The number of vessels employed on the north east shore is considerably increased. It is scarcely necessary to say that the latter mode of fishing is attended with less expense and risque and generally better success than the former. It is also more convenient to the planters as lying within the compass of their limited means and being connected with the seal fishing in the spring of the year, as both are carried on by the same vessels."[57] The cod fishery along the northeast shore—the "French Shore"—began about the middle of June.[58] "On this part the fish and bait are generally found in greater abundance." "The chief part of this fishery is carried on from Conception Bay[59] where the planters are more independent than in the other districts."[60] Families moved north, the men to use boats, and the women and children to split and dry the fish.

The seal fishery, the extension of the schooner fishery to the north, and the decline in cod prices, which fell from 27 shillings to 14*s.* 6*d.* a quintal in the four years ending in 1805, during the difficulties of the war period, meant increasing control by the resident merchant.

> It is impossible in so scattered a population with such amazing extent of fishing bank and shore, that he whose establishment is in St. John's or in one of the out-harbours or settlements, could attend to the large import and export trade upon which he subsists and at the same time employ himself or his clerks on a fish stage in twenty different places or in perhaps a hundred boats at sea. . . . He therefore uses at his need the planter, and as the fisherman must supply himself from his warehouses with winter food and with clothing he retains both planter and fisherman as his constant clients.[61]

Planters, i.e., residents or settlers, with families could "make" fish more cheaply than merchants or byeboatkeepers, who were compelled to pay

56. *C.O.*, 194:45. 57. *Idem.*

58. In 1797 St. John's and Trinity merchants sent a few large shallops north to Croc and adjacent harbors rather than to the Banks, with excellent returns. Poole merchants maintained establishments at Fogo Island, and, with Dartmouth and Jersey, at Harbor Breton on the south coast.

59. At Trinity Bay the merchants were fewer in number and each merchant had his own dependents. The winter or seal fishery, the salmon fishery, and shipbuilding contributed to make it "one of the best trades carried on from Britain." Reeves's evidence, *Third Report,* p. 170. Instances were cited of very large profits. John Jeffery had nine vessels, two of which made a double trip. Routh's evidence, *idem,* pp. 36–37. Thomas Street, a Poole merchant, also commented on the fur trade. *Idem.*

60. *C.O.*, 194:45.

61. R. H. Bonnycastle, *Newfoundland in 1842* (London, 1842), II, 165.

wages to servants. The principal merchants, relying on the planters, secured profits from the high price of goods and the low price of fish; in other words, by an extension of the truck system. "It is certain that the increased expense of every article required for the fishery and the low price of fish in the foreign market have obliged the merchants to have recourse to this mode; and it is perhaps the only means by which our fishery is enabled to maintain a competition against the rival fishery of New England."[62] "A custom prevails at St. John's for the merchants to meet together and settle the price of fish and oil, which is termed 'breaking the price.' This is done about the beginning of August after having received advices from Europe and ascertained the state of the markets." Prices were arranged "so as to enable the planters just to discharge their debts; but they never give a liberal price for fish unless, about the time of settling that point, several sack ships are arrived to buy, under favor of which the fish catchers make a good market, and that influences the rest of the season."[63]

By 1813, as a result of war conditions in Europe and America, the St. John's trade was very profitable, with its exports of cod brought from outlying centers and its imports of goods that would be distributed to them. "The principal mercantile men of this country, by monopolizing almost the whole of the external and internal trade, are thereby enabled to amass the most splendid fortunes with an inconceivable rapidity; whilst the middling and lower classes of fishermen may toil from year to year with patient and unremitted industry and yet find themselves in their old age many degrees worse off than when they first crossed the Atlantic."[64] Prices in the outports were lower since the merchants paid the cost of collection and the fish were less strictly culled. The St. John's merchant purchased at 32 shillings a quintal in the outports and sold at 40 to 46 shillings in Spain or Portugal, but the pur-

62. *C.O.*, 194:45 (1806). 63. *Idem.*

64. Lieutenant Edward Chappell, *Voyage of His Majesty's Ship Rosamond to Newfoundland and the Southern Coast of Labrador* (London, 1818). Irishmen were brought to Newfoundland under a bond. The man's parents or relatives went security for his passage money to the master of the trading vessel. This money was really paid by the employer in Newfoundland. "The slavery of the Newfoundland fishermen, thus commenced upon their first entering the country, is perpetuated by a system of the most flagrant and shameful extortion. . . . The prices are so enormous that the original debt due for the passage money of the emigrants instead of being diminished by the hardest and most faithful servitude, continues rapidly to increase." *Idem*, pp. 219–220. "Almost every merchant issues notes in lieu of cash. This paper currency is the principal circulating medium of the country." See also R. B. McCrea, *Lost Amid the Fogs* (London, 1869), and L. A. Anspach, *A Summary of the Laws of Commerce and Navigation Adopted to the Present State, Government and Trade of the Island of Newfoundland* (London, 1809); Charles Pedley, *The History of Newfoundland* (London, 1863), pp. 204 ff. and *passim*.

chases were paid for in provisions and supplies, sold at a high price.[65] Dried fish brought prices almost as high in Newfoundland as in England. As in the fur trade, the product tended to cost as much at home as abroad, but the established merchant gained through the prices obtained for the products exchanged.

In the closing years of the war,

> Great Britain possessed almost exclusively the fisheries on the banks and shores of Newfoundland, Labrador, Nova Scotia, New Brunswick and the Gulf of St. Lawrence [and] enjoyed a monopoly of supplying Spain, Portugal, Madeira, different parts of the Mediterranean coasts, the West Indies, and South America with fish; and our ships not only engrossed the profits of carrying this article of commerce to market but secured the freights of the commodities which the different countries they went to exported. [But this] was followed by a depression more ruinous to our fisheries than had ever before been experienced.[66]

Competition from the United States, France, and British North America, following the end of the war, made for difficulties.[67] Many things—the surrender of the shore to the French, their bounty-fed competition, the closing of their markets to Newfoundland, the preferential duties enjoyed by the French in Italy, the increased costs of taking fish, and the decline in consumption—united to bring about a fall in prices; and in 1816 cod dropped to 14 shillings a quintal. With that, there was a rapid increase in debt, and a severe and sustained depression.

In Italy duties were increased in 1815 from 4*s.* 9½*d.* a hundredweight to 8*s.* 10*d.*, and in Spain to 10*s.* 2*d.* in 1814, and to 10*s.* 6½*d.* in

65. The merchant's profits were derived from: (1) cod and oil secured from fishermen, (2) cod purchased from petty boatmasters along the coast, (3) gains made by supplying petty boatmasters with provisions, clothing, powder, shot, and salt at triple prices, (4) the products of a large salmon fishery, and (5) oil from the seal fishery.

66. McGregor, *op. cit.*, I, 213. Apparently the first shipments were made to Brazil in 1812. Prowse, *op. cit.*, p. 403.

67. NEWFOUNDLAND EXPORTS

	Dry Fish (quintals)	Sealskins
1821	903,892	
1822	884,647	306,982
1823	867,183	230,410
1826	969,216	292,067
1827	936,470	460,584
1829	924,237	245,408
1830	844,154	357,523
1831	726,881	601,742
1832	654,053	682,803
1833	663,287	501,436

1815.[68] The Portuguese and Brazilian duties were 15 shillings.[69] Losses because of bad weather in 1818 and the crisis of that year resulted in 1819 in the grant of a bounty, one of 3 shillings a quintal.[70] In 1823 cod in Lisbon brought only three dollars a quintal, and there was increased competition from the Norwegians. In 1827 the figures for this fishery were: 83 stations, 2,916 boats served by 124 "yachts," 15,234 fishermen, and a take of over 16,000,000 fish. In February and March[71] more than 20,000 men were employed at Lofoden with nets and trawls. As a result of this competition prices were low in Portugal, Brazil, and the West Indies. At the same time there was a very bad fishery in 1829.

The reoccupation by the French of the Petit Nord following the Napoleonic wars forced Newfoundland schooners to the Labrador.

> About one third of the schooners (of a total of 300) make two voyages, loaded with dry fish, back to Newfoundland during the summer; and several merchant vessels proceeded from Labrador with cargoes direct to Europe, leaving, generally, full cargoes for fishing vessels to carry to Newfoundland. A considerable part of the fish of the second voyage is in a green or pickled state, and dried afterwards at Newfoundland. . . . The Labrador fishery, (conducted largely with seines) has, since 1814, increased more than sixfold, principally in consequence of our fishermen being driven from the grounds now occupied by the French.[72]

68. *Report from Select Committee on Newfoundland Trade with Minutes of Evidence Taken before the Committee, and an Appendix, 1817.*

69. McGregor, *op. cit.,* p. 222.

70. Four serious fires also broke out in St. John's between 1816 and 1818. McGregor, *op. cit.,* pp. 167–168.

71. After the end of April fish were not prepared as stockfish but were cured and salted as klipfish; i.e., dried fish. See Camille Vallaux, "The Maritime and Rural Life of Norway," *Geographical Review,* October, 1924, pp. 505–518.

72. McGregor, *op. cit.,* pp. 184–187. According to a rough estimate of the total, in 1829, the Jersey Islands and England sent 80 vessels (4,000 men) and took 240,000 quintals; and New Brunswick and the Magdalen Islands sent 20 vessels (160 men) and took 8,000 quintals. Nine hundred and sixty passengers, employed in boats and shallops, took 16,000. Colonial shipping registered in Newfoundland increased from 20,548 tons in 1826 to 29,465 tons in 1830. Of a total of 900,000 quintals, 150,000 were taken on the Labrador coast by Newfoundland vessels. The resident fishery on the Labrador, including six or seven English establishments chiefly owned in Dartmouth and four or five Jersey houses, was of limited extent. Direct exports from these establishments to the Mediterranean probably totaled 50,000 quintals and from Newfoundland houses 25,000. Martin, *op. cit.,* III, 497. For an interesting account of the business of Messrs. Harrison, Slade and Company of Poole, at Carbonear and St. Mary's, see the story of Philip Gosse's apprenticeship, which began in 1827. Edmund Gosse, *The Life of Philip Gosse* (London, 1890), pp. 29 ff. About 70 schooners left Carbonear for the Labrador in that year. The firm supplied about 25 planters, men who owned schooners and prosecuted the seal and cod fisheries on credit. The planter shipped a crew of 18 hands who claimed one half of the gross product of the seal fishery and were entitled to advances up to one third or one half of their probable earnings. On returning from the seal fishery, from March to May, the schooners proceeded to Labrador and went back in October. Prices charged on account were about double those in England, not allowing for the difference between sterling and cur-

By the beginning of the 1830's "not more than eight or ten British vessels" were employed in the bank fishery. About Conception Bay there was a population of nearly 25,000, in a total of 58,000, with Carbonear and Harbor Grace as the most important centers. Trinity Harbor was a substantial settlement, Bonavista Bay had "some valuable fishing establishments," and Fogo Island had "several extensive mercantile establishments."[73] Settlements increased to the south of St. John's and along the south shore, particularly at Trepassey, St. Mary's Bay, Placentia, and Fortune, which also carried on a whale fishery.

The Maritime Provinces had become an important basis of support but not a substitute for the United States in the Newfoundland economy. Legislation was extended to facilitate imports from the United States and, by legislation enacted in 1813 and 1814, imports of provisions were permitted in any unarmed ship not belonging to France.[74] A growth of settlements in Newfoundland, particularly along the south shore at Burin, and the removal of restrictions on building enabled Nova Scotia to export increasing quantities of provisions, lumber, and Cape Breton coal.[75] The flexibility of an economy based on small

rency. Trade was carried on with the residents and with the shore fishermen. "The fish they [the latter] took were commonly of larger size, were better cured and commanded a higher price than the Labrador product, but the quantity was strictly limited." *Idem.* A memorial of the Chamber of Commerce at St. John's in 1825 stated that 60 to 70 vessels were fitted out from that port and nearly 200 from Conception Bay, employing a total of nearly 5,000 men in the Labrador fishery. Vessels were turning from the bank fishery to the Labrador. It was linked to the seal fishery. *P.C.,* III, 1232–1233. About one tenth of the Labrador product was fish of the best quality. For curing, it required more salt than Newfoundland fish. For an account of the technique of the cod fishery see McGregor, *op. cit.,* chap. ix; also W. G. Gosling, *Labrador* (Toronto, n.d.), I, chap. xvi, 385 ff.

73. THE NEWFOUNDLAND FISHERY, ETC., ABOUT 1830

	Bankers	Island Vessels	Foreign Vessels Trading	Tonnage	Men	Fishing Boats	Acres under Cultivation
St. John's	16	73	470	54,600	3,746	500	2,400
Bay Bulls	..	..	..	...	...	170	250
Ferryland	2	4	13	1,436	106	254	500
Trepassey and St. Mary's	..	2	3	340	30	50	150
Placentia	..	4	6	821	61	402	800
Burin and Mortier	..	5	43	4,279	362	129	70
St. Lawrence	..	1	11	1,185	61	55	30
Fortune Bay	..	4	30	4,285	275	494	300
Conception	..	167	77	18,603	1,614	420	3,000
Trinity	..	8	31	4,934	302	570	270
Bonavista and Greenspond	..	2	9	1,020	70	257	800
Fogo and Twillingate	..	31	34	5,334	257	496	200
TOTAL	18	301	727	96,837	6,884	3,797	8,770

74. 53 Geo. III, c. 67, and 54 Geo. III, c. 49.

75. "The fishermen's houses are one story high, built of wood growing on the island

schooners enabled merchants to take advantage of frequent changes in markets. Losses sustained by Nova Scotia in the West Indies trade were met by transfers to the Newfoundland carrying trade. Because of the depression in Newfoundland, labor migrated to Nova Scotia. In 1816 the schooner *Industry* took 150 passengers, chiefly laborers, from Newfoundland to Nova Scotia because of the distress due to the high prices of provisions.[76] The deficiencies of Newfoundland were deplored. "We are most thoroughly convinced the system practised in that island, however eligible it was formerly, is become as it were decrepit and that it will not suit these modern times. . . . Notwithstanding the two last unfavourable seasons in this province some timely aid was afforded from home besides feeding not less than six hundred emigrants from that Island, arrived in this province during the last winter, a convincing proof that the colonial is preferable to what is called in Newfoundland the *plantation* system."[77] In 1818 St. Andrews, Indian Island, Yarmouth, and Prince Edward Island sent lumber, flour, bread, beef, and cheese to Newfoundland.[78] Prince Edward Island complained of the depression there and its effects on the prices of oxen and horses.

Difficulties in the European markets resulted in exports from Newfoundland to the West Indies. Nova Scotia suffered from Newfoundland competition and became more concerned with the carrying trade. The effect of the colonial-trade acts of 1826 were apparent in 1827 in the importation from Newfoundland of 13,000 barrels of pickled fish

and covered with boards and shingles imported from Prince Edward Island, Cape Breton, Nova Scotia or New Brunswick." McGregor, *op. cit.*, I, 170. "The use of coal has of late years become general in parlours and even in kitchens; it is imported chiefly from Sydney, furnishing a profitable employment to the shipping until the fish is ready to be put on board for market." L. A. Anspach, *A History of the Island of Newfoundland* (London, 1819).

76. In the fall swarms of people "depart for Prince Edward Island, Nova Scotia and Cape Breton to procure a livelihood in those places among the farmers during winter. Many of them never return again to the fisheries but remain in those colonies." McGregor, *op. cit.*, p. 170. Seasonal unemployment became a serious problem with the increase in population. In one winter 500 people were fed by the government at St. John's. Attempts were made to prevent them coming in from the outports and "dieters, that is, those who entertain idle men in winter," were discouraged. Palliser's evidence, *Third Report*, p. 28. In 1789 a proclamation was issued "against fishermen coming from the out harbours to winter at St. John's." *Idem,* Appendix No. 8b, p. 204. Any person during the winter season harboring or entertaining "dieters" was subject to deportation, and the houses in which they were harbored or entertained were to be pulled down. *Idem,* Appendix No. 8a, pp. 202–203. See an order requiring a house to be pulled down, dated October 15, 1790. *Idem,* Appendix No. 11a, p. 207.

77. *Acadian Recorder,* June 14, 1817.

78. Patrick Campbell, *Travels in the Interior Inhabited Parts of North America in the Years 1791 and 1792* (Toronto, 1937), pp. 274–275. In 1822 lumber and livestock were exported from the gulf shore of Nova Scotia. Irish laborers from Newfoundland were accustomed to spend the winter in that vicinity.

and of 49,575 quintals of dry fish, chiefly "merchantable" for South America. The trade acts influenced imports as well as exports.[79] In a three-year period—from 1830 to 1832 inclusive—imports of lumber, cattle, and agricultural produce from the British North American colonies averaged £32,500, not including tea reëxported from Halifax amounting to £19,400. Imports of breadstuffs totaled 93,524 hundredweight, valued at £74,819, made up of 67,812 hundredweight from Hamburg and 25,712 hundredweight from British possessions, two thirds being from Great Britain. In detail these imports consisted of: flour, 37,552 barrels, valued at £52,573, 19,075 barrels being from foreign states, 18,477 from British possessions, and half of the latter quantity being foreign flour transshipped from England; pork, 22,594 barrels, valued at £73,430, 11,908 barrels being foreign, and 10,686 British, chiefly from the United Kingdom; and butter, 11,606 hundredweight, valued at £40,621, 3,119 hundredweight being foreign, and 8,487 British, chiefly from the United Kingdom. Produce imported directly from the West Indies totaled £47,500, foreign wines, spirits, and salt making £23,220 of this; and manufactured products from the United Kingdom were valued at £422,000.

The dominance of the resident fishery eventually brought about far-reaching changes in the institutional structure of Newfoundland. "The colonizing of Newfoundland is now going forward without the consent of government, consequently without proper regulations of law and order."[80] Palliser's Act added to the difficulties growing out of the American Revolution. A report, already much quoted, nevertheless stated "that the Newfoundland fishery ought to be carried on as much as possible by ships fitted out from Your Majesty's European Dominions; that by the yearly return of the sailors and fishermen to the said Dominions, Your Majesty may have it in your power to avail yourself of their services for manning your Royal Navy, when occasion may require."[81] Significantly, the report conceded something.

> Your Majesty's subjects who so resort annually to the said island of Newfoundland should be induced to and compelled by every wise and proper regulation to return to Your Majesty's European Dominions at the end of every fishing season . . . [but] from a change of circumstances it may now be beneficial to the fishery that a certain number of persons shall be suffered to remain on the said island after the fishing season, for the purpose of tak-

79. For an account of their effects on the trade of Poole see John Sydenham, *The History of the Town and County of Poole* (Poole, 1839), pp. 396 ff.

80. *C.O.*, 194:43.

81. *The Report of the Lords of the Committee of the Privy Council for Trade*, on the subject of the Newfoundland fishery, of March 17, 1786.

ing care of the fishing stages, boats, and other necessaries for the fishery, and to make preparation for the ensuing season, as our fishermen will be thereby enabled to commence their fishery at a more early period, and have in consequence thereof an advantage over other nations who are our rivals in the fishery.[82]

The act resultant upon the report (26 Geo. III, c. 26) reinforced the position of the fishing ships and increased the difficulties of adjustment.[83]

The fishery conducted by residents or boatkeepers on a large scale called for continual adjustments in spite of protests by West Country interests against the state of government that was then developing.[84] The problems to be faced showed themselves in wage disputes. Supplies were advanced by the merchant to the boatkeeper and in return fish and oil were delivered to the merchant. In general, accounts for supplies were settled and wages were paid by bills drawn for sixty days, which were easily negotiable. The danger that the servants might forfeit their wages was avoided by giving them, under Palliser's Act, a prior lien on the fish and oil.[85] This section of the act was also intended, however, to check all attempts to disrupt the fishery by seizures before the end of the season.[86] According to the same act it was possible for the employee to forfeit two days' pay, or five days' according to the act passed in 1786 (26 Geo. III, c. 26), for one day's absence.[87] The clause

82. *P.C.*, IV, 1868–1875.

83. *P.C.*, I, 273–286. "For the making of seamen I hold to be the first and principal object of the fishery." *Admiral Milbanke's Report upon the Judicature of Newfoundland*, December 31, 1789.

84. See *First Report, passim.*

85. As far back as 1749 Governor Rodney had issued an order prohibiting the seizure of fish, oil, etc. by creditors before wages had been paid. See also a decree of Governor Lloyd of 1754 at Trinity requiring the payment of wages before the payment of other creditors. *Third Report*, pp. 16–17. See the case of W. Compton at Petty Harbor in which servants were upheld in refusing to allow fish and oil to be loaded before their wages were paid. *Second Report*, Appendix No. 3, pp. 48 ff.

86. *Third Report*, Appendix No. 5, p. 192. Attempts were made to contravene the prior lien of servants by obliging them to contract with the merchants to forgo the right to seize the fish and oil before the merchants were paid. Newman argued that at Conception Bay he supplied 100 families without servants. The fishery was carried on as a common concern without a lien on the fish and oil for the payment of wages. He complained that when servants were paid wages they commonly refused to continue with the fishery after sufficient fish and oil had been taken to pay their wages, with disastrous results to the boatkeeper and the merchant. Graham pointed out that in refusing to lend to servants, and in limiting his loans to the families of fishermen, Newman was discouraging the byeboat fishery which he claimed he was anxious to encourage. In Conception Bay there were 15,000 people who "were much to be pitied and very much oppressed." Routh's evidence, *Third Report*, p. 42; also Reeves's evidence, p. 91.

87. *Admiral Milbanke's Report*, p. 21. Palliser was strongly in favor of severe pun-

was subject to grave abuse by the masters, and deductions were made when servants were not aware that they were neglecting their duties.[88]

Clause 13 of Palliser's Act required an employer to deduct 40 shillings from the wages of an employee, which sum was to be paid to the master conveying the employee to and from Newfoundland and thus enable him to leave the country. Admiral Milbanke pointed out that no authority existed to enforce the provision and that servants remained in the country while employers forfeited the 40 shillings.[89] At the same time there was "a rage for staying in Newfoundland,"[90] in spite of regulations against the formation of settlements with which were coupled instructions which advised, in effect, that "whatever they [the residents] loved to have roasted he was to give to them raw, and whatever they wished to have raw he was to give to them roasted."[91] Clause 14 forbade employers to advance to servants more than half of their wages in money, liquor, or goods.[92] It was claimed that this was disregarded,[93]

ishments and claimed that he had posted notices annually in all harbors that fishermen neglecting their duties would "be liable to make good damages to their masters to the utmost extent of their wages." He argued that "the neglect of one day's duty may be of very great prejudice, in certain periods of the fishery, to the interest of the merchant or employer," and suggested that the penalty should be extended to five days. *Report of the Committee of Privy Council for Trade etc.*, p. 6. See also Elford's evidence, *Second Report,* pp. 25–26, and Newman's evidence. Newman claimed to have had a ship detained because the splitter refused to work.

88. Graham's evidence, *Third Report,* pp. 17–19; see also the case cited by Admiral Milbanke. *Admiral Milbanke's Report,* pp. 28–29. A deduction of 25 per cent from wages because of a bad voyage, which was granted by a justice of the peace at Harbor Grace, was, however, the subject of a severe reprimand from the governor. *Third Report,* pp. 77–78.

89. "Copy of a representation of the Lords of the Committee of Privy Council for Trade and Foreign Plantations, on the subject of establishing a court of civil jurisdiction in the island of *Newfoundland,* dated 10th May 1790." *P.C.,* IV, 1876–1881.

90. Routh's evidence, *Third Report,* p. 41. Ougier held, to the contrary, that "he never knew a fisherman remain in Newfoundland the winter from choice," and that "no winter servant can earn as much for his master as the expense of his provision." *First Report,* p. 37.

91. Knox's evidence, *Second Report,* p. 17.

92. In 1787 a servant was engaged at Burin for one year for £26, and in the course of the year was advanced, chiefly in rum, brandy, gin, and molasses, £27 3*d.* His employer sought to collect £20 8*s.* from him for "neglect of duty and upholding and encouraging 2 men who ran away in my debt." The court refused to allow the claim, and held that the employer owed him £13. In 1789, in spite of his advances, an employer was held liable for one half the summer's wages, or £7. *Third Report,* Appendixes Nos. 3a and 3b, pp. 189–190; *idem,* Appendix No. 12a, p. 209. The *Report of the Committee of Privy Council on Trade* favored an amendment which would permit the advancement of £5 10*s.* to green men and boys, even though this was more than half their wages. Half of £7 10*s.*, the wages paid, was not adequate to buy clothing. In both cases payment was to be made on the return of the fishermen. *Idem,* pp. 4, 9. This was adopted in 26 Geo. III, c. 26. See an agreement dated at St. John's, June 11, 1790. *Third Report,* Appendix No. 4, p. 191.

93. Jeffery's evidence, *First Report,* p. 19.

as in many cases people leaving England had already been granted credit for one third of their wages in addition to advances made to their families. The number of young lads in the island increased rapidly, as half their wages was not sufficient to fit them out for the first year, and consequently they were hired for two years and were left in Newfoundland over the winter.[94] This made it unnecessary to pay half the wages on return or to pay their passage home at the end of the first season. The possibility of servants returning in ships other than those owned by their employer, and in some cases refusing to go on his ships, was an additional factor which did not encourage the employer to force them to return. In some localities no ship was available at the conclusion of the fishery,[95] and if the servant was given a choice "he would six times out of ten stay behind regardless of the loss of forty shillings."[96] In cases where the master was unable to secure him a passage, the wages and passage money were often paid to hucksters in return for "liquor and useless articles." The return had little attraction as fishermen attributed "their disorders to bad living in long passages, and it was no uncommon thing for him [Graham] to hear of very narrow escapes with life, after the greatest hardships had been endured, which the want of provisions and water could occasion."[97] The regulations in many ways had effects the opposite of those intended, and the number of residents increased.[98]

As a result of difficulties in carrying out the provisions dealing with wages there was testimony that "the merchants of late have taken care to remove themselves from the responsibility of paying the servants

94. Justice Reeves stated that men returned at the end of two, three, or four seasons and that the Irish generally stayed two summers and a winter.

95. See a letter from Graham to Greaves, St. John's, October, 1790, stating that if the master's vessel was not ready to go in reasonable time, or from six to eight days, the servant might demand his wages and go on some other vessel. *Third Report,* Appendix No. 13, p. 213.

96. On the difficulties of forcing people to return see Minute of Council, January 25, 1786; and Admiral Campbell's answer. *Third Report,* Appendix No. 6, p. 194.

97. Graham's evidence, *Third Report,* p. 87. The actual value of a passage was usually 24 shillings. John Jeffery of Poole (a fishery which was not represented at St. John's) stated that 40 shillings were not deducted in one case in six and that the men were brought back gratis. At Fortune Bay, Poole and Dartmouth conducted a bank and shore fishery and the men were taken back for nothing. This fishery had increased after 1786 in residents, adventurers, men employed, and fish taken. Over one half of the fish were caught by owners who returned to Great Britain. In 1791–92 it had four banking ships. As in the case of Conception Bay it was in a transition stage; some of the boats belonged to planters and some to merchants. J. Walden, for example, hired servants for planters, bought the fish from the planter, and charged him with the servants' wages.

98. *Third Report,* pp. 104–105; also Elford's evidence, *Second Report,* p. 15.

wages as far as possible."[99] The emergence of the planter as an intermediary brought in new problems. A petition of forty-five merchants for a court at Harbor Grace stated that the boatkeepers or residents (planters) were natives who "hire their own servants, and plan out their own voyages, independent of the merchant (except being supplied by him) which is not the case in many parts where master and crew are in fact servants to the merchant."[100] In April, or at the close of the first fishery, the planter contracted with the merchant for provisions, salt, clothing, canvas, cordage, and other necessaries, agreeing in return to give him "all the fish and oil." The advances ran up to £300, and the outlay prior to the first payment in September was extensive. "Besides being at the vast expense (contrary to the custom of other parts of the Island, where the planters bring their fish and oil to the merchant) of employing boats and servants to proceed to the different harbours of the district and collect the same," the merchant was forced to rely on the honesty of the planter, since in bad seasons he was likely to dispose of his fish elsewhere.

An increase of Irish laborers and the rise of the wage system accompanied the growth of trade, especially at St. John's, and raised further difficulties for the merchant. Ougier claimed that, since the master was forced to pay half the wages in bills of exchange, various hucksters' shops and public houses had come into existence which were willing to make advances on the endorsed bills of exchange thus due as wages. The Irish preferred dealing with shops owned by their own countrymen, and the English merchant who gave supplies on credit lost by it.

The fishing admirals had long since ceased to meet the demands of an increased population for the administration of justice, for they were generally illiterate, they spent much of their time on the Banks, and were prejudiced in favor of the employers.[101] As Palliser stated, "The prosperity of the fishery requires speedy, short and above all unexpensive issues, in a summary way. . . . The immense extent of the coast, the shortness of the season (for there is no communication by land either winter or summer) the necessity of every person being diligent during the season and [the fact that they] cannot be interrupted without great prejudice, all unite to render it [use of law courts] impracticable."[102] St. John's, with a growing population and increasing trade, had the greater need of courts, but the outports could not be neg-

99. Graham's evidence, *Third Report*, p. 106.

100. *Idem*, Appendix No. 15, pp. 215–217.

101. *Third Report*, p. 48.

102. Palliser's evidence, *Third Report*, pp. 28–29; also *Admiral Milbanke's Report upon the Judicature of Newfoundland*, p. 18.

lected.[103] The customs administration found it necessary to make seizures[104] for payments, and suffered because of legislation in 1792 which laid a prohibition upon customs officers' becoming justices of the peace.[105] *Admiral Milbanke's Report* in December, 1789, presented a thorough survey of the powers given under his commission in June, 1789, and pointed to their inadequacies. This report, together with further details in a letter from Admiral Milbanke to William Fawkener of February 20, 1790, led in 1791 to the passing of an Act for Establishing a Court of Civil Jurisdiction in the Island of Newfoundland for one Year (31 Geo. III, c. 29).[106] It was followed in 1792 by an Act for Establishing Courts of Judicature in the Island of Newfoundland and the Islands Adjacent (32 Geo. III, c. 46), an act which was renewed periodically until, in 1809, the Newfoundland Act (49 Geo. III, c. 27) set up a permanent court. The establishment of governmental machinery called forth vigorous protests and was preceded, in 1793, by an extensive investigation by the House of Commons.

The rise of the resident fishery meant not only the establishment of courts but also the introduction of government[107] on a more extensive scale. More elaborate equipment for carrying on the fishery, the growth of St. John's as a distributing center, an increasing dependence on larger vessels to bring supplies and provisions from Great Britain and Ireland, and a rise in the value of sites near which the fishery was conducted with greater efficiency were factors favoring the permanent occupancy of the land. The fishing ships' rooms disappeared in spite of legislation. The occupying of property for purposes other than the fishery had been prohibited in 15 Geo. III, c. 31 (1775). After hearing the principal merchants of Poole and Dartmouth and reviewing the

103. See Graham's evidence, *Third Report,* pp. 2 ff.

104. See Routh's evidence, *Third Report, passim,* but especially pp. 57–61; also *Third Report,* pp. 178–179; *Second Report,* Appendix No. 68, p. 61; also Newman's evidence, *First Report,* pp. 20 ff. Routh suggests that Dartmouth and Bristol were instrumental in this legislation, p. 43.

105. See a table of Revenues, Expenses and Seizures in 1781–1791. Routh's evidence, *Third Report,* p. 65; also *Second Report,* Appendix No. 58; also pp. 5 ff., and Appendix, pp. 35 ff. "The sheriff, with his accustomed inhumanity, declared to him, that he would readily sell his, this deponent's liver, if he had an order for it." Appendix No. 1, D. 35.

106. *P.C.,* I, 287–296.

107. There were repeated complaints of the frequency of court sessions during the fishing season, exorbitant fees, and the participation of judges in the trade. "If all the fishing conveniences are to become private property and if such kind of inhabitants as the present should increase in the same proportion as they have of late years . . . more justices will be necessary and even some kind of civil government." For an account of the beginnings of the courts see Ougier's evidence, *First Report,* pp. 43–45. See Reeves's evidence, *passim,* and especially John Reeves, *History of the Government of the Island of Newfoundland* (London, 1793).

evidence, the Lords of the Committee of Privy Council for Trade recommended in March, 1786, that "Your Majesty's subjects who may from time to time reside in Newfoundland ought never to be allowed to form themselves into a colony and with that view to possess in fee any landed property there."[108] It was also the advice of the Privy Council that those who had first taken possession of the shore should be entitled to continued possession provided they did not, for any season, neglect to carry on the fishery;[109] that no buildings should be erected, other than for preparing fish, within six hundred yards of high-water mark; and "that no right of property be acknowledged in any land or building even beyond that distance." The governor was given permission, however, to allow buildings, already erected, to remain, provided they did not interfere with the fishery. Admiral Campbell reported in 1786 that enclosures were being made and that "dwelling houses and retail shops along side the water" were infringements on legislation.[110] A proclamation was issued forbidding the construction of buildings and fences. But vested rights were already in possession, and grants to officers of the military corps and to members of civil establishments complicated the problem. Admiral Milbanke claimed credit for a decrease in the population of St. John's during the first year following the proclamation, but the check was temporary.[111]

An increasing number of buildings used in trade rather than the fishery helped to defeat the regulations. The fishing admirals, or the first three arrivals at St. John's, generally belonged to firms with ample fishing room, and, although entitled to a choice of the public fishing rooms, they never used them, but charged a rent.[112] The act of 10 and 11 Wm. III, c. 25 (1699), had been responsible for the encouragement of the byeboatkeepers since it had given them the privilege of building houses and fishing rooms to be held as their own property. With the increasing scarcity of timber these properties were "let at rent to any new adventurers for one tenth expense they could erect others."[113]

108. *P.C.*, IV, 1869.

109. See a grant of land to Anquetil and Company in Aquaforte, dated St. John's, October 8, 1788, to be void if neglected for one entire season. *Third Report*, Appendix No. 10, p. 206.

110. *Idem*, Minute of Council, January 25, 1786; and Admiral Campbell's answer, pp. 193 ff. Appendix No. 8a; Appendix No. 10, and Appendixes Nos. 11a, b, pp. 202 ff.

111. Graham's evidence, *Third Report*, pp. 79, 83–84.

112. *Idem*, pp. 20–21.

113. Ougier's evidence, *First Report*, p. 43. "From the situation of day labourers in England a great many have by their possessions (fishing rooms) in the latter part of their days had a comfortable subsistence." Fishing rooms in some cases, when including stage or covered wharf for handling fish, flakes or beach for drying, cookrooms for lodging men, and storehouse for fish were valued at from £10,000 to

There is scarcely a part in Newfoundland unoccupied, that adventurers from Europe would take possession of for the purpose of carrying on a fishery, without they could be allowed to do it as private property, and not under the restrictions of the 10th and 11th of William the IIId and 15th of George the IIId, and he [Graham] has known a great many adventurers of late years apply to the governor for grants of such places as they had fixed upon, saying unless they could be secured in the possession of it hereafter, they could not think of expending any money upon it. The difficulty and expense of clearing the ground and building upon it would be so great that it would not answer their purpose, nor had they wherewithal to carry on such an undertaking. It by no means follows that however inclined the government of this country may be, or whatever pains may be taken to encourage that branch of the fishery, that it ever could be carried on again in the manner, or any thing like to the extent, that it has been heretofore; for if they should have no other obstacle to counter with, the want of room alone, with proper conveniences, must render it difficult, if not altogether impossible. . . . It has for many years past been a common practice among the merchants particularly those who have engaged in the trade since the American war, to advance money, and encourage resident boatkeepers to build fishing rooms, and carry on fisheries upon them, depending upon the fish and oil with a mortgage of the fishing room taken as collateral security, for the repayment of the money so advanced. . . . Within his own knowledge [1779–91] a great deal of the shores of Newfoundland have in this manner been built upon, to the total exclusion of adventurers from Europe; and he thinks himself justifiable in saying, that six at the least out of every ten of such rooms have in the course of five years after they were first built upon by boatkeepers come into the possession of the merchants, from the incapacity of the boatkeeper to pay his debts, but by sale of his fishing room; and then he seldom sells it for more than a fourth part of the sum which it cost him in the building. . . . In this way it certainly may be said there is almost a complete monopoly made of the whole of the shores of that island, to the exclusion of adventurers from Europe.[114]

In spite of these trends the government persistently[115] refused to permit the ownership of property, and insisted on the return of the merchants and men from Newfoundland at the end of the season. But in 1798 Portland could write: "I am sorry to observe that the policy of

£15,000. "The greater part of what was formerly ships room is now occupied by individuals as private property." *C.O.*, 194:45. On the other hand, a letter from Sir Erasmus Gower to Sir Stephen Cottrell, June 9, 1806, referred to a resident fisherman occupying a vacant ship's room in St. John's who paid a sum of money to the first fishing ship arriving in the harbor from England for permission to remain undisturbed in his room, even though the ship belonged to a merchant who had rooms of his own and no use for the room occupied by the resident. *C.O.*, 195:16.

114. Graham's evidence, *Third Report*.

115. See the Instructions and Commissions to the Governors of Newfoundland, *P.C.*, II.

these laws has been completely defeated and that the population of Newfoundland has in consequence increased."[116] In 1806, William Fawkener described the situation clearly:

> The nature of the Newfoundland fishery and the mode of conducting it, have, from unavoidable circumstances, undergone so great a change that some of his Majesty's instructions as well as several provisions contained in the statutes now in force for regulating the said fishery and the trade between the island and the mother country are become wholly inapplicable to the present state of such trade and fishery and can no longer be expected to produce the effect and advantages with a view to which they were originally framed.[117]

The pressure from every side was steadily becoming greater. For the government, retreat was inevitable.[118] We find it foreshadowed in a letter from Governor Holloway to Castlereagh, written in November, 1808:

> I cannot but lament that it was ever recommended to His Majesty's Ministers by my predecessors to grant leases of land on that island; it was striking at the root of the law which had for so many years regulated this fishery as a nursery for seamen and meant to discountenance residency, the great and improper increase of which tends to colonization; it likewise gives the inhabitants a kind of sanction to claim from occupancy lands that are no longer used for the fishery, for which purpose alone they received their grants from the different governors agreeably to His Majesty's instructions; consequently [they] can have no legal right for to sell, mortgage, lease and transfer as is now become a daily practice.[119]

And in November, 1812, Governor Duckworth wrote to the Earl of

116. *P.C.*, IV, 1931.
117. *C.O.*, 194:45.

118. For an account see J. D. Rogers, *Newfoundland* (Oxford, 1911), pp. 146 ff. The handicaps of Newfoundland's people, in the matter of ownership, inability to raise their own products, and the restriction on imports from the United States have been pointed out. "From a system the first object of which is to withhold that principal of internal legislation which is acknowledged to be indispensable to the good government of every community—which restrains the building of comfortable dwellings in a climate exposed to the most inclement weather—which prohibits the cultivation of the soil for food and restricts the importation of it from the only market which the inhabitants have the power to go to—from such a system it is not surprising that the inhabitants of Newfoundland are not able to maintain a competition against American fishermen." *Considerations on the Expediency of Adopting Certain Measures for the Encouragement or Extension of the Newfoundland Fishery, 1805.* "The merchants of Dartmouth and Poole will ever oppose the shore fishery or [fishing] on the banks in small shallops; it would be a blow to their monopoly tho' a more general advantage to Great Britain." Thorne to Sullivan (February 26, 1803). *C.O.*, 194:43.

119. *P.C.*, IV, 1931. In 51 Geo. III, c. 45, certain ship's rooms were declared private property.

Bathurst: "There is a general concurrence in one respect only, that the fisheries of Newfoundland are decidedly sedentary and that the war has been protracted so long as to make it very uncertain whether any change of system would be produced by the return of peace." The increase in population could not be checked. "The wisest object of such revision [of the laws] would seem to be to remove from the sedentary fishery all unnecessary impediments, [namely] the provisions . . . by which all unoccupied places . . . are accounted fishing ships rooms, and the restrictions on cultivating the lands." "The merchants of St. Johns have formed themselves into a society and are making continual efforts for the acquisition of a power which ought not, in my opinion, to be vested in them."[120] On November 3 Duckworth complained of a merchant who insisted on erecting a building on his land. "This attempt was not that of an individual but was instigated and supported by the merchants in general who have created a fund, the real object of which is to oppose the measures of government and to establish the right of property upon a quiet possession of twenty years."[121] In 1813 provision was made for grants of land under a quitrent.[122] The Board of Trade declared in 1817 that the people could not be moved to Canada because of the depression, and that the fishing ship was already a thing of the past. "The great length of the late war and the increasing prosperity of the fisheries, gradually, and almost insensibly overthrew that system, and the resident population . . . is become extremely numerous . . . 40,000 to 50,000 persons. The Governor was a few years since authorized to make small grants of lands reserving a trifling quit rent, and my Lords see no reason why this plan should not be acted upon to a greater extent."[123]

The war had increased the hazardousness of the fishing-ship industry and contributed to its losses, and its place had been taken by a fishery carried on by the resident population. The winter and spring seal fishery encouraged the trend. Fishing ships from England had decreased from nearly 300 in 1792 to less than 50 in 1817; and in 1823 they numbered only 15.

> Thus the ship fishery has diminished to little more than a name, the result of the two systems being last year the production of 750,000 quintals of fish from the boat or island fishery while that of the ships made only 34,000 quintals. It was evident, therefore, that laws created as well for the encouragement of a ship fishery from England as with a view to discountenance settlement and a resident fishery were become nearly a dead letter; and that

120. *P.C.*, IV, 1933–1935.
121. *Idem*, p. 1936.
122. Prowse, *op. cit.*, pp. 398 ff.
123. *P.C.*, IV, 1938–1939.

some provisions on the other hand were wanting for the regulation of the latter [the resident fishery].[124]

The appointment in 1818 of Sir Charles Hamilton as governor and commander in chief in Newfoundland was generally conceded to mark the beginning of a policy of alleviation. Among the instructions which he proceeded to carry out was the making of grants of small lots of land at trifling annual rentals.[125]

In December, 1822, the residents took a more aggressive offensive. In a memorial from a committee of the citizens[126] of St. John's to the Earl of Bathurst, they complained of the bad administration of the laws, the inefficient taxation, the various bankruptcies, the lack of roads, the lack of education, and the uncertainty of employment. They made the charge that the policy of regarding "Newfoundland merely as a fishing establishment and a place of trade" was fostered by the merchants, who had a monopoly of trade and were anxious to check local development lest it should interfere with imports. They also held the merchants responsible for statements that the soil was barren, the climate severe, and the fishery important to the navy. They, the merchants, after accumulating fortunes of £50,000 to £300,000 by their annual profits of £20,000 or £30,000, retired to England, withdrew their capital, and kept the colony from making improvements.

In any case the commanding position of the merchants was being weakened by many things: by the internal and external competition of the Americans and the French, by an ever-increasing number of Newfoundlanders who were engaged in the fishery, and above all by the growing education of the people. Indeed, they had already made it felt in two general demands. One called for the encouragement of agriculture, if only as a means of relieving the colony of the disastrous effects of the severe fluctuations in the fishery. The other was financial. In the ten years ending in 1824 revenues had, in round numbers, totaled £140,000, and the expenses of civil establishment £60,000. The demand was, consequently, made for increased local expenditures and for increased control over finances.

Basically all this was a result of the increase in the resident popula-

124. Sir Charles Hamilton, February 2, 1824. *P.C.*, IV, 1941, and *P.C.*, II, 696.

125. See Royal Instructions authorizing the Governor of Newfoundland to make grants of Land from June 1, 1817. *P.C.*, II, 691.

126. See *Colonies, Observations on the Government Trade Fisheries and Agriculture of Newfoundland* (London, 1824), Appendix I. This pamphlet was printed in the newspapers and was given more effective publicity. See Lord Birkenhead, *The Story of Newfoundland* (London, 1920), chaps. vi, vii; also Prowse, *op. cit.*, chap. xiv.

tion. A charter was granted,[127] known as The Newfoundland Fisheries Act of 1824.[128] In the same year another act[129] established a supreme court and arranged for a system of registration. In 1832 a representative constitution was granted with elective suffrage and control over the "nett produce of all duties levied within the said colony."[130] Sir Thomas John Cochrane was appointed civil governor and in the royal instructions and dispatches sent to him on July 27, 1832, by Lord Goderich the system was made clear which was to be followed in the government of the ancient colony.[131]

127. *P.C.*, I, 300–323, and *P.C.*, II, 723–748.

128. 5 Geo. IV, c. 51.

129. 5 Geo. IV, c. 67.

130. 2–3 Wm. IV, c. 78.

131. "The fundamental principle," wrote Lord Goderich, "was to prevent the colonization of the island, and to render this kingdom the domicile of all persons engaged in the Newfoundland fisheries. The common interest or convenience of those persons virtually defeated the restrictions of the various statutes respecting them. . . . Notwithstanding the growing population and the wealth of Newfoundland," he continued, "no plan has hitherto been adopted for regulating such of the internal affairs of the colonists as demanded the enactment of laws specially adapted to their peculiar situation. Parliament indeed contemplated the erection of corporate towns with the power of making bye-laws, for remedying this inconvenience; but on attempting to carry this design into effect, unforeseen obstacles were encountered. It was found altogether impracticable to reconcile the contradictory wishes and recommendations of the parties who would have been more immediately affected by the measure; and it became evident that the boon which it was proposed to confer would be received by a great body of the inhabitants not as an act of grace but as an infringement of their rights into whatever form the intended charters might have been thrown. . . . Carrying with them from this kingdom the law of England, as the only code by which the rights and duties of the people in their relations to each other, and in their relation to the state, could be ascertained, it was obvious, as soon as the colony began to assume a settled form that the adaptation of that code to the various exigencies of the local society was a task demanding the exercise of much reflection and caution; that many of its provisions were entirely inapplicable to the wants of a population so peculiarly situated; and that many more could be applied only by a distant and uncertain approach to the original standard. Hence it occurred that in the administration of the law, the judges virtually assumed to themselves functions rather legislative than judicial and undertook to determine not so much what the law actually was, as what, in the condition of Newfoundland it ought to be. . . . In whatever related to police and internal improvements demanding the cooperation of different persons nothing could be carried into effect which any individual found an adequate reason for opposing, or which he opposed from mere caprice." *P.C.*, IV, 1954–1960.

The Newfoundland Fisheries Act was extended for two years and nine districts were delimited for electoral purposes. In spite of recommendations that, as in British Guiana, the Council and Assembly should be united in one house to avoid disputes between them, provision was made for the establishment of the two branches.

CHAPTER XI

NOVA SCOTIA AND CONFEDERATION, 1833–1886

NEW ENGLAND

The youth of the province are daily quitting the fishing stations, and seeking employment on board United States vessels, conducting them to the best fishing grounds, carrying on trade and traffic for their new employers with the inhabitants, and injuring their native country by defrauding its revenue, diminishing the operative class, and leaving the aged and infirm to burthen the community they have forsaken and deserted.

Journals of the Assembly, Nova Scotia, 1837

At the beginning of the half century from 1833 to 1885 New England had lost the markets of Europe. But she had increased the extent of her markets in the West Indies, even though they had been weakened by the emancipation of the slaves;[1] and she had an expanding internal market indicated by rising prices.[2] Enlarged markets contributed to the increase of her cod, mackerel, and other fisheries. Conflicts with Nova Scotia over fishing grounds and markets were avoided by the Reciprocity[3] and Washington treaties. But the increase of population in Nova Scotia and Newfoundland eventually brought about restrictions on New England. Moreover, after the Civil War, and with the opening

1. After the abolition of slavery the apprenticeship system was introduced and then, particularly after 1838, the wage system. Planters were compelled to import labor from China and India to meet the demands of soil exhaustion and the acquisition of land by freed slaves. The removal of differential duties on sugar from Cuba and Brazil, coinciding with the depression of the 'forties, added to the planters' difficulties. See J. R. McLean, "The Consequences of Slave Emancipation in British Guiana and Trinidad," B. Litt. thesis, Balliol College, *passim;* W. L. Burn, *Emancipation and Apprenticeship in the West Indies* (London, 1937).

2. The average prices of mackerel and cod per barrel on September 1, from 1842 to 1884, at Gloucester, were quoted as follows:

	No. 1 Mackerel	Cod
1842–53	$10.42	$3.08
1854–66	13.57	5.18
1867–72	14.16	5.05
1873–84	15.17	5.19

In 1830 the price of mackerel was $5 a barrel and in 1856, $19. Cod prices increased from $2.12 to $3.75 a quintal. *No Surrender,* Vol. 1, No. 1, Washington, December 17, 1887.

3. On the advantages anticipated for the New England fishery as a result of the advent of reciprocity see *The Report of Israel D. Andrews on the Trade and Commerce of the British North American Colonies* (Washington, 1853), *passim.*

of the West, markets for fresh fish displaced those for cured fish; and New England gradually withdrew from the more distant waters, and erected tariff walls to bar out fish from Nova Scotia.

Bounties[4] and duties[5] supported the expansion of the New England fishery. In 1833 a protective tariff abolished specific duties and introduced a 20 per cent ad valorem rate, and in 1842 restored the rates of 1816—namely a duty of $1.00 a quintal on smoked or dried fish, $2.00 a barrel on salmon, $1.50 a barrel on mackerel, and $1.00 a barrel on other varieties of pickled fish. The duty on salt was lowered from 20 cents a bushel in 1824 to 10 cents in 1832, to 8 cents in 1842, and was made 20 per cent ad valorem in 1846. The reduction on fish to 20 per cent in 1846, the Reciprocity Treaty of 1854, the abolition of the bounties in 1866, and the Washington Treaty, in force from 1873 to 1885, were indications of an efficient industry and the increasing importance of the domestic market.

The fishery was rapidly extended on the Banks and on the Labrador.

> The cheapness of their supplies enables the Merchants of Newport, a Town in the Eastern extremity of that Country, and bounding on New Brunswick, to enter extensively into the Labrador fishing, which in the Years of Eighteen Hundred and Thirty-two and Thirty-three, they have carried on with great success; their vessels are manned mostly with men from the western part of this Province particularly from Barrington and Argyle, to whom they pay higher wages than the owners of our vessels can afford, or that they can earn on board of vessels of this Province. And it is with extreme regret we hear, that from the competitions, difficulties and reversions before mentioned, and which our fisheries now labour under, many of the fishermen be-

4. The rates of allowance to vessels in the cod fishery were as follows: on vessels of from 5 to 30 tons, $3.50 a ton; above 30 tons, $4.00; above 30 tons, and having a crew of 10 persons employed at sea over 3½ months, but less than 4 months, $4.00. The allowance to any vessel was not to exceed $360. The mackerel fishery was protected by duties, but not bounties. See Lorenzo Sabine, *Report on the Principal Fisheries of the American Seas* (Washington, 1853), pp. 178 ff.; also Philip Tocque, *Newfoundland* (Toronto, 1878), pp. 289–292.

5. "Twenty-five millions of people agree to pay to 15 to 20,000 of their number, being fishermen, a protection equal to $1.25 to $1.50 for every barrel of fall mackerel they bring home. This is an inducement sufficient to stimulate a less enterprising people than such as inhabit the eastern harbors of the United States. The business has, consequently, increased until the number of their [American] fishing vessels in the Gulf of St. Lawrence, the past season, has been computed at from 1,000 to 1,200, and the result of their voyages has produced an inspection, in Massachusetts alone, of a total of 329,278 barrels; of which 90,411 barrels were No. 1, 102,364 barrels were No. 2, 136,089 barrels No. 3, and 412 barrels No. 4. This quantity of mackerel, together with the catch of the different ports of Maine and other parts of the United States, will, it is supposed, fully meet their consumptive demand for the year, and the prices of fall mackerel are from 1¼ to 1½ dollars per barrel less than at this period last year." *Journals of the Assembly, Nova Scotia,* 1852, Appendix No. 13.

longing to the ports before mentioned are induced to abandon the vessels of their own Country, and seek employ in a foreign service; thus transferring the benefits arising from their industry to the United States, by fishing on the same shores and taking the same fish they would otherwise do in vessels of this Province.[6]

Schooners of from 70 to 80 tons, manned by twelve men selected by the skipper, made one trip to the Banks between March 20 and May 20, one to the Labrador lasting until September 20, and finally one to the Banks again in November. On the Labrador, part of the fish was cured and sent directly to the Azores, Madeira, Portugal, Spain, and the Mediterranean. The green and refuse fish were carried home. Owners of ten or twelve ships were able to assort cargoes, send the poorest grades of fish to the West Indies, other grades to other markets, and the oil to England.[7]

The Gulf of St. Lawrence mackerel fishery was to a great extent complementary to the Labrador cod fishery.

Some of their vessels fish to the westward of the Straits, though most proceed to the vicinity of Sandwich Bay and Cape Harrison, where they generally complete their cargoes in about six or seven weeks. All their fish is salted down in bulk, a large portion of which is dried on their return to the Straits, from whence many fit out for the Mackerel Fishery in the Gulf. About the 10th of August is the average time of their leaving the Northward.[8] Fewer vessels, it is said, fished to the Northward this year [1852], probably not more than 150; many of their best hands are Nova-Scotians, receiving 20 dollars a month wages.[9]

Paul Crowell says in his Report:

In 1851 I was informed there were about one thousand sail of American

6. *Journals of the Assembly, Nova Scotia,* 1834, Appendix No. 31.
7. See John McGregor, *British America* (London, 1833), I, 219–222.
8. "The Northward" is the regular colloquial term for upper Labrador.
9. *Journals of the Assembly, Newfoundland,* 1853, Appendix, p. 128. "Having securely moored their vessels, they hoist out their boats, each vessel having three or four, and commence fishing, the Americans salting their fish in bulk, whilst the Newfoundland people carry them to some harbour on the coast, on the shore of which they have stages for drying their cargoes. Should the fish prove abundant, they remain there until they have completed their cargo; but if scarce they immediately proceed to sea, and grope their way to some other harbour, where the fish are more abundant. It is surprising how they manage to find their way among the numerous Islands and dangers which fringe this barren coast; and that during the dense fogs in which this part of the coast is sometimes enveloped, they are not more often wrecked, especially when they have neither chart, quadrant, or book of directions to guide them." *Journals of the Assembly, Nova Scotia,* 1841, Appendix No. 62. On July 4, 1857, 7 United States schooners were sighted at Bradore Bay, and in July, 1859, 15 at Mingan.

vessels, which with an average of 15 men would give fifteen thousand. Some of these vessels, I heard, made three trips in Chaleur Bay for mackerel. Some, after having made one or two trips or fares of codfish, proceed to the Bay de Chaleur, well fitted, taking sufficient barrels to cure their fish in. These are partly filled with menhaden and clams, which are considered the best bait for mackerel; others are filled with salt and water, which make ballast; when required for use they are emptied of their contents and filled with mackerel; this keeps their vessels in good ballast. They generally commence their fishing about Bradelle Bank, Shippegan, and follow the fish northerly, until the season advances, when they return to the north side of Prince Edward's Island and Cape Breton.[10]

Because of a growing scarcity in the regular mackerel grounds, mackerel fishermen had entered the Gulf of St. Lawrence about 1834 and they continued to fish in these waters up to the 'seventies. From about May 20 mackerel were followed along the Nova Scotia shore from Liverpool to the Gulf.[11] In the autumn, the fish ran close inshore and were followed by the Americans. The mackerel were attracted by feeding, and if the wind was offshore they followed the schooners; otherwise they drifted alongshore. "It is of first importance to have a smart weatherly vessel—the current and drift is usually off shore—the fish always make to windward. If you fall quickly to leeward, you lose the fish, therefore you must be continually stretching windward."[12] The Americans had "a new and superior class of vessels fitted with all the needfull appliances and variety of bait which a long experience aided by a well-endowed and careful nursery has taught them the value of." By the middle of the century small Chebaccos and "pinks" were being displaced by "jiggers" and by the "clipper schooner," which meant an increase in size of from about 45 tons to about 75 by 1885.[13]

As in the cod fishery, the share system was a powerful incentive. "American vessels for the Mackerel Fishery are fitted out in what is called a half lay, that is, the Men have half of the Fish caught, and the Natives of this Province are induced to sail in American Vessels because the value of the Fish is so much greater in the American Markets

10. Report of Paul Crowell, *Journals of the Assembly, Nova Scotia,* 1852, Appendix No. 25.

11. After a first trip off Cape Cod and Block Island, vessels shifted, from June 11 to 15, to the Bradley and Orphan Banks, and later to Chaleur Bay and the Gaspé coast. They returned to Prince Edward Island about September 10, and to Sydney about November 1. See Raymond McFarland, *The Masts of Gloucester, Recollections of a Fisherman* (New York, 1937).

12. *Journals of the Assembly, Nova Scotia,* 1852, Appendix No. 13.

13. Raymond McFarland, *A History of the New England Fisheries* (New York, 1911), chap. xviii.

than in Nova Scotia's . . . that their profits are thereby greatly increased."[14]

Reciprocity opened a market for the Canadian product, but it was not until the disturbances of the Civil War that the American fishery in Canadian waters declined. An approximate estimate of the number of American vessels in the Gulf of St. Lawrence shows that those in the cod fishery increased from 100 in 1853 to 160 in 1856, but fell to 100 in 1862.[15] In the mackerel fishery the number of ships rose from 100 in 1854 to 300 in 1856, but dropped to 120 in 1863.

As a result of the termination of the Reciprocity Treaty, the abolition of bounties in 1866, the imposing of duties by Newfoundland, and the increasing importance of the bank fishery, the American cod fishery on the Labrador disappeared in the 'sixties.[16] Following its disappearance, various factors such as the spread of the purse seine—due to the low prices in the depression of the 'seventies—the introduction of night fishing in 1874, and an increase in the "southern" mackerel fishery, where 50 vessels in the 'seventies had grown to 150 in the 'eighties, caused the inshore and the Gulf mackerel fishery to become of smaller concern.[17]

> Up to about 1870, the mackerel catch was made with the hook and line, as they could be used near shore. Often a considerable part of the catch was made there. Since the introduction into general use of the purse seine, in 1870, we find nearly all of the mackerel catch has been made . . . on the high seas, or more than three miles from shore. Of late years mackerel, in common with other varieties of fish that once were found in plenty near

14. Evidence of Charles Stewart, *Journals of the Assembly, Nova Scotia,* 1837, Appendix, No. 75.

15. *Sessional Paper* 71, 1864 (*Sessional Paper* is hereafter abbreviated to *S.P.*). See W. G. Pierce, *Goin' Fishin', the Story of the Deep Sea Fishermen of New England* (Salem, 1934); also S. E. Morison, *Builders of the Bay Colony* (Boston, 1930), chap. xix, pp. 375, 378. From 1837 to 1865 the Cape Ann tonnage in the fishery increased from 9,824 to 25,836, the value of the catch of cod from $186,516 to $839,675, and that of mackerel from $335,566 to $2,259,150. The number of Cape Cod ships declined from 359 to 314; but their tonnage increased from 21,280 to 50,166. The catch of cod rose from $392,772 to $976,326, and of mackerel from $490,638 to $1,169,074. Boston Bay tonnage decreased from 15,281 to 2,969; South Shore, from 11,302 to 5,360; North Shore, from 10,232 to 5,631; and Essex County, from 8,019 to 4,245. McFarland, *op. cit.,* chaps. ix, x. For an extensive mine of information see G. B. Goode, *The Fisheries and Fishing Industries of the United States* (Washington, 1887), sec. 5, Parts I and II; also *A Geographical Review of the Fisheries, for the Year 1880,* sec. 2; also Andrews, *op. cit.,* pp. 629–659.

16. W. G. Gosling, *Labrador* (Toronto, n.d.), p. 420.

17. For the returns of the American mackerel vessels from Port Mulgrave, in 1873 and 1874, see *Documents and Proceedings of the Halifax Commission* (Washington, 1878), pp. 222–229.

shore, are now seldom found in abundance within three miles of land, and oftener wide out, or on the more distant fishing banks. . . . The North shore of Prince Edward Island and Cape Breton are the localities in the inshore British waters which are now chiefly visited by American vessels in pursuit of mackerel. . . . Then too, the change in the method of fishing has, in recent years, led to the almost practical abandonment of the mackerel fishery in the Gulf of St. Lawrence. Occasionally a considerable fleet enters the Gulf; but, since the results have generally been unsatisfactory, there have been seasons when only a very few vessels went there. It is true, perhaps, that the mackerel being a remarkably erratic species, its movements cannot be predicted from year to year with any absolute certainty. It is an historical fact, now well established by the most accurate and careful investigation and inquiry, that the catch of mackerel in the Gulf of the St. Lawrence, not to speak of the inshore waters under British control, has been of comparative insignificance during the last decade. And even under most favorable conditions, when the catch there has been exceptionally large, as in 1885, the total product of the Gulf mackerel fishery did not amount to more than 8 per cent of the entire catch of the New England fleet. Of this, less than one-third was taken inside of the 3-mile limit.[18]

The large bank fishery grew in importance. From the Banks, fish were delivered to the owner of the vessel and dried on shares. The skipper and the other officers usually took one fish in every sixty-four, and credited every man with his individual part of the catch.

Every man on board has an interest in the returns of the voyage, and as they are all invigorated by an abstinence from spirituous liquors, the majority being members of Temperance Societies, they are beyond dispute a more efficient body of men than the Colonists can obtain for their crews. Being trained from their infancy to the pursuit, they are also more expert, the fishermen in Newfoundland being chiefly composed of Irish emigrants, who, though both apt and laborious, from having adopted the pursuit at an advanced period of life, never acquire except in rare instances, the same manual dexterity.[19]

From 1830 to 1850 Marblehead sent from 50 to 100 vessels of from 50 to 70 tons to the Grand Banks to engage in hand-line fishing. Its ships made two trips. They left in April and returned in late September or early October, and they obtained from 700 to 1,000 quintals a trip. After the middle of the century and during the Civil War this port declined. Gloucester suffered from the falling off of trade with the Medi-

18. *Fishing Interests of the United States and Trade with Canada. Reports from the Consuls of the United States* (Washington, 1887).

19. G. R. Young, *The British North American Colonies, Letters to the Right Honorable E. G. S. Stanley, M.P.* (London, 1834), pp. 71–72.

terranean. But in the 'fifties, with the introduction of trawl or setlines,[20] the increasing importance of the fresh-fish industry, and the accessibility of bait in Newfoundland, its tonnage began to expand rapidly, and reached a peak in the decade from 1880 to 1890. In 1880 Gloucester sent out 200 vessels, more than 100 to Georges Bank; and Provincetown sent a large fleet to Western Bank, Banquereau, and Grand Bank. In the 'seventies cod were taken on the Banks to an increasing extent for the preparation of boneless cod.[21]

THE herring fishery expanded as a result both of the increasing consumption in the home market and the demand for bait in the bank fishery. In 1839 nearly 150 American schooners of from 60 to 80 tons together "made" 100,000 barrels of pickled herring at the Magdalen Islands. This industry was very profitable in the late 'fifties but fell away in the early 'sixties. After 1830 Lubec and Eastport vessels proceeded to the Magdalen Islands for fish for the smoked-herring industry.[22] Bloaters were first prepared from Bay of Islands herring in 1859.The industry declined after 1870 because of the competition from Canadian herring made possible by the Washington Treaty.

Subsequent to the Reciprocity Treaty, frozen herring were purchased for bait for New England's fishery on the Banks and on the south coast of Newfoundland. Fresh bait was more effective than salt. Purse seines were used in Newfoundland after the Treaty of Washington. As the United States had been forced to withdraw from the dried-cod fishery on the Labrador, so she was compelled by Newfoundland to withdraw from the herring fishery. Newfoundland passed an act in 1876 declaring that "no person shall haul, catch or take herrings in a seine or such contrivance between the 20th October and 25th April in any year, or at any time use a seine or such contrivance for catching herrings except by way of shooting and forthwith hauling. Proviso: nets may be used set as usual, and not used for barring or inclosing herrings in a cove, inlet or other place." The same act prohibited hauling on Sunday. Americans accustomed to buying herring caught in gill nets by Newfoundlanders to sell in the New York market brought their own seines

20. "For years the French were the only fishermen that followed the set line fishing, but latterly the high prices given for large codfish in the United States has induced their fishermen to adopt it, and more recently many British fishermen have taken it up and are still doing so." *Journals of the Assembly, Nova Scotia,* 1861, Appendix No. 32. Americans were engaged in trawl fishing at the Magdalen Islands in the 'seventies. In 1858 hand-line dory fishing had been introduced at Southport, Maine, from whence it had spread to other ports.

21. The small fish, those under twenty-two inches, were sold to Newfoundland and Lunenburg for drying as "light-salted."

22. Sabine, *op. cit.,* pp. 191 ff.

and hauled herring for themselves. "The American seines are 30 fathoms deep and 120 fathoms long. These American seines are used for barring herring in deep water."[23] Herring were kept in the water until freezing weather and were not subject to loss because of warm weather as in the case of fish taken with gill nets. On January 6, 1878, Americans were forced to stop catching herring in seines at Long Harbor in Fortune Bay. In reply to their claims for damages, amounting to $105,305, Great Britain held that even if the act of 1876 was not applicable Newfoundland legislation of 1863 protecting the fishery and antedating the Washington Treaty should be regarded as valid, and that under the treaty Americans were not entitled to the use of the strand in hauling nets. The Americans replied that the indemnity paid under the Halifax Award of the Washington Treaty was of little value if their rights to the inshore fishery were to be whittled down in this fashion. Further difficulties arose when, on July 8, 1879, Gloucester fishermen were refused permission to take squid by "jigging" in Broad Cove, Conception Bay, and there was also trouble with the Americans at Aspy Bay in Cape Breton. An offer on the part of Great Britain to pay $75,000 was finally accepted. The arguments she advanced stressed the responsibility of the local legislature in the evasion of treaties, whereas the United States held that such matters were within the purview of the Senate. These difficulties contributed to the abrogation of the Washington Treaty. New England withdrew from Newfoundland, and the Bay of Fundy became the great center of the herring fishery.

The withdrawal of New England from the Labrador, the Gulf of St. Lawrence, and Newfoundland was hastened by improved transportation and the possibilities of marketing fresh fish. About 1837 live fish were brought to Boston in smacks and shipped by rail. In 1846 a railway was completed to Gloucester. By 1850 ships carried ice to Georges Bank and brought back haddock and halibut. In 1858 the first fish were sent packed in ice from Boston to New York. Increasing demands for fresh halibut and the exhausting of the nearer grounds, such as Georges Bank by 1850, led in the 'sixties to the development of the salt-halibut

23. "There were mackerel seines capable of taking 2,000 to 5,000 barrels and costing with boats $1,200. They were too expensive for the generality of Newfoundland fishermen and they would have no use for seines, only during the herring season, while we [Americans] can use them both summer and winter and thus make them pay for the great cost." *Correspondence Respecting Occurrences at Fortune Bay, Newfoundland, in January, 1878* (London, 1878); see also *Further Correspondence Respecting the Occurrences at Fortune Bay, Newfoundland in January, 1878* (London, 1880); *idem* (London, 1881); and *Further Correspondence . . . Newfoundland and other places* (London, 1881); also "The Alleged Outrage at Fortune Bay, Newfoundland," *House Executive Document* No. 84, 46th Cong., 2d Sess.

fishery in regions as far distant as Greenland. In the late 'sixties and early 'seventies Gloucester fishermen exhausted the halibut fishery along the Canadian Labrador and turned to deep-water fishing. About 1865 systematic halibut fishing began on the Grand Banks; and by 1880 the fleet totaled 50 sail. Fresh haddock from the La Have Bank became of marked importance in the 'eighties. The fresh-fish industry required faster vessels and expedited the introduction of cold-storage plants for the handling of both fresh bait and fresh fish; and the Boston T Wharf was developed as a fishing center in 1884. The 'eighties and 'nineties were also notable for the construction of fast fishing ships.

The increasing importance of the domestic market for fresh fish and the premiums on faster vessels were factors that narrowed the range of activity of American fishing vessels. The demand of the American market was for large fish. The introduction of improved vessels and trawls on offshore grounds made it easier to take them, and likewise lessened the importance of the factor of distance. The withdrawal of American ships hastened, and was hastened by, the increase of settlements and the expansion of the fishery in Newfoundland, the Gulf of St. Lawrence, and Nova Scotia.

NOVA SCOTIA TO CONFEDERATION

The Nova Scotian . . . is often found superintending the cultivation of a farm and building a vessel at the same time; and is not only able to catch and cure a cargo of fish but to find his way with it to the West Indies or the Mediterranean; he is a man of all work but expert in none.

THOMAS CHANDLER HALIBURTON, *The Old Judge*

To the expanding New England fishery Nova Scotia opposed a vigorous policy of restriction in British waters, in part to compel the United States to admit her own products. She was exasperated by the opening of the British West Indies to her rival, by the competition from a fishery supported by bounties and protected by duties; and she felt the consequent loss of markets and of men and protested vehemently.[24] She solicited the support of other provinces before the adoption of the Reciprocity Treaty; and after its abrogation she continued her aggressiveness, helped by Confederation and the policy of the Canadian government. Under the Washington Treaty, her success in the Halifax Award and in penetrating the United States market led to that treaty's termination.

The advantages possessed by New England, which were described in

24. *The Report of Israel D. Andrews*, pp. 553–571.

protests prior to the admission of the United States to the West Indies, were even more effective in the years that followed, as a result of the expanding American market and the increasing importance of the mackerel fishery. The reduction of British West Indian duties on foreign-caught fish and the increase in American duties on British fish "completely excluded our [Nova Scotian] fish from their markets while we have thrown open our colonial markets to their fish at less than half the rate of duty imposed by them."[25] Moreover, the Warehousing Act passed by the United States in 1846 permitted importation and reëxportation in bond, to the disadvantage of Nova Scotia in foreign trade.

What meant success in New England meant despair for Nova Scotians. They had poor equipment and complained of their inability to meet New England competition.

The American vessels which fit out for the hook fisheries are of a superior class from those in Nova Scotia. Their tonnage generally from 60 to 130 tons, very sharp built, well fitted in every respect; those they term the sharp-shooters are very superior sailing vessels. This enables them to reach the fishing ground and procure their cargo, while those of Nova Scotia are actually carrying sail to reach the fishing ground. Those vessels are likewise well manned, varying from 12 to 24 men.[26] They offer great opposition—a common threat among them is to run the Nova Scotia vessels down—they are usually prepared for this, their bowsprits are fitted large and strong and the end well ironed; they have double chained bobstays, and shrouds well bolted and geared. A number of them came armed for opposition. I have seen the arms on board of them. These vessels, with Nova Scotia masters, called white washed Yankees, are generally the worst.[27]

Nova Scotian arrangements between the owners and crew were less satisfactory.

I give it as my opinion, however, that the greatest reason why our Fisheries are not as productive as the Americans, arises from the difference in the way they are fitted out and owned; the greater part of our fishing vessels are owned by poor men; they get their out-fits on credit, at the highest possible rate; their hands are generally hired. His [the master's] own spirits are dull from a knowledge of the disadvantageous circumstances under which he has to labour. His hands have the same feelings, in some measure, with the additional one, of the uncertainty of being paid; thence their want of energy and the unprofitableness of our Fishing. The American Merchant owns the Vessel, fits her out at the cheapest rate, ships his Hands on Shares,

25. *Journals of the Assembly, Nova Scotia,* 1837, Appendix No. 75.
26. Report of Paul Crowell, *Journals of the Assembly, Nova Scotia,* 1852, Appendix No. 25.
27. *Idem,* Appendix No. 13.

from the Skipper to the Cook, according to what catches. An ambitious spirit is thus excited among them; this, and the liberal encouragement from their Government causes more active, enterprising men to embark in the Fisheries; consequently, they are generally more successful, and their Fisheries more productive; perhaps the encouragement from Government, more than any thing else, causes these good effects.[28]

The cost of outfitting was greater in Nova Scotia. "American fishing vessels are outfitted at cheaper rates than British; the difference consists chiefly in the price of provisions which is the principal item in the bill of outfits; other necessaries being equally as low, or even lower than can be procured by our Fishermen."[29]

As a result of these advantages labor migrated from Nova Scotia to New England, and its ranks were filled by fishermen from Newfoundland[30] compelled to migrate in their turn because of competition from the French.

They [New Englanders] are more expert Fishermen, and in most instances nearly every man in a crew is related by family more or less, also having shares in Vessel and Voyage, which naturally makes them take a deeper interest than the Servants of Nova Scotia Planters—In general the men that compose their crews are from Newfoundland and elsewhere—they, after serving a year or two, and realising a little money, proceed on to the United States, consequently two thirds of our crews are entire strangers every year.[31]

28. Evidence of G. R. Tucker, *Journals of the Assembly, Nova Scotia,* 1837, Appendix No. 75.

29. *Idem.* Agreement was general on the advantages but not on the details. "Pork, Bread and Flour, are quite as low here as in the United States; the American pays a duty upon his Salt of two Cents per 56 lb., and upon his Fishing Nets and Lines, five Cents per lb. Consequently, in those duties, he contributes largely towards the bounty he receives, which, in reality, to a large extent, is only a debenture. Upon those Articles our Fishermen pay no duty, and therefore, so far they both may be nearly equal; but the duty upon Foreign Fish is the Bounty and encouragement received by the American." Evidence of Joseph Allison. "The cost of out-fit for a Fishing Voyage varies every year. In general the Americans have the advantage over the British, their Provisions, Canvas, and some other Articles required, being cheaper than ours, while Iron, Cordage, Lines etc., are procured at lower prices in the Colonies. At the present time there is but little difference in the price of Provisions in the two Countries; but heretofore they have been from 15 to 20 per cent. cheaper in the United States." Evidence of D. and E. Starr and Company. It was claimed that domestic salt in the United States was cheaper than salt in Nova Scotia imported from Europe and the West Indies.

30. *Journals of the Assembly, Newfoundland,* May 9, 1851.

31. Evidence of Thomas Tobin, *Journals of the Assembly, Nova Scotia,* 1837, Appendix No. 75. "The crews of these vessels are nearly one-fourth belonging to Nova Scotia. . . . Some of these leave their homes in the spring of the year and take passage for the United States, for employment; others ship on board American vessels when they arrive in Nova Scotia. They may be a cause why American fishermen are found

In Nova Scotia "those engaged in the pursuit were persons of the poorest description who commencing without capital, without anything in fact but the power of bodily labour, had to procure credit in the first instance and then fight up-hill under an accumulation of debt for their fit-out, their annual equipment and their winter stores which keeps the greater part of them at this moment in arrear in the books of the merchant."[32]

The extension of the New England mackerel fishery to the Gulf of St. Lawrence facilitated the migration of labor, the growth of smuggling, and encroachments on British fishing grounds. It was said that in 1834 the larger part of the catch was sold to Americans, at a price one third higher than in Halifax, in exchange for flour, molasses, rum, tea, and tobacco; and it was recommended that cutters should be used to check

fishing within the limits. . . . But how will those do who sail in American vessels? When arriving in the United States they generally procure good wages, or should they ship on shares, their fish is taken to a market in the United States, free of duty or expense. As these vessels are generally bound to some port in Nova Scotia, those who are Nova Scotia men can take their little supplies for their families, and have them landed at their doors, nearly as low as they can be procured in the United States. When their voyages are accomplished, they either proceed on to the United States and receive their share, or, as the practise is in some places, a merchant supplies them with goods to the amount of their voyage. He then receives a draft, which is accepted by the owner of the vessel, payable in the United States. This answers the purpose of the fisherman, and likewise makes remittances for the merchant, who can step on board the packet and proceed to the United States, collect his drafts, make arrangements for a new supply for the coming season, and return. This appears to be the state of a large part of Nova Scotia at present." Report of Paul Crowell, *idem,* 1852, Appendix No. 25. Another account claimed that one half of the masters were Nova Scotians. "They know the coast well and are more at home in the harbors and can remain later in the gulf." "There were upwards of 200 men on board of them this season from the Straits of Canso alone. A large number of our western men were also with them. . . . One vessel had her whole crew, nearly, from Port LaBear; from Port Latour, every man capable of fishing was taken. In one of their vessels I saw three brothers (Nickersons), and on board of others their sons were shipped; indeed, it would be difficult to find one American vessel without a large part of her crew consisting of Nova-scotians—look at the number lost on board the American vessels in the gale at P.E. Island." *Idem,* 1852, Appendix No. 13.

32. *Acadian Recorder,* June 6, 1835. The fisherman depended "partly on land, partly on water for his subsistence instead of attending wholly to the one." Capt. W. S. Moorsom, *Letters from Nova Scotia* (London, 1830). "The Nova Scotian . . . is often found superintending the cultivation of a farm and building a vessel at the same time; and is not only able to catch and cure a cargo of fish but to find his way with it to the West Indies or the Mediterranean; he is a man of all work but expert in none." Thomas Chandler Haliburton, *The Old Judge* (London, n.d.), Preface, p. v. "In Nova Scotia we do not follow the fishing so exclusively as in Newfoundland nor lumbering as in New Brunswick. The bulk of our people are farmers. A large body living on the seacoast are fishermen, but not fishermen only. Having plenty of free timber when the fishery is unproductive, our men go into shipyards and build vessels for themselves or for their friends, and manning them go into the carrying trade or coasting business." Joseph Howe in a letter, June 5, 1854.

smuggling. Americans were accused of purchasing bait in Liverpool in return for pork, bread, and other articles "as, early in the season they cannot procure bait on the fish banks but must resort to the harbors for it."[33] Smugglers came into the harbors "regularly every night when the weather gets unsettled and the days shorten," and they purchased provisions, vegetables, herring, barrels, salt, etc., and sold tobacco, spirits, clothing, boots and shoes, and the like. A committee of the Assembly reported that they had ascertained "from sources on which they can rely with the utmost confidence, that an illicit Trade is now carried on in this province to a much greater extent than has heretofore been known; that they have reason to believe that it is not confined to any particular section of the Province; but it is doubtless conducted on a much larger scale in those parts where the facilities and inducements are known to be the greatest."[34]

Nova Scotia attempted on the one hand to reduce the costs of carrying on the fishery by pressing for a revision of legislation affecting the colonies and on the other to introduce effective measures for checking the New England fishery.

> Their canvas, rigging, all the outfits of the vessel, as well as their provisions, are afforded at cheaper rates than they can be commanded in the Colonies. It is unquestionable, that an American ship can be both manned and navigated at a lower charge than either in Britain or with us. In the latter, while we are subjected to the prices of articles, enhanced by British taxation, the charges of importation and the difference of exchange, they procure them

33. Evidence of John Barss, *Journals of the Assembly, Nova Scotia,* 1837, Appendix No. 75. For a description of the effects of the competition from New England in the Gulf of St. Lawrence in the shape of cheap supplies of breadstuffs, India goods, and tea, and of brandies, silks, and wines from "St. Peters [St. Pierre] . . . and the United States," see *Montreal Gazette,* September 22, 1824, and November 12, 1829.

34. *Journals of the Assembly, Nova Scotia,* 1834, Appendix No. 15. "The harbours are inundated with a supply of smuggled goods and cheap manufactures, the best fish are seduced from the British fishermen, and the trade of the coast withdrawn from its native and natural channels, to increase the commercial marine, the foreign connections, and the elements of naval power of our great national rival." G. R. Young, *op. cit.,* p. 59. "By this system of bold and open invasion of its laws the local revenue of the province is not only lessened, but is affected, for the future, by the illegal abstraction of those resources which furnish the materials of that branch of the foreign trade, from which the largest amount of revenue is derived. The local merchant in the outport is, in addition, induced to expand his ready money in the purchase of a stock of goods which he procures at cheaper rates; and thus, while the revenue is subjected to large losses the fair trader is deprived of his profits and of the property pledged to him on every principle of justice. The colonial and Irish agriculturist, and British manufacturer are exposed to a competition, against which the violated law has raised protection. In this view of the question I find it difficult to persuade myself that the manufacturing interests of Great Britain have no concern in this inquiry." *Idem,* pp. 52–53.

in the home market at the simple cost of production. It will form a natural inquiry how it arises that we do not obtain these articles at the nearest market, in the place of importing them from Britain . . . but the tariff of protecting duties imposed by the Imperial Act, 6 Geo. IV, c. 114, for the protection of the British manufacturer, and of which we are not disposed to complain, will furnish a very satisfactory answer. Upon these articles an impost is exacted, varying from 15 to 30 per cent ad valorem.[35]

The colonies were restricted in methods of encouragement. "Parliament has long since repudiated the principle of bounties." With the imperial act (3 and 4 Wm. IV, c. 59) which imposed a duty of 12 shillings a hundredweight on salted provisions and 5 shillings a barrel on flour although these commodities were free in Canada a provincial act (4 Wm. IV, c. 1) reduced the duties on articles used in the fishery.[36] But the reductions resulted in abuses and were abolished in 1841. Adjustments of differentials between colonial duties and imperial duties were made reluctantly. The difficulties imposed by Great Britain intensified the aggressiveness of Nova Scotia in pressing for the exclusion of American vessels from British waters.

Legislation in 1826 (7 Geo. IV, c. 4) exempted from duties goods bought in return for exports of fish "to encourage our own fisheries, and the export of products in Vessels owned and registered in the Province, or belonging to British Merchants engaged in, and carrying on, the Fisheries within the same." The reduction was found to be open to abuse; "and in 1838, by the 2d clause of 1st. Vic. Chap. 9, it was subjected to certain restrictions requiring the Foreign Goods, exempted from Colonial Duty, to have been shipped in some port or place in South Amer-

35. G. R. Young, *op. cit.*, pp. 71–72.

36. An Account of All Articles Entered for the Use of the Fisheries for the Year 1839*

Article	Unit	Quantity	Article	Unit	Quantity
Beef and Pork	(Barrels)	108	Lard	(Half Barrels)	10
Boots and Shoes	(Boxes)	78		(Kegs)	250
	(Pairs)	36	Molasses	(Gallons)	167,132
Boats	(Number)	4	Oil Clothes	(Bundles)	20
Cordage	(Coils)	566		(Suits)	1,303
Corkwood	(Tons)	6	Oakum	(Bales)	100
Flour	(Barrels)	26,095	Pitch and Tar	(Barrels)	1,826
Fish Hooks	(Boxes)	1	Varnish	(Barrels)	2

* For a list of goods imported during the five years ending in 1839 see *Journals of the Assembly, Nova Scotia,* 1840, Appendix No. 25. In 1834 bounties had been proposed for mackerel taken with the hook between Cape Sable and Cape Canso, and on tonnage engaged for at least four months in the cod and mackerel fisheries, the outlay to be met by the imposition of a duty on the fishermen's flour. *Idem,* 1834, Appendix No. 31; also *idem,* 1835, Appendix No. 34. Again in 1846 a bounty of 10 shillings a ton for three years on all vessels of more than 20 tons engaged in the mackerel fishery was recommended. *Idem,* March 12, 1846. A grant of £2,000 was made in 1851, and 75 vessels, totaling 3,378 tons and representing 699 men, applied for the bounty.

ica or in Europe; and the Fish or Fish Oil to have been exported to some port or place in South America, or in Europe, and there sold."[37] Great Britain objected. "The act," it was said, "exempts from Colonial Impost Duties, all Foreign Goods purchased with the proceeds of Fish and Fish Oil, the produce of the Colony. . . . This provision is so objectionable in principle, and so open to imposition, that I shall be under the necessity of advising Her Majesty in Council, to disallow any Acts which shall be passed containing a similar Clause."[38]

In the end Great Britain conceded the handicaps of Nova Scotian trade, and the Assembly could rejoice in

> the substitution of a uniform, ad valorem duty of Ten per cent for the present Duties of Thirty, Twenty, and Fifteen per cent., which, on many articles, amounted to a prohibition, and are inclined to hope that the Government will gradually come to the conclusion, that the protective Duties imposed by the Imperial Acts, while they cramp the energies and retard the expansion of Colonial industry, are of no real advantage to the British produce; and as our Legislature would impose no Duties, except only for purposes of Revenue, that we would consume more largely of Home Manufactures, and become more valuable Customers of the Mother Country, were the regulation of our Trade left in our own hands, subject always to the control of the Government. Such would be the effects, as the members of this Committee have long thought, of a perfectly free and unconstrained intercourse with all the world, relieved of the old restrictions.[39]

The colonial system was gradually relaxed to the advantage of the Maritime Provinces.

> With respect to the eastern ports of New Brunswick, your Committee are not of opinion that the trade between that portion of the above Province and

37. *Journals of the Assembly, Nova Scotia,* 1840, Appendix No. 1.

38. *Idem,* April 20, 1838. "Indeed nothing can be more obvious than that the policy of the Imperial Parliament was to preserve, in certain cases, a distinctive Duty which should be beyond the power of Colonial interference; the mode adopted for enforcing that policy is clear and effective in itself, and is fortified by the general enactment of the 56th Section, which declares to be null and void any existing or future Law in any of the British Possessions in America, in any wise repugnant to that Act; and it follows that in a matter like this, touching the General and Commercial relations of the Empire, and thus carefully regulated, no power constitutionally exists in a Colonial Legislature either to repeal or indirectly to defeat the enactments of the Imperial Parliament." *Idem,* 1839, Appendix No. 25 (see also June 28, 1843). Great Britain permitted colonial duties imposed by legislation (7 Vic., c. 16). "Looking to the moderate rate of the Imperial Duty on Foreign Wheat, Flour, and Molasses, imported into the Province, and to the disadvantageous effect on the Revenue of Nova Scotia, which, the exemption of these articles from duty, when supplied for the use of the Fisheries, appears to produce, no objection will be raised to the proposed Colonial Duty, such Duty being equal to the Duty levied upon Foreign Wheat, Flour and Molasses, by Imperial authority." *Idem,* June 19, 1844; also *idem,* April 8, 1847.

39. *Idem,* 1841, Appendix No. 79.

Canada has materially increased within the last few years; and with respect to the trade with ports in the Bay of Fundy, regret to say that it all but ceased, which your committee attribute to the changes in the Imperial laws, more specially the Act passed in 1842, generally called Gladstone's Act; before the passing of which all American provisions by passing through the Canadas, were allowed to take the privileges and character of Canada produce, and imported into our sister-colonies as such, but with that change all inducements to receive their supplies from this ceased, as the proximity of those ports to Boston and New York, and the cheapness of bread-stuffs, and provisions in those markets, offered superior advantages; and the result has been as stated; the same remarks apply, to some extent, to Halifax and other ports in Nova Scotia, where merchants, from their large increasing trade with Boston, by shipments of coals, plaister, etc., are enabled to take advantage by the return vessels of very moderate rates of freights, and a selection from a comparative cheap market. With Gaspé the trade has been very gradually increasing.[40]

Final concessions were granted (9 and 10 Vic., c. 94) empowering British North American legislatures "to repeal differential duties in favour of British produce, imposed in these colonies by former Imperial acts."[41] It was hoped that the provinces would combine to establish "a common system of custom-house duties and divide the revenue which these duties produce." These hopes were[42] realized only by degrees; and it was the Reciprocity Treaty and Confederation that eventually fulfilled them.

The struggle to secure control over the country's commercial policy was accompanied by a struggle for control over the customs,[43] and also

40. See *Journals of the Assembly, Nova Scotia,* March 12 and March 31, 1847.

41. *Journals of the Assembly, Nova Scotia,* 1848, Appendix No. 12. See also 8 and 9 Vic., c. 90, and 8 and 9 Vic., c. 93.

42. See a protest against high duties on agricultural products by Earl Grey to Sir John Harvey: "I apprehend . . . that duties of this kind are likely to prove injurious to the real interests of all classes, and that they are moreover calculated to produce dissatisfaction in the neighbouring Colonies, from which Nova Scotia might be expected to derive a considerable supply of these articles." *Idem.*

43. It was recommended that an act passed March 31, 1834, and continued with amendments to March 31, 1836, providing for a combined collection of duties should be allowed to expire and that the act for raising revenue in the following year should embrace provincial duties only. *Journals of the Assembly, Nova Scotia,* 1836, Appendix No. 19; also *idem,* 1837, Appendix No. 5. Revenue from Imperial duties for 1835, 1836, and 1837 averaged £11,570 sterling, and from provincial duties £30,629. *Idem,* 1840, Appendix No. 1. In merging the customs offices, "The second main difficulty is, the aversion of the Mercantile body, especially in Halifax, to any change, however beneficial in other respects, which would subject them to the unrestricted action and control of the Customs. A degree of liberality, involving some personal responsibility and almost looseness in the practice of the office, is extended by the Excise Department to Importers worthy of its confidence, which is productive of no loss to the Revenue or the Public, but could not be expected in, and would not be shown by, Imperial Officers belonging to the Customs, and in some degree independ-

for freedom of the ports.[44] Under 3 and 4 Wm. IV, c. 59,[45] Halifax, Pictou, and Sydney were made warehousing ports for foreign ships, and Yarmouth and Liverpool warehousing ports for those of Great Britain. The Assembly represented interests demanding an extension of such facilities.

The importation of Foreign Articles into Nova-Scotia, in Foreign Vessels, is confined to the Ports of Halifax, Pictou, and Sydney. These are denominated Warehousing Ports, and the articles thus Imported, may be exported to the other Colonies and else where. Into Liverpool and Yarmouth Foreign Articles in British bottoms, may be Imported and Warehoused, and thence also Shipped elsewhere from the Warehouse. By Warehousing, the Importer obtains time for the payment of the Duties imposed on such articles, until they are taken out for Home Consumption. The trade outwards is carried on from all the Ports of the Province, thus from Windsor and Douglas, Gypsum; from Cumberland, Grindstones, Gypsum and Wood; from Pictou, Lunenburgh, Arichat and others, Fish, Lumber, Agricultural Produce; and other products of labor are exported. The official returns shew the trade of the Out-Ports to be considerable and rapidly increasing. But while Foreign Vessels can go and come freely into Halifax, Pictou and Sydney, while our own Shipping can bring the returns of their Sales into Yarmouth and Liverpool, vessels from all the other Ports of the Province, are compelled either

ent of each other." *Idem,* March 16, 1840. Customs salaries were revised in 1834, and the department of local revenue was merged with the customs in 1839–40. Four years later all positions in the customs except that of the collector were filled by Nova Scotians. See Marion Gilroy, "The Imperial Customs Establishment in Nova Scotia," *Canadian Historical Review,* September, 1938, pp. 264–277.

44. "It is in the ports on the Continent of South America, where we are subjected to the full measure of its reaction. The Merchantable fish, caught and cured on our own shores, produced by our native industry and belonging of right to the Colonial Merchant, meets his vessel in the foreign market, in the hands of a foreign competitor, and under circumstances which place him in a position of most disadvantageous inferiority. From climate, and from the less advanced condition of the agriculture of the Colonies, (I speak now in reference to Nova Scotia and Newfoundland, where the fisheries are carried on to the greatest extent,) we are yet unable to supply any saleable production, with which to assort our cargoes. Although our ports enjoy the advantage of the free warehousing system, and we are allowed to tranship flour from bond, the double freight and the expenses of unloading and transhipment, increased by an absurd regulation in the Colonies that the property must be actually landed, render the opportunity of competition of no practical utility. The American shipowner, on the other hand, can and does assort his cargo of fish, with flour and those coarse manufactures which are required in the South American markets, and in which the Americans are now able to compete successfully with the mother country. This brings his lesser quantity of fish within the means of a larger circle of purchasers; and while the flour and manufactures assist in selling the fish, the fish lend their aid to sell their companions, so that the system, like the double-edged sword, cuts equally against the manufacturer at home, as it does against the Colonist abroad." G. R. Young, *op. cit.,* pp. 69–70.

45. See *Journals of the Assembly, Nova Scotia,* January 11, 1838. See page 266.

to return in Ballast to enter into the favoured Ports, or to smuggle their return Cargoes into their own ports. In many instances they adopt the last alternative. Yarmouth and Liverpool are desirous that Foreign Vessels may be allowed to enter into their ports as into Halifax, because they have reason to apprehend that the United States will exclude their Vessels, unless the privilege of entering these ports is given to the Vessels of the United States. The other portions of the Colony pray that in their own ports they may be permitted to bring back from Foreign Countries, such articles as are now admissable into Halifax, in return for their own produce, and there pay the Duties imposed by the Imperial Parliament.[46]

The problem became part of the struggle for responsible government. The Assembly agreed that, in the infancy of the colony,

its whole Government was necessarily vested in a Governor and Council, and even after a Representative Assembly was granted, the practice of choosing

46. *Journals of the Assembly, Nova Scotia,* 1835, Appendix No. 1. Also see, for demands for an extension of free ports, *idem,* March 5 and March 29, 1834; also January 7, 1835, for requests for free ports on the Bay of Fundy and the requests of a large number of ports, January 3 and January 15, 1835. "And whereas, her Statesmen, originally adhering to the same wise line of policy, have cautiously guarded her Coasting and Colonial Trade, by partially admitting Foreign Shipping to her Colonies and totally excluding them from the privilege of carrying Freights to and from international Ports, whereby the whole Inland Navigation is engrossed exclusively by Colonists, and the Foreign Trade only thrown open to competition. And whereas, the extension of the Foreign Trade of this Province (unless barriers are interposed to prevent other Nations from entering for commercial purposes, the waters thereof,) will be pregnant with mischief to the Inhabitants of the Province, injurious to the Fisheries, and destructive of the Coasting and Inland Trade, now affording profitable occupation to a numerous and useful class of people. And whereas the Establishment of a few Free Ports in eligible situations will secure the payment of Duties, on Foreign Productions, to the Revenue of this Province, and insure to its people prompt and continued supplies of such Merchandize for which they can barter and exchange their own Exports without encountering the formidable rivalry of other Nations in the Coasting and Carrying Trade, etc." *Idem,* January 5, 1835. The Assembly favored the addition of Digby, Lunenburg, and Arichat as free warehousing ports, but were opposed to a more liberal policy. *Idem,* January 6. In 1838 and 1839 the Assembly was "most anxious that all the ports should enjoy the same privilege of free access and entering" as the five ports in possession of those privileges. Yarmouth had nearly doubled its shipping since 1834, and trade with Halifax and the demand for British goods had increased. A free port at Digby would check payment of duties at Saint John. Cumberland, with exports of grindstones of a value of £10,000, and Windsor, exporting from 120,000 to 150,000 tons of plaster of Paris, and Cornwall, potatoes or hay, could import flour, corn, tobacco, and other American goods at lower costs. Free ports would check illicit trade, increase revenue, and augment resources. "Our exports being chiefly of bulky articles, such as Coal, Gypsum, Lumber, Granite, Paving Stones, Grindstones, Agricultural Produce, Fish, Fire Wood, &c., cargoes are often of small value, and if the returns must be carried for entry to a distant Port, the profit is swallowed up in the extra expense and delay. The great Towns in the United States are the natural and the only markets for many of our exports." *Idem,* July 8, 1839. A petition from Hants on the other hand held that the opening of ports

Members of Council, almost exclusively from the heads of Departments, and persons resident in the Capital, was still pursued, and, with a single solitary exception, has been continued for the last thirty years; that the practical effects of this system have been, in the highest degree, injurious to the best interests of the Country, inasmuch as one entire branch of the Legislature has generally been composed of Men, who, from the want of local knowledge or on account of their official stations, were not qualified to decide upon the wants or just claims of the People of this Province, by which the efforts of the Representative branch were, in many instances, neutralized or rendered of no avail; that among the many proofs that might be adduced of the evils arising from this imperfect structure of the upper branch . . . it is only necessary to refer to the unsuccessful efforts of the Assembly, to extend to the Out-Ports the advantages of Foreign Trade and . . . to the enormous sum which it was compelled, after a long struggle, to resign, for the support of the Customs's Establishment.[47]

Halifax was represented by the Council in its opposition to free ports and to the lowering of duties.[48] It claimed that the creation of free ports to attract United States trade from the West Indies had not been justified and that the best interests of the colony would not be served by imports of foreign products. Revenues were not increased with the addition of free ports and Halifax returns had declined because of the increases at Sydney and Pictou. Smuggling actually increased with free

to foreign trade would injure agriculture, the coasting trade, and the merchants because of the introduction of American manufactures. *Idem,* January 9.

47. *Idem,* February 27, 1837.

48. "That from the County of Hants, to the Eastward, there is most pronounced difference of interest, in many particulars, from that which prevails to the Westward, is a fact of which there can be no dispute. To the Eastward, Halifax, as the general market place, draws to her the Inhabitants of all the adjacent Counties, and it is her interest to engross as much of the Colonial Trade as possible. On the contrary, the Counties of Hants, King's, Annapolis and a greater part of Shelburne, being Agricultural Counties, the almost contiguity, by easy Water Carriage on the Bay Fundy, to St. John, in the sister Province of New Brunswick, affords them a most advantageous Market, of which the Inhabitants avail themselves. . . . Your Petitioners humbly beg leave to submit, as one instance of Legislative discouragement of this Trade, that (on the Shores of the Bay Fundy, and the Rivers thereinto falling, Ship Building, for the English Market, being carried on to a large extent) such was the enormous impost duty during the last year upon Cordage and other Naval Stores imported from New Brunswick to this Province, that the Builders were obliged to transport the Hull of all ships built, to be rigged in the Sister Province. How powerful then must be the influence of the Trade of Halifax in the Legislature (for your humble Petitioners can attribute the Impost above alluded to among others, to no other motive than that of a wish to benefit such Trade) when it has even led to a tax upon Articles which otherwise must have employed a large portion of a most useful class of the Community, in fitting them for Articles of Export." *Journals of the Assembly, Nova Scotia,* 1836, Appendix No. 1, p. 4. See also *Journals of the Assembly* for March 14, 1839.

"An Account of the number of Vessels Cleared on a Fishing Voyage at the Port of

ports; for example, between Sydney and St. Pierre and Miquelon. Foreign products could be obtained at the outports cheaper than at Halifax.[49]

The Lords of the Committee of Privy Council for Trade and Plantations on May 24, 1835, followed the advice of the Council and decided that the difficulties could be met by allowing imports of flour and provisions into Nova Scotia "upon the same free terms upon which they are admitted into the Canadas. The Commissioners of the customs," they believed, "would be enabled without the appointment of the free ports which have been applied for, to make arrangements under which every material object which the inhabitants of the province have in view may be effected."[50]

Halifax and other Ports in this Province, during the year of 1853,—together with the Total Amount of Tonnage of the said Vessels, and the number of Men employed in the said Fisheries:

Port	Vessels	Tonnage	Men
Halifax	149	5,816	1,240
Yarmouth	54	1,982	400
Lunenburg	23	1,130	244
Windsor	1	14	4
Liverpool	13	585	106
Pictou	6	316	63
Guysborough	11	382	76
Digby	4	97	23
Sydney	8	204	53
Arichat	44	1,155	152
Annapolis	1	16	4
Clements Port	22	23	8
Port Medway	5	152	33
Pugwash	5	380	60
New Edinburg	6	282	30
Cape Canso	24	861	174
Argyle	7	193	49
Sheet Harbor	2	57	7
Pubnico	7	206	55
Canada Creek	1	26	5
Gates Breakwater	3	43	15
Westport	16	422	94
Ragged Islands	27	952	218
Ship Harbour	10	283	52
St. Marys	3	75	19
Port Hood	3	139	36
Barrington	17	413	116
Church Point	3	72	19
	455	16,276	3,355"

Reports of Committees of the House of the Assembly of Nova Scotia on the Subject of the Deep Sea and River Fisheries of the Province (Halifax, 1854).

49. *Journals of the Assembly, Nova Scotia*, 1835, Appendix No. 6, p. 15.

50. *Idem*, 1836, Appendix No. 4.

The demands of the Assembly eventually prevailed in the achievement of responsible government, and in 1839 the ports of Windsor, Parrsboro, Cumberland, Shelburne, and Lunenburg were opened to foreign trade. Finally, the abolition of the Navigation Acts placed the ports under provincial control.

With these modifications of the colonial system which had been achieved by Nova Scotia there were linked others which involved losses. The West Indies finally succeeded in escaping from the burden of the preferences given to the British American colonies.[51] New Brunswick and Halifax vessels sold their cargoes of lumber and fish in the British islands for cash and proceeded to Cuba just as New England, in the time of the first British Empire, had proceeded to the foreign islands. With only 5 shillings a hundredweight protection on sugar the British West Indies insisted on the elimination of the privileges given to British North America. "Let them go where they will for their sugar, but let us too purchase our lumber, our fish, our flour, and other daily wants, at the Cheapest market, unshackled by restrictive duties from which we derive no reciprocal advantages." As to the duty of 5 shillings, "It might have been a protective duty as between foreign slave sugar and British slave sugar but it is an utter mockery as between foreign slave sugar and British free sugar."[52]

The Assembly of Nova Scotia protested against the rescinding of preferences.

51. Pine and spruce were exported for building and for the heads of casks for sugar, molasses, and coffee; and oak and birch in blocks 3½ feet long to be split into pieces 4 inches wide and one inch thick for hogsheads (for sugar), puncheons (for molasses), and tierces (for coffee). The straight-grained, more carefully prepared wood of the United States was preferred to that of Nova Scotia. The West Indies also preferred American or Irish salt pork and beef. Other articles exported included oil, salmon, handspikes, spars, shingles, white-oak staves, hoops, potatoes, apples, cheese, butter, lard, chocolate, flour, rice, tobacco, snuff, tea, oats, and, experimentally, fresh provisions in ice in winter. In return rum, molasses, and sugar were most important, also coffee, cocoa, and hides. Rum purchased in the West Indies at 20 pence a gallon, or £9 10*s*. a puncheon, paid an excise of 1*s*. 6*d*. 3 farthings a gallon. It cost the importer 3*s*. 9*d*. freight included, the retailer 4*s*. 1*d*., and the dram drinker 1½ pence a glass, or 9 shillings a gallon. If one half of the imports were sold to dram drinkers, the remainder was sold at 4*s*. 6*d*. a gallon or, diluted one eighth, brought, for the same original quantity 5*s*. 2*d*. A great proportion of the rum trade was carried on by agents of nonresidents who purchased cargoes for export in vessels from Bermuda, had a large portion of carrying trade, and drew cash out of the province; 2,190 puncheons of rum purchased in the province involved a total additional cost to the consumer of £68,320. Sugar was purchased by the more wealthy class and molasses by fishermen, "the most profitable labourers we have for assisting to carry on our commerce." *Acadian Recorder*, February 25, 1837.

52. *Trinidad Standard*, July 16, 1839; see an argument to the effect that free labor was cheaper than slave labor, Sir Francis Hincks, *Reminiscences of his Public Life* (Montreal, 1884), chap. xviii.

An extensive and valuable Trade has sprung up in the transhipping of American Flour, Beef, Pork and Lumber to the West Indies, which will be annihilated by the reduced Duties on these Articles. Upwards of 60,000 barrels of Flour, and large quantities of Beef, Pork, and Lumber of Foreign production passed through the ports of this Province, and paid a freight to our Vessels during the last year; and our exports to the West Indies will be henceforth confined almost wholly to Dry and Pickled Fish, Lumber, and other articles, the produce of the British Colonies. The substitution of moderate Duties on the importation of Foreign Fish, in place of the entire prohibition which has hitherto prevailed, will expose our Fisheries to a new and formidable class of competitors, who are enabled, by the immense bounties offered by their own Governments, to undersell us, having no such advantages.[53]

The American Warehousing Act of 1846 capitalized the advantages accruing from the relaxing of the British colonial system and assisted Nova Scotia in her fishery, if not in her direct trade with the West Indies. Return trade from the West Indies was also hampered by Canadian restrictions and was subject to protest. Higher duties were imposed by Canada on West Indian produce from Halifax than from other ports. Cuba and Porto Rico were able to enter their sugar more cheaply in Canada than in Nova Scotia. In 1850 a steamship service from Newfoundland to Halifax and from Halifax to the West Indies, subsidized by the United Kingdom, served as a partial remedy.

Attempts to lighten the burden imposed by the colonial system were accompanied by efforts to restrict the New England fishery. The depression in the 1830's intensified smuggling activities while encroachments on British waters led to demands for higher tariffs to secure funds for protective purposes and, in turn, to increased smuggling and encroachments to avoid the higher tariffs. Nova Scotia, compelled to concede the loss of West Indian markets to the United States, proceeded to take drastic measures to exclude American vessels.[54] In 1835 the *Java*, *Independence*, *Magnolia*, and *Hart* were seized and confiscated, and in 1836 the Hovering Act (6 Wm. IV, c. 8) was passed, which permitted revenue officers to board vessels within the three-mile limit.[55]

53. *Journals of the Assembly, Nova Scotia,* April 8, 1841.

54. For an account of these conflicts see Sabine, *op. cit.;* also Wallace Graham, "The Fisheries of British North America and the United States Fishermen," *Collections of the Nova Scotia Historical Society,* XIV, 215 ff.

55. "Your Majesty's subjects in this Province have experienced great inconvenience and loss in this branch of Industry, by Foreign interference, and the Province is injuriously affected by the Illicit Trade carried on by Vessels ostensibly engaged in the Fisheries, who hover on the Coast, and, in many cases, combine Trade with the Fishery—a traffic, prejudicial alike to the Revenue, the importation of British Manufactures—the honest Trader, and the political and moral sentiments, habits and man-

H.M.S. *Champion* was sent to Chaleur Bay, Gaspé, and the North Shore, and H.M.S. *Wanderer* to the Bay of Fundy. In spite of this protection complaints were made of smuggling, that the trade was completely destroyed along the eastern shore, and that

> the Mackerel Fishery, carried on from the United States in the Bay Chaleur, where the Fish resort for the purpose of spawning, is destructive to the net and seine Fishery on the shores of Nova Scotia; for the Fish being detained in the Bay by the food thrown to them from the Vessels, till the season of their feeding on the shore of Nova Scotia is past, they pass to the westward, at a distance from the shore too great to permit their being taken with nets.[56]

An address of the House of Assembly in 1837 read in part as follows:

> The paralysed state of our fisheries . . . ought to afford a valuable export and constitute the staple of Nova Scotia, and, although we admit that the past season has been an unfavourable one, we are compelled to attribute the

ners, of the people. To prevent the continuance and extension of such evils, the Legislature of this Your Majesty's loyal Province of Nova Scotia have embodied in an Act such Regulations and Restrictions as they conceive will most effectually prevent such interference in the Fishery and the Illicit Trade connected with it, and thereby secure the Rights and Privileges recognised by the Treaty, and intended to be guarded by the Statute. This course has become the more necessary, as the Act of the Imperial Parliament contemplates the further Regulations of the Fisheries by some such means, of which all persons concerned will be bound to take notice. Many of the irregularities complained of may have taken place from the want of such Regulations. There is no intention of intimating that the Government of the United States approves of, or sanctions any interference with a branch of the Fishery which they have expressly relinquished." *Journals of the Assembly, Nova Scotia,* February 24, 1836. See official correspondence from the years 1827 to 1872, inclusive, showing the encroachments of United States fishermen in British North American waters since the conclusion of the Convention of 1818. *Documents and Proceedings of the Halifax Commission* (Washington, 1878), Appendix H.

56. Evidence of John Barss, *Journals of the Assembly, Nova Scotia,* 1837, Appendix No. 75. A petition was presented "complaining of the encroachments, by Foreigners, upon the Fisheries of the Province, and *particularly* of their Forestalling the Herrings, and other Bait for the Cod Fishery, which they purchase in large quantities from Persons who are thereby induced to sweep for the same, with seines and nets, in the Rivers and Creeks on the Coast of this Province; and praying that the same may be remedied by the passing of an Act, imposing a penalty upon the selling, bartering or giving of such Herring or other Bait to American, or other Foreign Fishermen." *Idem,* March 23, 1837. A petition from Parrsboro complained of three years of crop failure, and of a failure of potatoes in 1836 and of the diminished catch of herring and cod, "in consequence of encroachments made by foreign fishermen on our coasts and shores." *Idem,* March 14, 1837. A petition from Guysborough complained of "the great injury resulting to the revenues and prosperity of the fisheries . . . from the encroachment of foreigners upon the fishing grounds of the province and the infraction of existing treaties and praying that relief may be afforded by the fitting out of armed vessels to protect the British fisheries on the coasts of this province from such repeated aggressions." *Idem,* March 28, 1837. See especially Appendix No. 75, 1837; also *Acadian Recorder,* April 10, 1839.

continued decline of this valuable branch of industry to repeated infringements of existing treaties by the citizens of other nations.[57]

Those concerned in the fishery thought

> the employment of additional Capital in the Cod and Mackerel Fisheries, by parties living in convenient places for conducting them, an essential point, and that Larger Vessels, suitable for the Bank Fishery, should be more generally employed. It is notorious that our Fishermen on the shore do not follow the business with that energy which is requisite to ensure success, but by dividing their time between Coasting, Farming, and Fishing, they fail in producing any good result. To induce Capitalists to embark in the business, it is of the first importance to restrain Foreigners from fishing within the limits of Treaties, and thus secure an undoubted and undivided right to the in-shore Fisheries to British Subjects.[58]

A joint address from the Council and Assembly, "complaining of the habitual violation by American citizens of the treaty subsisting between Great Britain and the United States on the subject of the fisheries and praying for additional naval protection to British interests," was laid before the Queen with the result that a letter from Lord Glenelg to Sir Colin Campbell of November 5, 1838, made this announcement:

> It has been determined for the future, to station, during the Fishing Season, an armed Force on the Coast of Nova Scotia, to enforce a more strict observance of the provisions of the Treaty by American Citizens; and Her Majesty's Minister at Washington has been instructed to invite the friendly co-operation of the American Government for that purpose. The necessary directions having been conveyed to the Lords Commissioners of the Admiralty, their Lordships have issued orders to the Naval Commander in Chief on the West Indian and North American Station, to detach, so soon as the Fishing season shall commence, a small Vessel to the Coast of Nova Scotia, and another to Prince Edward Island, to protect the Fisheries. The Commander of these Vessels will be cautioned to take care that, while supporting the rights of British Subjects, they do not themselves overstep the bounds of the Treaty. You will of course afford them every information and assistance which they may require for the correct execution of this duty. I trust that these measures will prove satisfactory to the Legislature of Nova Scotia.[59]

A code of regulations was introduced in 1840, and it was held after two years' experience with the revenue schooners[60] that they had

57. *Journals of the Assembly, Nova Scotia,* February 1, 1837.

58. Evidence of Joseph Allison and Company. *Journals of the Assembly, Nova Scotia,* 1837, Appendix No. 75. See *Acadian Recorder,* January 2, February 11, and February 18, 1837.

59. *Journals of the Assembly, Nova Scotia,* 1839, Appendix No. 9.

60. The *Papineau* and *Mary* were seized in 1840 for purchasing bait. See a list of

"proved an efficient check on illicit trade and have repressed foreign encroachment on the reserved fishing grounds of the colony, whereby the domestic fishery has rapidly increased and that of the republic declined."[61]

During the past season the Fishermen of the Republic have not intruded to any great extent on the Fishing Grounds of Cape Breton; previous to the adoption of restrictive regulations upwards of 160 sail annually infested those waters, and bore away upwards of 30,000 barrels of Pickled Fish. The gain to the Province by their exclusion must be great, when the Port of Halifax alone shows such an enormous increase in 1842 over the year 1839, a year immediately anterior to the employment of Revenue Cutters, and the Committee have reason to think that the Fishery would have been more productive if the tempestuous weather of last autumn had not occasioned such severe loss in nets.

Because of the restrictive measures it was claimed that the American tonnage engaged in the fishery declined from 61,082 in 1835 to 11,775 in 1844.[62]

vessels detained and seized, *Journals of the Assembly, Nova Scotia,* 1841, Appendix No. 27. On March 27, 1841, the United States protested against the enforcement of the Hovering Act. An interpretation of the Convention of 1818 was asked of the Crown authorities on April 28, 1841, and a reply made to the United States on May 8, 1841. See complaints of Lunenburg fishermen against American violence, October 18, 1839. *Idem,* 1840, Appendix No. 85.

61. *Idem,* 1842, Appendix No. 75.

62. *Idem,* 1844, Appendix No. 68. "In the Eastern Fishery from the entrance of the Strait of Canso, including the Island of Cape Breton, the Inhabitants of Nova Scotia engaged as operative Fishermen equals 5,000 men, having upwards of 120 shallops, and 1,700 Boats; and computing that an equal number are employed in the Western and other Fisheries of the Province, an aggregate of 10,000 Fishermen, 240 or 250 Shallops and 3,400 Boats may be assumed as a fair statement of the Fishing interest of Nova Scotia; and taking an average of the catch of pickled Fish for three years, selected so as to prove the utility of employing small vessels to repel encroachment on our Fishing Grounds, the Committee are gratified in being authorized to report that the experiment has been successful." *Idem,* 1843, Appendix No. 74.

EXPORTS FROM THE PORT OF HALIFAX*

(Barrels)

	Mackerel	Pickled Fish
1839	19,127	60,810
1840	25,010	60,495
1841	35,917	64,649
1842	54,118	84,879
1843	71,854	95,875
1844	50,698	70,192
1845	38,230	97,577
1846	82,645	136,448

(continued on next page)

The Committee are of opinion that there is more energy evinced by the class of the community engaged in this useful occupation, which contributes a more valuable staple for trade and commerce than heretofore, which is attributable to demand in foreign markets, in particular in the United States, to which from Halifax alone, between March 1847 and [March] 1848, there have been exported of pickled salmon and mackerel one hundred and twenty-four thousand five hundred barrels, besides alewives, herrings, dry codfish and pollock.[63]

The policy of restriction was pursued too aggressively. In 1843 the *Washington* was seized, and in the same year Americans were kept from landing on the Magdalen Islands. But as a result of American protests Great Britain was forced to intervene. Lord Stanley suggested a relaxing of regulations in a letter of May 19, 1845,[64] but was opposed by

62 (*continued*).

EXPORTS OF FISH FROM NOVA SCOTIA IN THE YEAR ENDING 5TH JANUARY, 1841

Destination	Dried Fish (Quintals)	Green Fish (Barrels)	Smoked Fish (Boxes)
Great Britain	56	140	119
British N.A. Colonies	12,555	11,262	14,250
British W. Indies	232,541	38,393	11,547
United States	....	13,182	1,637
Foreign W. Indies	14,065	1,001	62
Brazils	17,063	...	...
Mauritius	274	108	...
Africa	42	27	140
Foreign Europe	5,335	...	...
Western Islands	288	1	...
From Cape Breton†	44,807	7,562	...
TOTAL	327,026	71,676	27,755

* *Idem,* 1847, Appendix No. 75.

† "The Custom House Returns from Arichat and Sydney . . . shew the exportation [of] 41,328 quintals dry Fish, 10,794 barrels pickled Fish, 270 casks of Oil; and the following quantities are fair estimates of the catch in other parts of Cape Breton, where no Customs Officers are stationed: Strait of Canso, 2,500 quintals; Port Hood, 500; Mabou, 2,000; Marguerite, 5,000; Cheticamp, 8,000; Bay St. Lawrence, 3,000; Cape North, 4,000; Inganiche and Lowpoint, 8,000; Bras d'Or, 3,000; Mainadieu, 4,000; Louisburg, 5,000; L'Ardoise, 6,000, making 51,000, clearly evincing that this valuable branch of industry, under every disadvantage, is furnishing an export equal to a million annually, while the internal consumption of the Province, with a population exceeding 200,000, (many, from pious feelings, and more from choice or necessity, making this an article of food) may be fairly estimated at 300,000 quintals." *Journals of the Assembly, Nova Scotia,* 1840, Appendix No. 85.

63. *Idem,* 1848, Appendix No. 89.

64. *Idem,* July 2, 1845; *idem,* 1846, Appendix No. 11. In 1841 Nova Scotia submitted the question of the right of American vessels to use the Strait of Canso to Crown lawyers for an opinion and secured a favorable decision. *Journals of the Assembly, Nova Scotia,* 1841, Appendix No. 6. "Between 700 and 800 sail of American vessels belonging to the Republic of the United States pass through that Strait annually, and usually return with average freights amounting to nearly half a million

Nova Scotia in a reply of June 16. On September 17, 1845, Great Britain insisted on a policy of relaxing in the Bay of Fundy but strict adherence elsewhere.

Effective limitation of the New England fishery was further weakened by the lack of regulations in the other provinces. "The good work is only partially accomplished while Foreign Shipping is allowed to Fish within the prescribed limits on the Coasts of Prince Edward Island and the shores of the Magdalen Islands, where Herring spawn, and a system of the most destructive character is annually in full operation for taking them."[65] The coöperation of New Brunswick was sought in a resolution granting £500 to protect the fisheries in 1837; but it was not until 1853 that New Brunswick introduced legislation (16 Vic., c. 69) similar to the Hovering Act. Prince Edward Island introduced such legislation in 1843 (6 Vic., c. 14). The necessity of coöperation with the other provinces was imperative.

The course to be pursued to prevent foreign vessels from trespassing on the grounds reserved for British subjects, requires more talent and experience than I have, to decide. However, with the information which I have received, and the little experience I have, it appears that it would take a larger amount than the legislature of Nova Scotia would grant for the protection of the fisheries, when we take into consideration the extent of the coast on Nova Scotia and Cape Breton, which, in the latter part of the season, is completely lined with American vessels, from Cape Gaspé to Cape North, in Cape Breton. These vessels I have been informed, often fish within half a mile from shore, paying little or no regard to the limits stated in the national convention. In fact the day on which I seized the "*Tiber,*" there were sixty or seventy sail in sight, which were nearly all within limits; but as these are fast

of quintals, taken in British waters; they hitherto have made this Strait a resting place where they procure wood and water at one third of the price in their own markets, which induces them to leave home scantily supplied, and encourages our people to engage in an unprofitable employment, to the neglect in some measure of Agricultural pursuits, and the fostering of illicit trade; this state of things retards investment of capital in our Fisheries, and accounts for an extent of Coast exceeding 400 miles, only furnishing 5,000 fishermen, prosecuting their calling in Boats, whereby the Province sustains heavy annual loss, from the limited means of this hardy and industrious class of society." *Idem,* 1843, Appendix No. 74. "The legal control of the passage in question is vested in this Government; and the Committee recommend that such regulations should be adopted with respect to foreigners, as will compel payments of dues for its use, conceiving that it is wise policy to meet the high tax imposed by the Tariff of the United States on fish taken by British subjects, by a corresponding tax on their vessels using said passage, in proceeding to or returning from the Gulph, such regulations to be modified so soon as their tariff is ameliorated." *Idem,* 1848, Appendix No. 89.

65. *Idem,* 1842, Appendix No. 75; 1843, Appendix No. 74; and 1844, Appendix No. 68. *Idem,* 1840, Appendixes Nos. 85, 86, for a discussion of the difficulties of controlling Newfoundland as well.

sailing vessels, if they once get the start, and are out of gunshot, they feel quite secure. Were the British Colonies united, or each colony equally interested in the fisheries, and would all come forward to protect the fisheries, it would be of great consequence. The coast cannot be protected from encroachment by foreigners, by sailing vessels, unless there are three or four in number.[66]

In 1851 arrangements were made with Canada and New Brunswick for the protection of the fisheries.

Mr. Howe, having called the attention of his excellency and the council to the importance and value of the gulf fisheries, upon which foreigners largely trespass, in violation of treaty stipulations, and Mr. Chandler having submitted a report of a select committee of the house of assembly of New Brunswick, having reference to the same subject, the government of Canada determines to co-operate with Nova Scotia in the efficient protection of the fisheries, by providing either a steamer or two or more sailing vessels to cruise in the Gulf of St. Lawrence and along the coasts of Labrador. It is understood that Nova Scotia will continue to employ at least two vessels in the same service, and that Mr. Chandler will urge upon the government of New Brunswick the importance of making provision for at least one vessel, to be employed for the protection of the fisheries in the Bay of Fundy.

(Signed) Jos. Bouret, Joseph Howe, E. B. Chandler[67]
Toronto, June 21, 1851.

In 1852 the committee of the Nova Scotia Assembly recommended the selection of four fast sailing vessels to seize all foreign fishing vessels within the three-mile limit. New Brunswick supplied two, Prince Edward Island one, and Great Britain a small fleet of steamers. An order was issued on August 28, 1852, which asseverated that "no American fishing vessels are entitled to commercial privileges in provincial ports but are subject to forfeiture if found engaged in traffic. The colonial collectors have no authority to permit freight to be landed from such vessels which, under the convention, can only enter our ports for the purposes specified therein and for no other." It was recommended "that the rights of the province in reference to the fisheries should be strictly and rigidly enforced and that no participation in them should be conceded to any foreign power; but that the Colonial fishermen should be invested with the exclusive rights to fish in the waters adjacent and belonging to the province."[68]

66. *Idem,* 1852, Appendix No. 25.
67. *Idem,* p. 169. See J. S. Martell, "Intercolonial Communication, 1840–1867," *Canadian Historical Association Report,* 1938, pp. 41 ff.
68. *Reports of Committees of the House of Assembly of Nova Scotia on the Subject of the Deep Sea and River Fisheries of the Province* (Halifax, 1854).

The Reciprocity Treaty was supported by Great Britain[69] as a means of avoiding conflict in the fishery and was proclaimed on September 11, 1854.[70] It brought numerous difficulties to an end by Articles 1 and 2, which permitted United States vessels to fish in Canadian waters and Canadian vessels to fish in American waters. Duties were removed in the case of fish, flour, and many other products. Neither country was to participate in the salmon, the shad, and the river fisheries of the other.[71] Three commissioners were appointed to adjust difficulties.[72] The effectiveness of the American fishery was maintained, in spite of the removal of duties on fish from Nova Scotia, until the outbreak of the Civil War and the consequent rise in American prices.[73]

69. See Palmerston to Sir H. Bulwer, November 1, 1849, *The Elgin-Grey Papers 1846–1852* (Ottawa, 1937), IV, 1482–1483. D. C. Masters, *The Reciprocity Treaty of 1854* (London, 1937), chaps. i–iii; also L. B. Shippee, *Canadian-American Relations, 1849–1874* (New Haven, 1939), chap. iii; J. M. Callahan, *American Foreign Policy in Canadian Relations* (New York, 1937), chap. xi.

70. A request from Marcy made on August 4, 1854, asking that American fishermen should not be molested "should they at once attempt to use the privileges secured to them by the treaty, although Great Britain and the provinces may not have passed the laws required on their part to carry it into complete effect" received an answer on August 18, 1854, to the effect that "it is the desire of Her Majesty's government that this wish of the government of the United States should be acceded to; and that American fishermen may be immediately allowed the use of these privileges." The Executive Council of Nova Scotia on September 14, 1854, insisted on the remission of duties in the United States in return for the privileges granted American vessels, and the United States on October 10, 1854, proposed a plan by which duties should be refunded, and it was made public in a circular issued on October 16, 1854. Nova Scotia worked out a plan for similar and reciprocal arrangement for the refunding of duties on November 13, 1854, and issued regulations to that effect on December 1, 1854. *Journals of the Assembly, Nova Scotia,* 1854–55, Appendix I.

71. "It is understood that there are certain Acts of the British North American Colonial Legislatures, and also perhaps, executive regulations intended to prevent the wanton destruction of the fish which frequent the coasts of the Colonies, and injuries to the fishing thereon. It is deemed reasonable and desirable that both United States and British fishermen should pay a like respect to such laws and regulations, which are designed to preserve and increase the productiveness of the fisheries on the coasts. Such being the object of these laws and regulations, the observance of them is enforced upon the citizens of the United States in the like manner as they are observed by British subjects. By granting the mutual use of the inshore fisheries, neither part has yielded its right to civic jurisdiction over a marine league along its coasts." W. L. Marcy to Peaslee, March 28, 1856.

72. See P. E. Corbett, *The Settlement of Canadian-American Disputes* (New Haven, 1937), pp. 28–30. Howe was appointed a fishery commissioner in 1862. See J. A. Roy, *Joseph Howe* (Toronto, 1935), pp. 320–322.

73. "Any protest they [the Maritime Provinces] could have made at the time the bargain was concluded would have been drowned by the stronger voices of Imperial and Canadian policy; Canada proper having little to lose in the way of fisheries and much to gain in the matter of a market for raw material. The fisheries suffered first by our idiotic reciprocity treaty and secondly by the blockade of the southern ports. Since . . . the northern states have enforced a blockade of the southern ports the chief privilege allowed our colonists by this treaty has been done away with while

The abrogation of the Reciprocity Treaty in 1866 and the reimposition of a duty brought about the reëstablishment of a protected market for the New England fishery. Pressure for the exclusion of Americans from Nova Scotian fishing grounds was renewed, particularly as Newfoundland began to exclude Nova Scotia and the United States. On November 4, 1865, Hastings Doyle wrote to Edward Cardwell, the Colonial Secretary:

> There can be no doubt that the right to enjoy these fisheries was one of the leading inducements which actuated the Government of the United States in the arrangement of the Reciprocity Treaty, and that there is no way in which they can be made to feel the impolicy of its abrogation more effectually than by the rigorous exclusion of their fishermen from the fishing grounds to which they have had the right to resort during the past ten years. Independently of the want of any naval force in these provinces, the duty to be performed in the protection of these fisheries will be one of great responsibility and delicacy, in the discharge of which the most serious questions of national interest are not unlikely to arise. Having reference, therefore, to the character and importance of this service, I trust that her Majesty's Government will be enabled to adopt such measures in this important matter as will effectually protect these fisheries from intrusion, and from the outset assure the Government and people of the United States that they cannot withdraw the commercial advantages conceded to these colonies without losing the privileges which were extended to them in exchange.[74]

On March 19, 1866, the Hon. Charles, later Sir Charles, Tupper, as Premier of Nova Scotia, issued a proclamation which declared "that hereafter all vessels and boats belonging to any Foreign Country pursuing the fisheries within the territorial jurisdiction of her Majesty, in the province of Nova Scotia, are by law subject to forfeiture, and the parties engaged therein to penalties, and that the law will be rigorously applied to all cases of trespass on the fishing grounds of Nova Scotia. Of which all parties will take notice and govern themselves accordingly."[75]

The Province of Canada[76] suggested to the Imperial government that

they are still subject to the rivalry of the Yankees on their other coasts. The sufferings of our fishermen in America can hardly be realized." Francis Duncan, *Our Garrisons in the West* (London, 1864). See S. A. Saunders, *Studies in the Economy of the Maritime Provinces* (Toronto, 1939), pp. 103–159. An estimate of American vessels fishing in Canadian waters during the years of the Reciprocity Treaty gives an increase from 234 in 1854 to 476 in 1856 and a decline to 235 in 1863. E. C. Gould, "Relations between Nova Scotia and the United States," Master's thesis, University of Toronto, 1934.

74. *Journals of the Assembly, Nova Scotia,* 1866, Appendix No. 21, p. 1.

75. *Idem,* p. 6.

76. In 1791 the old Province of Quebec became Canada, divided into Upper and Lower Canada. Under the Act of Union in 1840 they were united and were known as

it might be best to impose license fees on American vessels rather than exclude them. But this suggestion was made "without preconcert with the other colonies to be affected by the proposed arrangement . . . in a matter so vitally affecting the rights and interests of the maritime provinces," and Nova Scotia resented it.

The Council, after the most serious deliberation, and with a view to meet the wishes both of the Imperial Government and the Government of Canada, are compelled to state that they are of the opinion that any concession at this moment of the admitted rights of British subjects to the exclusive use of the inshore fisheries of British North America, would be most impolitic and disastrous to the interest of British North America. The privilege of using these fishing grounds has been deliberately abandoned by the Government and Congress of the United States, and abundant notice was given to the people of that country by the official announcement made more than a year ago, which abrogated the Reciprocity Treaty.

If under these circumstances, when the United States are exhausted by a four years' war, and paralysed by an oppressive debt, any indecision is exhibited in the maintenance of these undoubted and admitted rights, and a temporizing policy substituted, which will be certain to be misconstrued, the Council believe that the prospect of obtaining a fair reciprocity treaty will be diminished; that the most injurious results will follow, and that the difficulties to be encountered a year hence in dealing with the question will be vastly enhanced.[77]

Great Britain, however, insisted on the acceptance of the arrangement,[78] and in 1866 a tonnage duty of 50 cents was imposed.

Nova Scotia also sought to limit smuggling from France.

In the Bras d'Or Lake, for several years past, quantities of French Goods have been introduced from the Islands of St. Pierre and Miquelon by French

Canada West and Canada East. Under Confederation in 1867, they became the provinces of Ontario and Quebec. Until Confederation, therefore, Nova Scotia and the other Maritime Provinces did not consider themselves as in any sense a part of Canada.

77. *Idem,* Appendix No. 18, p. 9. Canada suggested licenses for one year at a moderate fee, the proceeds of which would be used to maintain a joint maritime police, force American vessels to realize that reciprocity was at an end, and assert Canadian rights. Minutes of the Executive Council, Montreal, March 23, 1866. Consideration of a despatch from the Governor-General of Canada, April 4, 1866, by the Council, May 9, 1866.

78. "Downing Street, 26th May, 1866.

"Her Majesty's government learn with great regret the opinion entertained by your government with respect to a policy which her Majesty's government consider extremely calculated to facilitate an arrangement with the United States of a question affecting the foreign relations of this country.

"Her Majesty's government trusts that on further consideration, and when the Executive Council are informed that there are *reasonable* grounds for hoping that be-

Vessels arriving from those Islands in quest of Live Stock and Agricultural Produce, and the same illicit trade was carried on by British Vessels from Newfoundland, whence Foreign Articles were brought, which had been entered in that Island free of Duty. . . . In the Harbor of Lingan or Bridge-

fore next season permanent arrangements may be made with the government of the United States, they will feel themselves at liberty to withdraw their objections to a temporary arrangement for the year which has received the cordial approval of her Majesty's government.

"I must distinctly inform you that on a matter so intimately connected with the international relations of this country, her Majesty's government will not be disposed to *yield their own opinion* of what is reasonable to insist on, nor to *enforce* the strict rights of her Majesty's subjects beyond what appears to them to be required by the reason and justice of the case.

(Signed) EDWARD CARDWELL"

Idem, p. 10.

The Council at Halifax replied on June 21, 1866: "The Council not only failed to perceive how the issue of licenses for one year would promote the object in view, but regarded that policy as fraught with greater difficulties and complications than the moderate and temperate enforcement of the exclusion of American fishermen from privileges which they had voluntarily surrendered, and for which the government of the United States was unwilling to give any adequate consideration. The difficulty of carrying out the proposed licensing arrangements, the Council considered obviously greater and more likely to cause unpleasant collisions with American fishermen than the judicious enforcement of the treaty of 1818; as in the latter case no foreign fishermen could enter the prohibited waters, while in the former a constant and irritating, and frequently repeated search must be made by the numerous vessels belonging to the various provinces and to her Majesty engaged in compelling the American fishermen to respect the licensing regulations.

"The Council feared that the uninterrupted enjoyment of the fishing privileges acquired by the reciprocity treaty would prevent the government and people of the United States from appreciating the loss to themselves, caused by the abrogation of that treaty, while a year hence the withdrawal of these privileges will be our act instead of their own.

"It is not, however, necessary now to expand the numerous objections entertained in this province to the proposed Canadian policy. Suffice it to say that the Council, entertaining the opinion that policy would be most disastrous in its effects upon British interests, felt it their duty respectfully to submit their opinions for the consideration of her Majesty's Government.

"After giving this important question the most careful consideration, the Council regret that they cannot change the opinions at which they had arrived; but they fully appreciate the necessity of meeting the view of her Majesty's Government, so strongly expressed in Mr. Cardwell's despatch of the 26th ultimo, and accordingly withdraw their objections and agree to grant the licenses for this year as desired." *Idem,* p. 11.

Nova Scotia protested again in 1867: "As to the system of granting fishing licenses to American fishermen, adopted and practised during the last year by the Governments of this and the adjoining Provinces . . . the committee agree with the petitioners in their expressions of deep regret, that the adoption of such an arrangement had become, or was considered necessary. Nothing could more injuriously affect the fishing interests of this Province; and the committee cannot in terms too emphatic express their disapproval of the injustice done to our industrious and enterprising fishermen, in allowing American fishermen, upon nearly equal terms, to fish in our waters side by side with the former, while the American market is virtually closed by a high tariff to their products."

"If," in the words used by the Colonial Secretary in the correspondence laid before

port a practice prevailed among the Americans of landing, clandestinely, Brandy and Wine, which they brought from the French Islands, when calling at Lingan for a cargo of Coal, on their way home.[79]

Nova Scotia likewise resisted the attempts of Newfoundland to restrict her fishery. Regarding Newfoundland's imposition in 1846 of an export tax on bait she had made this protest:

The Committee have enquired into the statements made by the Merchants and Inhabitants of Isle Madame, complaining of a duty on fish exported from Newfoundland and find the sum of three shillings sterling is demanded on every 100 lbs. weight of pickled fish exported in bulk, and 2s.6d. per barrel on fish exported without inspection. This Law appears partial in its operation, applying to the Southern coasts of Newfoundland, but not extending to the Bay of St. George. The policy which induced the Legislature of that Colony to pass an Act so restrictive in its nature, may have been to break up a trade in baits with foreigners; but its application to British subjects trading with British possessions, is oppressive and unwise. The Herring Fishery of Fortune Bay produces from 30 to 40,000 barrels annually, chiefly taken in nets by the inhabitants, and sold to traders or exported to the French Islands. The effect of the Law is to compel a sale to the resident Merchants or Traders, at their own price, and to limit the markets. This Fishery is carried on between November and June, a period of the year when the Fishery of Nova Scotia is interrupted, and the fishing population of the Eastern parts of the Province resort to those waters, and have contributed extensively by their enterprise to develop the value of that fishery, and the continuance of a law imposing an export duty equal to 6s. sterling per barrel on fish caught by nets in vessels which hold no communication with the shore, by British

the House, "motives of forbearance and good policy still demand the exercise of this privilege, the committee earnestly recommend that, instead of levying a pecuniary license fee therefor, steps be taken to arrange, if practicable, with the American market free, or under a more reduced tariff than that now imposed."

"The consideration received for the privilege would thus accrue to the benefit of our fishermen as a class, who alone are entitled thereto as being the parties immediately injured." *Journals of the Assembly, Nova Scotia,* 1867, Appendix No. 33, p. 2.

79. *Journals of the Assembly, Nova Scotia,* 1840, Appendix No. 52. *Idem,* 1841, Appendix No. 62, gives a general statement of the importance to Nova Scotia of keeping the French from engaging in the Newfoundland fishery: "Large quantities of fish have been imported into this province from the French island of St. Pierre, greatly to the prejudice of our fishermen, as in consequence of the large bounty paid by the French government to that branch of industry, our fishermen are unable to compete with them; and as the French fishermen do not exchange commodities, but take only specie in return for their fish, your committee would recommend the imposition of a duty on all fish caught and cured in foreign countries which do not reciprocate with this province." *Idem,* 1859, Appendix No. 40, p. 504. An address was forwarded by Nova Scotia to Great Britain to be sent to France to ask that steps should be taken to prevent depletion of the Banks by the use of boulters. The French "expressed their opinion that it is more practical and therefore preferable to leave each government to take such measures as they may judge most suitable for the preservation of the fisheries." *Idem,* 1862, Appendix No. 7.

subjects, in British waters, is impolitic and oppressive if not arbitrary, illegal and unconstitutional, the repeal of which ought to be sought by Address to Her Majesty or the action of the Executive Government of this Province in England and Newfoundland, in such way as may be most likely to afford redress.[80]

In 1849 Newfoundland afforded the redress here called for by making special provision for the British colonies. Later, the interests of Nova Scotia in the Labrador fishery were endangered by Newfoundland's insistence upon the collection of duties. In 1857 nearly 150 Nova Scotian schooners were reported at Blanc Sablon. Trade in agricultural products on the Labrador increased but was checked by the effective collection of duties on Nova Scotian goods.

As a result of the difficulties with New England, France, and Newfoundland, Nova Scotia turned to Confederation for support.

NEW BRUNSWICK, QUEBEC, AND PRINCE EDWARD ISLAND TO CONFEDERATION

Her Majesty's Government although desirous not to sanction any unnecessary deviation from that policy which regulates the Commerce of this Country, are still disinclined to prevent those Colonies, by the interposition of Imperial Authority (and especially pending the negotiations with the United States of America for the settlement of the principles on which the Commerce with the British North American Colonies is hereafter to be carried on) from adopting the policy which they may deem most conducive to their own welfare and prosperity.

SIR JOHN PAKINGTON, 1853

NEW BRUNSWICK, Quebec, and Prince Edward Island were, like Nova Scotia, interested in the exclusion of New England from the fishery, and were also being increasingly restricted by Newfoundland. On the New Brunswick shore in 1832 the Jersey firm of William Fruing and Company* had 60 boats engaged in fishing at Shippigan[81] and 20 at Point Miscou. The industry was carried on at various points, the largest establishment being Caraquet, with 200 boats. But this included the station of Charles Robin and Company at Point Miscou, with 40, and another, at Petit Rocher, also with 40. In 1832 in this area 903 men and

80. *Idem,* 1846, Appendix No. 87.

81. Le Bouthilier Brothers had a "room" at Point Miscou where "all the settlers . . . complained bitterly of their poverty and state of bondage." The cod fishery was on the decline on the New Brunswick coast. Tracadie Gully was abandoned in 1844. M. H. Perley, *Report on the Sea and River Fisheries of New Brunswick* (Fredericton, 1852), pp. 30 ff.; Tocque, *op. cit.;* W. H. Ganong, "A History of Miscou," *Acadiensis,* April, 1906; "The History of Shippegan," *idem,* April, 1908.

*See Notes to Revised Edition, p. 509.

250 boys were employed. They used 432 boats for their fishery, and had a catch of 24,050 quintals.

On the Gaspé Peninsula, Channel Islands firms extended their fishery.[82] Vessels from Mal Bay and Eboulements participated in the Gulf fishery for the first time in 1857, and apparently introduced the trawl system about 1862. As a result of the activities of Robert Christie, the representative of Gaspé; the expiration of the lease of the "King's posts" held by the Hudson's Bay Company; the introduction, in 1852, of the joint protective service,[83] shared by New Brunswick and Nova Scotia; and difficulties arising from competition with Newfoundland on the Labrador—Canada passed an act in 1853 (16 Vic., c. 92) which permitted all British subjects to participate in the fishery along the Labrador coast, "several of whom concerned in those fisheries," the act said specifically, "have been of late years by strong hand prevented by persons residing on or frequenting the Labrador or north shore of the said gulf from making [fish] on the coasts thereof and islands contiguous thereto."[84] And this act, to remain in force for three years, was continued from May 1, 1856, by further legislation. A second act, the Fishery Act of 1857 (20 Vic., c. 21), was designed to encourage the Gulf fishery by the introduction of numerous regulations.[85] It was followed by additional legislation (22 Vic., c. 86) which provided for the establishment of a Fisheries Division in 1858.

82. The merchants let boats to fishermen and paid so much per draft, or 238 pounds, of green fish "from the knife," that is, when headed and cleaned. It was claimed that advances were made good by August 15, and after this date men were allowed to fish for themselves, to sell their catch for cash or to send it to Quebec. Three quintals of green fish made one of dry. The trade price of fish exceeded the cash price by about 20 per cent. According to M. H. Perley 300 pounds "fresh from the knife" or 252 pounds that had been split, salted, and left to lie for one night were regarded as a quintal. The curer was given one tenth. A barrel of flour costing $6 at Quebec was sold for $7 cash or $10 in trade. The merchants made about 25 per cent profit on the year's transactions. See *Documents and Proceedings of the Halifax Commission* (Washington, 1878). See also a description of operations at the fishing station of Percé, on July 14, 1873, in *My Canadian Journal*, 1872–78, by the Marchioness of Dufferin and Ava (London, 1891), pp. 90–91. Also Antoine Bernard, *La Gaspésie au soleil* (Montreal, 1925), pp. 218 ff.

83. See the "Seventh Report of the Committee for Managing the Affairs of the North American Colonial Association," which gave promise of assistance in resisting encroachments upon the fisheries by the United States. *Montreal Gazette,* July 22, 1837.

84. "The district of Quebec suffers as much from the young fishermen as the district of Montreal from the Shantymen." *Journals of the Assembly of Canada,* July, 1850, pp. 196–197.

85. E. T. D. Chambers, *The Fisheries of the Province of Quebec* (Quebec, 1912), pp. 157 ff. The fisheries department disappeared after Confederation, but the Supreme Court decision in *The Queen* vs. *Robertson* gave the provinces the fishing rights in water bounded by Crown lands and a new division was set up. *Idem,* pp. 177–178.

"Since the passing of a law permitting British subjects to take possession of any portion of beach unoccupied, a great number of fishermen from the Bay of Chaleur and the coast of Gaspé have made establishments at various points about the King's posts and in the seigniory of Mingan."[86] In 1856 fishermen went to Natashkwan, Magpie Bay, Sheldrake, and Seven Islands.[87] In the following year 13 families were reported as permanently settled at Natashkwan where "the schooners going to fish on the north shore for cod commence their operations." The settlement comprised 120 people with 16 boats, and made 1,700 quintals. At Seven Islands six new cod-fishing establishments were started in 1857, the largest of which, owned by Hamilton of New Carlisle, had 12 boats and 30 men. Within three years the Thunder River had five establishments with 13 boats and 40 men, and made 2,200 quintals. Magpie Bay had eight establishments with 29 boats and 103 men. Many of its hands came from Bonaventure. It made 4,810 quintals. Sheldrake had seven establishments with 30 boats and 100 men. It made 4,590. Bradore Bay[88] had 30 men working for large establishments concerned with the seal fishery, and Bradore Basin had three families. It was estimated in 1857 that along the coast from Godbout to Blanc Sablon there were 1,225 employed in the fishery and 300 fishing vessels. They took, in all, 33,060 quintals of cod that brought $3 a quintal; 2,235 barrels of herring, worth $4 a quintal; 700 barrels of mackerel, worth $10; 1,200 barrels of salmon, worth $18; 300 barrels of cod oil, worth $30; and 5,730 seals, worth $6 apiece; or a total of $186,100. In 1876 it was claimed that 17 firms engaged in shipping fish from Gaspé had 30 establishments on the North Shore.[89] Stations were also

86. See the valuable information included in the annual reports of Pierre Fortin, commanding the expedition for the protection of the fisheries in the Gulf of St. Lawrence. The Appendixes in the *Journals of the Legislative Assembly of Canada,* 1853 and later.

87. "Below the mouth of each fishing stream proves to be the best station for cod-fishing as there the fish accumulate to feed on the fry which runs into the river, and to deposit spawn which they follow to sea after this as soon as the fry make off from the rivers to deep water." M. R. Audubon, *Audubon and His Journals* (London, 1848), p. 380.

88. "Bradore is the great rendezvous of almost all the fishermen that resort to this coast for cod-fish. We found here (July 26, 1833) a flotilla of about one Hundred and fifty sail, principally fore and aft schooners, . . . mostly from Halifax and the eastern portions of the United States." Audubon, *op. cit.,* p. 413. See *Journals of the Legislative Assembly of Canada,* 1859, Appendix No. 30. For an excellent account of the industry on the Labrador see also H. Y. Hind, *Explorations in the Interior of the Labrador Peninsula* (London, 1863).

89. For a list of companies trading at Gaspé and on the North Shore see *Documents and Proceedings of the Halifax Commission,* pp. 883–884; also Chambers, *op. cit.,* pp. 134 ff.; J. C. Langelier, *Esquisse sur la Gaspésie* (Quebec, 1884), chap. vii; Tocque, *op. cit.,* pp. 316–318.

established to which ships could go directly from the Channel Islands. The Touzel firm, from Jersey, established a post at Sheldrake in 1851, and the firm of La Parelle, also from Jersey, set up a post at Natashkwan in 1857.

Below Natashkwan and Eskimo Point, known after 1924 as Havre St. Pierre, the character of the coast changes from one of sandy beaches to a bleaker aspect. The technique of the fishery adapted to Gaspé was less suited to the lower coast, and, below Harrington, settlements contained a good many Newfoundlanders. At Bonne Esperance and in the vicinity of the Straits of Belle Isle near the Canadian border, large firms from Newfoundland and the Channel Islands carried on operations in spite of a Canadian tariff. From Harrington to Blanc Sablon, traders from Halifax competed with traders from Quebec in the retailing of supplies other than flour and pork and in the purchase of fish from settlers. Vessels from Halifax could vie successfully with those from Quebec, particularly as the latter were less advantageously affected by the Reciprocity Treaty and were hampered by the tariffs of 1858.

A free-port policy was introduced and a free port was established at Gaspé in 1860, to the end

> that the fisheries would be materially encouraged by the bounty which the purchase of supplies free of duty would give—that an extensive commercial mart would be established at Gaspé Basin to which foreigners engaged in the fishery would resort for their supplies, at which a market would be established for the purchase and sale of fish, and which would be frequented by foreign shipping for the interchange of foreign produce for the produce of our fisheries.

Three years later, however, the results were found to be disappointing. "The free-port policy has failed to accomplish the objects aimed at." It failed, it was claimed, because of the monopoly control by "a few merchants who command large capital and with whom it is useless for small capitalists to attempt competition." "The business [is] done by barter and the outfit is advanced before the fish are caught which are to pay for it. The result is that practically a *quintal* of fish will not bring to the fishermen any greater quantity, either of necessaries or luxuries, than it did formerly. . . . The chief advantage has gone to the principal merchants."[90] The free-port system was abandoned, but the conditions which developed as a result of the varying tariff policies of the provinces contributed to the movement for Confederation.

In 1857 Le Bouthilier Brothers had establishments at Bonaventure—

90. *S.P.* No. 37, 1865. See also *Select Documents in Canadian Economic History*, ed. H. A. Innis and A. R. M. Lower (Toronto, 1933), II, 698 ff.

where they had 40 boats—at St. Anne des Monts, Isle au Bois and Forteau. They sent 3,000 quintals to Civitavecchia from St. Anne and, altogether, 12,000 quintals to the Mediterranean. They sent 5,000 tubs of cod to Brazil;[91] 750 quintals, 80 tons of oil, and 1,200 barrels of herring to England and Jersey; and 1,000 quintals of cod to Quebec. L'Esperance had 18 vessels, and 40 men at Grand Etang, and sold more than 3,000 quintals to Italy and Spain. The firm of Robin and Company developed its export trade to Brazil and the Mediterranean, and handled an average of 30,000 quintals chiefly from their headquarters at Paspebiac.[92] The Gaspé coast sent about 25,000 barrels of pickled cod to Quebec and Montreal.

The Magdalen Islands interests concerned themselves more largely with pickled fish for the American market, and were the gainers by the Reciprocity Treaty. In 1833, 160 families were annually supplied by one or two vessels from Quebec. In 1854 various fishing establishments sent cargoes of cod and seal oil to Halifax; green fish, cod, herring, and mackerel to Quebec and Montreal; and cod to the United States. There was an American establishment at Amherst Harbor. Fish were sold in bulk, heavy-salted, at 3 shillings per 200 pounds, or in barrels at from 10 shillings to 12*s.* 6*d.* In 1859, 12,429 hundredweight of dried and smoked fish, 101,380 barrels of pickled fish, and 27,971 gallons of oil were exported from Amherst.[93] They sent 21 schooners to the seal fishery. They were an important center for the herring fishery, and use was made of large seines—110 fathoms by 8½ in depth. Nova Scotian and other schooners came to the islands to fish for bait.

In spite of this expansion of trade, the small-scale fisherman was dissatisfied with the high duties and with a monopoly that controlled both land and trade. He was attracted by the opening of the Labrador,[94] and migrated to the North Shore along with the fishermen from Gaspé. In 1856, six families moved to Natashkwan and two to Eskimo Point. In 1858 the Abbé Ferland[95] reported the further migration, within

91. Tubs were used in the Brazil trade. "Each tub contains 128 pounds of well-dried fish. The packing is done by means of an iron screw worked by three men, the fish being thus pressed in the tub and forced into the smallest possible space."

92. In 1876 this firm had 16 stations: 3 in Cape Breton, 7 on the Gaspé coast, and the remainder in New Brunswick on the North Shore. It exported 80,000 quintals in that year.

93. Tocque, *op. cit.*, pp. 315–316.

94. Fishermen went with schooners to purchase goods from Jersey houses on the Labrador. A more rigid collection of customs and higher duties after 1858 increased the price of Quebec goods and encouraged competition from Halifax. In 1860, 38 schooners, 232 boats, and 574 nets were reported on the islands, but the schooner fishermen preferred the more protected regions of Newfoundland and the Labrador to the adjacent waters of the Magdalen Islands themselves.

95. See Hind, *op. cit.;* A. S. Packard, *The Labrador Coast* (New York, 1891);

three years, of 21 families to Eskimo Point and, in two years, 15 to Petit Natashkwan. In 1863 large numbers left the Magdalen Islands. "More than three hundred heads of families have left the Islands and have gone to establish small French centers at Kegashka, Natashkwan, and Eskimo Point."[96] But later, island fishermen ceased to migrate as a result of the help given by Canadian bounties after 1882, the abrogation of the Washington Treaty, the growing settlements on the North Shore, and the increasingly important American bank fishery, with its demand for bait.[97] A steamship service to Pictou was started in 1856. And, due to more favorable conditions and the occupation of the North Shore, population began to increase.

The mackerel fishery brought numbers of American vessels to Prince Edward Island to obtain supplies and provisions and to hire labor. Reciprocity resulted in increased sales of agricultural products to the United States. With the growth of the settlements went a growing interest in the fishery. The increase in shipbuilding and the system of bounty support for the vessel fishery introduced in 1851 made for expansion;[98] and in 1859 many ships in and about the Straits of Belle Isle

L'Abbé Ferland, *Opuscules* (Quebec, 1877); Paul Hubert, *Les Iles de la Madeleine* (Rimouski, 1926), pp. 111 ff.

96. *Report of the Special Committee on the Magdalen Islands and the Western Part of this Province above Lake Huron* (Quebec, 1853); *Journals of the Legislative Assembly of Canada*, 1859, Appendix No. 20, especially pp. 47–74; G. Sutherland, *The Magdalen Islands* (Charlottetown, 1863); Faucher de Saint-Maurice, *De tribord à babord* (Montreal, 1877), p. 204; also V. A. Huard, *Labrador et Anticosti* (Montreal, 1897), p. 440. Agriculture was limited. "Like most fishing coasts where communication by water is easy, and the making of roads difficult and demanding labour at the time fishing is profitable, the roads on these islands have been too much neglected."

97. Hubert, *op. cit.*, pp. 132 ff. See also *Reports of Special Committees on the Magdalen Islands in 1853 and 1859*, and of the Legislative Assembly of Quebec in 1872 and 1875; and Antoine Bernard, "Les Iles de la Madeleine," *Transactions of the Royal Society of Canada*, 1934, sec. 1, pp. 15–45.

98. "Downing Street, 26 May, 1853.

"With regard to the question of promoting the Fisheries of the British Colonies by the means of bounties, Her Majesty's Government although desirous not to sanction any unnecessary deviation from that policy which regulates the Commerce of this Country, are still disinclined to prevent those Colonies, by the interposition of Imperial Authority, (and especially pending the negotiations with the United States of America for the settlement of the principles on which the Commerce with the British North American Colonies is hereafter to be carried on) from adopting the policy which they may deem most conducive to their own welfare and prosperity. Entertaining these views it is the intention of Her Majesty's Government to advise the Queen to give Her assent to an Act passed by the Legislature of Prince Edward Island in the Session of 1851, for the promotion of its deep sea Fisheries, and they will be prepared to give favourable consideration to any Acts for a similar purpose which may be passed by the Legislatures of the other North American Colonies.

"JOHN S. PAKINGTON
"Secretary of State for the Colonies."
Journals of the Assembly, New Brunswick, 1853.

were from Prince Edward Island. "It is only in the last five or six years that the inhabitants of the island have entered into the fishing operations on a large scale. But they now possess a fleet of fishing vessels quite as well equipped as those of the United States and Nova Scotia."

With the abrogation of the Reciprocity Treaty, vessels were sold and the boat fishery increased. Customs regulations introduced and enforced in 1868 by the new Confederation government in New Brunswick, Nova Scotia, and Quebec compelled American fishermen to resort to Prince Edward Island.[99] Maintaining control over its own activities, it was able to take advantage of the freedom from the legislation imposed by more powerful regions; and it was more profitable to handle transshipments of American fish and to engage in American trade than to engage in the fishery.

> During the continuance of the Reciprocity Treaty, and even during the season of 1867, a very large and lucrative trade and business, extending a distance of 25 miles interiorly from the Strait of Canso, had existed between the merchants and inhabitants of the County of Guysborough, and the American fishermen passing through the Strait. This trade and business consisted in the sale to the Americans of very many thousands of barrels manufactured by the people of that county; in the sale of salt, bait and necessary fishing and other supplies, in the storage of the cargoes and materials of such vessels, and in the refitting of the same. This trade and business has rendered the western side of the Strait of Canso (embracing three convenient harbors and forming a portion of the County of Guysborough), the constant resort of American fishing vessels, and a very prosperous and progressive section of the province. During the present season the Department of Customs, through its officers, by a strict construction of the Treaty between Great Britain and the United States have put a stop to all commercial intercourse between the American fishermen and the constituents of Mr. Campbell, in consequence of the refusal by the former to pay the tonnage dues now exacted from them. The effect of this prohibition has been to transfer to Prince Edward Island the whole of the advantageous trade heretofore subsisting, and as a natural consequence a very serious depression at this moment exists in Guysborough.[100]

On August 23, 1870, resultant upon a memorandum presented on August 4, following a request by Canada to Great Britain of May 21, 1870, instructions were issued to customshouse officers in Prince Edward Island to refuse admittance to United States vessels. The Execu-

99. See an account of a visit by an American delegation to Prince Edward Island in 1868 with a view to a reciprocity arrangement. D. C. Harvey, "Confederation in Prince Edward Island," *Canadian Historical Review*, June, 1933, pp. 123–136.

100. *S.P.* No. 12, 1869.

tive Council of the island protested in a minute of September 2, 1870;[101] and, as a result of support from the Imperial government,[102] the instructions of August 23 were withdrawn on November 17 in a notice which directed "that such vessels shall be permitted to enter, tranship their cargoes of fish, and obtain supplies in the ports of this colony—from this date and until further orders." This was followed by demands for insistence on a strict interpretation of treaty rights by Great Britain.[103] The divergent interests of Prince Edward Island as regarded

101. ". . . Second. Lest it should be supposed that the people of this Island alone of all the Maritime Provinces of British North America, have deliberately and with the connivance of their Local Government, carried on an illegal but lucrative business, the Council remind your Honor, for the information of the Secretary of State, that the practice referred to in your Honor's Minute, has until a recent period, been permitted in the Strait of Canso, that the New Brunswick Railway has transported large quantities of fish of foreign take, and that Her Majesty's Representatives could not fail to be cognizant of the practice of transhipping cargoes, and supplying foreign fishing vessels. Moreover no attempt at concealment thereof was made in the summer of 1869, during the visits of the two Vice-Admirals and several Commanders of Her Majesty's Ships to Charlottetown Harbor, consequently, it is not surprising that merchants and traders in this Colony should regard the practice referred to without suspicion of its illegality. . . .

"5th. Having thus acquitted themselves of their duty and caused the law to be carried into effect, though at a sacrifice to their fellow Colonists, which will be little appreciated elsewhere, the Council feel bound to protest against the policy now re-adopted. . . .

"Fairly stated, the old policy revived, demands from the people of Prince Edward Island, the exclusion from their harbour of their best customers—customers who have employed the colonial marine in importing salt for their use; the colonial mechanics in manufacturing their barrels; customers who have purchased their clothing, their provisions and their sea-stores in the Island markets. These men are to be expelled until the forty million citizens of the United States succumb to the pressure put upon them by four million colonists, and consent to concede reciprocity in exchange for free access to the fishing grounds and harbors of the colonies." *S.P.* No. 12, 1871.

102. Lord Kimberley instructed the lieutenant governor to relax the restriction on American fishermen. "The transhipment of fish and the obtaining of supplies by the United States fishing vessels in the ports of the colony cannot be regarded as a substantial invasion of British rights." *Idem.*

103. "Downing Street, 16th January 1871.

"It appears from the correspondence before them, that the Government of Prince Edward Island, while admitting the correctness of the Canadian interpretation of the Treaty, is disposed to make concessions, with the avowed object of fostering a trade with the American trespassers, which is advantageous to individuals who have no interest in the Fisheries. Her Majesty's Government may not be aware that the inhabitants of Prince Edward Island have engaged in the Fisheries to a very limited extent, and that Charlottetown has been the headquarters of the American trespassers. The Committee of the Privy Council readily acknowledge that there are persons in Canada who would be very desirous that an illegal traffic, by which they would derive profit, should be encouraged. There have always been persons interested in smuggling and in poaching, who although not immediately engaged in such pursuits, have nevertheless profited by them and indirectly encouraged them. The Committee of the Privy Council have only, in conclusion, to express their firm conviction that Her Majesty's Government will adhere to the stipulations of the Treaty of

the fishery contributed to her delay in entering Confederation even as they contributed to the failure of Newfoundland to enter it.

CANADA, 1867–1886

Canada considers inshore fisheries her property and that they cannot be sold without her consent.

SIR JOHN MACDONALD TO SIR CHARLES TUPPER, APRIL, 1869

THE change in the Quebec Resolutions at the London Conference in December, 1866,[104] which gave the federal government control over seacoast and inland fisheries, reflected the importance of the new instrument of Confederation as a means of resisting New England. With the ending of reciprocity, Nova Scotia, New Brunswick, and Quebec took refuge behind an organization more efficient for the checking of encroachments on British fishing grounds, the prevention of smuggling, and bargaining for a new treaty. An act was passed in 1868 incorporating the main features of earlier maritime legislation. Some of them were as follows:

1. The Governor may grant licenses to fish within three miles of the coast.
2. Any one of a number of specified officers may go on board of any vessel within any harbor of Canada, or hovering (in British waters) within three marine miles of any of the coasts, bays, creeks, or harbors in Canada, and stay on board as long as she may remain within such place or distance.
3. If such vessel shall be bound elsewhere and shall continue within such harbor or so hovering for twenty-four hours after it shall have been required to depart, the officer may bring her into port, search her cargo, and examine the master on oath touching her voyage and cargo; if the master do not truly answer the questions put to him, he shall forfeit four hundred dollars; if the vessel be foreign and have been found fishing, or preparing to fish, or to have been fishing, within three marine miles of any of the coasts, bays, creeks, or

1818, which, in their judgment, cannot be abrogated without the consent of Canada. It appears to the Committee of the Privy Council that if the Government of the United States should make any complaint of the stringency of the regulations for the protection of the British Fisheries, Her Majesty's Government will be enabled to state in reply that they have learned from the reports of the Naval Officers on the North American Station, that there has been systematic trespassing by American fishing vessels in the water in which they expressly renounced all right of fishing by the Treaty of 1818; and that Her Majesty's subjects in British America have good reason to claim a strict adherence to Treaty rights, when the abandonment of such rights would obviously encourage the illicit trade which is openly carried on." Dispatch from Sir John S. Pakington, Secretary of State for the Colonies. Privy Council Chambers, February 17, 1871. *Idem.*

104. J. A. Maxwell, *Federal Subsidies to the Provincial Governments in Canada* (Cambridge, 1937), p. 19.

harbors of Canada not included within the above mentioned limits, without a license, the vessel, stores and cargo shall be forfeited.

Provision is then made for the proceedings upon seizure. Customs regulations introduced and enforced in 1868 paralleled these regulations for the fishery.[105]

License fees were increased from 50 cents to $2 a ton in 1868,[106] on the insistence of Sir Charles Tupper, representing Nova Scotia,[107] the result being that the larger American fishing vessels accepted the challenge to run the risk of capture and smuggling increased.

The strong interest that both the resident British traders and the United States fishermen have in maintaining the trade would in my opinion render its suppression extremely difficult even were it thought judicious to continue the attempt; whilst the combination between these two to evade British law, and the sympathies arising therefrom must be very undesirable. The sympathies of the inhabitants are entirely with the Americans as the American schooners are principally manned by men who are natives of the Strait of Canso. The store-keepers secure large profits from their intercourse with the crews. Every facility is given in the ports of the island to foreigners for obtaining and replenishing their stock of stores and necessaries for fishing.[108]

105. See *S.P.* No. 12, 1869, and No. 101a, 1885.

106. Three hundred and sixty-five American vessels, totaling 19,355 tons, paid licenses of 50 cents a ton in Nova Scotia, New Brunswick, and Canada in 1866. The fee was raised to $1 per ton in Nova Scotia in 1867, and to $2 in 1868. A cruiser was employed in 1867.

Province Issuing License	Number of Vessels Licensed 1866	1867	1868	1869
Nova Scotia	354	269	49	16
New Brunswick	1	..	..	2
Quebec	10	..	7	7
Prince Edward Island	89	26	5	6

For a list of licenses issued to American vessels, see *Documents and Proceedings of the Halifax Commission* (Washington, 1878), pp. 197–218; also *S.P.* No. 12, 1869.

107. Sir Charles Tupper to Sir John Macdonald, April 18, 1868: "I think I also satisfied his Grace that assent ought to be immediately given to raising the fishing licenses to two dollars, and doing away with the present arrangement as to notices. . . . Lord Stanley's policy is evidently one of abject dread of the United States, and to give them anything British American they ask. I have presented in the strongest terms the fact that the licensing was only assented to by the colonies for a single year and that the plan proposed is practically to abandon the fisheries altogether, and keep up the existing restrictions on trade and promote continued difficulty with the United States. That the policy we propose would lead to a renewal of reciprocity and settle the whole question permanently." On May 9, 1868, Tupper was informed "that Her Majesty's government have assented to the adoption this year in the Dominion of Canada of a fee of two dollars a ton on licenses to vessels to fish in Canadian waters." See *The Life and Letters of the Right Honourable Sir Charles Tupper, Bart., K.C.M.G.*, ed. E. M. Saunders (Toronto, 1916), I, 164, 166.

108. Report of Vice Admiral Fanshawe, November 22, 1870. *S.P.* No. 5, 1871.

From Cape Sable to St. Mary, Nova Scotians were anxious to sell stores, bait, ice, and frozen herring. "As a consequence of the continued indulgence towards the Americans [after the abrogation of the Reciprocity Treaty] very few colonial fishermen are engaged in fishing owing to the almost prohibitory tariff imposed by the United States on fish imported in colonial vessels and colonial fishermen therefore in considerable numbers man American vessels."[109]

The advantages of the Nova Scotian policy of rigidly excluding Americans from British fishing grounds were conceded in the discontinuance of the Canadian policy of license fees, and in the issue of an order in council on January 8, 1870, which gave notice that "henceforth all foreign fishermen shall be prevented from fishing the waters of Canada." Special instructions were issued on May 14, 1870, to officers in command of government vessels engaged as marine police to protect the inshore fisheries.[110] Americans were forbidden to fish in bays less than six miles broad at the mouth but, following a suggestion from Great Britain, the narrower interpretation of the treaty was not insisted upon and new instructions were issued on June 27, 1870. A fleet of six cruisers costing nearly a million dollars was employed, some 400 vessels were boarded for trespass, and 15 were condemned.

Partly as a result of the difficulty[111] of enforcing these instructions, a

109. W. F. Whitaker, *Report on the Fishery Articles of Treaties between Great Britain and the United States* (Ottawa, 1870). Shipping interests were involved along with fishing interests. "These fish are cured and packed on board the fishing vessels directly they are caught, it being necessary to do so in order to preserve them. As the nearest principal markets for the sale of fish are New York and Boston, and as there is a tax in the United States on fish landed from British vessels, but none from American, and as fish are more plentiful near the coast, it follows that the business on this coast is most profitable when it can be carried on close to the shore, and under American colours. Also, as fishing is apparently most profitable under American colours, and there is an American law which prevents a British built ship from ever being able to sail under American colours, it follows that the ship-builders in the United States have a better market for their fishing schooners than the British. On the other hand, it is a disadvantage to the United States Government for vessels to fish under their flag, for they lose the tax which they would get were the fish to be landed at their seaports from British vessels instead of American." Written by the Commander of H.M.S. *Sphinx* at Halifax, November 15, 1870.

110. American vessels were seized and confiscated on the following grounds: (1) fishing within the prescribed limits; (2) anchoring or hovering inshore during calm weather without any ostensible cause, having on board ample supplies of wood and water; (3) lying at anchor and remaining within bays to clean and pack fish; (4) purchasing and bartering bait and preparing to fish; (5) selling goods and buying supplies; (6) landing and transshipping cargoes of fish. The provision allowing vessels to remain in the harbor or to hover for twenty-four hours was struck out.

A United States circular of May 16, 1870, was revoked following a Canadian protest, May 31, 1870, and a revised circular issued on June 9, 1870.

111. See a voluminous literature: W. H. Kerr, *The Fishery Question or American Rights in Canadian Waters* (Montreal, 1868); *The Fishery Question, Letters from the New York Herald's Special Commissioner* (1870); *Review of President Grant's*

Joint High Commission was appointed by Great Britain and the United States, and Sir John Macdonald[112] became a member of it in February, 1871. As in Newfoundland responsible government was recognized as implying control over fishing rights in the dispute with France, so in Canada responsible government and Confederation implied recognition of control in the dispute with the United States. The President of the United States had stated that "the Imperial government is understood to have delegated the whole or a share of its control or jurisdiction of the inshore fishing grounds to the colonial authority known as the Dominion of Canada and this semi-independent but irresponsible agent has exercised its delegated powers in an unfriendly way." The Canadian government asked for an opinion from London, and in answer received a cable from Lord Kimberley which said in part: "We think the right of Canada to exclusive fisheries within the three mile limit beyond dispute and only to be ceded in return for an adequate consideration."[113] The Assembly of Nova Scotia adopted a resolution "protesting against a transfer of the fisheries or sacrificing them to Imperial or Canadian interests." Sir Alexander Galt introduced resolutions in the Canadian House of Commons to the effect "that this house regards the inland fisheries and the navigation of the inland waters of the Dominion as especially within the powers conveyed to the parliament of Canada under the British North America Act and will view with the utmost concern and apprehension any proposal to alter or diminish the just rights of the Dominion in these respects without their consent."[114] Sir Francis Hincks, representing the Canadian point of view, wrote to Sir John Macdonald: "We have no object in refusing the fisheries and the St. Lawrence—on the contrary the fisheries are a mere expense—but we can't yield the fisheries without at least free importation of our fish and free or low duty on coal, lumber and salt, particularly the first."[115] The

Recent Message to the United States Congress Relative to the Canadian Fisheries and the Navigation of the St. Lawrence River (1870); *Return Correspondence between the Government of Canada and the Imperial Government on the Subject of the Fisheries, 20th February 1871; Report on the Practice which Prevailed Previous to the Reciprocity Treaty Respecting United States Fishing Vessels Trading in Provincial Ports* (November 5, 1870).

112. Sir Joseph Pope, *Memoirs of the Right Honourable Sir John Alexander Macdonald, G.C.B.* (Toronto, 1930), chap. xx.

113. For a discussion of the negotiations see G. A. Smith, *The Treaty of Washington,* master's thesis, University of Toronto, 1934; also Allan Nevins, *Hamilton Fish* (New York, 1936), chap. xx, and pp. 917–920; Shippee, *op. cit.,* chaps. xiv–xv; Callahan, *op. cit.,* chaps. xiv–xv; Hincks, *op. cit.,* chap. xx.

114. These resolutions, after achieving their purpose—that of stimulating discussion—were, at the suggestion of Tupper and Howe, withdrawn as unnecessary.

115. See *House of Commons Debates,* May 18, 1868, and April 29, 1869; *S.P.* No. 36, 1866; *S.P.* No. 12, 1868; and *S.P.* No. 18, 1872, for suggestions that Canada hoped to use the fisheries to obtain a renewal of reciprocity.

position of Nova Scotia, as represented by Sir Charles Tupper, was a dominant factor in the negotiations. Macdonald wrote to Tupper:

> My impression was that it would be out of the question for Canada to surrender for all time to come her fishing rights for any compensation. . . . Any surrender must be for a term of years renewable by either party. . . . I spoke also of the means which the exclusive enjoyment of the fisheries gave us of improving our position as a maritime power and that were our fishing grounds used in common by our own and American fishermen the latter would enjoy the same training school as ourselves. Would you put the question concerning an equivalent for the fisheries to the Council?

Tupper wired in response: "Dominion cannot agree to the sale of fishing rights for money consideration." Macdonald instructed Tupper to cable the Colonial Office the message quoted at the opening of this section, that "Canada considers inshore fisheries her property and that they cannot be sold without her consent." To which this reply was made: "Her Majesty's government never had any intention of disposing of the fisheries of Canada without her consent." Sir John wrote: "As the inshore fisheries are admittedly the property of Canada no sale or lease can properly be made by England of our property without our consent. Without such consent the treaty would be binding on England, and the United States would claim our fisheries but it would be a wrong done to Canada without the same plea of necessity which has justified France in agreeing to the cession of Alsace." To this Lord Lisgar replied: "If you can procure in return for free access to the inshore fisheries the free admission into the United States market of fish, and other products of the sea, and an annual tribute of $100,000, you will have drawn yourself handsomely out of the difficulty." Macdonald telegraphed to Tupper: "My colleagues press strongly the necessity for agreeing to grant fishing privileges for a money consideration for a term of years." He "anticipated a good deal of trouble in the fisheries matter from the desire of England to settle with the United States at any price." The Council wired Macdonald: "Neither the government nor the people of Canada will ever consent to concede fishing privileges for even ten years for money consideration, as such sale even though period limited would be regarded by the Canadian people as equal to parting with a portion of the territory of the Dominion." In reply to a request asking, "How much money will Canada take if coasting trade, fish, coal, lumber or salt made free?" Tupper answered: "Council will concur . . . for a term of years and $200,000 per year." In further negotiations, the Council wired: "Canada cannot concede fishing privileges for free fish, coal, salt and lumber." Macdonald made a determined stand against his fellow commissioners, who recommended "the granting of the fisheries for a term of years with notice in exchange for the four free articles." Further pressure by his colleagues on the commission brought a telegram: "Council cannot entertain the proposition to accept a sum of money, or free fish and a sum of money to be settled by arbitration for the inshore fishery for ten years. Either of these proposals would be promptly rejected by Parliament and cause incalculable mischief by creating the impression that the rights of Canada had been sacrificed to Imperial interests."

In the end, however, and in spite of such strong opposition, Great Britain prevailed. She authorized the commission "to negotiate on the basis of free fish and arbitration for an additional sum." Macdonald wrote to Tupper: "My first impulse was to hand in my resignation."[116] "The Queen's government," wrote Sir George Etienne Cartier, "having formally pledged themselves that our fisheries should not be disposed of without our consent, to force us now into a disposal of them, for a sum to be fixed by arbitration and free fish would be a breach of faith, and an indignity never before offered to a great British possession." But, as a result of pressure from Great Britain, the treaty was signed on July 4, to become effective on July 1, 1873, or, in Newfoundland, on June 1, 1874.[117] It provided for the participation of Canadians in the inshore waters of American fisheries and of Americans in the inshore waters of the Canadian fisheries in return for freedom of entry into the United States for Canadian fish, and a sum to be determined by arbitration. Howe, in his "Comedy of Errors speech" in Ottawa early in 1872, which Macdonald described as "more untimely than untrue," referred to England's recent diplomatic efforts "to buy her own peace at the sacrifice of our interests." Protests against the treaty enabled Macdonald to capitalize the discontent by a demand for, and a grant from, Great Britain of a guaranteed loan of £2,500,000.

A stubborn resistance had emerged with the growth of commercial interests in Halifax. It became vocal with the organization and petitions of the Society of Merchants, beginning in 1804 and extending through protests preceding and following the Convention of 1818. It was consolidated by the enactment of legislation and by attempts to enforce rights in the fishery, in the extension of control over Cape Breton, and

116. Macdonald also wrote to Tupper: "I think it well to keep the case of Canada separate—if we come to any satisfactory treaty I shall endeavour to have it limited to the Dominion of Canada so that if Prince Edward Island and Newfoundland desire the advantages of the treaty they must come into Confederation."

117. Caleb Cushing, *The Treaty of Washington* (New York, 1873), Appendix; also Shippee, *op. cit.,* chaps. xvi–xvii.

in arrangements for coöperation with New Brunswick and Canada. And it persisted both as a factor favorable to Confederation and in the determined efforts of Canada to protect rights in the Treaty of Washington and in the later arbitration proceedings. Considering the exposed position of Tupper, and the fact that Howe was a member of the Cabinet, Macdonald as commissioner was compelled to resist strenuously attempts to belittle the importance of the fisheries.

Upon the coming in of a new federal administration under Alexander MacKenzie, an attempt was made to secure the free admission of Canadian products to the American market in lieu of the monetary payment to be determined by arbitration. George Brown and Edward Thornton entered into negotiations with Hamilton Fish, Secretary of State, and drew up a draft reciprocity treaty; but it was rejected by the United States Senate. Galt was then appointed a member of the Fisheries Commission to arbitrate the payment. On May 31, 1875, following his appointment, Macdonald wrote to him, "The value placed by the Maritime Provinces on these fisheries is so large that no award is at all likely to satisfy them and a terrible howl will be raised from that quarter." The Fisheries Commission, sitting at Halifax in 1877, settled upon a final majority award of $5,500,000 as payment for the excess value of the Canadian inshore fisheries over those of the United States.[118] Newfoundland received $1,000,000 and Canada $4,500,000; and $4,000,000 of the latter sum was invested to provide interest for the payment of bounties (45 Vic., c. 18, 1882) at the rate of $150,000 per year, a sum that was later increased to $160,000.

Discontent with the award and the payment of Canadian bounties from its proceeds to increase competition with the American fishery[119] combined with difficulties in Newfoundland to strengthen a demand in the United States for the abrogation of the Washington Treaty. New England complained of the competition from Canadians, "with their soft-wood schooners built at half the cost of the American; the fish at their own doors, with light taxation and free salt, and the great American market free." "The treaty," it was claimed, "gives foreign fishermen a monopoly of the business; it throws us completely out of the

118. See O. D. Skelton, *Life and Times of Sir A. T. Galt* (Toronto, 1920), chap. xvi; also *Correspondence Respecting the Halifax Fisheries Commission* (London, 1878); *Correspondence Respecting the Award of the Halifax Fisheries Commission* (London, 1878); *Further Correspondence* (London, 1878); *Documents and Proceedings of the Halifax Commission* (Washington, 1887); *Record of the Proceedings of the Halifax Fisheries Commission, 1877;* "The North American Fisheries and the Halifax Commission," *Blackwood's Magazine,* 1878, pp. 287–304; Corbett, *op. cit.,* pp. 31–34; and Shippee, *op. cit.,* chap. xx.

119. For a survey of bounties paid see *S.P.* No. 22, 1909, Appendix No. 1.

traffic—it will work to our utter ruin and destruction. The Canadians will clamber into fortunes on our necks."

The expansion of the French fishery owing to the advantage given by the trawl system, bounties, and protected markets, and the growth of the fishery of the United States due to a protected and expanding market provoked measures of defense in Nova Scotia. Losses in the fishery and the growth of smuggling continued to mean a loss of revenue and trade, and led to a rigid enforcement of the Convention of 1818. The Reciprocity Treaty, although opening the markets of the United States to Nova Scotian products, chiefly before the Civil War, made easier the development of the United States fishery, and of smuggling as well. After reciprocity had come to an end, renewed and vigorous efforts to check encroachments from the United States resulted in the Treaty of Washington, which remained in force for twelve years. But the increasing importance of the fresh-fish industry, the decline of the mackerel fishery, the growing part played by the purse seine in the 'seventies, and less dependence on inshore fishing were factors which limited the interest[120] of the United States in any further extension of the treaty. In Nova Scotia's insistence on her rights she turned more and more toward Canada. As a result, a joint resolution of March 3, 1883, gave notice of the termination, after two years, of Treaty of Washington Articles XVIII to XXV which had to do with the fishery.

Confederation and the Treaty of Washington involved not only the opening of the American market but also the expansion of the Canadian.[121] Steamships to Quebec and Montreal and the completion of the Intercolonial Railway contributed to the development of the fresh-fish industry.[122] Lunenburg ships[123] followed American vessels in withdraw-

120. Certain American commercial interests, on the other hand, were favorable to a continuation of a free entry for Canadian fish. The Boston Fish Bureau said, in a statement of September, 1885: "We rely entirely upon the provinces for our stock of fat herring and for the larger part of the cheaper grades of herring, both pickled and smoked, of alewives, salmon, trout and shad. We need the hard-dried codfish of Newfoundland and the choice slack-salted codfish and pollock of Nova Scotia." The large, fat mackerel were obtained chiefly from Nova Scotia and Prince Edward Island. Vessel owners, on the other hand, were in favor of duties. *Correspondence Relative to the Fisheries Question 1885–87* (Ottawa, 1887), pp. 17–18.

121. "Fish is not only a staple article of commerce but also an article of extensive home consumption. The farmer therefore justly considers the annual catch highly important to his interest." Abraham Gesner, *Industrial Resources of Nova Scotia* (Halifax, 1849).

122. See Langelier, *op. cit.,* chaps. vii, x; Richard Nettle, *The Salmon Fisheries of the St. Lawrence and Its Tributaries* (Montreal, 1857).

123. R. F. Grant, *The Canadian Atlantic Fishery* (Toronto, 1934), pp. 9–11. See L. Z. Joncas, *The Fisheries of Canada* (Commercial Union Document No. 6, New York) for a general description of the industry in 1883.

ing from the Labrador; and, in the 'seventies, they adopted the technique of trawl-line fishing on the Banks. As a result of the rise of the fresh-fish industry in the United States, the latter withdrew from the dry-fish markets of the West Indies. The Nova Scotian fishery reached its peak in the early 'eighties as the extension of steamship services brought disaster to the sailing vessels. Nova Scotia profited by the inauguration of a steamship service to the West Indies[124] and Brazil in 1881, the development of the bank fishery, and the payment of bounties which began in 1882.

Having to compete with the New England fishery, supported as it was by bounties and tariffs, and suffering from the loss of preferences in the West Indian markets, Nova Scotia had early turned to the development of increasingly effective machinery designed to check American encroachments on the fishery and on trade. The necessary defensive measures had called for an increasing control over governmental machinery, as had been made plain by the successful struggle of the Assembly against the Council, and by the victory which gave the Assembly control over the customs revenues in the 'forties. The mechanism had in the end involved a realignment of the British colonial system, cooperation with the other provinces, reciprocity, Confederation, and the Treaty of Washington. The success of Nova Scotia, when supported by Canada, coincided with the part played by New England in erecting American tariff barriers. For Newfoundland, the struggle against France, the United States, Nova Scotia, and Canada meant isolation;[125]

124. With the prospect of losses following the abrogation of the Reciprocity Treaty, representatives of the provinces had studied the possibilities of Caribbean trade. *Report of the Commissioners from British North America Appointed to Enquire into the Trade of the West Indies, Mexico and Brazil* (Ottawa, 1866), *S.P.* No. 43, 1866. The steamship service subsidized by the United Kingdom in 1850 terminated in 1886 and was taken over by Canada.

In the period from 1874 to 1883 the foreign West Indies became increasingly important as a market for codfish, and surpassed the British West Indies in 1882 and 1883. Salt and molasses were purchased chiefly in the British West Indies, but sugar most largely first from the foreign West Indies, and later, in 1882 and 1883, from Brazil. *S.P.* No. 67, 1884. Halifax served as an entrepôt for Newfoundland cod for re-export to the West Indies and elsewhere in spite of high charges. These imports met the demand of "vessels not owned in the province, resorting here for cargoes; and such must continue to be the case until more efficient means are adopted to increase the catch of this staple export and protect the rights of our fishermen."

125. The attempt to include Newfoundland in Confederation met with overwhelming defeat. See *Addresses from the Two Houses of Parliament of Canada Praying for the Admission of the Colony of Newfoundland into the Dominion of Canada,* August, 1869. "Newfoundland possessed an independent government and could not bear the idea of becoming a mere appendage to the government of Canada." Thomas Talbot, *Newfoundland* (London, 1882), p. 51. The long hostility of Newfoundland to the

and for Nova Scotia, the struggle against New England, France, and Newfoundland meant Confederation.[126]

The fisheries were inevitably connected with the frontier of controversial constitutional territory represented by the three-mile limit. The controversy was intensified by the effect of American encroachment on the business of the production and exporting of fish, and also by the possibilities of smuggling, particularly, for Nova Scotia, in the case of the mackerel fisheries. She assumed a position in the forefront of the struggle for control over legislation dealing with the fishery,[127] partly as a result of her ancestral and New England tradition of assertiveness which was in turn based on the fishery, and partly as a result of the inherent peculiarities of her own fishery. Her position was strengthened because the American Revolution made flexibility imperative, and this enabled her to press successfully for modifications in the British commercial system. The intensely competitive character of the industry was a driving force which demanded revisions that would place Nova Scotia on a position of equality, if not of superiority, in the legislation of the mother country. The imperialistic outlook of the statesmen of Nova Scotia when dealing with the fishery and world trade was destroyed by the breakdown of the commercial system in Great Britain,[128]

French continued to make itself felt in its attitude toward the French Canadian. Natural resources were "the cornerstone of provincial finances." In the British North America Act, "the resolutions of the Quebec conference of 1864 expressly provided in the case of Newfoundland that, in the event of that colony entering Confederation, its crown lands, mines and minerals would be surrendered to the Federal Government, and the province would receive in consideration of this surrender an annual subsidy of $150,000. But Newfoundland decided to remain outside Confederation and the result therefore was that all original provinces did retain their natural resources." *Report of the Royal Commission on the Transfer of the Natural Resources of Manitoba* (Ottawa, 1929), p. 11; also Chester Martin, *The Natural Resources Question* (Winnipeg, 1920).

126. See W. M. Whitelaw, *The Maritimes and Canada before Confederation* (Toronto, 1934), *passim.*

127. See *Life of Sir Charles Tupper,* chap. 13.

128. Adam Smith favored an Imperial Federation to satisfy the ambitions of public men. "Instead of piddling for the little prizes which are to be found in what may be called the paltry raffle of colony faction; they might then hope, from the presumption which men eventually have in their own ability and good fortune, to draw some of the great prizes which sometimes come from the wheel of the great state lottery of British politics." *Wealth of Nations* (New York, 1937), p. 587. The inability of Howe to secure Imperial posts was, for example, in part a tragedy which grew out of the movement toward responsible government and free trade. "If monarchial institutions are to be preserved," said Howe, "and the power of the Crown maintained, the leading spirits of the empire must be chosen to govern provinces; and the selection must not be confined to the circle of two small islands, to old officers or broken down members of parliament." Under Confederation, Howe was lieutenant governor of Nova

but revived when given the larger scope of Confederation. The sailing vessel with its world outlook was displaced by the steamship and the railway with their contacts on the continent.

Scotia for a brief period and Tupper as a younger man became minister of railways. Howe predicted a transcontinental railway, but it was for the younger man to fulfill the prediction. See J. A. Roy, *op. cit.* For Howe's conception of the place of the fisheries in the empire see *The Speeches and Public Letters of Joseph Howe,* ed. J. A. Chisholm (Halifax, 1909), II, 288, and a reply in Hincks, *op. cit.,* chap. xi.

CHAPTER XII

COMMERCIALISM AND THE NEWFOUNDLAND FISHERY, 1833–1886

EXPANSION OF THE FRENCH FISHERIES

It is on our fisheries that . . . repose all the most serious hopes of our maritime enlistments. . . . The fisheries give employment to a great number of men, whom a laborious navigation under climates of extreme rigour speedily forms to the profession of the sea. No other school can compare with this in preparing them so well, and in numbers so important, for the service of the navy.

M. ANCEL, QUOTED IN *Report of Israel D. Andrews,* WASHINGTON, 1853

AFTER the Napoleonic wars, the elaborate system of bounties developed by France to stimulate recovery and to build up her navy was further extended. In 1846 the apportioned totals were as follows: to fishermen, 558,110 francs; on exported fish, 3,903,910 francs; and on imported cod oil, 19,511 francs. In 1851 bounties were increased.[1] Drawbacks were given in the case of duties on salt used in the curing of the fish and on outfits for the fishery, including vessels. Duties of 7 francs a quintal

1. BOUNTIES PAID TO SUPPORT THE FRENCH FISHERY

Per man, for those outfitted for the fishery on the Newfoundland coast, St. Pierre and Miquelon, and the Grand Bank,	50 *frs*
Per man for the fishery in the Iceland seas, without drying fish	50 "
Per man, on the Grand Bank, without drying	30 "
Per man, on the Dogger Bank, without drying	15 "

BOUNTIES ON EXPORTED FISH
per Metric Quintal

Dry codfish, of French catch, exported directly from the coast of Newfoundland, St. Pierre or Miquelon; or warehoused in France and exported to the French colonies or to transatlantic ports having a French consul,	20 "
Dry codfish, not warehoused, exported from French ports,	16 "
Dry codfish carried directly from fishing regions to ports of France, Portugal, and Spain, or to other foreign ports in the Mediterranean,	16 "
Dry codfish carried directly to ports of France, and thence to Sardinia and Algeria,	12 "

See Léon de Seilhac, *Marins pêcheurs* (Paris, 1899), pp. 115–116; *Journals of the Legislative Assembly of Canada 1859,* Appendix No. 20; *Report of Israel D. Andrews* (Washington, 1853), pp. 661–685; *Journals of the Assembly, Newfoundland,* 1850, Appendix, pp. 111 ff.; G. R. Young, *Letters to the Right Honorable E. G. S. Stanley, M.P.* (London, 1834), p. 66; also *Montreal Gazette,* April 26, 1834. For a detailed account see Henry Schlacther, *La Grande pêche maritime* (Paris, 1902), *passim.*

on fish in 1845 were lowered to 3 francs in 1860. The effectiveness of government assistance[2] was apparent in the increased use of large-scale methods of operation such as the setline, or the trawl or "bultow" (boulter).

> The method of fishing with set lines is as follows. A small-sized rope or stout cod line, according to the depth of water, is provided, varying in length from half a mile to as much as five miles, according to the ground on which the line is to be set out. These lines are termed by the French bultow lines, and by the American fishermen thrott or throat lines. Each vessel following this method of fishing has a number of lines, according to her size. To the line, about every three feet, there is attached a "guaging" or lanyard. These are about a foot and a half or two feet in length, and the cod hooks are made fast to them. With all hooks baited the line is run out by the men in their boats and sunk to the bottom, buoys being fastened to it at different distances to denote its locality, and make it possible to haul it up. It is allowed to lie there a certain length of time, generally determined by how the cod are running. Boats from the vessel go to the buoys, haul up the line, take off what fish are on the hooks, re-bait them, and let the line down again, and so continue while there are fish to be taken in quantity, or till the "fare" is made up. The writer has been informed, incredible as it may appear, that some of these lines have as many as ten thousand hooks fastened to them.[3]

It was estimated that a vessel of 300 tons with a crew of 40 men and four or five large boats was capable of laying out 5,000 fathoms of baited setline.

Encouragement by bounties and improved methods of catching fish led to a rapid extension of the fishery in the region of the Banks. It was extended from the Grand Banks to more distant banks, either, as alleged, because of the depletion of the old grounds, or from a more aggressive development of the industry.

> The French have so seriously injured the fishing grounds of the Grand Bank and other places where they have had the privilege of fishing that they have been forced to search for other grounds; and for a number of seasons past several of their largest vessels have been fishing on the Sable Island Banks, up to the very western edge of them. Bank Quereau, to the north-east of Sable Island, one of the best fishing banks, found a few years since, has

2. "Set line fishing was, I believe, first practised, to any extent, by the French fishermen in the Newfoundland fisheries, and principally on the Grand Bank, and is one of the evils produced by a bounty system—the natural offspring of a vile parent. Set line fishing, there can be but little doubt, was induced by the enormous bounty of ten francs paid by the French Government for every quintal of fish caught by their fishermen." *Journals of the Assembly, Nova Scotia,* 1861, Appendix No. 32.

3. *Idem.* See also M. G. Massenet, *Technique et pratique des grandes pêches maritimes* (Paris, 1913), chap. iii.

been completely ruined by set line fishing, first practised on it by French and latterly by United States fishermen.[4]

Vessels came out in the spring to St. Pierre and on the first trip took herring bait.[5] Captain Milne reported on June 21, 1840, that 80 or 90 French vessels of from 120 to 300 tons were arriving at St. Pierre from the Banks and discharging their cargoes of fish in preparation for the second fishery. Capelin bait was used for this; and these French ships returned in September and made a third "fare" with squid bait. Jukes,[5a] in October, 1839, reported that ships and fishermen were returning to France for the winter, leaving a small number of residents. They employed large vessels carrying regular crews, with every man numbered, and a regular system of catching and curing. Fish were dried in St. Pierre or taken green to France to be dried and exported to the West Indies and the Mediterranean.

Of the large outfitters for the bank fishery, four were located at Granville, two at St. Servan, four at St. Malo, and one each at Morlaix, Bordeaux, and Bayonne. Bordeaux[6] occupied an increasingly important

4. *Journals of the Assembly, Nova Scotia,* 1861, Appendix No. 32.
5. Captain Erskine, *Report on the Newfoundland and Labrador Fisheries* (St. John's, 1875).
5a. J. B. Jukes, *Excursions in and about Newfoundland during the Years 1839 and 1840* (London, 1842).
6. BORDEAUX IMPORTS OF COD
in 1,000 kilos

	From Banks		Total from All Sources
	First Fishery	Second Fishery	including Iceland
1875	2,388	7,315	13,030
1876	2,052	9,323	13,667
1877	2,177	9,501	13,939
1878	2,291	12,339	17,054
1879	2,873	12,707	18,179
1880	3,282	10,768	17,671
1881	2,319	10,975	15,721
1882	2,993	15,599	22,292
1883	3,466	18,250	24,417
1884	4,739	20,157	28,318
1885	4,711	24,957	32,794
1886	4,420	28,754	36,456

French vessels arrived at Bordeaux from the Banks from June to the end of the year, with fish in salt bulk. Twenty-five drying grounds, each with an area of 75,000 square feet, were available at Bègles; and with three days of drying weather the fish were ready for market. The cod lost about 25 per cent in weight, and was not dried as much as Norwegian or Newfoundland fish. After curing, it was packed in 600-pound casks for the French West Indies and in 100-pound barrels for European markets. The operations were carried out by the Syndicat du Commerce de la morue, which consisted of thirty-one merchants, twenty dryers, and six consigners.

See *Report on the French Fisheries on the Great Bank of Newfoundland and off*

position as a drying center because of its facility of access to the West Indies, Portugal, Spain, Northern Africa, Italy, and Greece. "About 150 vessels varying from 100 to 350 tons came annually from France, which, with 50 schooners and 500 boats belonging to St. Pierre and Miquelon, employ about 12,000 men, more than a half of whom are enrolled seamen."[7] The tonnage fitted out from France for the fishery ranged from 18,000 to 19,000 in the 'forties and increased to nearly 30,000 by 1886.[8]

The islands experienced a steady increase of population. In 1820 the figure was 800; in 1831, 1,100; in 1848, 2,130; in 1860, 2,910; in 1870, 4,750; in 1880, 4,916; and in 1887, 5,929. By 1858 twelve mercantile houses were in operation at St. Pierre and Miquelon, of which one half were connected with the fishery, the largest being the General Maritime Company of Paris.

The expansion of the setline, trawl, or bultow system meant a marked increase in the demand for bait and extensive purchases from the English on the south shore of Newfoundland. Trade in bait was accompanied by smuggling.

> The French from St. Pierre carry on and encourage the English in an illegal traffic in bait, from Fortune Bay, Burin, and Placentia Bay, and it is supposed that the sum of £20,000 is annually paid by them to the people on these coasts for bait alone.[9] . . . Since my residence in Lamaline the price of herring has been as high as 25 francs per barrel, filled loose, or 20s. 10d.; and the average may be taken at 10 francs, [or] 8s. 4d., which itself is a high price for fresh herring, as the French prefer taking them as they are taken from the seine or net. Late in the season, from the 1st to the 20th May, the price is generally low, being from 2 to 8 francs. The price of caplin has been from 20 francs per hogshead of two barrels, to 5 francs; but the average may be estimated at about 9 francs. The quantity of herring annually sold for bait is at least 21,000 barrels; and that of caplin 20,000 hogsheads; which, at the average price I have stated, would amount to £16,200.[10] British subjects from Placentia and Fortune Bay supply herring for the Bankers to the extent of 57,130 barrels, and receive . . . about 15 francs per barrel on

Iceland, Diplomatic and Consular Reports, No. 492, Miscellaneous Series (London, 1899).

7. For an excellent account of a trip from Granville in 1876, see *Sur le Grand-Banc. Pêcheurs de Terre-Neuve, récit d'un ancien pêcheur* (Paris, 1905); also Le Comte A. de Gobineau, *Voyage à Terre-Neuve* (Paris, 1864), chap. ii.

8. From 1880 to 1886 inclusive the number of French vessels on the Grand Bank, with their total tonnage, was as follows: 1880, 147 vessels totaling 23,588 tons; 1881, 137 vessels and 21,083 tons; 1882, 156 and 23,824 tons; 1883, 160 and 23,292 tons; 1884, 178 and 28,140 tons; 1885, 174 and 28,281 tons; and 1886, 187 and 20,337 tons.

9. *Journals of the Assembly, Newfoundland,* 1851, Appendix, p. 151.

10. *Idem,* p. 165.

an average, making £39,276; and about 60,000 hhds. caplin at 6 francs per hhd. would give £16,550, making a total of £55,826, paid by the French for bait alone. On my return to St. Peters [St. Pierre] in August (1856), I saw 80,000 squids sold at 40 francs per M., brought from Placentia Bay. . . . Persons at St. Peter's[11] supply many in Placentia and Fortune Bays, with cordage, canvas, tea, tobacco, spirits, &c., to a very considerable extent, and take fish and oil in return, which latter articles are shipped direct from Newfoundland in British and American bottoms to the United States and British Colonies. Much is sent without clearing at the Custom House, and the goods so supplied pay scarcely any duty in this Island, on account of such articles being supplied to fishing boats carrying bait, which boats do not enter at any Custom House; nearly all the boats are supplied with cordage and canvas, on account of the cheapness at St. Peters.

Suppliers at St. Peter's will ship more largely from Newfoundland this present year to the American market; last year they shipped from Burin alone 4,000 qtls., and 6,000 from other places on our shores, and their trade will considerably increase on our coast, and cause a corresponding decrease with St. John's while inducements are found at St. Peter's, and those inducements arise principally from parties purchasing and landing goods without paying duty. Firewood to the amount of £2,500 Stg. is sold at St. Peter's by persons from Newfoundland.

At St. Peters new cordage is sold at from 6d. to 7d. per lb. It cannot be bought in any of the Bays for less than 1s. 3d., and it is also high at St. John's;—here then is great inducement to deal there largely in these heavy articles. The reason why the French can supply cordage cheaper is because of the exportation from France having been prohibited during the Russian War, except to their own colonies. This has kept the price low in France, hence St. Peter's is able to furnish these articles so cheap.[12]

On the French Shore of northern Newfoundland, expansion followed an ample support by bounties.

The French resort to this coast in Spring, bringing out with them in their vessels (which are chiefly brigs of 100 to 200 tons) a cargo of salt for curing their fish, as also their implements for fishing. Having moored their vessels in security, they commence the repairs of their salting-houses, drying stages, and likewise the huts for the abode of their crews. The former are covered

11. English fishermen defrauded their merchants in St. John's and Conception Bay by selling both bait and fish to the French. "The practice of supplying the French with Fish has now become so common that the French merchants give provisions, &c., in advance for bait and Codfish; the latter is collected at the end of the season, and sent either to Halifax or St. John's, Newfoundland, and thence to Europe, the French not being allowed to land dried fish procured from English Fishermen at St. Pierre." *Idem,* 1856, Appendix, p. 91.

12. "The outports of Newfoundland are thus in part supplied with French spirits and manufactures which are introduced of course free of duty. The Magdalen Islands are a celebrated depot for this contraband traffic." *Idem,* 1857, Appendix, p. 357.

with canvass. The ship's bell is landed, and shipped in its former cranks near the superintendent's house. The large boats, which were hauled up in places of security at the close of the last year's fishing, are launched, repaired, and fitted. In all their arrangements and mode of prosecuting the fishery there was displayed system and neatness which we observed nowhere else. Each vessel has from six to ten of these boats, according to the number of their crew. They are of large dimensions, being about 25 to 30 feet long, with a great beam, and all rigged alike, with two lug sails. Their crew consists of two men and a boy: they start in early morning to their Fishing Grounds, which are generally at the entrance of the harbour, and continue to prosecute their avocations with hooks and line until they procure a cargo, when they return to their establishments: the crew are then relieved, and the fresh hands immediately commence throwing the fish into the salting-house where the process of splitting, boneing, cleaning, and salting is prosecuted with amazing quickness. . . . The fish are not laid out to dry until about three weeks before their departure for France, so certain are they on this northern coast that no fogs which are so prevalent and dense on the south shore of the island, will continue to interrupt the process of the fishery. . . . It is on this account that the French fishing ground is considered so much more valuable than the English. Besides the hook and line, cod-seines are also used,[13] with which they take immense quantities of cod.[14]

From all that I could learn respecting the French fisheries, I was led to the belief that they were more of a governmental affair than of private mercantile enterprise; and indeed the people do not deny it, but freely acknowledge that the intention of their government in sustaining the fisheries on the coast of Newfoundland by the bestowment of liberal bounties, is solely to secure a sufficient number of seamen. They are shipped at a fixed rate of wages, averaging from two to six hundred francs, and the boys, who chiefly compose the shore crews, receive from fifty to sixty francs each; they derive no advantage from the bounty, nor the catch of fish, with the exception of those men who are induced to venture into British waters to fish; to them the employers make some allowance, but I could not correctly ascertain in what shape or form this gratuity was given. The vessels employed in the shore fishery are of three classes and chiefly large tonnage, some of them exceeding four hundred tons. By law the first or largest class are required to employ fifty men and two seines, the second, thirty-five men, and the third class, twenty-five men, with one seine to each vessel, but they are not restricted as regards an increase in the number of men to each vessel, only as regards the number of seines. The number of vessels harboured between Cape St. John and Cape

13. They were described as "chiefly very large nets which are nearly all 150 fathoms long and 30 fathoms wide. Nearly forty men [are] required to handle them successfully. They are very costly." "Report of Pierre Fortin for 1852," *Journals of the Legislative Assembly, Quebec,* 1853, Appendix iiii.

14. *Journals of the Assembly, Nova Scotia,* 1841, Appendix No. 62. Philip Tocque, *Newfoundland* (Toronto, 1878), pp. 294–296.

St. George this year [1852] is one hundred and fifty-nine, and bankers, within the same range of coast, about one hundred, employing nearly fourteen thousand men and boys; the increase over last year being seventeen vessels and about one thousand men.[15]

The French fisheries are conducted upon a principle entirely different from our own in every respect; rules for their guidance in the minutest particular are laid down, adopted by the merchants interested and approved of by the Ministers of the Marine Royale. French vessels cannot fish on their own coast where they please, as our own can; each harbour is classified as fit for a vessel of the first, second, or third class, according to their tonnage; and when the harbor is large it is divided into parts; each harbor or part is drawn for in France by the different merchants every four years, and they have no right in any other.[16] Codroy Harbor is an exception to this rule, it being always reserved for the craft belonging to the islands of St. Pierre and Miquelon; there are certain harbors either too poor to be drawn for, or, if drawn, not suitable to carry on a successful fishery, if the vessels were confined to these alone; these are at present carried on, trusting for the completion of their catch to the Labrador; and I doubt much if any of these vessels would continue to come, provided decisive measures were taken in reference to this latter fishery.[17] The French have ruined their fishery by the abuse of the use of cod-seines, and it is the intention of those interested to get a law passed to do away with them if not altogether, at least for a certain number of years, so as not to scare off the fish to the northern coast of Labrador; and in the meantime, to use Belle Isle for cod-seines, and the Labrador to make up their voyages.[18]

The survey of Captain Alexander Milne in 1840 reported 10 or 12 French brigs at La Scie, and 5 vessels and four French establishments at Pacquet.[19] A later survey in 1848 described the French at Croc as having two rooms with 37 boats and 130 men. Of the men, 100 worked

15. *Journals of the Assembly, Newfoundland,* 1853, Appendix, pp. 138–139.

16. In the decree of 1852 vessels were divided into classes: (1) more than 149 tons and 30 men, (2) 90 to 149 tons and 25 men, and (3) less than 90 tons and 20 men. They had to draw lots every five years for the various allotted posts. These regulations restricted the development of the ports. Émile Hervé, *Le French-Shore* (Rennes, 1905), pp. 23–24; also *Journals of the Assembly, Newfoundland,* 1858, Appendix, pp. 419–435; and *idem,* 1859, pp. 619 ff.

17. *Idem,* Appendix, p. 159.

18. *Idem,* 1851, Appendix, p. 157. "In 1852 the question concerning Bultows was seriously debated in the General Assembly of the Ship owners. The use of these lines was adopted by twenty-nine against twenty-five for the Eastern Coast, and unanimously for the Western Coast." *Idem,* 1860, Appendix, pp. 473 ff.

19. "There was a peculiarity noticed in the French boats at Pacquet Harbour which I never observed elsewhere, namely, both oars being pulled on the larboard side, and no oar at all on the starboard; but the third person used an oar fitted into a crutch placed broad on the starboard quarter, with which he pulled the stern round against the power of the two larboard oars, and thus steers her course." *Journals of the Assembly, Nova Scotia,* September 25, 1840.

with boats and 30 on shore. They also had 6 boats for taking capelin and herring for bait. These boats were manned by eight-man crews who used seines principally. Other harbors included Rouge, St. Julien, Goose Cove, Cremaillere, Brehat, and Quirpon.[20] Belle Isle was alleged to be in the hands of the French and to have furnished 30,000 quintals.[21] Even the English area on the Labrador between Forteau and Red Bay was visited by from 1,000 to 1,500 fishermen in 200 vessels, and it was claimed that the grounds were being exhausted. The French hired staging from the English for the cod fishery in return for the livers. On the French Shore from Cape St. John to Cape Ray it was said that 11,000 were employed. French schooners left St. George Bay with cargoes of fish in November and returned in May. The fishery was conducted by interests from Granville, St. Malo, Paimpol, Binic, Havre, Nantes, and St. Brieuc, and they were said to have much greater weight and influence than those in the southern fishery; but there was evidence of decline by the 'seventies.[22]

The total French exports of cod from St. Pierre and Miquelon increased from 8,305,765 kilos of dry fish and 2,085,303 kilos of green fish in 1850 to 11,198,342 kilos of dry fish and 35,042,475 kilos of

20. In 1852, Quirpon had seven establishments, chiefly from St. Servan and St. Malo, with 18 ships, of from 200 to 500 tons. Fish were sent to the Bourbon Islands and Mauritius. "There was one French brig, the *Concorde*, of St. Brieuc, in St. Lunaire Bay, and there are two French rooms and stages for curing and drying fish. . . . I counted fifteen French vessels, brigs and barques, anchored in the different creeks and bays between St. Lunaire and Cape Bauld, and off the Maria Rocks. I was informed these vessels are from 150 to 300 tons burthen, and have from 30 to 60 men each, ten boats and two cod seines to the larger, and six boats and one cod seine to the smaller vessels, besides herring nets and capelin seines. The larger of these vessels generally take away about 3,000 quintals of cod, and the smaller about 2,000 quintals." *Journals of the Assembly, Newfoundland,* 1851, Appendix, p. 145.

21. "Five or six French brigs had been fishing this year at Green Island but had left before my arrival. When the fishing slackens on the Newfoundland shore they generally encroach on the Labrador side." *Idem,* p. 150. "The fishery on the coast of Labrador is carried on in boats, the vessels lying at anchor in the harbours. The fish are chiefly taken near the shore, say within a mile. The French vessels lying on the north side of Newfoundland have Shallops which they send to the Labrador Coasts, but chiefly fish in deep water in the Straits where they catch the largest fish." *Journals of the Assembly, Nova Scotia,* 1837, Appendix No. 75. At Black Bay, French vessels from St. Pierre had been employed in the early 'fifties in taking capelin for the Grand Banks fishery.

22. In 1875 the French had one schooner, 16 men, seven small boats, six stages, and 450 people on the mainland at Codroy Island. There were engaged in the fishery at Pond's River, Kepple's Island, three vessels and 200 men; Port au Choix, five vessels and 550 men; St. John's Island, four vessels and 450 men; Englée Cove, Cremaillere Bay (unoccupied nine years), Cape Rouge Harbour, seven French rooms, four vessels, and 500 men; St. Anthony, two French rooms, three vessels, with an estimated loss of 4,000 to 5,000 francs a vessel; Rouge Harbour, eight rooms and 300 men. The usual difficulties arose with the English over the salmon fisheries. Captain Erskine, *op. cit.* See also De Gobineau, *op. cit.*

green fish in 1886.[23] The importance of various markets after the collapse of the Norwegian fisheries in 1880 is indicated in the table below.[24]

The expansion of the French fishery was a result of bounties, and of improved technique as illustrated in trawl fishing. This expansion meant the growth of a smuggling trade and likewise competition in fish exports to foreign markets. In 1848 it was claimed that nearly all the bank fish were sent to the West Indies, two thirds going to Martinique and Guadeloupe with a bounty of 11 francs per quintal. In Newfoundland it was held that the average price of Labrador fish, "which is more especially competed with by French bank fish," did not exceed that amount. "The bounty . . . and differential duty on St. Pierre fish entering Spain under the most favoured nation clause in their tariff amount to 12½ francs . . . or to more than the whole value obtained by our fishermen for Labrador fish." The bounty was particularly exasperating during periods of depression and low prices for fish. The British consul at Leghorn reported that "whereas as recently as 1883 it [French fish] was only consumed in small quantities," in the thirteen months ending January 31, 1887, 63,500 quintals had been imported. As a result of the bounty of 16 francs a hundred kilos on exports to Leghorn—for Genoa the bounty was 14 francs and for Naples 11—fish were sent to Leghorn for reëxport to other centers, and this "had a very baneful effect on our own Newfoundland trade." In 1884, because of quarantine regulations, bounties were paid on shipments transferred by rail as well as on fish imported directly by sea in French vessels and from French fishing stations; and the British consul at Naples stated that French imports to Genoa increased from a few barrels in 1884 to 5,300 quintals in 1885, or to 13 per cent of the total imports; and to 19,800 quintals, or 44 per cent, in 1886. Whereas Labrador cod commanded 14 shillings in 1886, French bounty-fed fish brought 20 shillings. French fish were also driving the Newfoundland product out of Valencia, Malaga, and Alicante, the distributing centers for Madrid, Saragossa, and elsewhere. French shippers in Bordeaux "have actually offered and sold fish to Spain for nothing," the Spanish importers paying the duties and cost of carriage. The encroachments of French fish

23. Prowse, *op. cit.*, chaps. xv–xvi.

24. THE FRENCH FISHERIES, 1881–1885

	Cod Fish Taken by French (kilos)	Exported with Bounty (kilos)	Remaining in France (kilos)	Shipments from St. Pierre (quintals)
1881	27,378,700	9,482,171	17,896,529	374,017
1882	25,419,697	8,673,222	16,746,475	411,986
1883	34,395,000	11,653,332	22,741,668	530,045
1884	36,517,000	15,886,818	20,630,182	632,005
1885	53,055,815	19,606,230	33,449,585	820,350

on the markets for Newfoundland fish resulted in a lowering of the price of fish and pressure on markets such as Brazil and other tropical countries not handicapped by prohibitive duties.[25] Competition from French fish in foreign markets led Newfoundland to press for measures calculated to handicap the French industry.

The Newfoundland fishery was subject not only to French but also to Norwegian competition. The average annual exports of cod from Norway increased as follows: 1846–50, 537,450 quintals; 1851–55, 605,737; 1856–60, 666,076; and 1861–65, 751,382. By 1877 it was estimated that production had increased to 1,400,000 quintals. The fishery at Lofoten employed 21,287 men and 4,567 boats.[26] Norwegian stockfish, "dried in the sun until it is as hard as a stick . . . easily portable and carried by mules into the interior of Spain," made substantial inroads in the market for Newfoundland fish.[27] A respite followed the bad year in 1880.

The Portuguese attempted to build up a fishery on the Banks. In 1835 a Lisbon company, the Compania Piscarias, purchased seven English schooners of 100 tons each and shipped Devonshire and Portuguese fishermen in equal numbers. On the first two trips the fish were taken to Fayal to be dried, and taken green to Lisbon on the third trip. By the middle of the century the company had practically disappeared; but duties of 8 shillings in currency and of 15 per cent ad valorem at Oporto and Lisbon, and 10 per cent at Figueria, were still in force.

NEWFOUNDLAND'S DEFENSE MEASURES

It is, therefore, no matter of surprise that the Council so constructed, and influenced by such objects, should have opposed Free Trade with the United States, the introduction of Responsible Government, and the passing of any measure of Representation that would have the effect of obliging them to relinquish their ill-used authority, or submit to the Constitutional checks secured by Responsible Rule. But to attempt to cover their intentions with the assumed garb of Religion is, I do not hesitate to assert, without wishing by any means to wound their pious sensibilities, little more than hypocrisy.

Journals of the Assembly, Newfoundland, 1854

The expansion of the French fishery under the bounty system and through the use of trawls led to the adoption of defense measures in

25. English exports from Newfoundland in 1870 totaled 970,176 quintals, of which 211,222 went to Spain, 167,589 to Portugal, and 249,425 to Brazil.

26. Tocque, *op. cit.,* pp. 330–331.

27. *Journals of the Assembly, Newfoundland,* May 28, 1861; also 1868, Appendix, pp. 837 ff.

Newfoundland which comprised the rise of responsible government, control of the fisheries, the adoption of conservation measures including bait legislation, and the extension of a customs administration on the Labrador. The growth of the Newfoundland fishery and the increasing importance of resident commercial interests brought to an end the influence of the West Country ports, even in the Labrador, and established government machinery designed to stimulate agriculture,[28] industry, and trade by such developments as steamship services, the beginnings of railway construction, and the adoption of tariff protection. The influence of St. John's was extended and consolidated.

French competition in foreign markets and the increase of smuggling from St. Pierre and Miquelon both had serious results, especially when there was an expanding population.[29]

> There appears to have been no decline in the cod fishery; for the average quantity of Cod Fish exported from the first five years to the Return, viz., from 1840 to 1844, inclusive, is 944,372 qtls., and of the last five years, viz., from 1858 to 1862, inclusive, 1,075,687 qtls.; although, as will always be the case, more or less have been caught in certain years; but, it must not be lost sight of, that, although the average quantity caught appears not to have decreased, a great increase has taken place in the population, and, consequently the produce has to support a much larger number of fishermen and their families; and should the population continue to increase, and trust to the fishery for their subsistence, the natural results must inevitably follow.[30]

The limited possibilities of increase in local agricultural[31] production necessitated an expansion of trade based on cod.[32] With the modifi-

28. Revenues obtained from duties on imports were spent as subsidies for ocean steamships, in the construction of lighthouses and roads, and for local steamships and other improvements. In 1881 a contract was signed for the construction of a railway from St. John's to Hall's Bay with branches to Brigus and Harbor Grace. See a statement on the tariff in 1875. Tocque, *op. cit.*, pp. 345 ff. On communications, also see Tocque, *op. cit.*

29. In 1832 the population was 60,008; 1836, 75,094; 1845, 98,703; 1857, 124,288; 1869, 146,536; 1874, 161,374.

30. *Journals of the Assembly, Newfoundland,* January 28, 1864. See also *Report of Israel D. Andrews,* pp. 573–603.

31. "An acre of sea is worth a thousand acres of land." W. F. Rae, *Newfoundland to Manitoba* (New York, 1881). For a description of the agricultural possibilities see R. M. Bonnycastle, *Newfoundland in 1842* (London, 1842). Meat and bread were expensive and much higher in price than fish and biscuit. Horses and cattle were imported from Nova Scotia and Prince Edward Island. See also Joseph Hatton and Moses Harvey, *Newfoundland* (Boston, 1883), pp. 333–344; and Tocque, *op. cit.*, chap. xvii and *passim.*

32. "Such of the inhabitants as are not engaged directly in this trade are so indirectly by supplying the fishermen with the necessaries, and in a good season, with the luxuries of life." The people of St. John's were employed in "fish hauling, varied with fish curing, and a noisome way of extracting seal oil by putrefaction."

cations of the colonial system the trading organization centering in St. John's became of more importance in the matter of imports and exports.

> In former times all Foreign Countries were excluded from this Commerce, but the liberal policy of Great Britain, after the year 1820, admitted the vessels and the produce of Foreign Countries, without distinction, to supply the fishery stations at Newfoundland. The Hanse Towns, and especially Hamburg, have taken an active part in this Commerce; 31 vessels were bound from Hamburg [in 1853] to these British Possessions, carrying principally provisions, as breadstuff, butter, cheese, salted meats, pork, &c.[33]

The dominance of the British mercantile group[34] was steadily weakened as a result of the growth of a more flexible organization.

> The close of the war and consequent fall of the price of fish led to the breaking up of the large mercantile establishments, others failed from various circumstances, the increasing and more stable population drew people with smaller capital to set up stores in a smaller way and opened the door to competition, and the larger houses concentrated their business in St. John's, or a few of the principal places and supplied to merchants in the outports, or any persons who would pay for their goods either in cash, fish or oil. Lastly the number of small peddling schooners trading along the coast, frequently stepping in between the merchant and his planter, and buying the fish from under his nose as it were, acting in concert with the other courses, gradually broke up the old system while political and religious differences completed the alienation between fishermen and the merchant. The fisherman [in the Avalon peninsula] may carry his fish to any one he chooses and though he cannot fix the price at which it shall be sold, as the merchants fix that by common consent from the state of the foreign markcts he has still the great benefit of competition in the choice of the provisions and goods he is to buy.[35]

The fishermen became more bold and self-assertive. At Carbonear and Harbor Grace they met on January 9 and again on February 9, 1832,

> with the avowed object of combining together in order to compel their merchants and employers to adopt a different mode of dealing with them from that which had theretofore constantly prevailed, as much respecting the prices to be given for the fish and oil caught and made by their fishermen, as

33. *Journals of the Assembly, Newfoundland,* 1858, Appendix, p. 428.

34. "The merchants are chiefly of English birth and as the island has no attractions for them it is only tolerable on account of the means it affords of acquiring the wealth whereby they are enabled to live in luxury and magnificence at home." "They must either live in that city [St. John's] or have trust-worthy agents there for the transaction of their business." John Mullaly, *A Trip to Newfoundland* (New York, 1855), p. 47.

35. Jukes, *op. cit.*

to those to be paid by the fishermen for the supplies furnished to them by their merchants and employers. . . . Manifestly [it was] against the law of the land for any number of persons to compel others to receive and pay for their services at such price or rate as the persons so combining may think fitt to prescribe and dictate.[36]

Governmental institutions began to reflect more adequately the influence of the fishermen and the class of small local traders. "The administration of Newfoundland was in a great measure an exclusively mercantile or trading government; which as Adam Smith very justly observes 'is perhaps the very worst of all governments for any country whatever.' "[37] The establishment of representative government in 1832 was followed by the imposition of a 2½ per cent import duty in 1835 to finance expenditures for capital improvements, and this was regarded as an attack on the merchants.[38] In 1838 the Council, representing their

36. *C.O.* (*Colonial Office*), 197:1. For an account of the influence of O'Connell in Newfoundland and of the troubles at Carbonear see Edmund Gosse, *The Life of Philip Henry Gosse* (London, 1890), pp. 81–82. "There existed in Newfoundland in 1827 among the Protestant population of the island, an habitual dread of the Irish as a class, which was more oppressively felt than openly expressed, and there was customary an habitual caution in conversation, to avoid any unguarded expression which might be laid hold of by their jealous enmity." *Idem,* p. 43. Increasing difficulties with the Irish had its part in the decline of West Country influence. By 1838 the firm in which Gosse had been employed had disappeared. *Idem,* p. 105. Irish Catholics ate their way into the fishery more effectively than Protestants.

37. John McGregor, *British America* (London, 1833), I, 158. The "fishocracy" comprised, in descending order: (1) the principal merchants, high officials, and some lawyers and medical men; (2) small merchants, important shopkeepers, lawyers, doctors, and secondary officials; (3) grocers, master mechanics, and schooner holders; and (4) fishermen. Tocque, *op. cit.,* p. 86.

38. "The policy of the merchants was to keep the attention of the people altogether fixed upon the fisheries and to repress every movement that had a tendency to encourage agriculture or any other pursuit which clashed with their own interests. All improvements would induce increased taxation, that is increased duties on imports, and such increased duties would tend to a diminution of their profits, because although they charged these duties (which they themselves paid in the first instance) upon the provisions and goods supplied by them to the fishermen, yet the fishermen might not always be in a position to pay back those charges." Thomas Talbot, *Newfoundland* (London, 1882), p. 43. It was claimed that the merchants discouraged shipbuilding. "We are spending £60,000 annually in purchasing ships of foreigners to keep up our mercantile fleet." "A Nova Scotia shipbuilder brings a schooner ready rigged and found in all things necessary for business, and sells her in our own ports free of all duty. This is an advantage to the merchant but an injury to the colony." Stephen Marsh, *The Present Condition of Newfoundland* (St. John's, 1854). On the opposition to expenditure on roads, see *A Sketch of the State of Affairs in Newfoundland* (London, 1841). For an account of the importance of stimulating agriculture as a method of supporting the shore policy to combat the French, see Patrick Morris, *A Short Review of the History, Government, Constitution, Fishery and Agriculture of Newfoundland* (St. John's, 1847); also J. S. Buckingham, *Canada, Nova Scotia, New Brunswick and the Other British Provinces in North America, with a Plan of National Colonization* (London, 1843), chap. xxiv.

interests, refused to grant bills of supply. Early in the same year a petition of merchants, traders, and other residents of St. John's protested that of St. John's total population of 75,000[39] about one half were Irish Catholics, and that the priests had secured control of the Assembly. The petitioners also held that the occupation of a dwelling for one year as an electoral qualification was much too broad. They objected to making magistrates, constables, and subordinate functionaries dependent on the Assembly for salaries, and they claimed that "the merchant and fisherman have but one common interest and are bound together by one tie of mutual dependence." Petitions from the merchants of Liverpool bearing 119 signatures, from the merchants of Bristol bearing 29, and from the merchants of Poole with 21, received in September and October, 1838, expressed a fear of possible disturbances during the winter, considering the large number of unemployed, and asked that troops should be sent out and that the constitution should be revised. "It is indispensably necessary that the nature of the elective franchise be revised."

The Assembly protested on October 25 that

> the policy heretofore pursued by the parent government . . . at first to forbid residence, then to restrain settlement, anon to deny agriculture; in fine to fetter the resources and cramp the energies and blast the prospects of the people, has produced the natural result. Native gentry there is none; a resident landed proprietary there does not exist, and consequently society is reduced to two classes, the one mercantile composed not of native "merchants and adventurers" and indeed to a considerable extent even these, non-resident, to whom may be added the officers of the government, all strangers too; the other the humble fishermen, whose destinies are riveted to the soil of their nativity. . . . The interests of a mercantile class of society in Newfoundland by no means bear that intimate analogy with those of the fishermen, which, in the provinces referred to, subsists between the different classes of people. The native inhabitants of Newfoundland are sighing for the promotion of agriculture for the full development of the internal resources of the country. The merchant sees in the accomplishment of their wishes the grave of his monopoly, for if agricultural produce be raised in the country the profits of the merchant in the importation of provisions must proportionably decline. . . . While the mercantile portion of the council support the official in passing the Bill for the civil salaries and expenditures, the official support the commercial in rejection of Bills contemplating public improvement, defraying the cost incurred in seeking redress of grievances, or the just and legitimate remuneration for services honestly and zealously performed, because performed by persons selected by the representatives of the people.

39. 1837: Roman Catholic, 37,376; Church of England, 26,748; Dissenters, 10,636. See Charles Pedley, *The History of Newfoundland* (London, 1863), chap. xiv. St. John's and Conception Bay returned seven of fifteen members in the Assembly.

On December 11, 1838, a petition signed by 2,626 clergymen, magistrates, merchants, traders, and others desired that this be known:

That some of the merchants of Liverpool, connected with the trade of this country should, although totally unacquainted with the present state of the colony, address your Majesty's government to adopt measures of coercion against the people of Newfoundland, appears to your Majesty's petitioners only as part of a general hostility to the growth of free institutions in Newfoundland manifested by a portion of the mercantile body at all times.

On December 18, a petition bearing 1,520 signatures from a similar group in Harbor Grace and Carbonear claimed that with 30,000 people in Conception Bay there was almost a total absence of crime, and that with the support of the Assembly agriculture was increasing.

This was answered on February 20, 1839, by a petition from 79 merchants, traders, and shipowners of Conception Bay.[40] In the preceding December a petition of the Chamber of Commerce of St. John's, signed by thirteen "mercantile men" elected by ballot from the "general commercial society," had argued to the same effect and stated that St. John's had annual imports of £400,000 or £500,000, chiefly manufactured goods from Great Britain, and that nearly 800 vessels averaging 100 tons cleared at the customs.

Upon its trade therefore and upon it alone does the very existence of the colony and its value to the Crown of Great Britain depend. . . . In this country there are not as there are in most others any persons of education residing unconnected with business—the population therefore with the exception of the learned professions, consists entirely of the merchants, possessing capital and the means of giving employment to the fishermen, and to the fisherman whose wealth consists in his labour, who is not attached to the soil by any tie of family or possessions and who is prepared to migrate to the United States . . . upon the first symptoms of a depression in trade or upon the first suggestion of caprice.

Debt had increased under the House of Assembly until in 1838 one sixth of the entire net land revenue was absorbed by it. The value of property had depreciated and the rates of insurance had increased. "We do not desire nor can we patiently endure that persons who have no property in the country, and who can contribute nothing to the revenues, shall exercise unlimited power over and rule with a rod of iron, those who do

40. They claimed that a trade employing 250 vessels, of which more than 150 were British, and imports exceeding £150,000 annually were endangered by the difficulties which had followed the establishment of the Assembly. There had been a marked emigration to Canada and the United States, and the empty places were filled by Roman Catholics from western and southern Ireland.

possess property and who mainly contribute to the support of its government." In February and March, 1839, 18 merchants of Liverpool, 17 merchants of Dartmouth, 19 merchants of Teignmouth, and certain merchants of Torquay stated in their several petitions that they were about to send out fishermen, capital, and supplies for the year, and asked for protection and the abolition of the local legislature. The Teignmouth merchants viewed "with great suspicion and alarm the encroachments on our liberties in the island and the imposts levied on the articles of life necessarily imported there to carry on the fishery, which have been made and done by the House of Assembly."

As a result of the conflict, the constitution was suspended and the Council and Assembly were amalgamated in 1842 (5–6 Vic., c. 120). The disastrous fire of June 9, 1846, in St. John's and the gale of September 19, 1846, contributed to the general unrest and added strength to renewed demands for responsible government. In 1848 the old constitution was revived, and the struggle broke out anew. On February 25, 1854, a memorial of the Chamber of Commerce of St. John's to the Duke of Newcastle voiced opposition to responsible government "until all classes of its population are fairly represented in the Assembly,"[41] and resented the charge by delegates of the Assembly that they, the members of the Chamber of Commerce, "were being actuated only by a spirit of monopoly, and, combined to fix a price below its value on the staple produce of the country, and to establish and confirm a credit and truck system ruinous in its effects upon the operative population." The speaker of the Assembly stated on March 24 that the leaders of the Executive Council "have invariably opposed the introduction and progress of free institutions in this colony and, by the system of trade they have long pursued, reduced our operative population to a deplorable degree of nursing and dependence." Delegates of the Assembly lodged a protest[42] on July 28 that the Council had secretly repealed the fishermen's charter which guaranteed that their wages should be paid from the proceeds of the voyage. "Now," they said, "if the voyage should fall short of the amount of the outfits given to the planter, with their enormous overcharges, the unfortunate fisherman is deprived of his wages and thrown on the government to be supported as a pauper." The

41. Marsh, *op. cit.*

42. "It is no uncommon occurrence in Newfoundland for a planter to fell, and bring out of the forest, timber and other materials necessary to construct a vessel, to build her from keel to topmast, and afterwards to take charge of and navigate her in prosecuting the trade of the colony. Surely then such men are not to be supposed devoid of that intellectuality which would qualify them to become the recipients of a system of constitutional rule under the enjoyment of which they observe their sister colonies thriving." *Journals of the Assembly, Newfoundland,* 1854, p. 41.

refusal of the Council to pass a bill remedying this situation and the Assembly's refusal to pass supply bills precipitated a deadlock which led to responsible government.

Her Majesty's Government had come to the conclusion that they ought not to withhold from Newfoundland those institutions and that civil administration which under the popular name of responsible government had been adopted in all Her Majesty's neighboring possessions in North America; and that they were prepared to concede the immediate application of the system as soon as certain necessary preliminary conditions had been acceded to on the part of the legislature.[43]

The decline in the influence of British merchants at St. John's was accompanied by a persistent cutting down of their activities. "Shopkeepers, as a respectable class," it could be said in 1861, "are only *now* gaining ground in St. John's; while almost the only attempt elsewhere takes the form of a petty barter trade, carried on between the more successful fisherman and his poorer neighbours, in which the illicit sale of ardent spirits forms the most marked characteristic."[44]

An active trading organization in St. John's and smuggling from France and the United States led to encroachments upon monopoly control on the south coast. Competition from French bounty-fed fish caught by trawls baited with Newfoundland herring and capelin brought about demands for restriction of trade.[45] As a result, an export tax of 3 shillings per hundredweight was imposed on bait fish in 1845 (8 Vic., c. 5),[46] and in 1846 and 1847 a cruiser was employed to check smuggling and encroachments on British waters. The effects were apparently negligible.

It may be concluded that the operation of the Export Duty on Bait in 1846, had no effect whatever upon the Outfit for the French Bank Fishery. . . . Either showing that the outfit for 1847 was not affected by any apprehension of the consequences of our export duty on bait, arising from the experience of its operation in the previous year, or leading to the alternative conclusion that if the working of that duty had any practicable or appreciable influence upon the supply of bait and the catch of cod upon the Banks; that injurious result was more than neutralized by some other cause, prob-

43. *P.C.*, II, 749 ff.

44. Tocque, *op. cit.*, p. 191.

45. "Many of the Fishermen are driven to such illicit means of traffic, to enable them to maintain a livelihood—Provisions and Clothing at St. Pierre being from 60 to 70 per cent. cheaper than English traders are willing to supply them." *Journals of the Assembly, Newfoundland,* 1856, Appendix, p. 91. See pages 403–404.

46. In 1786, English subjects in Newfoundland had been forbidden by law to sell bait to foreigners (26 Geo. III, c. 26), and in 1833 trade in capelin was made illegal under 3 and 4 Wm. IV, c. 59.

ably by the large bounties paid by the French Government, upon grounds of national policy, to all concerned in the cod-fishery.[47]

French vessels visiting Newfoundland ports were displaced by Newfoundland fishermen who carried bait from Fortune, Grand Bank, Lamaline, Burgeo Islands, and other points to St. Pierre or to French vessels.

As the price of bait carried into the French ports is extremely fluctuating, those who might be inclined to pay the duty on exported bait would sometimes not realise the amount of duty paid; and were the duty reduced, there are other causes on account of which those who are in the habit of carrying bait would still seek to evade its legal export; so long as that evasion continues possible, those who would leave the ground where the bait is caught to sail back to a distant port of entry, instead of sailing direct from the ground to St. Peters [St. Pierre] or Miquelon, not only run the risk of having such a perishable article rotten on their hands, but they could in any case be undersold in the French ports by the less scrupulous evaders of the law, who would, in all probability, also have the advantage of being first in the market; and besides which, it is commonly the desire of all such parties to conceal their dealings with the French from the merchants by whom they are generally more or less supplied, and by whom such traffic is as far as possible discouraged.[48]

After the French increased their bounties in 1851 there were strong protests

That your Petitioners regard with much alarm the intelligence lately received of the alterations made by the Government of France in the terms on which they award bounties to the Fishermen of that country, on Fish caught on the Coast of this island and sent to the Markets of Europe.

That your Petitioners have learned that in consequence of the increased encouragement thus offered, the French Fisheries are to be prosecuted this season with increased vigour.

47. *Idem,* 1857, Appendix, p. 227.

48. Thomas E. Gadon, November 1, 1850. *Journals of the Assembly, Newfoundland,* 1851, Appendix, pp. 160–161. "Under the present system of High Duties the exporter cannot afford to pay it, as he goes to an uncertain market, as it often occurs that bait varies in price several times in a week; but should a low rate be imposed of 9d. to a 1s. per barrel for herring, and the same per hogshead for caplin, the owner would not run the risk of losing his boat for the amount of duties he would have to pay; though, of course, there would be some exceptions to this, as some would avoid paying anything if they could, were it only 1d. per barrel. Low duties, in all probability, would induce the French to come to Lamaline and other places, and purchase themselves, which would materially benefit the really poor man, as then he would be enabled to sell what he could catch; but under the present system, it is only the man who is comparatively well off and in good credit, that can carry on this traffic, as it requires large sized boats, good seines and nets, which the poor man has not the means of procuring." *Idem,* p. 166.

That your Petitioners have reliable information that the French fishermen catch from 200,000 qtls. to 300,000 qtls. of fish, at Cape John, Belle Isle, and Labrador, within limits in which they have no right to fish, according to the Treaties between Great Britain and France.

That the fish so caught enters into competition in the Markets of Europe with that exported from this colony, causing a reduction of price ruinous to our trade, which is not stimulated by a bounty as the French Trade is.[49]

I cannot close this report without bearing my strongest testimony against the suicidal traffic in bait, at present carried on with the French at the Labrador . . . the effects of which will sometimes, even at the height of the fishery, leave our own people without a sufficient supply, and which traffic, independently of other evils, furnishes such an easy opportunity of encroachments upon our fishery grounds, when, as is but too often the case, the French grounds should fail in consequence of their destructive mode of conducting the fishery by bultows and large seines.[50]

A strong plea was made for protection in the north and south.

For many years after the peace, the produce of the French fisheries was not greater than the requirements of their own Home markets; and while this continued, we experienced but the primary loss of the best portions of our fishing coast. Of late, however, the increasing growth of their operations has given them a large surplus above what the French markets require, and this finds its way into places which formerly were supplied by our produce. From some of our oldest markets we have been driven altogether; and in most of those on which we chiefly rely, our interests are weakened to a degree that menaces the integrity and foundations of our trade. The evils of this unequal competition have been progressively developing themselves for some years; but in the past season we experienced them to an alarming degree, a large quantity of our fish having been disposed of in the European ports at *one-half its actual cost*. . . . Great as the French competition would be, even if the terms of the Treaties were adhered to, the ruinous increase of their rivalry that we now experience is mainly attributed to their audacious intrusions on the Western Coast in quest of Bait, and on the Labrador Coast, to which, when the fishing has ceased or failed on the French Shore, they resort in great numbers. And whilst the interests of the subjects of France are completely guarded throughout the whole season by War Steamers and other

49. *Idem,* April 30, 1852.

50. *Idem,* 1867, Appendix, p. 721. "I think that it is most necessary that the question as to whether it is legal or not for the French boats to come on the coast of Labrador to bring bait from the English fishermen, should at once be settled decisively, as it is the cause of continual quarrels and complaints amongst the fishermen. There are two parties, one of whom is in favour of selling bait to the French, the other most decidedly against it, as they argue that it is with that bait sold from our coasts that the French are enabled to bait the bultows laid down in the Straits of Belle Isle, and which they declare do so much damage on the coast of Labrador." *Idem,* August 20, 1865. See page 401.

armed Government Vessels peculiarly suited to the service, we are in effect, wholly unprotected, and hence the daring intrusions to which we refer.[51]

Great Britain was asked to send a steam vessel to do police work.

To put down the traffic decisively will require that a War Steamer be placed in the locality in question, for the winter and spring months. The knowledge that such means of prevention are to be used (which will be impressed on the minds of the people by the presence of the ship during the winter months) will deter the major portion of them from preparing to engage in the pursuit; and when the time arrives, the work of protection would be comparatively easy, as few will hazard the serious risk it will then involve. Nor do the Committee apprehend that even the people generally of that part of the island from which this trade is now carried on, will be injured by its prevention, for if, as is generally admitted, the Western cod fishery is severely injured by the activity of the French fishery on the Banks, it necessarily follows that the withdrawal of the present supplies of bait, in weakening the means by which the vigour of their operations is now sustained, would lead to a marked improvement in the summer voyages in the Western bays by which all the population would be beneficially affected. The new vents now being found for our herring trade is an additional reason to negative the presumption that any injury can be consequent on the measure in question.[52]

To this Great Britain replied on June 29, 1853:

It will not be in the power of Her Majesty's Government to comply with the wishes of the House of Assembly by sending a Steamer, during the winter, to the West Coast of Newfoundland; but they would suggest that the Colonial Government should fit out a Schooner for the prevention of the illicit traffic complained of—such vessel being placed under the immediate direction and control of the Admiral commanding on the station—an arrangement which has been approved by Her Majesty's Government with respect to the Colonial vessels employed in protecting its Fisheries by the neighbouring Province of Nova Scotia.[53]

51. *Idem,* May 28, 1852.

52. *Idem,* March 16, 1853. "The reasons for recommending that the colony should be at the expense of finding the house boats for the protection of the south coasts, are:

"First—That the duty imposed on exporting caplin is sufficiently high to cover the expenses of the protection the colony requires; and I consider the colony would be considerably enriched by this protection.

"Secondly—That by treaty, the fishing within three miles of the shore, and to halfway between it and St. Pierre and Miquelon, belongs to the English exclusively. As the Americans have the right of fishing on the coast of Labrador, I propose that the protection there required should be at the expense of the Home Government, more especially as a large number of those engaged in the Labrador fishing come from ports in England and the Channel Islands, and there is no revenue to be derived by the legal export of bait as might be from Lamaline to St. Pierre and Miquelon." *Idem,* 1851, Appendix, p. 153.

53. *Idem,* 1854, Appendix, p. 65.

The necessity for control over the expenditures required for the needed protection hastened the grant of responsible government and its introduction was followed by attempts to check French encroachments.[54]

Great Britain attempted to solve the problem through the negotiation of a treaty with France. The provision of the treaty of 1783 forbidding the establishment of settlements[55] had been disregarded by both English and French—the French encouraging the English to settle and look after stores and fishing establishments during the winter.[56] A convention for the settlement of all fishing rights on the coast of Newfoundland and adjacent regions was arranged with France, and in January, 1857, an agreement was signed which gave French subjects exclusive rights to certain sections of the coast. But this was resisted with determined hostility as an encroachment upon the control over natural resources assumed under responsible government.

> It may be said that our concession to the Americans [in the Reciprocity Treaty] justified the conclusion that the like privileges may with security be ceded to the French. We respectfully submit that the commercial policy of the two nations is entirely different. The Americans are pursuing a similar commercial policy to that of Great Britain. The small bounty allowed their fishermen is only equivalent to the duty on the salt used in curing their fish. The French nation engaged in the Fisheries, not so much as a commercial pursuit, as a means of fostering and extending their national power, and the large bounties they grant from national motives would completely destroy the position of British Fishermen, sustained only by private enterprise; that the concession to the Americans under the Reciprocity Treaty of 1854 was contingent on the assent of the Local Legislature, who, seeing the reciprocal advantages likely to result to our trade, accepted the terms of the Treaty;

54. *P.C.*, III, 1283–1284.

55. The problem of fishing rights had been covered in earlier correspondence between the two governments. In a note of July 10, 1838, Lord Palmerston wrote to Count Sebastiani: "The British government has never understood the Declaration [of 1783] to have had for its object to deprive the British subjects of the right to participate with the French in taking fish at sea off that shore provided they did so without interrupting the French cod fishery. . . . In none of the public documents of the British government . . . does it appear that the right of French subjects to an exclusive fishery either of cod fish or of fish generally is specifically recognized." *Journals of the Assembly, Newfoundland,* 1857, Appendix, pp. 178–179. A dispatch from Lord Stanley, the Colonial Secretary, to the governor, Sir John Harvey, dated July 29, 1843, endorsed the Earl of Aberdeen's opinion that "Great Britain is bound to permit the subjects of France to fish during the season in the districts specified by the treaty and declaration of 1783, free from any interruption on the part of British subjects; but if there be room in these districts for the fishermen of both nations to fish without interfering with each other this country is not bound to prevent her subjects fishing there." *Idem,* p. 252.

56. Half of the people of St. George on the west coast were French and the remainder English and Jersey Islanders. A complaint was made in 1852 that the construction of a plant for grinding cod bones was an infringement of the treaty.

and it should be remembered that the Americans have now a right in common with the British Fishermen to fish on our Coasts. The extension of such a privilege to any other power would considerably complicate this right, and would tend to engender conflicts between the fishermen of the three nations, and disturb the peaceful relations happily existing between those powers.

Conscious of the many and great disadvantages we labor under, when brought in direct competition with the French, either in the pursuit of our Fisheries on the Coast, or in the disposal of our staple products in foreign markets, we have just reason to be alarmed for our very existence as a British Colony, dependent entirely, as such, on our Fisheries, should any further Fishery privileges be conceded to the French Government.[57]

Henry Labouchere, in a dispatch of March 26, 1857, conceded that

the rights enjoyed by the community in Newfoundland are not to be ceded or exchanged without their consent, and that the constitutional mode of submitting measures for that consent is by laying them before the colonial legislature; and that the consent of the community of Newfoundland is regarded by Her Majesty's government as the essential preliminary to any modification of their territorial and maritime rights.

Newfoundland's control over her natural resources, as confirmed in the defeat of the convention with France, led to the development of local legislation designed in the interests of conservation and the exclusion of foreign fishermen. For example, in 1858 an act (21 Vic., c. 14) provided "that the size of the mesh, &c., shall not be less than 2⅜ inches from knot to knot; [and] that no herring or other Bait shall be taken for exportation between 20th April and 20th October, within one mile of any settlement situate on the coast between Cape Chapeau Rouge and Point Boxey." An extensive investigation of possible conservation measures was carried out by a Select Joint Committee of the Assembly and the Council in 1862 and 1863. It was brought out in evidence that large-scale methods of operation had been introduced from the French. As early as 1845 the boulter method had been used in Bryant's Cove.

57. *Journals of the Assembly, Newfoundland,* February 3, 1857. Newfoundland was supported by the other colonies, as indicated in the following resolution:

"Resolved, That a delegation, comprised of Messrs. Kent and Carter, Members of this House, having been sent last Spring to the British North American Colonies to solicit their aid in resisting the Convention entered into, January, 1857, between Her Majesty's Government and the Emperor of the French, on the subject of the Fisheries of this Colony; and these gentlemen having received the promptest and most cordial co-operation of the Legislatures and people of our sister Colonies; and the said Convention having been withdrawn, with an emphatic recognition and declaration of the territorial and maritime rights of the people of this Colony:

"Resolved, That the Speaker do communicate the warmest thanks of this House to the Legislatures of the several Colonies to whom we are so deeply indebted for their influential aid and sympathy." *Idem,* April 14, 1858; also *idem,* 1857, pp. 45-52.

"Within the last thirty years the trade and population of Ferryland have very much declined." The roomkeepers of Bay Bulls petitioned as follows:

We, the undersigned memorialists in our time have carried on a hook-and-line fishery for a great number of years and for want of fish to pay the expense of such fishery were compelled to abandon the same and adopt the cod seine fishery, although in that time our shore fisheries were much increased by a large quantity of fish caught in the offing, which now is not to be found. We . . . most humbly pray your honourable house will allow [your petitioners] the prerogative of catching fish in their usual way, as [they] are of opinion that any alteration would be injurious. Large seines must be used in Bay Bulls, or none, in consequence of the depth of water in that locality. Should [they] by any enactment be prevented from using those seines, it would be most ruinous.[58]

The use of seines had increased on the Labrador.

By permitting the use of seines on the Labrador, we prevent our fishermen from obtaining a supply of herring for bait, and to dispose of, for the benefit of Nova Scotians, Americans, and others, as very few seines are owned by Newfoundlanders. The net fishery is conducted by fishermen who have rooms, the seine fishery mostly by those who follow it in vessels; still I think it would be unfair to those who have laid out their money to prosecute the seine fishery, to at once abolish their use.[59]

With large-scale operations went poorer methods of curing. "The people in this part of the District [Old Perlican] pickle their fish; this

58. *Journals of the Assembly, Newfoundland,* 1863, Appendix, p. 511. A petition opposing the use of seines stated that "we, your petitioners, are fishermen who fish by the hook-and-line; and the best of us find it most difficult to support our families by the fishery, as the amount of pauper relief transmitted to this place the previous winters can bear testimony; And we attribute the cause mainly to the custom adopted, of using Cod seines along the shore here, which runs almost in close proximity with our fishing ledges, impeding and proving a complete obstacle to our mode of fishing; and, so much so, that should Cod Seines continue as those past years, to haul in the contiguous neighbourhood of our fishing ledges, it will ultimately result in general pauperism." *Idem,* p. 509. Another petition told the same story. "We have reason to regret and to complain of the method and practice of catching fish in this part of the Bay with spilliards for the past five years; up to that period there was always a fair average catch with the hook-and-line, but since the commencement and increase of the spilliards with the few that use them, the many of us who use the line and hook have not the smallest chance; we could use the spilliards as well as them, but we solemnly protest against the use of the spilliards; and from experience and information from the oldest fishermen, if they are allowed by law to be continued, we shall become as so many paupers; and we believe them to be the cause of leaving many families destitute the coming winter." *Idem,* p. 512. As late as 1899 fishermen complained of the use of trawls at Pouch Cove. Abraham Kean, *Old and Young Ahead* (London, 1935), pp. 117–118.

59. *Journals of the Assembly, Newfoundland,* 1863, Appendix, p. 589.

is ruinous to the merchant; and the heavy losses on some of the fish shipments are occasioned by this practice, and if it could be put a stop to, the earlier the better for the country. Fish salted in bulk takes less time to cure, will stand in cargo much better, turn out well, and realise a better price at market."[60]

A "Fishermen's Society" made the following recommendations:

First: They are impressed with the absolute necessity of preventing the hauling of caplin for agricultural purposes, inasmuch as it is a practice detrimental to the interests of the fishermen. Those who haul bait for the purposes of manure, resort early in the morning to the places frequented by the caplin, so that when the fishermen come for their supply they frequently are obliged to go without any. This was not the case some twenty years ago, when the fishermen could, at any time in the course of the day, obtain whatever quantity they required, there being no obstruction, comparatively, to their pursuits in this respect. . . . Third: They are impressed with the conviction that the use of bultows is highly injurious to the general interests of the Cod fishery, and ought to be prohibited within a distance of five miles from the line of coast. . . . Fifth: They are strongly of opinion that the mooring of herring seines in any waters within the jurisdiction of the Government of this Colony should be strictly prohibited, as being exceedingly injurious to great numbers of fishermen of the country. . . . Seventh: They are thoroughly satisfied that the indiscriminate use of the cod seine is prejudicial to the interests of the fisheries; and that it ought not, therefore, be permitted to interfere with the hook-and-line men in any part of the country. Eighth: The use of cod nets they cannot but regard as injurious to the interests of the fishermen; and therefore they would urge the prohibition of them as a wise and judicious step towards the improvements of the fisheries. Ninth: They would suggest, as a useful measure, tending in the same direction, the prohibition of the jigger, from the 1st of June to the 1st of October. Tenth: The necessity of protecting and promoting that useful branch of our fisheries, the catching of herring, suggests the propriety of preventing herring being caught, except for bait for our own fishermen, from the 20th of March to the 20th of May, which is considered their season of spawning.[61]

A report of the Select Joint Committee was presented on March 16, 1863, and its chief recommendations were enacted in legislation.

60. *Idem,* p. 444. "The cure of Fish, your Committee believe, may be much improved, by washing it immediately after it passes from the splitting-knife, and salting it while fresh; cleanliness, with such judgment as our people possess, would, in the workings of it, secure at all times, (except when unfavorable weather prevents) the desired improvement. The Committee are of opinion that the practice of pickling Cod-fish, instead of salting it in bulk, after the old fashion, is very pernicious. The fish so cured will not keep in humid climates, or stand a long sea-voyage; and they believe the only remedy to correct these evils, and to secure a well cured and marketable article, rests with the purchaser, in making a suitable distinction in the price." *Idem,* March 16, 1863.

61. *Journals of the Assembly, Newfoundland,* 1863, Appendix, pp. 496–497.

When framing the proposed statutes, the Newfoundland government was advised by Great Britain that it should have regard to the following considerations:

1st. That if any misconception exists in Newfoundland respecting the limits of the colonial jurisdiction, it would be desirable that it should be put at rest by embodying in the Act a distinct statement that the regulations contained in it are of no force except within three miles of the shore of the Colony.

2nd. That no Act can be allowed which prohibits expressly, or is calculated by a circuitous method, to prevent the sale of bait.

3rd. That all fishing Acts shall expressly declare that their provisions do not extend, or interfere with any existing treaties with any foreign nation in amity with Great Britain.[62]

A concession of the right to introduce conservation measures in the fishery was followed by the demand for control over the land on the French Shore. A dispatch from Great Britain on December 7, 1866, prohibited "the issue of grants of land in that part of the island." Resolutions by the Newfoundland legislature insisted on "authority to issue grants within the island for mining, agricultural and other purposes."[63] In May, 1872, a further resolution called for the removal of restrictions "affecting the territorial rights of the people of the island." And, in the end, Great Britain in 1881 conceded territorial jurisdiction over the French Shore to the Newfoundland government.

As a result of this concession, in 1882 a commission of two members was appointed, one from France and one from Great Britain, and an agreement was signed on April 26, 1884 (an additional clause being added on November 14, 1885), which provided that the "superintendence and police of fisheries shall be exercised by the ships of war of the two countries," and that the British would not "interrupt in any manner the fishery of the French by their competition during the temporary exercise of it which is granted to them upon the coasts of Newfoundland." The French were also given the right to purchase bait free of duty after April 5. But the Newfoundland government refused to ratify the agreement because of the latter concession.[64] Following this refusal the French issued a protest on June 21, 1886, and gave new instructions to French commanders "to seize and confiscate the gear belonging to

62. The Duke of Newcastle to Governor Bannerman, August 3, 1863, *idem*, 1864, Appendix, pp. 607–608.

63. *Journals of the Assembly, Newfoundland*, April 9, 1867; August 24, 1868; April 23, 1874.

64. See *Correspondence relating to an Agreement between Great Britain and France Respecting the Newfoundland Fishery Question* (London, 1886); also *Journals of the Assembly, Newfoundland*, 1873; and *Report of the Council of the Royal Colonial Institute on the Newfoundland Fishery Question, November, 1875* (London, 1875).

foreigners resident or non-resident." The French also protested against the construction of buildings or the working of mines by Newfoundland, disregarded the jurisdiction of local magistrates on British territory, and were determined to protect their citizens in the development of the lobster fishery. Newfoundland introduced a bait act in 1886 but it was disallowed.[65] Further protests were made and a new act was passed in 1887 (50 Vic., c. 1). It was amended and clarified in 1888 and 1889.

The struggle to restrict the French fishery was accompanied by attempts to narrow the trade from the United States, Nova Scotia, Canada, and England. English firms continued to exercise monopoly control both in the north and south. They migrated to new areas[66] as they were excluded from territory tributary to St. John's. Garland's moved from Trinity to Greenspond[67] and Bloody Bay and became Robinson, Brooking and Garland; and Slade's moved from Trinity to Twillingate.[68] In addition to West Country and Jersey firms, trading schooners came to the Labrador from the United States and Nova Scotia.

Expansion was greater to the north, partly as a result of the greater expansion of the French to the south. In 1840 large numbers of vessels proceeded from Carbonear[69] to Labrador in spite of a series of failures extending over eight years. The Labrador fishery "employs annually upwards of four thousand persons, the greater part of whom come from Trinity and Conception Bays, in Newfoundland. The principal fishing stations are Henley Harbour, Battle Harbour, Cape Charles, Deer Island, Seal Islands, and Long Islands."

At Red Bay there are from 20 to 25 fishing boats of different sizes, employing from 2 to 3 men each. Ten families reside here during the winter,

65. *Despatch from the Secretary of State for the Colonies to the Governor of Newfoundland in the Subject of the Reserved bill of the Newfoundland Legislature Entitled "An Act to Regulate the Exportation and Sale of Herring, Caplin, Squid and other Bait Fishes"* (London, 1887).

66. See Tocque, *op. cit.*, pp. 147–148.

67. There were two large branch houses of London and Poole merchants at Greenspond in 1860. Julian Moreton, *Life and Work in Newfoundland* (London, 1863).

68. At this point in 1860 two brigs were loaded for Lisbon and made a return voyage in seven weeks, ending September 13. At Change Island Tickle, schooners were loading fish for St. John's. The Funk Islands, as in Cartier's time, exported boatloads of eggs. *Idem.*

69. *P.C.*, III, 1281–1283. See Talbot, *op. cit.*, for an account of the fishery from Carbonear and Harbor Grace in 1838 and a description of the contrast in Newfoundland between that date and 1882. See also W. Chimmo, "A Visit to the Northwest Coast of Labrador during the Autumn of 1867 by H.M.S. *Gannet*," *Journal of Royal Geographical Society*, 1868, pp. 258–281; also H. Y. Hind, *Explorations in the Interior of the Labrador Peninsula* (London, 1863); A. S. Packard, *The Labrador Coast* (New York, 1891); Hatton and Harvey, *op. cit.*, pp. 285–303, also Appendix IV for a survey of the fishery in 1883; and Erskine, *op. cit.*, is particularly valuable in describing the technique of various regions. [See also Notes to Revised Edition, p. 509.]

and about 100 persons of both sexes come here annually from Carbonear, in Conception Bay, to fish during the season. These are called freighters, and are brought in a vessel belonging to Mr. Penny, of Conception Bay, and return in the same at the end of the fishing season. About 40 English boats [at Lance au Loup], under the direction of Mr. Crockwell, were fishing for the firm of Stabb & Co., of Newfoundland. . . . They receive from 3*s*.9*d*. to 5*s*. per hundred fish. The fishermen are provided with money orders on the firm at St. John's, the value of their provisions being deducted from the sum due to them.[70]

The Cod Fishery in the Straits of Belle Isle is a shore Fishery: that generally conducted to the Northward of Cape St. Louis is by means of vessels, which follow the fish to wherever they happen to have struck in. Having completed their cargoes, which are salted down in bulk, they are taken to various ports to be cured, some of which are about Francis Harbor, others in Newfoundland. I should suppose that the total number engaged in the Cod Fishery, from Cape Charles, Northward to Esquimaux Bay, was about 6,500; of these a great number are women, wives and daughters of fishermen, and are employed to clean the fish, and also nominally to clean the vessels; they are engaged at small wages, and are said to do as much work as the men. A stronger and more healthy set of men and women I have rarely seen. Of the extraordinary quantity of fish taken to the Northward, I was told by Messrs. Larmour and Daw, at Gready Harbor, that two tons and a half of round fish was caught with hooks and lines, per man, per diem; that is equal to about fifteen quintals dry. The fish, it must be observed, are caught about a fathom from the surface. Five hundred quintals of round fish, and sometimes much more, are hauled in seines, though the average may be considered sixty. The fish are generally small not averaging more than four pounds.[71]

The past few years, I should say from 150 to 200 boats from Conception Bay, Trinity and Bonavista Bays, as well as Twillingate, Fogo, and other northern harbours, as soon as caplin shoal is over, go to Labrador, and return with trips of green fish; if the French were not in possession of Belle Isle, many of them would remain there; at any rate would not leave to go so far North as they now do.[72]

The merchant establishments sold necessaries and purchased fish, and were operated by superintendents who came from St. John's, England, or elsewhere to prosecute the fishery for the season. At Lance au Loup in 1850 "there was one brig belonging to the Hudson's Bay Company, one Jersey and one St. John's schooner; and at Schooner Cove one brig belonging to St. John's and two Jersey schooners." At Blanc Sablon, the Jersey firm of De Quetteville "generally brings out from 150 to 200

70. *Journals of the Assembly, Newfoundland,* 1851, pp. 146–148.
71. *Idem,* 1853, pp. 128–129.
72. *Idem,* 1851, Appendix, p. 157.

persons from Jersey to be employed in the fisheries, of whom 120 are at Blanc Sablon and the Isle au Bois—some paid on shares of one third, some on wages from 10s. to 40s. per month, and found in everything. They generally load six vessels for ports in the Mediterranean and Jersey, with from 10,000 to 12,000 quintals of cod."[73] The firm of De Quetteville also purchased fish from surrounding areas—from Little St. Modest for instance—"at 9s. per quintal, payment in truck. Others disposed of their fish to traders from Halifax and Yarmouth in Nova Scotia."

On the northwest side of Forteau Bay there are three extensive establishments of Jersey merchants, viz.; Boutillier, De Quetteville, and Dehaume, two smaller ones, one belonging to Ellis at English Point, the other to Buckle at the mouth of the river; and another to Davis, at L'Ance Amour, on the southeast side of Forteau Bay. . . . The establishment of De Quetteville and Brothers, of Jersey, bring about 50 men every year from Jersey, and engage 5 or 6 more with their boats, to fish during the season, at payment of 4s. per hundred fish, with firewood and spruce beer, the parties finding themselves in everything else. These hired men only averaged 100 fish a day per man this year. The men who come from Jersey are found everything, and a free passage out and home again. Six boats and twelve men are employed as sharemen, and get one third of the fish they take, and one third of the oil. Six more boats and twelve men are employed on wages from 15s. to £2 per month, and found in provisions. Twenty-five men are employed in splitting, curing, and other work, and return to Jersey every winter. Last year this establishment exported about 2,500 quintals of fish, 8 tons of oil, and 60 barrels of herrings; this year they expect to export from 1,800 to 1,900 quintals of fish only and about 7 tons of oil, and had only cured 15 barrels of herrings to this date; but the latter promised to be abundant. Dehaume's establishment bring out, and carry back to Jersey every year, from 30 to 40 men, some paid on shares at one third, others on wages from 15s. to £2 per month, finding them in provisions; 12 boats and 24 men are generally employed in fishing, the rest in splitting and curing fish. They generally arrive about the middle of June and leave again in September or October. The sharemen's time generally ends about the 10th of September, after which they, and the wages men also, are employed in curing fish, and loading the vessels for the voyage. . . . At Green Island there are three Jersey establishments of La Bruile, Savage, and Mallett, who send their fish over to Point Ferolle in Newfoundland, to be dried, as there are no stages on that island. At Isle au Bois are two establishments belonging to Leboutelier, and De Quetteville, exporting between

73. Carroll's Cove is described as "a small fishing station . . . which employs about 30 men, 6 of whom remain for the seal fishing in winter, and take about 300 seals each year. Their average take of cod fish is about 2,000 quintals. . . . At Great St. Modeste, there are eight boats and eighteen men." *Journals of the Assembly, Newfoundland,* 1851, pp. 146–147.

2,000 and 3,000 quintals of cod each. At Grand Point there are two small establishments belonging to Lefevre, and Syout, who export generally about 1,500 quintals of cod each. The number of persons employed fishing in this vicinity is in fact increasing, and the working portion is said to have increased one third in the last ten years. The scattered inhabitants settled along the coast to the westward generally sell their fish to Americans who, the Jersey men informed me, traded extensively, to the injury of the English trader; but I suspect, much to the advantage of the poor fishermen.

Boutelier's establishment employs about 22 boats and 44 men in fishing, besides 36 in splitting and curing. They catch and export about 3,000 quintals of cod every year to ports in the Mediterranean, and Jersey, besides from 10 to 12 tons of oil to England, about 100 barrels of herring to Jersey and Quebec, worth about 15s. per barrel, and about 30 barrels of caplin to Jersey, worth there about 20s. per barrel. The fishermen and splitters of this establishment are brought from the Bay of Chaleur[74] every year about the middle of June, and are sent back there again about the middle of August, the fishing being then over. The fishermen are paid from 4s. to 5s. per hundred fish; payment is made at the establishment of the same firm in Chaleur Bay on their return, half in cash, half in goods.[75]

In other parts of Newfoundland English firms endeavored to maintain effective control. The south coast was largely dominated by Jersey Island and London firms. "Complaints were many of the monopoly control exercised by the two firms of Newman and Company, and Nicolle, and of the curse of absenteeism in Fortune Bay." Nicolle and Company of Jersey had branches at Burgeo and La Poile in 1848; and in 1850 they exported about 10,000 quintals of cod from the establishment at Jersey Harbor, "and also upwards of 150 quintals of salmon at 18s. or 19s. a quintal, principally to Halifax and Quebec." The Jersey firm of Falle and Company at Burin purchased fish from fishermen who had "80 boats of 4 men each, and 60 punts with 80 men." "They receive 11s. per quintal for the best fish, and 10s. for the general run for it." This firm, along with St. John's merchants and Nova Scotia traders, purchased fish from Great and Little St. Lawrence, which had "35 small schooners[76] and 100 punts, employing upwards of 200 fishermen—for

74. About 1840, Le Bouthilier (spelled variously as above) and De Quetteville began bringing labor from the Magdalen Islands. A. P. Hubert, *Les Iles de la Madeleine* (Rimouski, 1926), p. 112.

75. *Journals of the Assembly, Newfoundland,* 1851, Appendix, pp. 148–150.

76. On the "western coast," that is, west of St. John's, or actually on the south coast, from Channel to Hermitage Bay, Garnish and Placentia, where the fishery could be carried on all the year round, a western boat of from 22 to 28 tons capable of staying out three or four weeks involved an outlay of about £350 for six men for six months, and took some 400 quintals. Because of their smaller English vessels and their lighter crews, the English bank fishery was obliged to use the ordinary trawl line.

which payment is in truck at about 10s. 6d. the quintal." The same firm exported

about 25,000 quintals annually, to Spain and Portugal; and the inferior to the West Indies. Mr. Falle also sends to Halifax and St. John's about 35,000 quintals in addition to what is sent foreign. He also exports about 300 barrels of salmon to the United States and British North America, at about 45s. a barrel. Since November last 3,000 barrels of herrings were exported to Halifax and Boston, at about 10s. per barrel. In Mortier Bay there are about 100 fishermen. The fish taken in the Mortier Bay and the Flat Islands is sent to Mr. Falle at Burin, and included in the quantity exported by him.

In 1848 the firm of Newman, of London, had branches at Burgeo and Gaultois. The Burgeo Islands had a population of 650 with 10 vessels. From Harbor Breton, the headquarters of the firm,[77] 10,000 quintals were exported annually to Vigo, Oporto, and Brazil.

In Harbour Briton I could only hear of there being 3 large boats, and 9 punts, the former employing about 12, and the latter 15 men, whose average take is from 1,800 to 2,000 quintals of cod; for which they receive 10s. and 11s. per quintal, payment in truck, from the agents of Newman of London, and Nicolle of Jersey, resident here. From Hermitage Bay, Belloram, Grand Bank, (300 people); Fortune, (240 people, 25 vessels); Lamaline, (400 people, 70 boats); and other small coves in Fortune Bay, a considerable quantity of fish is sent to Harbour Briton, to the firms of Newman of London, and Nicolle of Jersey, who export principally to Spain, Portugal, and Brazil. The exports of last year were—Cod, 55,186 quintals, at 10s. to 11s. per quintal; salmon, 296 barrels, at 36s. per barrel; oil, 120 tons, at £23 per ton; herrings, 3,050 barrels, at 8s. per barrel. . . . [Newman's] constantly employ upwards of 100 men, on wages from £20 to £30 per annum, and employ women and children occasionally, on wages from 1s. 6d. and 2s. 6d. per diem, in curing and drying fish.[78]

In 1840 Great Lawn had a population of 120 to 150 and fishermen brought small schooners and boats from Fortune and Grand Bank.

There are five small schooners, manned with 3 or 4 men, capable of carrying about 100 qtls. each; and 20 punts, with two men in each; the average take of fish is about 80 qtls. per man during the year. . . . Payment is made in truck, 3 quintals of green fish at 3s. 3d. per quintal, livers included, counting for one quintal of cured fish; some of the fishermen complained of being compelled to deal in this manner, as the truck agents of Nicolle of Jersey were not allowed salt to sell them. Some of the old and poorer classes of fish-

77. Newman and Company moved from Fortune Bay to Little Bay and then to Harbor Breton. Tocque, *op. cit.*, pp. 178–179.

78. *Journals of the Assembly, Newfoundland,* 1851, Appendix, p. 142.

ermen complained also, that others whose boats were larger and better equipped, haul the caplin and start off to St. Pierre with it, leaving those who have only small punts very often without bait. The fish from this place is sent to firms at Burin and Harbour Briton.[79]

A merchant named Thorne, probably of the American firm of Atherton and Thorne which did a substantial business in St. Pierre, had an establishment at Little St. Lawrence in 1840. The port had a population of 200 with 60 small boats, as well as small boats attached to schooners from Fortune Bay.[80]

The legislation in 1849 exempting the British colonies from the export tax on bait imposed in 1846 was followed by a rapid increase in trade in herring.

The section of 12th Vic., c. 7, which allows the free export of herring under bond to any part of the British dominions, has been productive of signal benefit to the fishermen of the district by affording them employment during the winter and early spring, and a market for their surplus herring, and enabling them to procure in return, provisions, clothing and other requisites at a low rate from the schooners in which the herring is exported to Halifax and other British Colonial Ports.[81]

79. *Idem,* 1851, Appendix, pp. 143–144. "The causes of poverty are many; and first, is the oft-repeated failure in the fishery . . . the potato disease, which destroyed a large item of the people's food; the Green Fish System, which persons in business have been obliged, in self-defence, to adopt towards those who are extravagant and careless when supplied with necessaries for the fishery, and the number of unlicensed pedlars who encourage the poor to barter with them the proceeds of their voyage, thereby destroying the confidence of the supplier who is left unpaid for the fisherman's outfit, and in succeeding seasons such persons will only barter for the Green Fish." *Idem,* 1855, Appendix, p. 277.

80. "The Inhabitants complain of the Fishermen of Fortune Bay coming to the Harbour of St. Laurence, with large Schooners, (which they cannot afford to procure for themselves,) each having two or three punts with them, for the purpose of the in-shore Fishery, while the Schooners are employed in the offing: by which means the fish are prevented from coming in shore, or are driven off the coast, before they [the inhabitants], in their small Boats, are able to catch them; and when the fish are scarce, or will not bite, these Fortune Bay Fishermen haul Caplin on their Shores and Bays, with which they load their Schooners and proceed to St. Pierre, dispose of their Cargo, and again return to prosecute the Fishery, at a more favourable period." *Journals of the Assembly, Nova Scotia,* 1841, Appendix No. 62.

81. *Idem,* p. 161. In St. George Bay and the vicinity, "there are said to be over 1,000 inhabitants, English, French, and descendants of Canadian, and Indians, of whom there are about 300 in the settlement of Sandy Point. The principal occupation is herring fishing, which usually commences in May, and lasts from two to three weeks, during which about 20,000 barrels of herrings are taken by the inhabitants; and from 5,000 to 6,000 more by vessels that touch there. About 300 barrels of salmon, and from 300 to 400 quintals of cod-fish are usually taken each season after the termination of the herring fishery. There are from 8 to 10 schooners trading from St. George's Bay, mostly to Halifax, which carry the fish to that market, and bring goods back. The herrings generally sell from 9s. to 11s. per barrel, and the salmon from £2

The lack of a certain determination of Canadian and Newfoundland jurisdiction was responsible for confusion. In 1850 Newfoundland traders complained that "twenty-seven trading vessels visited the harbors of this district previous to 1st August, with cargoes for barter, averaging £500 each, say, £13,500, none of which paid any duties, and all received drawbacks, either from Canada, Nova Scotia or United States." A Newfoundland merchant, W. H. Ellis, as "a severe sufferer" complained on behalf of persons residing in St. John's or anywhere in Newfoundland:

Labrador from Blanc Sablon to the Hudson Bay is within the jurisdiction of the Government of Newfoundland. I fit out from this island to carry on business on that coast, and if I do a large trade, I pay £200 duties, there being no drawback allowed. I commence under this disadvantage against Nova Scotians, Canadians, or Others, who receive the drawback on all goods. The voyage completed I want to realize the produce of it. Canada offers an excellent market for my herring, seal oil, and a portion of my fish; but here again I have 12½ to 25 per cent. duty staring me in the face because I happen to be within the jurisdiction of Newfoundland, while the others enter these parts free. Thus because Newfoundland does not reciprocate, or throw off the Labrador, these persons doing business there, though unconnected with St. John's or the outports are made liable to these advantages, [*sic*] and as no persons pay any Customs on that coast, Newfoundlanders labour under a double system of taxation.[82]

The number of vessels belonging to the United States, as well as to the neighbouring provinces, every year engaged in trading with the people of the French shore and coast of Labrador, is immense, and their dealings to an almost incredible extent. The resident population upon these coasts amount to several thousands, and from the traders the chief part of the supplies are drawn, whilst the transient fishermen have an opportunity to dispose of their surplus produce with great advantage to themselves. These adventures have now monopolized the entire trading business, especially upon the coast of Labrador; they pay neither duties nor taxes of any description, although they unquestionably come within the jurisdiction of this government. By this system of traffic the merchants and traders of this colony are greatly injured, being driven out of the trade by unjust and illegal competition; to correct this evil is not so unsurmountable a difficulty as might be imagined and, (with all due deference) should not be lightly regarded. . . . The protection of the fisheries, in my humble opinion, is of paramount interest to the colonies

to £3 10s. according to the season and demand in the market. After the termination of the fishery the inhabitants employ their time in procuring hoops and staves, and making barrels for the next season; each individual who is ordinarily industrious can make from 100 to 125 barrels, and fill them in the short fishing season, besides keeping their nets and boats in order." *Idem,* p. 150.

82. *Idem,* p. 160.

generally, and, however extraordinary it may appear, has a very powerful bearing upon the principles of free trade; and I have no doubt whatever that, should the fisheries be rigidly guarded throughout the whole of the British Provinces for another year or so, the principle of reciprocity would be more fully developed, and the true interest of all parties better understood. Then free trade might be established between the United States of America and these colonies upon more satisfactory terms than at the present period.[83]

With the introduction of reciprocity, effective in Newfoundland on July 7, 1855, there was an increase in trade with Canada, the United States, and Great Britain.

By the great number of persons resorting to the Coast of Newfoundland on account of the Fisheries, and supplying themselves there with provisions, a considerable Commerce with provisions is carried on to that coast. . . . The recent change of the Newfoundland Tariff, introduced in consequence of the above mentioned Treaty, threatens to put an end to this commerce, as the provisions imported to Newfoundland from the Hanse Towns will have to pay the duty, those from America and Great Britain entering duty free. I beg leave to state, that by the present Tariff of Newfoundland (Revenue Act, 21st July, 1855) the following Articles are admitted free of duty, when they are the produce of the United States or of Great Britain and the British Colonies, viz., breadstuffs, smoked and salted meats, butter, cheese, tallow, lard; whereas the duty on importation from other countries is as follows, for bacon, hams, smoked beef, the cwt., 7s., beef, salted and cured, the brl., 2s.; biscuit, the cwt., 3d.; butter 3s.; cheese, 5s.; flour, the brl., 1s.6d.; pork, 3s. These duties have the effect of differential duties, to the prejudice of the other countries, rendering their competition nearly impossible.[84]

In 1856 large mercantile houses from Jersey and Poole and vessels from Canada and the United States were carrying on an extensive trade without paying duties. It was estimated that ninety sailing vessels, many from Halifax and the Magdalen Islands, were on the coast in 1858. "This system of course operates unfairly upon those merchants resident in Newfoundland who also enter into the Labrador trade but whose goods have been subjected to a duty at the Colonial custom house, but its more palpable injustice lies in the fact that many thousands of those who maintain the trade are inhabitants of Newfoundland who migrate to the Labrador during the fishing season." Trade also increased on the south coast.

I believe a very considerable amount of illegal trading to be carried on along the South Coast, by Nova Scotians and American vessels, not entering

83. *Idem,* 1853, Appendix, pp. 139–140. 84. *Idem,* 1858, Appendix, p. 128.

at the Custom House, for the District, by which the Colony is defrauded of considerable revenue, and the merchants of the fish to which they are entitled, after supplying the people with the outfit, to enable them to prosecute the fishery; and a very demoralizing system consequently arises on both sides, the merchants charging largely to cover bad debts, and the people knowing it, evade payment when they can, by disposing of their fish to the illegal traders, (who, giving no credit, cannot loose) and pleading a bad catch to the merchant who has supplied them with their outfit. As a proof, I give the price of pork and flour at the outports and at St. John's—

	Outports	St. John's
Pork	£6 10	£3 10
Flour	2 8	1 15[85]

Customs regulations and conservation measures were alike disregarded.

In the presence of the Sub-Collector, they violated the Act, by using seines. I imagine that he is not in possession of the Act, otherwise the destruction would have been prevented. Unless a stop is put to the use of seines, Fortune Bay will be ruined. There has been at Bay-de-North, this winter, 60 large seines; at Rencontre, 40; total, 100 [with] 10 men to each seine. Not less than 25,000 barrels have left the Bay this winter, (many vessels without clearing.) To compute the number shipped, with the quantity thrown away, shews a destruction of 100,000 barrels. It is the opinion of every owner of a schooner or boat that I have been speaking to, if something is not done by the Government to enforce the laws for the protection of the Herring fishery, the inhabitants of Fortune Bay will become as dependent upon the Government for relief as Placentia and other Bays to the eastward.[86]

In 1853 herring was taken chiefly for bait, and a small quantity was carried home. Ten years later we find that at least 30,000 barrels of herring, caught on the Labrador by the people of Conception Bay in the past season, had found their way directly into the United States markets.

The markets of Canada were fully supplied early in the autumn, those of Ireland were overstocked; so that there would have been no vent for them, had the markets of the United States been closed against us. These fish were shipped in Colonial vessels, manned with our own people, and we received return cargoes of flour, pork, &c., thereby aiding and assisting our maritime interests.

The herring fishery has been of vast importance, the past season; they formed a valuable adjunct to our cod fishery at Labrador, the catch of which in many places on that coast was very partial, compensating, in a great measure, for the deficiency in our chief staple, and has been the means of

85. *Idem,* 1863, Appendix, p. 396.

86. *Idem,* January 31, 1866.

providing ample supplies for many who would otherwise have been destitute. The large quantity of herring caught, the past season, at Labrador, has had more to do with the present prosperity of this Bay, than is at first sight apparent, and has been chiefly instrumental in doing away with the necessity for pauper relief, the present winter, that great drain on the revenue of the Colony, thus effecting a great saving of the public funds.[87]

Salmon were taken in large numbers, particularly about Sandwich and Esquimaux Bays by English firms.[88] In 1863 the catch of salmon was unusually large,

particularly at Chateau and Sandwich Bays; in the latter 1500 tierces were caught, principally in Eagle River where 34,000 lbs. have been preserved fresh. Messrs. Hunt have also established two other posts for preserving, at Paradise, and Divers Island; and next year they intend commencing at Cape St. Francis. It is a valuable branch of the fisheries, giving much more employment than simply catching and pickling the fish, as, during the winter, the tinmen are employed making tinware, and other men making boxes and preparing firewood for the boilers; they also make, at Eagle River, large quantities of tinware, which is much sought by the Newfoundlanders who fish in the vicinity, as being far superior to anything of the sort they can get at St. John's. . . . Messrs. Hunt have a branch establishment in Davis' Inlet, about 120 miles beyond Cape Harrison, and the Hudson's Bay Company one at Kypococke, 70 miles beyond it, where they trade with the Esquimaux for seal skins, oil and salmon. . . . The Hudson's Bay Company employ Orkney men, and Messrs. Hunt, West of England men; many of them remain out when their period of service has expired, and being steady and sober, are valuable settlers.[89]

The cod fishery on the Labrador coast continued to expand to the north in the 'sixties.

87. *Idem,* February 24, 1864. The fishery was confined chiefly to the region between Blanc Sablon and Indian Tickle. See also Hatton and Harvey, *op. cit.,* pp. 266–276. "For ten days any quantity could be barred or netted, even from 1000 to 1500 barrels could be barred in a few minutes. From Battle and Sizes Harbors to Cape Charles, there must have been 25,000 barrels barred at one time. This is the only way the herring fishery can be profitably carried on, as herrings barred in large seines will keep good for twenty days, can be taken from the net and cured as required. A great many Nova Scotia vessels visit this coast late in the fall laden with fresh beef, pork, poultry, and all kinds of vegetables. The residents and others can thus find a ready market for their herring, and are thus supplied with the necessaries of life. After six years' experience, I would respectfully suggest to the Legislature, the propriety of repealing that part of the Herring Act which refers to what is termed 'barring,' so that every man may have free access to those shoals of wealth when they come within reach." *Idem,* 1866, Appendix, p. 719. See Hind, *op. cit.,* I, 327 ff.; and Packard, *op. cit.,* pp. 126 ff.

88. *Journals of the Assembly, Newfoundland,* 1853, Appendix, p. 130.

89. *Idem,* 1864, p. 470; see also *P.C.* (*Privy Council*), iii, 1286 ff.

The fisheries of Labrador have been increasing for several years past, until now they have attained an extent of such importance that Newfoundland could not sustain itself without them. It appears that the cod fishery has for some time declined on the southern part of this coast, so that many of our fishing vessels and crews have gone further North, until they have reached beyond Hope Dale, the Southern Moravian settlement.[90]

The seal fishery increased as an important joint industry[91] for the schooners engaged in the Labrador fishery. In 1834 St. John's had 125 ships and 3,000 men seal fishing; Conception Bay, 218 ships and 4,894 men; and Trinity Bay, 19 ships. Legislation enacted in 1827, which kept ships from leaving before March 17, was changed in 1840 to permit their departure on March 1, to meet competition from European sealers.[92] In that year and in 1841 the fishery was successful, but in

90. *Journals of the Assembly, Newfoundland,* 1868, Appendix, p. 549C. See W. G. Gosling, *Labrador* (London, 1910), pp. 412–414. For an account of trade at Bonne Esperance see W. A. Stevens, *Labrador* (Boston, 1884), chaps. v, viii. Legislation was introduced in 1881 prohibiting the crowding of schooners to the Labrador. Gosling, *op. cit.,* p. 442. A precursor of the "trap" appears to have been tried in 1840. "This year a Schooner from St. John's is prosecuting this Fishery with great success, having brought several large Nets, which are laid straight out from the points in the Harbour with anchors, in which the fish mesh themselves as they run along shore; and the number daily taken is from twenty to Forty, which are salted for exportation." *Journals of the Assembly, Nova Scotia,* 1841, Appendix No. 62. About 1877 the cod trap was invented by W. H. Whiteley at Bonne Esperance and spread rapidly to other districts. Gosling, *op. cit.,* p. 458. "A trap is a square net, generally fourteen fathoms by twelve fathoms and ten fathoms deep. From the corner of one side of the net the mesh runs in towards the centre on an angle, thus forming a small door. A leader or mooring line runs through this door, and through the trap, one end being attached to a stake on the shore about thirty fathoms away, the other to a grapnel outside. The distance from the shore is governed by the depth of the water; it may be more or less than thirty fathoms. At each of the four corners another mooring line is attached and indicated by buoys, these are securely fixed to grapnels. The cod swim through the converging space made by the angles and enter the trap, from which few escape. The law fixes the size of the mesh at four inches but the law is sadly violated. The net is laid in leaves one hundred meshes in a leaf. A good trap costs about $400. It is this method of catching cod in wholesale quantities that is depleting these waters." G. F. Durgin, *Letters from Labrador* (Concord, 1908), pp. 26–27.

91. See Warren, *op. cit.* S. G. Archibald, *Some Account of the Seal Fishery of Newfoundland and the Mode of Preparing Seal Oil* (Edinburgh, 1852); Michael Carroll, *The Seal and Herring Fisheries of Newfoundland* (Montreal, 1873); also Hatton and Harvey, *op. cit.,* pp. 247–266; W. H. Greene, *The Wooden Walls among the Ice Floes* (London, 1933); *Chafe's Sealing Book* (St. John's, 1923), 3d and subsequent editions; H. de la Chaume, *Terre Neuve et les Terre Neuviennes* (Paris, 1886); Kean, *op. cit.,* pp. 132–133; Basil Lubbock, *The Arctic Whalers* (Glasgow, 1937).

92. Tocque, *op. cit.,* pp. 304–307. Jukes described a sealing voyage in 1840. Schooners and brigs of 80 to 150 tons were employed. His ship, with a crew of 36 men, started on March 3 and reached the seals on March 12. The ship had nine four-oared punts with three men and a master. The crew paid berth money of about £4 currency after which the returns of the voyage were divided, half for the men and half for the vessel. The seals were killed with a spiked hook at the end of a stout pole, six

1842 sealers complained of high prices for berths. In 1846 it was estimated that 12,000 men were earning an average of £15 a season by five or six weeks' work. In 1850, 229 vessels (20,581 tons) and 7,919 men took 440,828 seals, and their skins and oil had a total value of $298,796. About five sevenths of the catch was brought to St. John's. In 1852, 367 vessels (35,760 tons) and 13,000 men took more than 500,000 seals in spite of a disastrous loss of vessels; and in 1857, 370 vessels of from 80 to 200 tons and 13,600 men took 530,000 seals valued at $425,000. In 1860 complaints were numerous regarding heavy losses of seals through making the pans too large.[93] In 1863 the fishery was a failure in spite of the introduction of the first steamers.[94] The firm of Baine, Johnston and Company purchased the *Bloodhound,* of 216 tons gross and 40 horsepower, and Walter Grieve and Company, the *Wolf,* of 270 tons, 30 horsepower. In the following year S. March and Sons employed the *Osprey.* In 1866 Walter Grieve and Company added the *Hawk,* and Ridley and Sons of Harbor Grace, the *Retriever.* In the same year, 177 sailing vessels and 5 steamships were employing 8,909 men.

With this expansion of the Labrador fishery the demand for the placing of restrictions upon American, English, Canadian, and Nova Scotian traders became more persistent. Legislation introduced by Canada

to eight feet in length, and the skins were hauled by several yards of stout cord to the vessel. A voyage varied from two or three weeks to two months; and many vessels made two trips and some three in a season. A catch of a thousand seals was regarded as profitable, but generally three and occasionally five thousand were taken. The skins were landed and the fat removed, an expert skinner handling 300 to 400 pelts daily. The skins were dry-salted and sent to England. The fat was placed in strong wooden cribs 20 to 30 feet square and 20 to 25 feet high, which were set up in strong wooden pans three to four feet longer than the cribs and three feet deep. The oil melted and ran into the pans, the first runoff being known as pale seal oil. The process covered a long period, some six months, and involved a horrible stench. After the destruction of the vats in the fire in 1846 they were rebuilt on the opposite side of the harbor. Archibald claimed to have invented the steam process which rendered the oil in 12 hours, produced a better grade, and used both old- and young-seal fat successfully. Steam was introduced and plants used with a capacity of 4,000 pelts in 24 hours. In 1852, 2,000 tons of oil were sent to the United States; but the market had been handicapped by the strong smell of the oil, for it was used for lamps and in a comparatively warm climate. Seal fats were of different weights, varying for the kind and the age of the seal. Some weights, per barrel, are given below along with the oil and residue produced per barrel.

	Fat (Pounds)	Oil (Gallons)	Residue (Pounds)
Old harps' fat	228	22½	73
Young harps' fat	225	22	52
Hooded harps' fat	230	21	80
Old and bedlamers' fat	246	21	103

93. *Journals of the Assembly, Newfoundland,* 1860–61, pp. 526 ff. The skins were hauled to large piles, called "pans," preparatory to loading on the ships.

94. Steamers left on March 10.

to check the confusion on the Labrador in the 'fifties met with opposition from Newfoundland. On June 10, 1862, notice was given that duties imposed by the annual revenue acts would be collected, and in 1863 a court of limited civil and criminal jurisdiction was established (26 Vic., c. 2 and c. 3).[95] In the same year the customs collector reported that Messrs. Hunt & Henley, an English firm, refused to pay duties.[96] As a result of the enforcement of the payment of customs duties this firm was displaced by Newfoundland firms: "Messrs. Hunt & Henley, having abandoned the supplying business at Long Island and Gready, the Planters and fishermen of said places . . . [have] been chiefly supplied by Mr. Larmour, who obtained goods from St. John's duty paid. Another circumstance to be observed is, that traders knowing that they are likely to be met with by the Revenue Cruiser, have been induced to call at Blanc Sablon, to enter and pay duty."

The old English mercantile establishments at Labrador, have of late years contracted their business, so that the direct importations of supplies from abroad are not now so great as they were formerly. The intercourse between Newfoundland and Labrador has greatly increased, but of course the supplies for the Newfoundland fishermen employed at Labrador during the summer, are imported into the former country, and there pay the Customs' duties.[97]

95. *Idem,* October, 1862. An attempt to collect duties on spirits sold by Nova Scotians and Americans in 1840 was successful in that duties were paid under protest by all except the Messrs. Slade at Battle Harbor. *Copy of papers and correspondence between the Colonial Office and the Government of Newfoundland relating to the levy of customs duties on the coast of Labrador and reports of the collectors, judges or other officers sent to that coast last season March 17, 1864.* Also Gosling, *op. cit.,* pp. 416–419. "The illegal sale of spirits by American, Nova Scotian and Canadian traders, is a great source of demoralization on the Labrador, (the people purchasing spirits with fish which ought to go to their suppliers,) and if it is the intention of the Newfoundland Government to establish Custom Houses there, I would beg to suggest to your Excellency the necessity of those vessels being warned by their own authorities to comply with Art. 21, Act 8 & 9 Victoria, c. 93, as otherwise they will plead ignorance of the establishment of Custom Houses. I have also heard that it is probable large quantities of goods procured from English and other Colonial traders are likely to [be] smuggled into Newfoundland this year." *Journals of the Assembly, Newfoundland,* 1863, Appendix, p. 400. See also *P.C.,* III, 1423 ff.

96. See copy of report of James Winter, Esq., of his proceedings as collector of revenue on the coast of Labrador under the act, 26 Vic., c. 2, sec. 9. *Journals of the House of Assembly, Nova Scotia,* 1865, Appendix No. 42, p. 4. By an impressive show of authority, duties were collected from the branch houses of De Quetteville and Company and Lebouthilier of Blanc Sablon at Forteau, from an English and an American vessel at Lance au Loup, from a Nova Scotia trader at Henley Harbor, from Messrs. T. & D. Slade of Poole at Battle Harbor and Venison Tickle, from Henley and Hunt of London at Cartwright, Long Island, and Gready, Warren at Indian Tickle, and from King and Larmour at Gready. Four Canadian trading vessels were found at Cape Charles and four Canadian and American vessels at Henley Harbor. For documents on the collection of revenue in the Labrador, see *P.C.,* III, 1488 ff.

97. *Journals of the Assembly, Newfoundland,* 1868, Appendix, p. 537.

The firm of De Quetteville failed in the crash of 1873 and the bankruptcy of the Banque Union in the Island of Jersey.

At last the fishing-ship tradition was killed, and its protests in death were typical of its protests throughout the history of Newfoundland.

We found here [at Gready] the *Escort* (Messrs. Hunt & Henley's) discharging cargo (salt). I boarded her and ordered the work to be stopped, which was done, the captain not being on board. Soon after, the agent of this place came on board, and ordered the men to proceed discharging, to which I objected, till the vessel was entered. Mr. Goodridge, the agent at Cartwright, then came on board, (and with more authority than the other,) gave orders to go on discharging, to which I again objected; he then demanded my authority, when I produced and read my commission, in presence of the captain. I then left an officer aboard the *Escort,* and went on board the *Volant.* On returning I found the men discharging, in spite of the officer, and again stopped the work. I then had a warrant issued for the arrest of the captain. The warrant not being served that day, nothing more was done.

At sunset, as was customary aboard the *Volant,* a gun was fired and the colours lowered; immediately the gun was fired, the port of the *Escort* was opened, and a gun was fired several times, a number of guns were also fired from the establishment and in different parts of the harbor. From the hill, near the dwelling house of Messrs. King and Larmour, a heavy cannon was fired, and the firing was kept up incessantly all over the harbor, for more than an hour, evidently for the purpose of intimidating me in the execution of my duty.

The next morning they again commenced discharging, which attracted the attention of the Judge, who immediately sent the bailiff, who arrested the captain and brought him on board. He was then sentenced to pay a fine or imprisonment, he chose the latter. Mr. Goodridge, the agent, seeing the decided steps that had been taken, consented to pay the duties, produced all the papers, invoices, &c., and gave a bill for the full amount of duties, (under protest). I consented to the release of the captain.

The abrogation of the Reciprocity Treaty led to the enforcement of the regulations against Canada and the United States. "Many Nova Scotia and Canadian vessels come down to Labrador for cargoes of herring; they now generally purchase them, in barter, from our fishermen; and do not so much as formerly catch them on their own account. This is an arrangement satisfactory and beneficial to our people. Very few vessels from the United States of America now come to Labrador for any purpose." This was in 1867. The American fishery on the Labrador is said to have ceased in 1869.

Under the Treaty of Washington of 1871, the export of herring in the frozen state for the New York market which began under reciprocity expanded rapidly. As with Nova Scotia, discontent with the Halifax

Award and conflict with Newfoundland in the Fortune Bay dispute led to demands for the termination of the treaty. With the end of the Treaty of Washington in 1885 and the loss of the American market[98] Newfoundland became more aggressive toward France, and the Bait Acts were finally passed.

The appearance of a trading organization centering about St. John's was an aid to the growth of responsible government and the exclusion of France, England, and continental America from trade and the fishery. It failed, however, to develop effective control over external trade. Spain's discrimination in favor of her own shipping was linked to expansion of the French fishery, but it was more difficult to combat.[99]

> The duty on Cod Fish in Spain in British bottoms is 10s.7d. currency per 100 lb.; in Spanish vessels, 8s. currency, direct importation, and 11s.11d. currency per 100 lb. indirect importation. The case of Spain is, of course, that which is open to the more serious objection—as in addition to the primary heavy tax the discriminating principle is there upheld—a principle which has so detrimentally affected the interest of British Shipping that British Vessels are rarely employed in the trade, except during the short period of the year when Spanish Ships are not available.[100]

In 1866 Newfoundland retaliated with a differential duty, but it was dropped because of objections from Great Britain. On September 19, 1866, Lord Carnarvon wrote to Governor Musgrave:

> This enactment is inconsistent with the stipulations of existing Imperial Treaties, which provide that the produce and manufactures of particular Foreign Countries, when imported into any British Possessions, shall be liable to no other Duties than are imposed on similar articles of British origin. In virtue of these Treaties, Belgium and the Zollverein by express stipulations, and indirectly all Countries having Treaties containing "most favored

98. "Newfoundland was anxious for trade relations" the disturbance of which would be attended with inconvenience and injury. The opening of new markets would be "a work of time and possible difficulty and meanwhile losses on shipments might reasonably be apprehended." See *Correspondence Relative to the North American Fisheries 1884–86* (London, 1887).

99. "We also desire to bring under the notice of your Majesty the change made in the duties in Spain on fish imported into that country. This change, while increasing the former heavy tax on our staple, also effects an increase in the previous difference of charge on fish imported in British ships, as compared with that payable on the article when brought in by the ships of Spain. Spanish vessels entering the ports of this colony enjoy all the immunities that are incident to British Ships, and bring their produce into our markets on equal terms with ourselves; while in return we are met by a Tariff, the old hostility of which has been further aggravated to a degree which must end in depriving us of the markets which that country has long afforded us." *Journals of the Assembly, Newfoundland,* April 24, 1850.

100. *Idem,* May 2, 1860. Prowse, *op. cit.,* pp. 453–473.

nation" articles, may at any time claim the exemption accorded by this Act to British Fish, and should any of these Countries, at any future time, demand admission for their fish on the same terms as British caught fish, their claims could not be disputed.

As the Act is one under which the Revenue of Customs for the current year has for some time been in actual course of collection, and as it expires on the 20th of May next, I will not expose the Island to the inconvenience which would follow from advising its disallowance. But Her Majesty's Government cannot concur in Commercial Legislation which is inconsistent not only with the established policy of this country, but with express engagements made with Foreign Powers; and must therefore instruct you that you will not be at liberty to assent to any renewed enactment of differential duties.[101]

In reply, Newfoundland argued as follows:

The system of bounties given by the French Government, and the differential duties imposed by the Spanish Government, for instance, in favor of fish imported under the Spanish flag, render it of great importance to our trade that French fish should not be brought into our Ports for exportation under Foreign flags. No differential duty imposed upon it will, however, operate to prevent this without the provisions of the 3rd. Section of the Act, which enacts that no fish not exempted from duty shall be warehoused without payment of duty, or shall be allowed the usual drawback on exportation. This Section has formed part of every Revenue Act for several years past, having been first inserted in the Act of 1859, and together with the exemption of British fish from duty in all Acts since the beginning of the Reciprocity Treaty with the United States, has had the effect of giving that protection against French Fish, without any ostensible differential duty, which [it] is now sought to retain, though the mode of doing so is open to objection, and as far as the differential duty alone would operate, is without effect. But no action of the Reciprocity Treaty with the United States, and the contemporaneous fiscal regulations with regard to importation from the neighbouring Provinces, or the same articles as those to which that Treaty referred, has been more clearly defined than the establishment of a differential duty in favor of United States and British produce as against that of the rest of the world. United States and British American fish being admitted free under these provisions, it was only necessary, in the late Revenue Acts, to impose a duty on "fish" without discrimination, to obtain what in operation was a protective duty.

It is obvious that the Newfoundland Trade suffers seriously from a competition supported by artificial protection at their expense in this manner; and the Legislature may be excused for regarding the article of fish as entitled to be excepted, under these circumstances in their case, from the application of the general principles of Free Trade. If French Bounties were abolished, and British caught fish admitted into France on the same terms

101. *Idem,* 1867, Appendix, p. 894.

as French fish, I believe there would not be the least desire to impose any discriminating Customs Regulations.[102]

With the support of agricultural development in the Maritime Provinces, population had increased in Newfoundland. Commercial interests had become increasingly concentrated in St. John's and enforced the withdrawal of the interests of France, the West Country, the United States, Nova Scotia, and Canada. Responsible government had come into existence because of the demands of Newfoundland interests; and responsible government, in its turn, had meant support for the customs, defense of territorial rights, and bait legislation as a defense against fishery and trade competition by other regions. The aggressiveness of commercial interests in Newfoundland had accentuated and paralleled the activity of Nova Scotian commercial interests against New England. An expanding domestic and protected market and increasing American demands for mackerel from the inshore waters had been met by the increasing assertiveness of Nova Scotia. The possibility of conflict had been checked by the Reciprocity Treaty, and the acute difficulties following its abrogation were checked by the Treaty of Washington. Nova Scotia had been strengthened in her demands by her entrance into Confederation with Canada and New Brunswick. New England became more concerned with the demands for fresh fish taken in relatively adjacent territory, and the Treaty of Washington was terminated. The iron steamship had proved its superiority over the wooden sailing vessel. The enforcement of regulations for the control of trade and the fishery became effective with steamship patrols, and hastened the decline of smuggling and encroachment.

Whereas Nova Scotia entered into Confederation in part as a measure of protection against the United States, Newfoundland continued in close relation with Great Britain as a measure of protection against

102. Governor Musgrave to Carnarvon, October 25, 1866, pp. 895–896. "The fish must be exported exclusively in French vessels; but it is not required that it should be the same vessels which are employed in the fishery; and the bounty is paid without regard to the distance of the foreign port where the fish is sold, from the French fishery grounds. So that a cargo landed in St. John's, Newfoundland, or in Sydney, Cape Breton, or Halifax, will get the same bounty of 20 francs per metric quintal, as if it was landed in a United States or a Brazilian port. In former times, the law named the foreign ports where the cargoes could be carried; but now the only conditions required, is that a French Consular Agent (whatever may be his rank) should reside in the Foreign Port where the fish is landed and sold.

"It is contemplated now by some French Fish Merchants to make arrangements with Spanish owners to send their vessels to Halifax, where they will receive their cargo from a French vessel direct from St. Pierre on the French shore of Newfoundland; and thus the shippers will profit of the double advantage of the French export bounty and the Spanish differential import duty." *Idem,* 1867, Appendix, p. 902.

France. The attempt of France to increase production in spite of her restricted grounds involved conflict with the expanding population of Newfoundland. Trawls, with their demands for bait, were combated by bait legislation. Treaties between France and England which threatened Newfoundland's interests were resisted successfully. And the increase of the English-speaking population threatened the rights of France in its own ominous fashion.

In the period that began with 1783, Nova Scotia had emerged as a focal point in the realignment of the colonial policy of Great Britain. In the period after 1833, the increase in population and the growth of a commercial group in Newfoundland were aids in the development of responsible government; and in her struggle with France she achieved a status of independence in the British Empire which precluded the signing of treaties without her consent. Through her struggle with France, her contributions to the development of the empire paralleled the contribution made by Nova Scotia in her struggles with the United States. What Newfoundland accomplished in external affairs when dealing with France was paralleled by the significance in Imperial affairs of the inclusion of Nova Scotia in the Canadian federal system.

CHAPTER XIII

FROM COMMERCIALISM TO CAPITALISM, 1886–1936

NEW ENGLAND

In the very short interview afforded by your visit I referred to the embarrassment arising out of the gradual emancipation of Canada from the control of the mother country, and the consequent assumption by that community of attributes of autonomous and separate sovereignty, not, however, distinct from the Empire of Great Britain.

The awkwardness of this imperfectly-developed sovereignty is felt most strongly by the United States, which can not have formal treaty relations with Canada, except indirectly and as a colonial dependency of the British Crown, and nothing could better illustrate the embarrassment arising from this amorphous condition of things than the volumes of correspondence published severally this year, relating to the fisheries, by the United States, Great Britain, and the Government of the Dominion.

The time lost in this circumlocution, although often most regrettable, was the least part of the difficulty and the indirectness of appeal and reply was the most serious feature, ending, as it did, very unsatisfactorily.

It is evident that the commercial intercourse between the inhabitants of Canada and those of the United States has grown into too vast proportions to be exposed much longer to this wordy triangular duel, and more direct and responsible methods should be resorted to.

SECRETARY OF STATE BAYARD TO SIR CHARLES TUPPER, MAY 31, 1887

THE end of the long period of commercial strife between areas engaged in the fishery came in sight with the spread of machine industry and the substitution of the steamship and the railway for the wooden sailing vessel. The vigorous and aggressive economies which had characterized the fisheries regions collapsed and sought shelter in protected markets.

The spread of industrialism evident in urbanization, improved transport, and refrigeration had profound effects on an industry that had its life in a commodity which depended on salt as a preservative if its product was to be sold in distant and tropical countries. In New England the trawler increased the supply of fresh fish from near-by waters; and the widening of the market made possible by improved transportation led to imports of fish from Canada and Newfoundland. In Canada refrigeration supplied the markets of Ontario and Quebec. In the case of France and other European countries the trawler of the fresh-fish industry was also used in the salt-fish industry. Obsolescent craft such as the schooner moved from Nova Scotia to Newfoundland and were displaced by the steamship in both local and external trade. After the war a mounting production of soft-cure fish accompanied increasing industrialism. Mine sweepers were converted into trawlers; and trawler-caught fish from Iceland and elsewhere pressed on the markets of Newfoundland. The overhead costs of large-scale equipment in the fresh-fish industry tended to force dried cod into the position of a by-product.[1] Pushed from the European market, Newfoundland relied to a growing extent on the West Indies and South America, with serious consequences for the Canadian product.

The increasing concentration of New England[2] on the fresh-fish industry, as reflected in the termination of the Washington Treaty, continued after the exclusion of American ships from Canadian waters; and it was a concentration that was intensified by the growing demands for fresh fish, especially in the urban markets of the Atlantic seaboard, by the development of refrigeration, and by the increasing use of the trawler.[3]

After the expiry of the Washington Treaty, Canada entered upon "a vigorous license and cruiser policy."[4] To enforce treaty rights, Canadians boarded some 700 vessels in 1886, and 1,362 in 1887. In 1886 ves-

1. See *The Proceedings of the Nova Scotia Fisheries Conference, Halifax, July 13–14, 1938*, p. 42.

2. E. R. Johnson, *The History of Domestic and Foreign Commerce of the United States* (Washington, 1915), Vol. II, chap. xxxiii; *The Fisheries of the United States in 1908* (Washington, 1911); and *New England's Prospect, 1933* (New York, 1933), pp. 247–278. For an appreciation of the significance of the collapse of the era of the wooden sailing vessels see Sarah Orne Jewett, *The Country of the Pointed Firs* (Boston, 1896).

3. "Fishing for the fresh-fish markets not only, as a usual thing, pays better but it involves less work for the fisherman and his family. Hence the trend wherever practicable and to the extent it has been possible has been away from the production of dried fish to fishing for the fresh-fish markets." *The Proceedings of the Nova Scotia Fisheries Conference, Halifax, July 13–14, 1938*, p. 19.

4. Revised Statutes of Canada 94 and 95, 1886. *Correspondence Relative to the North American Fisheries 1884–86* (London, 1887). "No attempt has ever been made by the Parliament of Canada, or by that of any of the provinces to give a 'construction' to the treaty, but the undersigned submits that the right of the Parliament of Canada, with the royal assent given in the manner provided in the constitution, to pass and act on this subject to give that treaty effect, or to protect the people of Canada from the infringement of the treaty provisions, is clear beyond question. An act of that Parliament, duly passed according to constitutional forms has as much the force of law in Canada, and binds as fully offenders who may come within its jurisdiction, as any act of the Imperial Parliament.

"The efforts made on the part of the Government of the United States to deny and refute the validity of colonial statutes on this subject have been continued for many years, and in every instance have been set at naught by the Imperial authorities and

sels were seized by the Canadian government and a furious agitation arose because of the "inhumanity and brutality with which certain Canadian officials treated defenseless American fishermen." Canadian vessels, as in the case of the *Scylla* at Lunenburg in 1887, were penalized for supplying ships with provisions beyond the three-mile limit. "The admitted purpose of the colonial authorities throughout the controversy," it was asserted, "has been to compel the United States to grant trade concessions as the price of uninterrupted enjoyment of privileges, claimed by the United States as a right under the treaty. . . . The fishery question has come at last to be an undisguised attempt on the part of the Dominion government to harass the United States into revising the tariff in the interest of Canada." The bitterness that arose following the rigid enforcement of the Convention of 1818, which forbade American fishing vessels to enter port, transship crews, purchase bait, or ship fish in bond to United States markets, led to the passing of a nonintercourse act authorizing the President to deny entry of Canadian vessels to United States ports and to prohibit the entry of fish or any other product or goods coming from the Dominion.[5] The proclamation of a Canadian act to enforce customs regulations[6] issued on December 24, 1886, led Congress to pass a bill giving the President power to retaliate, and it received his approval on March 3, 1887.

by judicial tribunes." Report of Hon. J. S. D. Thompson, Minister of Justice, July 22, 1886; *idem*, p. 181. See also *S.P.* (*Sessional Papers*) No. 101r, 1885; *Special Report on the Fisheries Protective Service of Canada, 1886* (Ottawa, 1887); and *Correspondence Relative to the Fisheries Question 1885–87* (Ottawa, 1887); *United States Senate Executive Document, No. 221,* 49th Cong., 1st Sess.; "Rights of American Fishermen in British North American Waters," *House Executive Document No. 19,* 49th Cong., 2d Sess.; "Report of the Committee on Foreign Relations in Relation to the Rights and Interests of American Fisheries and Fishermen," *Senate Report No. 1683,* 49th Cong., 2d Sess.; *Senate Report No. 1891, idem; Senate Misc. Document No. 54, idem; Senate Executive Document No. 55, idem; House Executive Document No. 153, idem.*

5. See *S.P.* No. 16, 1887; No. 36a, 1888; and J. G. Bourinot, *The Fishery Question, Its Imperial Importance* (Ottawa, 1886); also Charles Isham, *The Fishery Question, Its Origin, History and Present Situation* (New York, 1887), for a valuable summary. Part of the bitterness arose from the claim that, following the proclamation of 1830 and the removal of discriminating duties in 1849, fishing vessels were placed on the same footing as commercial vessels, but that after the Washington Treaty a distinction had been rigidly enforced. *Testimony Taken by the Select Committee on Relations with Canada, United States Senate, submitted by Mr. Hoar, July 21, 1890* (Washington, 1890), p. 869.

6. A report of the committee of the Privy Council of January 15, 1887, insisted that "United States fishing vessels come directly from a foreign and not distant country, and it is not in the interests of legitimate Canadian commerce that they should be allowed to enter our ports without the same strict supervision as is exercised over all other foreign vessels, otherwise there would be no guaranty against illicit traffic of

The danger of friction led to a search for compromise.[7] Sir Charles Tupper was appointed High Commissioner, and, together with the British Minister at Washington and Mr. Thomas Bayard, Secretary of State, in 1887 entered upon negotiations for a new treaty. A *modus vivendi* agreement was arrived at on February 15, 1888, pending the ratification of the treaty.[8] The determined protest of Massachusetts

large dimensions to injury of honest trade and the serious diminution of the Canadian revenue." *Further Correspondence Respecting North American Fisheries 1886–87* (London, 1887), p. 84. The smuggling of bait, fresh vegetables, fresh pork, salt, and barrels was difficult to check. "The American skippers pay Gloucester prices in American currency for their purchases whereas in dealing in their local markets the shore fishermen and farmers are usually obliged to accept high-priced goods in barter for their wares. The Canadian fishing-vessel owners are, it is believed, the only persons in the provinces who object to the prosecution of this commerce. They complain that it facilitates the operations of United States fishermen, who, by replenishing their stores . . . are relieved of the necessity of returning to the United States for a fresh supply, and that the immediate effect is to augment local prices and increase the cost of Canadian cargoes to that extent."

7. On November 15, 1886, Secretary of State Bayard wrote to Mr. Phelps* suggesting that such measures should be taken by the respective governments "as will prevent the renewal of the proceedings witnessed during the past fishing season in the ports and harbors of Nova Scotia and at other points in the Maritime Provinces." *Idem,* p. 4. On April 2, 1887, the Marquis of Lansdowne wrote Sir Henry Holland: "In view of the fact that owing to the action of the Government of the United States in terminating the fishery clauses of the Treaty of Washington, a large body of American fishermen have suddenly found themselves excluded from waters to which they had for many years past resorted without molestation, and that the duty of thus excluding them has been thrown upon a newly constituted force of fishery police, necessarily without experience of the difficult and delicate duties which it is called upon to perform, there would be no cause for surprise if occasional cases of hardship or of overzealous action upon the part of the local authorities engaged in protecting the interests of the Dominion were to be brought to light. . . . It is the desire of my government to guard against the occurrence of any such cases, to deal in a spirit of generosity and forbearance with United States fishermen resorting to Canadian waters in the exercise of their lawful rights and to take effectual measures for preventing arbitrary or uncalled for interference on the part of its officials with the privileges allowed to foreign fishermen." *S.P.* No. 16, 1887, p. 238; see also "Canadian Non-Intercourse," *Report No. 4087, House of Representatives,* 49th Cong., 2d Sess.

* E. J. Phelps, American Minister in London.

8. "1. For a period not exceeding two years from the present date, the privilege of entering the bays and harbors of the Atlantic coasts of Canada and Newfoundland shall be granted to United States fishing vessels by annual Licenses at a fee of $1½ ton—for the following purposes: The purchase of bait, ice, seines, lines, and all other supplies and outfits. Transhipment of catch and shipping of crews.

"2. If, during the continuance of this arrangement, the United States should remove the duties on fish, fish-oil, whale and seal oil (and their coverings, packages, &c.), the said Licenses shall be issued free of charge.

"3. United States fishing vessels entering the bays and harbors of the Atlantic coasts of Canada or of Newfoundland for any of the four purposes mentioned in Article 1 of the Convention of October 20, 1818, and not remaining therein more than

brought about its defeat in the Senate,[9] and the *modus vivendi* agreement was continued until 1918.[10]

The relative decline of the fishery in the United States and the expansion of the industry in Canada meant the most active attempts on the part of Canada to capture the American market and strenuous efforts on the part of the fishermen of the United States to protect themselves from Canadian competition. The rigid enforcement of legislation based on the Convention of 1818, which had previously resulted in the Reciprocity and the Washington treaties, now called forth demands for retaliation and an increase in the tariff.[11] The American fishery retreated behind a protective tariff and was supported by an expanding domestic market and improved technique.[12] The Canadian fishery could no longer be used to secure reciprocity, and the moderating influence of New England participation in the Canadian fishery on the American tariff declined. On March 3, 1883, a new tariff imposed a duty of 50 cents a hundred pounds on dried or smoked fish not in barrels or half barrels, or 84 cents a quintal on dried fish; 1 cent a pound on mackerel, ½ cent a pound on pickled or salted herring, and 1 cent a pound on pickled salmon or other fish; and the rates became effective upon the abrogation of the Washington Treaty. In 1890 rates were raised to ¾ of a cent a pound on dried or smoked fish and on frozen or fresh fish packed in ice. Fresh herring paid a duty of ¼ cent a pound, frozen herring ½ cent in 1894 and ¾ of a cent in 1897. A duty of 1 cent a pound was imposed on fresh mackerel, halibut, and salmon in 1897. Improvements

twenty-four hours, shall not be required to enter or clear at the custom house, providing that they do not communicate with the shore.

"4. Forfeiture to be exacted only for the offences of fishing or preparing to fish in territorial waters.

"5. This arrangement to take effect as soon as the necessary measures can be completed by the Colonial Authorities." *Further Correspondence Respecting North American Fisheries 1887–88* (London, 1888), pp. 7–8. See also "Message from the President of the United States," *Executive Document No. 113,* 50th Cong., 1st Sess. (1888); "The Fisheries Treaty," *Misc. Document No. 109, idem.*

9. See the "Fisheries Treaty" speech of the Hon. G. F. Hoar of Massachusetts in the Senate of the United States, July 10, 1888; also J. I. Doran, *Our Fishery Rights in the North Atlantic* (Philadelphia, 1888); John Jay, *The Fisheries Dispute* (New York, 1887); W. L. Putnam, *The Fisheries Treaty* (Washington, 1888).

10. *The Report of the Royal Commission Investigating the Fisheries of the Maritime Provinces and the Magdalen Islands* (Ottawa, 1928), pp. 60 ff.

11. The American Fisheries Union formed in 1884 took a very active part in the campaign. The controversy was also muddied by the Irish dispute. See *Testimony Taken by the Select Committee on Relations with Canada, United States Senate, submitted by Mr. Hoar, July 21, 1890* (Washington, 1890), for ample material indicating the views of the chairman and others on the fishery.

12. For an interesting account of the schooner fishery on the Banks see Rudyard Kipling, *Captains Courageous* (London, 1932).

in refrigeration and increasingly rapid transportation enhanced the importance of the fresh-fish industry and lessened that both of the dry fishery and of those on the more distant banks.

Following the stimulus given by the mackerel fishery to the development of Gloucester, it became an increasingly important center;[13] but Boston increased more rapidly, thanks to the fresh-fish industry. In 1888 the Grand and Western Banks fleet of the United States totaled 339 vessels, and the Georges Bank and the New England shore fleets numbered 284.[14] From 1900 to 1910 the average number of ships going to the banks had declined to 60, and they produced 36 per cent of the total catch. In the total landings at the principal New England ports salt fish had dwindled to 1 per cent. It has been estimated that 85 per cent of the New England catch is now sold in the fresh-fish markets.

The broadening of the market in the United States, together with refrigeration and improved communications by telephone, telegraph, and radio, brought about improved facilities for handling fresh fish. The New England Fish Exchange was established in 1908, the Boston Fish Market Corporation organized in 1910, and the Boston fresh-fish pier completed in 1915. Increased urban demands led to a lowering of the tariff, and in 1913 the duty on fish skinned or boned was reduced from 1¼ cents to ¾ of a cent, while other classifications were admitted free. The introduction of the filleting process in 1921 and the marketing of packaged fillets reduced the weight of fish and extended the market. Filleting plants increased from 40 in 1924 to 128 in 1930.[15] Packaged fish increased to 85 million pounds in 1929. With the expansion of the fresh-fish industry there was a concentration on haddock. Of the total trawler landings in 1931, 80 per cent was haddock, and of the packaged fish in 1929, 85 per cent.

A rapid increase in trawlers accompanied an expanding market. The first trawler, the *Spray*, was introduced by the Bay State Fishing Company in 1905. Schooner draggers were introduced in 1919 and had increased to 198 in 1929. Diesel engines heightened the efficiency of

13. See *Testimony Taken by the Select Committee on Relations with Canada* (Washington, 1890), pp. 804 ff., 844 ff.

14. For statistics on the annual take of cod from 1886 to 1909 showing the increasing importance of the northeast shore and Georges Bank, and the decline of the Grand and Western Banks, see Raymond McFarland, *A History of the New England Fisheries* (Philadelphia, 1911), p. 371.

15. Fish-meal plants were built to handle the waste. One hundred pounds of fish as purchased will yield, in the form of fillets, from 25 to 40 pounds, depending on the species of the fish and the season of the year. It takes about five tons of raw material to make one ton of fish meal. *Proceedings of the Nova Scotia Fisheries Conference,* p. 85.

trawlers and contributed to a marked increase in the catch, especially after 1928. In 1931, 58 per cent of the fish landed were caught by trawlers and draggers. Trawlers concentrated on the nearer banks.[16] In 1929 Georges Bank furnished 42 per cent of the fish landed by vessels of over 5 tons; South Channel near Georges Bank, 21 per cent; Brown's Bank, 5 per cent; and the shore grounds, 12 per cent. Fish taken from Canadian and Newfoundland fishing grounds totaled 5 per cent. With the decline of the fishery[17] on Georges Bank in 1931 there was a sharp increase on the other banks, which became more accessible due to more rapid steam and motor ships.

The immediate postwar years meant financial difficulties for the large-scale capital organizations essential to a mechanized industry. The Atlantic Coast Fisheries Company emerged from a difficult period, launched a frozen-fish program, and constructed a processing plant at Groton in 1922. The development of rapid-freezing processes by H. F. Taylor, as well as the invention of Clarence Birdseye in 1923, was followed by expansion and the acquisition of other companies, including the largest Nova Scotia organizations. The plant at Groton was closed down in 1931. The company operated 12 trawlers in 1936. In 1929 the General Sea Foods Corporation was organized as a subsidiary of the General Foods Corporation to develop the rapid-freezing high-pressure Birdseye process. After difficulties during the depression, it was operating 5 trawlers in 1936 and reopened operations at Halifax. The Bay State Fishing Company developed the production, on a large scale, of fresh fillets; and although it avoided heavy investments incidental to freezing processes it was exposed to the market fluctuations of high-priced commodities. It operated 15 trawlers in 1936. One effect of the drought in the western states and the rise in the price of meat was the expansion of the industry in 1936. The radio has been of help by providing immediate information bearing upon the location of fish and the most promising markets. Mechanization offset the losses due to the exclusion of Canadian labor under the quota immigration law.

Tariff protection for the processed-cod industry was increased to 2½ cents a pound on skinned and boned fish and 1¼ cents on dried fish in 1922. In the depression of the 'thirties the duties on dried fish were increased to 2½ cents, on pickled fish with less than 43 per cent moisture, by weight, to 1¼ cents, and over 43 per cent, to ¾ of a cent; and,

16. R. F. Grant, *The Canadian Atlantic Fishery* (Toronto, 1934), pp. 119–120; also R. H. Fielder, *The Fishing Industries of the United States,* United States Department of Commerce, Bureau of Fisheries (1929).

17. On the general decline of the halibut, mackerel, and haddock fisheries see E. A. Ackerman, "The Depletion in New England Fisheries," *Economic Geography,* July, 1938, pp. 233–238.

if skinned or boned, 2 cents in 1930. As a result of the opposition of New England fishing interests to a lowering of the tariff, the Reciprocity Treaty of 1935 brought no change, and the agreement of 1938 reduced the rate on fish with over 43 per cent moisture from 3/4 of a cent to 3/8. Duties were reduced 25 per cent on boneless and fillets, under a quota. Exports of fresh or frozen mackerel from Canada to the United States increased from 2,658 hundredweight in 1935 to 26,776 hundredweight in 1938, fresh or frozen halibut from 26,205 to 55,576, fresh or frozen cod from 44,261 to 75,209, and green-salted or pickled cod from 108,126 to 112,355.

CANADA

Since the abrogation of the fishery clauses of the Washington Treaty and the reïmposition by the United States of a heavy duty on our fish, many of our best fishing skippers and men have gone to Gloucester each season to fish in United States vessels and thus get the benefit of the higher prices ruling in that market by reason of the exclusion of foreign-caught fish. Indeed so large has this annual exodus become, it has been asserted that fully half the Gloucester cod-fishing fleet is manned by Nova Scotians. As far as concerns these fishermen and their families, they are probably making a better living by this change of base than they could make at home, but other interests suffer. . . . We lose the building and outfitting of the vessels, the curing and packing of the fish, and the profit of selling them.

R. R. McLeod, *Markland or Nova Scotia*

The expansion of the American domestic market and restrictions upon imports of Canadian fish to the United States[18] had resulted in the migration of labor. It was estimated that, in 1886, of a total of 13,938 men employed in the New England fisheries, 2,254 were from the Canadian provinces,[19] and that the wages were $125 to $190 a month in contrast with $75 to $82 in Nova Scotia. The tariff was a powerful weapon, for it enabled the United States to dominate the fishery; and labor as always was a mobile factor. It moved from low wages to high, and from the truck system to the cash system; that is, from Newfoundland to Nova Scotia and from Nova Scotia to New England.

18. For the significance of the collapse of the wooden sailing vessel in Canada, see S. A. Saunders, *The Economic Welfare of the Maritime Provinces* (Wolfville, 1932); *The Maritime Provinces in Their Relation to the National Economy of Canada* (*Dominion Bureau of Statistics, 1934*); and *The Cambridge History of the British Empire*, VI, 659–671. The subject covered in this section is considered in detail in R. F. Grant, *The Canadian Atlantic Fishery* (Toronto, 1934).

19. Of 5,193 employed in Gloucester, 1,102 were Canadians. *Testimony Taken by the Select Committee on Relations with Canada*, p. 1209; also pp. 815, 832.

The decline of the dried-fish and pickled-fish industries was a result of the increasing demands of the fresh-fish industry and of the falling off of the dried- and pickled-fish demand both from the Negro regions in the southern states and from foreign urban populations, especially in New York. Their decline was further hastened by the abolition of slave labor in the Spanish colonies in 1880. The increasing importance of the beet-sugar industry brought about a drop of 60 per cent in sugar prices in the 'eighties. The Canadian fishery was injured by the increase in duties in Porto Rico and Cuba as a result of their release from Spain and their preferential arrangements with the United States. The steamship and the railway continued to effect far-reaching changes in the fishing industry, particularly because of the decline of the wooden sailing vessel. Its gradual disappearance in the carrying trade involved its disappearance from the fishery and the decay of the small ports. The fleets of Arichat and Cheticamp fell away. Their vessels were lost or sold to Newfoundland. In 1887 it could be said that "a large volume of Canadian production reaches the West Indies by foreign steamers via foreign ports."[20]

The dried-fish industry has become in part a by-product of the fresh- and frozen-fish industry, and in part a product of highly specialized areas.* The advantages of large-scale organization in the dried-fish industry made itself felt in that supervision of curing[21] which large-scale

20. See a report on the trade relations between Canada and the West Indies, *S.P.* No. 43, 1887. A subsidized steamship service to the West Indies was started in 1889. *S.P.* No. 26A, 1891. "The cod fish caught along this coast [Gaspé] find their principal market in the Mediterranean, and while a few years ago this traffic moved altogether by water direct from Caraquet or via the Gaspé ports, whither it was sent in small vessels and transhipped, it is being gradually diverted to the New York route, and during 1907 no less than 80 carloads were shipped over the Caraquet Railway (Intercolonial Railway) and connections to New York, where it took direct steamer to Mediterranean points. The time occupied in transit from Caraquet to destination being some 24 or 25 days, while sailing vessels take anywhere from 30 to 50, or even 60 days, in making the voyage. The principal factor, however, in turning this traffic to the New York route is not the difference in time the shipments are in transit, but the fact that the banks will make advances on shipments made by rail via New York while they will not do so on shipments made by sailing vessels, so that the rail traffic in this commodity will assuredly continue to increase and to move via New York until such time as direct and frequent communication is established between Canadian ports and the Mediterranean, when we can reasonably expect to divert it to Canadian channels." *S.P.* No. 65, 1909.

21. After catching, splitting, and washing, the fish were put into three-quintal tubs along with two gallons of salt per draft. (A "draft" and a half of fresh fish, or 238 pounds, made about one quintal of fish, or 108 pounds dried.) On the first day fish were added to the pickle formed by the salt. After four or five days they were washed in the pickle, put into the "water horse," allowed to drain, were spread on the floor for six hours, and then put on the flakes or drying racks, with the flesh side up. At night they were turned skin up, and the following day turned flesh side up until

*See Notes to Revised Edition, p. 509.

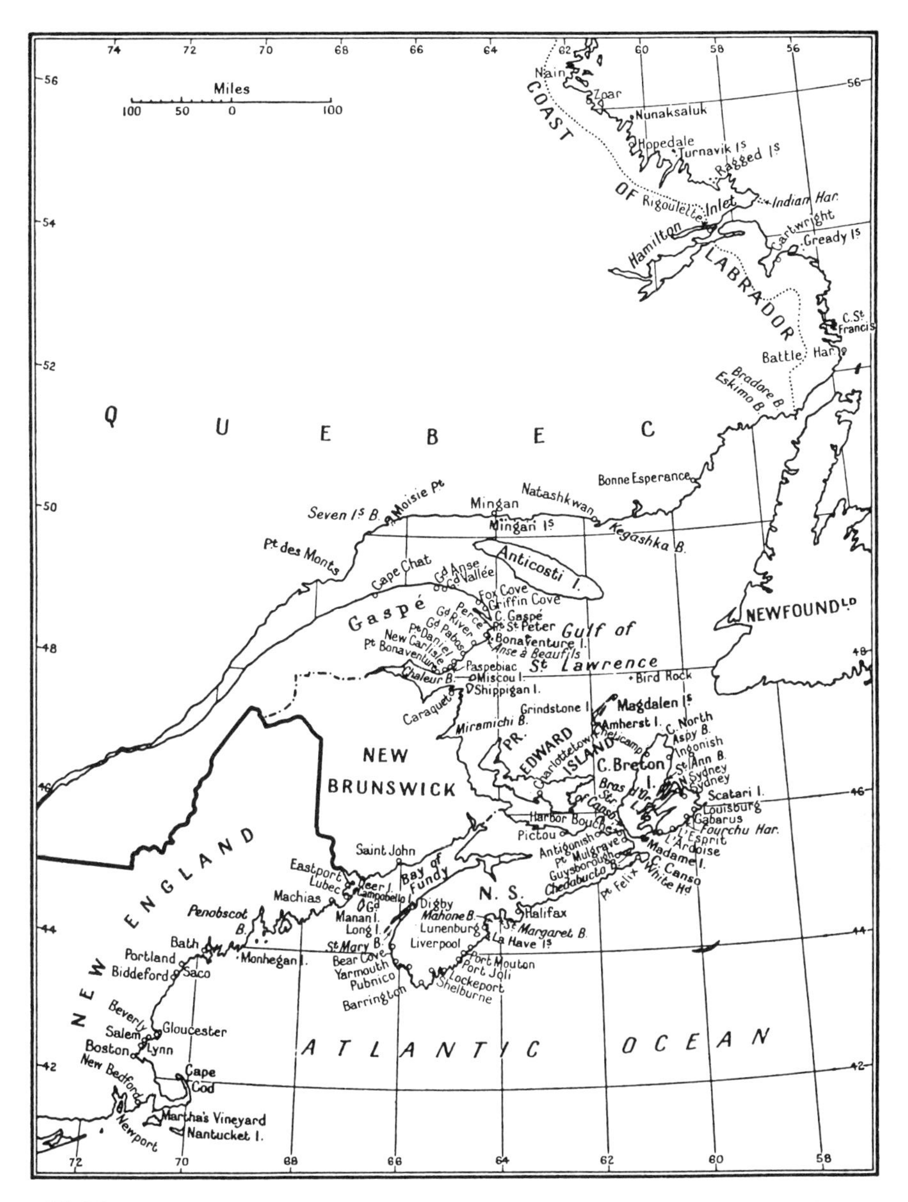

Fishing Ports of Labrador, Quebec, the Maritimes, and New England

organization could offer in its effective control of the product until marketed, and in its ability to compete in a wide variety of markets, with their varying grades. Channel Islands firms continued to maintain their system of apprenticeship which made easier the operation of several stations under one control. Around the Gaspé Peninsula, at Mal Bay, Barachois, and Caraquet small craft fished at considerable distances from the shore and even on the Orphan and Bradelle Banks; and they brought in the fish at less frequent intervals. Independent fishermen were able to give smaller quantities of fish better supervision, and to cure a better product; but differences of skill and capacity in the individual and fluctuations of weather and catch inevitably produced wide variations. The risks that lay alike in extended credit, the careless grading of fish, and the intense competition in foreign markets made for both the disappearance and the amalgamation of companies. The Jersey firms of J. and E. Collas united with Charles Robin and Company in 1892 and became the Charles Robin Collas Company. As such, in 1895, the new company had thirty-four stations on Chaleur Bay and the North Shore. In 1889, a member of J. and E. Collas severed relations and formed a new firm, Collas Whitman and Company, with A. H. Whitman. The latter succeeded as partner, and had adapted an apple-drying process to the drying of fish. In 1904 further amalgamation followed, and the Charles Robin Collas Company became the C. Robin Collas Company, while its head office was moved from Jersey to Halifax. In 1910 this firm, which sold its Canso plant to the Maritime Fish Company, acquired the Lunenburg plant of the Atlantic Fisheries Company, Ltd., Black Brothers (producers of boneless fish), and A. G. Jones and Company, salt-fish merchants in Halifax, and so formed the firm of Robin, Jones and Whitman. In 1935 it had stations at eighteen points in Quebec, at Caraquet and Lameque in New Brunswick, at Cheticamp in Cape Breton, and also wharves and warehouses at Halifax. It handled the light-salted Gaspé cure, the medium Nova Scotia cure, the heavy-salted Lunenburg cure, and the Canadian Labrador cure, and used them to combat the uncertainties of fluctuating markets. It was compelled to conduct a strategy of retreat. Along the North

early afternoon. On the second night they were placed in piles of three or four, skin up, on the following day spread out flesh up, and on the third night put in piles of twelve or thirteen, heads out, tails in, with a large fish on top skin up. On the fourth day they were put in round piles of fifteen or twenty quintals to sweat. After "transpiling" on the fifth and sixth days they were left for seven days, put back on the flakes for three hours' sun, and repiled for eight to ten days. This process continued for three or four weeks. Decline in skill was a serious factor and the increase in the production of pickled cod had serious effects in the dry-fish industry.

Shore of the Gulf of St. Lawrence[22] it suffered as a result of the scarcity of fish, the increase in settlements, the growing importance of the fresh-salmon industry, and because of competition from Quebec merchants, such as the Clarke Trading Company, who had the help of a steamship service. Along the lower Labrador to Blanc Sablon the increasing importance of traps, the poorer grades of fish, and the importance of the West Indian market made easier the competition from Halifax by shipping and trading firms such as the Halifax Fisheries, Ltd., and Rawlings, which absorbed the fish business of Farquhar and Son in 1933.

The high-grade product of the Gaspé area met difficulties in the Italian market and a prohibitive tariff in Brazil. In 1922 the Quebec government assumed direct responsibility for the administration of the industry in that area; and, from 1935 on, with exchange restrictions and sanctions against Italy, they pursued an aggressive policy in develop-

22. At Pointe aux Anglais the Abbé Huard noted in 1895 that 25 boats were engaged in the industry. At Seven Islands fishermen sold the dried product to Halifax or preferably to agents of a branch house of the Charles Robin, Collas, Company at Moisie. At the latter point, fish were bought by the draft, as at Gaspé, from men fishing with the company's 21 boats. Other boats—12—were owned by independent fishermen. The firm of Halliday et Frère shipped fresh salmon from the Moisie River to Quebec. At Sheldrake, Robin, Collas, and Company purchased fish by the draft as did also Touzel, the latter selling the dried product to the former firm. Bouthilier brought over about 120 men each season from Paspebiac and Bonaventure of whom about 80 prosecuted the fishery at Thunder River. Independent fishermen sold by the draft, or dried their fish and sold them to the company or, at the end of the season—about August 20—sent them green to Quebec. There was a total of 85 boats valued at about $100 each. At Richepointe, Robin, Collas, and Company had 35 boats. Every season the two firms brought over about 250 men from Chaleur Bay to Magpie. At St. Jean, Sirois employed 33 boats bringing about 100 men from Chaleur Bay and selling the fish to Bouthilier. In 1895 the Robin, Collas, Company had employed 45 to 50 boats and brought over 120 men. Nova Scotia schooners continued to fish with dories in the vicinity. Vibert, a Jerseyman, had come to Long Point in 1871 and prepared cod to be sold to Charles Robin and Company, but the fishery declined from more than 300 boats to 18. At Eskimo Point the number of schooners had increased from 12 in 1865 to 26, in 1882, but had declined to 12 in 1895. As in the case of the "floaters" in Newfoundland, they fished along the coast—schooners of from 40 to 55 tons carrying 8 men, 3 boys, and 3 boats, and those of 25 to 35 tons 6 men, 2 boys, and 2 boats. The boats were purchased from the owners of Nova Scotia schooners. At the end of the cod-fishing season, schooners used seines for taking herring. The fish were sold to traders or taken green to Quebec. At Natashkwan the fishing declined, especially after the failure of 1885. The house of La Parelle was purchased by Robin, Collas, and Company. In 1895 it had 7 boats engaged in the fishery, while independent fishermen had 21 boats and 5 schooners. In Anticosti waters the fishery was conducted on a small scale at Baie des Anglais and L'Anse-aux-Fraises. The firm of Robin, Collas, and Company had a total of 105 boats on the North Shore and brought 480 men from Gaspé. The fish were taken to Paspebiac and exported, chiefly to Brazil. V. A. Huard, *Labrador et Anticosti* (Montreal, 1897), *passim*. See also Raoul Blanchard, *L'Est du Canada Français* (Montreal, 1935), I, Part III.

ing markets for fresh fish, especially in the summer and fall.[23] The Caraquet "fall cure" has been sold in the Italian market in New York. Quebec continued to follow the policy of the French regime,* leasing concessions on salmon rivers and licensing traps, in contrast with the free fishery of Newfoundland and its numerous disputes as to trapping berths.

The advantages of Nova Scotia in the West Indian market for lower-grade fish, especially after the West Indies Agreement in 1913, enabled firms to purchase substantial quantities of west-coast fish from Newfoundland for reëxport. Regular lines of steamers made it possible for the consumer in the West Indies "to get his fish in small quantities fresh from the northern clime, and the Canadians were enabled to obtain higher prices for fish than they could do if their steamship facilities were not so good, or if they were compelled to ship in cargoes. . . . Newfoundland houses have branches in Halifax . . . and Halifax houses have branches in . . . Newfoundland."[24] "The subsidized lines to the Windward Islands absolutely cut out all competition."[25] In 1922 it was claimed that Halifax controlled the fish trade at Havana, Kingston, Port of Spain, and to a large extent in Barbados, where Halifax fish commanded a premium of $1.50 over that of Newfoundland.

With the abolition of the preference for Canadian fish by Jamaica in 1924, and later on, during the depression, Newfoundland and other production areas began to displace Nova Scotia in the West Indian markets. It was claimed that the Canadian–West Indies Treaty of 1925[26]

23. Louis Bérubé, "Le commerce du poisson frais de la Gaspésie," *Le Canada Français,* April, 1936, pp. 741–768; Pierre Asselin, "L'industrie de la pêche en Gaspésie et ses possibilités de développement," *Etudes Economiques,* IV, 105–131; also *Proceedings of the Nova Scotia Fisheries Conference, Halifax, July 13–14, 1938,* pp. 32, 46, 71. See also Blanchard, *op. cit.,* Part I.

24. *The Royal Commission on the Natural Resources, Trade and Legislation of Certain Portions of His Majesty's Dominions: Newfoundland* (London, 1915), p. 38.

25. *Idem: The Maritime Provinces* (London, 1915), pp. 130–131. Nova Scotia had the added advantage of being able to use the New York lines in competition.

26. The rigidity of Canadian National steamship rates, and preferences to Newfoundland fish, were regarded as further handicaps. Newfoundland and other rates to Brazil were lower than Halifax rates. *Proceedings of the Nova Scotia Fisheries Conference, Halifax, July 13–14, 1938,* pp. 65–71, 88, 90–93. See O. F. MacKenzie and F. H. Zwicker, *Reports on the Markets for Dried and Pickled Fish* (Ottawa, 1938), pp. 17–20; also *Reports of the Nova Scotia Economic Council* (Halifax, 1938), I, 6–10; II, 11–14; III, 43–83. The grant of preferential treatment in 1898 and later years by Canada to the British West Indies revived the struggle between the sugar islands and the fishing regions. Weakened markets in the West Indies led Canada to attempt to secure a permanent share by granting a preference on sugar. Sugar interests have become more aggressive and the fishing interests have become active in protesting against the effects of a sugar monopoly. See "Memorandum on the Effect of the United Kingdom Colonial Sugar Preference on Canada's Trade," *Imperial Economic Conference, 1932, Secret.*

transferred the purchase of sugar from Cuba and the Dominican Republic to sugar-growing regions of the empire, and this led to tariff reprisals on their part, and to the sharp decline in exports of fish from Canada. The lower American duties imposed in 1930 on high-moisture-content fish imported by Puerto Rico increased the advantages held by Newfoundland and Labrador fish.[27] Norway dominated the Argentine market, and the United Kingdom that of central Brazil, while Newfoundland fish stood first in northern and southern Brazil. British Guiana imported largely from the United Kingdom by means of English reëxports of Iceland fish, under the British preference, and by cheaper freight rates. Dutch Guiana also purchased Iceland fish, chiefly pollock. Exports of spring mackerel increased, those of herring declined, and those of alewives improved. Havana has purchased imports directly from Norway and Iceland, and Santiago has increased its imports from Newfoundland at the expense of Nova Scotia. An association of Nova Scotian exporters has capitalized the demands of Trinidad for Lunenburg bank fish. The decline of New York brokerage houses has weakened the markets in the foreign West Indies and Central America for Nova Scotia fish accessible to New York routes.

The removal of the American tariff in 1913 brought about a sharp increase in exports of pickled fish. "Half a million dollars of Gloucester money will be distributed on this coast this year [1914]."[28] "Gloucester takes the very choicest of our catch; Gloucester will pay the biggest price for the best."[29] This market suffered from the increases in the tariff after the war. In 1936, as a result of expanding markets in the United States and depression in the dried-fish trade, supplies of pickled fish for the production of boneless cod and codfish cakes became a product of the domestic fresh-fish trade, and more than 50 per cent of the supply was imported under the low-moisture-content classification. Boneless cod was imported chiefly from western Nova Scotia, its nearness to the market giving it an advantage in spite of the high tariff.[30]

The dried-fish industry has been profoundly influenced by mechanization—for example, by the trawler in Iceland—and by the consequent

27. MacKenzie and Zwicker, *op. cit.*, pp. 29, 49, 59.

28. *Royal Commission on the Natural Resources, Trade, etc.: The Maritime Provinces*, p. 131.

29. *Idem*, p. 129. The Magdalen Islands became an important source of supply. The Eastern Canada Fisheries, Ltd., which acquired rights from the Coffin interests, failed in 1924 and its successor, William Leslie and Company, got into difficulties in the depression.

30. *Proceedings of the Nova Scotia Fisheries Conference*, pp. 61, 116. See G. V. Haythorne, "Canada-United States Trade Agreement and the Maritime Provinces," *Public Affairs*, March, 1939.

encroachments of Newfoundland fish on Nova Scotian markets in the West Indies. The effects of the depression have been more lasting in tropical countries, the result being that recovery in North America has brought about a rapid shift to the United States markets, and pickled cod have been put up and exported to it, rather than dried fish. The falling away in the dried-fish market has also meant an extension of the fresh-fish industry. But when larger quantities of fish were offered to this industry prices were depressed. When skilled labor has been driven from the dried-fish industry, it has been found increasingly difficult to get it back even with the revival of markets.[31] Mobility within the industry depends on such factors as the character and variety of fish available on grounds adjacent to ports, the distance to the fishing grounds, familiarity with them, and on technique and personnel.

With restrictions on imports from Canada to the United States and the decline in the markets for dried fish went a growth of the fresh-fish industry to meet the demands of the industrial centers of Canada. The expansion of coal mining and the iron and steel industry in Nova Scotia and the rise of an urban population increased the local demands for fresh fish and attracted labor from the dried-fish industry. Improved rail and steamship transportation from Nova Scotia, Gaspé, and the North Shore of the Gulf of St. Lawrence meant shipments of fresh salmon to the urban centers of the St. Lawrence. Trawlers were introduced in 1908,[32] and the difficulties inherent in Nova Scotia's distance from her markets were overcome by the imposing of a tariff of one cent a pound on imports of American fish and by a policy of subsidized fast-rail service which began in that year and continued until 1919.[33] The increased consumption of fresh fish stimulated by food propaganda during the war also accelerated the development of the home market.

The expansion of the fresh-fish trade has come with the emergence of special processes. In 1899 the Dominion government had initiated a policy of subsidizing the construction of bait freezers by fishermen's bait associations, and in 1908 forty-five freezers were in operation, thirty-seven being in Nova Scotia. The uncertainty in the demand for bait

31. Mr. Zwicker, in 1938, describes the effect of "easy markets" during the war in lowering the quality of the product. Fish were not properly salted and not dried sufficiently. MacKenzie & Zwicker, *op. cit.*, p. 60.

32. An order in council of September 9, 1908, prohibited trawler fishing within the three-mile limit. Small trawlers for the finnan-haddie industry were introduced as early as 1902. R. R. McLeod, *Markland or Nova Scotia* (n.p., 1903), p. 281.

33. See R. F. Grant, *op. cit.*, pp. 131–133; also R. A. McKenzie, *The Fish Trade of Southern Ontario* (Ottawa, 1931). In 1937 improved refrigeration cars made possible shipments of fresh fish from Halifax to Omaha.

weakened the position of the associations and contributed to the development of a freezing industry in private hands for the production of frozen fish for domestic consumption. The industry tended to concentrate at points less suited to the export of fresh fish, iced, and to suffer from competition in the nearer markets. Demands for capital in the fresh-fish industry have given rise to large-scale organizations. The increasing importance of the American market, of improved technique developed in the United States, and of cheaper Canadian and Newfoundland labor have brought about American control. At North Sydney, cold-storage facilities occupied a strategic position owing to their accessibility both to the fishing grounds and to abundant supplies of herring for bait. These facilities were acquired by the Leonard Fisheries, Ltd., of St. John, New Brunswick. Arthur Bouthilier of Halifax attempted to develop the fresh-fish trade locally and in central Canada, and formed the Halifax Cold Storage Company which was taken over by the North Atlantic Fisheries Company with plants at Halifax and Hawkesbury. The latter plant was sold to the Leonard interests during the war. In 1914 Bouthilier formed the National Fish Company which took over the fish business of the North Atlantic Fisheries Company at Halifax, and later included a plant built at Hawkesbury during the later war years and a plant rented at Harbor Breton in Newfoundland. In 1921 the National Fish Company went into liquidation and was reorganized. After the death of Bouthilier in 1928, control was acquired by the Atlantic Coast Fisheries Company of New York. At Canso, A. and M. Whitman sold cold-storage facilities to the Atlantic Fisheries Company which, in turn, sold to the Maritime Fish Company. The latter, founded in 1910 with the support of Montreal capital, also purchased a fresh-fish plant at Digby. This company purchased and operated trawlers from Canso, but difficulties with ice and an order in council of October 30, 1929, which imposed taxes on trawlers, led to its withdrawal from the district. Fresh fish were sold from Canso and Digby. In 1929, the Maritime Fish Company also came under the control of the Atlantic Coast Fisheries of New York. At Canso the fishery was restricted to exports of fresh fish to inland markets and to the production of pickled fish for export to Gloucester and processing into boneless cod. The Leonard Fisheries, Ltd., operated a wholesale distributing organization at Montreal with plants at Hawkesbury and North Sydney. Upon its failure in 1934, the Hawkesbury plant was purchased by Ralph P. Bell. The North Sydney plant, purchased by D. J. Byrne, was later sold by him to Ralph P. Bell. In western Nova Scotia a Lockeport plant shifted from a salt- to a fresh- and frozen-fish industry, under a Mr. Hodge from Boston.[34] After his death the property was acquired by Bell. In 1936 the latter disposed of his interests in the Hawkesbury plant. He also sold the North Sydney plant and the Lockeport plant to W. C. Smith and Company of Lunenburg. This company, formerly concerned largely with the salt-fish industry, prepared for the shipment of fresh fish, in 1926, by the construction of cold-storage facilities. An additional plant came under their control with their purchase of the Nickerson Brothers' interests at Liverpool.[35]

The fresh-fish industry has largely come under the control of two large organizations.[36] The Atlantic Coast Fisheries controlled the National Fish Company and the Maritime Fish Company, together with such smaller companies as the Pioneer Steam Trawling Company and the A. H. Brittain Company, through the Maritime National Fish, Ltd., with plants at Digby, Halifax, Hawkesbury, and Canso; and W. C. Smith and Company had plants at North Sydney, Hawkesbury, Lockeport, Liverpool, and Lunenburg. Being closer to Halifax as a terminal point for transportation to the interior and possessing dominance in the bank fishery, the latter firm was able more effectively to combine the frozen- and fresh-fish with the salt-fish industry. This diversity has been extended by the acquisition of plants in the eastern and western parts of the province. The larger organization had the advantage of connections with St. John at Digby which became the center of a varied industry including salt fish, especially hake, fresh fish, scallops, smoked herring, and, in 1934, a fish-meal plant. Halifax, however, was the chief center of its frozen- and fresh-fish industry. It has operated trawlers, and, since their numbers were reduced by Dominion regulations, Lunenburg power schooners, for supplies of fish to be handled fresh, frozen, and as fish meal. The Dominion government subsidized, under the Department of Agriculture, a large cold-storage plant in 1926, and later—in 1932—acquired it under the Harbour Commission.

34. In 1914 Lockeport had about 21 small schooners ranging from 20 to 60 tons, 2 schooners of 100 tons each, about 150 boats ranging from half a ton to 3 tons, and 4 small steamers from 50 to 60 feet in length overall, engaged in the lobster, herring, and cod fisheries. "In the summer months it is a port of call for many American fishing schooners seeking bait. During a recent summer, over $20,000 worth was sold to American and local interests. Lockeport's shipping is valued at $150,000. These small vessels make from forty to sixty trips a year.

35. Nickerson Brothers started a salt- and fresh-fish business but sold out to the Seven Seas Fisheries. Later, the latter went into liquidation and came back into the hands of Nickerson Brothers.

36. Among smaller firms should be mentioned H. B. Nickerson at North Sydney, the Hensbee Company and R. E. Jamieson at Canso, Sweeny at Yarmouth, and the Swim Brothers at Lockeport.

Space was leased to the National Fish Company and more recently to a second American organization, the Sea Food Company, controlled by General Foods and General Sea Foods, for the production and storage of fish frozen under the Birdseye patents. The latter company bought out Mitchell and McNeill in 1929, but ceased operations with the depression until 1936. The expansion of the frozen- and fresh-fish business, aided by rapid transportation, has involved large-scale capital equipment and increasingly centralized control.

The trawler controversy is complicated by problems of cold storage and transportation. The schooner fishery implies the use of trawls and an extensive demand for bait which in turn calls for cold-storage facilities and the employment of equipment and labor to take bait. Lunenburg and Lockeport have been engaged in the bait trade for their own ships as well as for those of New England. Trawlers, while not needing bait, require an abundance of coal and ice. They can support with greater dependability, and under a variety of weather conditions, a market demanding larger quantities of fish on certain days of the week and during certain seasons of the year. The greater variations in the catch make necessary cold-storage equipment and provide a greater quantity of material for by-product industries, including the production of dried fish.

The large-scale capital investment[37] now essential to the fresh-fish industry—that is, an investment in cold-storage equipment, packing equipment, and by-products plant, extending in some cases to the ownership of mills for the production of lumber—demands a continuous supply of raw material. The suitability of haddock[38] for filleting purposes, because of its uniform size, has led increasingly to concentration on that variety. Herring can be sold fresh, smoked, or frozen, for human consumption, or it can be sold for bait. The variation in the size of cod makes possible a fresh iced-fish, notably cod steak, a pickled-fish, or a

37. For an account of the burden of the tariff see the *Royal Commission Economic Enquiry, Nova Scotia, 1934*, Appendixes. In the *Report* the writer committed himself to a position opposed to trawlers. The grounds were not stated but it may be said that an increase in trawlers would seriously endanger the prospects of a lower tariff in the United States. Moreover, so long as the fisheries are involved in a conflict of jurisdictions in which the Dominion determines policies and the province is compelled to face the result of those policies in unemployment, resistance to trawlers is justified. It was an essential part of the report, and of the position, that the province should be given control over the fisheries.

38. A. W. H. Needler, "The Migrations of Haddock and the Inter-relationships of Haddock Populations in North American Waters," *Contributions to Canadian Biology*, New Series, VI, No. 10; and A. W. H. Needler, *Statistics of the Haddock Fishery in North American Waters* (Ottawa, 1929).

dried-fish industry in the case of smaller sizes. Variations in the size of fish, the different kinds,[39] varying demands, based in part on religious beliefs, and the problem of devising storage for meat products in tropical regions all limit the possibilities of large-scale production and make heavy demands on managerial skill and ability to adjust the materials to the demands. An increasing centralization permits an increasing specialization for a greater variety of markets, and admits of a growing utilization of facilities. Lunenburg schooners[40] have become less interested in the dried-fish industry and have been engaged in taking catches for the fresh-fish market, particularly in the summer season.[41] The limitations of centralization appear in the costs of transportation and in the inevitable decentralization involved in exploiting widely separated areas.

American restrictions on imports of Canadian fish have made plain the need, particularly in areas adjacent to the American market, of depending on exports of high value, on which the burden of the tariff is lighter. It is claimed that one of the reasons why Canadian fish can be sold in the United States is found in its high quality. This is because of the comparative nearness of the fishing grounds to the Nova Scotia coast as compared with their distance from Boston.[42] Probably a third of the fresh fish, chiefly frozen, is shipped to the United States.

39. In 1934, 212,127 hundredweight of cod were taken offshore in Nova Scotia and 794,546 hundredweight inshore. Of this grand total, 369,566 hundredweight were taken by Lunenburg County, of which 43,553 were used fresh; 82,012 by Halifax County, of which 46,079 became fresh fillets; 19,781 by Shelburne, and 13,956 by Halifax; 94,366 green salted (Guysborough 24,871, Shelburne 18,254, Halifax 10,663), and 132,635 dried (102,902 hundredweight Lunenburg). Haddock taken offshore in Nova Scotia totaled 148,858 hundredweight and inshore 192,648 hundredweight (Halifax 157,429 hundredweight). It was marketed as fresh 87,534 hundredweight (45,275 hundredweight Halifax, 13,838 Lunenburg), fresh fillets 47,366 hundredweight (31,864 hundredweight Halifax), smoked 26,851 hundredweight. Halibut totaled 24,254 hundredweight (24,232 hundredweight used fresh). Of these important classes,—cod, haddock, and halibut—155,341 hundredweight was marketed fresh and 93,445 hundredweight as fresh fillets. Large-scale production of fillets meant the introduction of fish-meal plants. Production began in 1922 with exports to Germany. During the depression Canada has absorbed the total output. *Fisheries Statistics of Canada, 1934* (Ottawa, 1935). See *A Summary of the Report on the Marketing of Canadian Fish and Fish Products* (Ottawa, 1932). About 70 per cent of the fresh fish handled was frozen and 30 per cent packed in ice. For descriptions of the drying of fish at La Have and Lunenburg, see Clara Dennis, *More about Nova Scotia* (Toronto, 1937). See D. B. Finn, "Recent Developments in Processing Fish," *Public Affairs,* June, 1939, pp. 164–167; S. Bates, "The Economic Problems of Nova Scotia Fisheries," *idem,* pp. 1–5.

40. In 1937 Lunenburg sent out 27 vessels which took 103,725 quintals as compared with 79,550 quintals in 1936.

41. When the internal-combustion engine came into use schooners, instead of laying up in winter, turned to fishing for the fresh-fish markets. *Proceedings of the Nova Scotia Fisheries Conference,* p. 20.

42. *Proceedings of the Nova Scotia Fisheries Conference,* p. 50.

The lobster fishery in the Maritimes has been conspicuous in its demand for labor and in contributing to the difficulties of the dried-fish industry. It increased with the depletion of the American lobster fishery, the difficulties of the fishing industry,[43] and the development of steam navigation.

The pioneers of the lobster canning industry in Canada were either United States citizens, who had been engaged in it along the northern shores of the United States or Canadians who had learned the methods of the industry from our neighbours and saw the wonderful opening which our own waters offered for its continuance here. This was hastened by the fact that the fishery was already being exhausted wherever lobster canneries were operated from Massachusetts to Maine.

As a result of the rapid spread of the lobster-canning industry and the growth of the live-lobster trade, the lobster fishery probably surpassed the cod fishery in value in Nova Scotia by the end of the century. The number of one-pound cans packed in that province declined from 5,263,780 in 1900 to 1,959,888 in 1924, but increased slightly after that date.

In the beginning lobsters were invariably bought by count, and the price ranged as low as thirty cents per hundred. As the run of lobsters became smaller undersized ones were taken at two for one; but when they got smaller still the custom changed, and payment is now generally made by weight, market fish being still bought by count. The history of this change in the method of payment, together with the nearly universal practice of narrowing the slats and making the trap more of a jail than it was originally, offers the best possible proof of the decrease in the average size of the fish and the methods adopted, all round, to capture, hold and dispose of the undersized fish to the packer.[44]

In 1928 it could be said that "about three times the number of traps required ten years ago are needed today to take one hundred pounds of

43. High tariffs on herring in the United States and in Haiti have contributed to the increasing importance of lobster fishing at Grand Manan. The Grand Manan Smoked Herring Board was established under New Brunswick legislation to increase the control over marketing.

44. Report of Commander William Wakeham, *S.P.* No. 22a, 1910. See also Report of the Canadian Lobster Commission, 1898, *S.P.* No. 11C, 1899; *Dominion Shell Fishery Commission,* 1912–13 (Ottawa, 1913); and "Evidence Taken before the Marine and Fisheries Committee Respecting the Lobster Industry," *Journals of the House of Commons* (1909), XLIV, Appendix No. 3; McLeod, *op. cit.,* pp. 273 ff. In 1910 United States firms operated 71 canneries in Canada. The Portland Packing Company operated 21; Burnham and Morrell, 30; H. C. Baxter and Brothers, 8; H. L. Forhan, 5; D. W. Hoegg and Company, 6; and The Snow Flake Canning Company, one. *S.P.* No. 22, 1910, Special Appended Report 11.

lobsters." The total shipments from Nova Scotia had increased to 326,313 hundredweight by 1898 but declined to 87,321 in 1908, and to 49,435 in 1918. Live-lobster exports to the United States became conspicuous in the western part of the province subsequent to 1881. The region from which live lobsters have been shipped has broadened with improved transportation. In 1930 a lobster-transport service between points in eastern Nova Scotia and Boston was subsidized by the government. The live-lobster trade was extended as far as the Magdalen Islands. In 1934 Canada exported 97,485 hundredweight of fresh lobsters, of a value of $1,500,452, almost entirely to the United States, and 52,938 hundredweight of canned lobsters, of a value of $2,499,372, of which about one half went to the United Kingdom. Conservation measures have been steadily extended.[45]

The decline of the dried-fish industry, the increasing importance of large-scale organizations, the demands of capital equipment, and the expansion of the American market have been followed by displacements of labor and serious maladjustments.[46] The Canadian Fisheries Association, formed in 1915, was chiefly confined to producers and distributors. The active promotion of fishermen's organizations was begun as a result of the report of the Royal Commission in 1928 and the appointment by the Department of Fisheries in 1929 of the Reverend M. M. Coady as organizer. Nearly 150 various organizations and the United Maritime Fishermen then came into being. The latter had become sufficiently strong in 1937 and 1938 to protest against the fishing companies in the strike of Lunenburg fishermen.[47] It has interested itself increasingly in

45. See *The Report of the Royal Commission Investigating the Fisheries of the Maritime Provinces and the Magdalen Islands* (Ottawa, 1928), pp. 9 ff.; also p. 119.

46. See Diamond Jenness, "Canada's Fisheries and Fishing Population," *Transactions of the Royal Society of Canada,* 1933, sec. 11, pp. 41–46.

47. The "lay" on schooners varies in the different ports and firms. Schooners purchase supplies on credit from the outfitter, and operate on a fifty-fifty basis—the owner supplying provisions, salt, gear, and vessel, and the crew labor, half the bait, and half the ice. The returns are divided equally among the men. Ownership is divided into shares up to 64 parts. Firms prefer not to take shares, but ownership tends to pass to their hands. In the fresh-fish business, slight modifications have been introduced at Lockeport; for example, "the one-quarter lay"—the vessel taking one quarter, and the crew three quarters. From gross stock first allowances are made for gear, which is paid for by the owner and the men. The crews supply the food, fuel oil, bait, and lubricating oil. The owner meets the cost of the upkeep of the vessel, the insurance, and other items, and it is claimed that one quarter is not sufficient. "The one-fifth lay" allows 20 per cent clear for the boat, the crew taking all the risk on gear. A strike for an increase from 2½ cents a pound to 2¾ cents lasted through January and February, 1938. It was carried on by the crews and captains of vessels owned by Maritime National Fish, General Sea Foods, Lockeport Cold Storage, and Lunenburg Sea Products; that is, a sharp break with large-scale capital organizations was in evidence.

commercial work, particularly in the exporting of lobsters for individual coöperatives, and in the purchase of supplies such as rope and twine for local branches in eastern Nova Scotia and Cape Breton. Prince Edward Island formed a separate union in 1934. Local organizations were developed along coöperative lines and were assisted in their buying and selling by the central organization at Halifax.

Under the Reverend M. M. Coady and Mr. A. B. MacDonald, the Department of Extension in St. Francis Xavier University has actively fostered educational work and the growth of the coöperative movement by mass meetings, study clubs,[48] and the dissemination of informative material through libraries and a regular publication, *The Maritime Cooperator*. The department, according to its report in 1936, "has from the very beginning fostered the idea of economic coöperation." The first step is generally the formation of a buying club, and is followed by the opening of a coöperative store. Ten stores "have come into existence during the past four years and many more are in process of formation." Following an act passed in Nova Scotia in 1932, the first credit union was formed in December of that year and many others have rapidly been added to it.[49] The table below and those on page 440 may give some idea of the growth of the movement and how its activities are divided.

Seventeen cooperative lobster factories, fostered by the Extension Department, are now operating in eastern Nova Scotia and serve the fishermen of

48. UNIVERSITY OF ST. FRANCIS XAVIER, EXTENSION ACTIVITIES

	General Meetings	Attendance	Study Clubs	Membership
1930–31	192	14,856	173	1,384
1931–32	280	20,476	179	1,500
1932–33	380	23,000	350	5,250
1933–34	500	25,000	950	7,256
1934–35	450	27,000	940	8,460
1935–36	470	43,000	860	8,000
1936–37	...	....	1,013	10,000
1937–38	...	....	1,100	10,000

49. CREDIT UNIONS IN NOVA SCOTIA (1936)

Counties	Members	Borrowers	Assets
Antigonish	590	118	$ 3,115.35
Pictou	782	248	7,963.50
Colchester	50	...	150.00
Digby	50	4	125.02
Inverness	1,034	307	6,782.46
Victoria	245	45	883.30
Halifax	259	12	3,285.27
Guysboro	245	74	1,998.42
Richmond	378	149	2,996.00
Cape Breton	6,552	3,496	162,944.03
GRAND TOTAL	10,185	4,453	$190,243.35

seventy-five communities. The fishermen in eleven communities have organized societies for the sale of their fresh and cured fish, particularly herring, cod and haddock. A number of fishermen's cooperatives canned blueberries and fox berries as well as salmon, mackerel, etc. The people of three communities own their own mills for the sawing of rough lumber.[50]

The success of the movement has been partly a result of the able leadership of Catholic priests,[51] more recently also that of Protestants, working in Scottish communities in areas that have suffered acutely from the

50. COÖPERATIVE LOBSTER FACTORIES

Limited Companies in the Organization	Capital	Date of Incorporation
The Tor Bay Canning Company, Larry's River	$ 5,000	March 29, 1934
The Maryville Cooperative Cannery, Little Judique Ponds	2,700	April 14, 1934
The Northumberland Cooperative Packers, Arisaig	5,000	Nov. 4, 1933
The Cheticamp Fishermen's Cooperative Society, Cheticamp	3,000	Dec. 7, 1933
The Richmond Shorefish, Petit de Grat	10,000	Aug. 24, 1933
The La Pointe Fishermen's Cooperative Society, Plateau	2,000	July 13, 1933
The South Ingonish Cooperative Cannery, South Ingonish	16,000	May 18, 1933
The Blue Ribbon Canners, Little Dover	1,200	March 30, 1932
The Fishermen's Cooperative Canners, Port Felix	3,000	Feb. 9, 1932
The Cape Rouge Fishermen's Cooperative Society, Cape Rouge	1,500	Dec. 1, 1932
The Capeview Cooperators, Main-à-Dieu	5,000	March 9, 1936
The Bayview Cooperative Cannery, Judique South	5,000	Oct. 23, 1935
The St. Georges Cooperative, Ballantynes Cove	5,000	Sept. 26, 1935
The Havre Bouche Cannery, organized in 1932	Operated through locals of the United Maritime Fishermen; not incorporated on share-capital basis.	
The Little Bras d'Or Cannery, organized in 1934		
The Grand Etang Cannery, organized in 1930		

FISHERMEN'S COÖPERATIVES FOR PROCESSING AND MARKETING FISH

New Harbor Cooperative Fisheries, New Harbor	1,600	April 14, 1934
Charlos Cove Cooperative Fish Society, Charlos Cove	150	Feb. 21, 1934
Mabou Cooperators Fisheries, Mabou Harbor	5,000	March 9, 1936
Amet Sound Packers, Tatamagouche	5,000	Dec. 9, 1935

In 1937, sixteen fishermen's coöperatives with 760 members had total assets of $67,976, liabilities $17,546, paid-up capital $14,849, surplus and reserves $35,580, and total sales of $251,144. *Report of Cooperative Associations . . . 1937* (Halifax, 1938).

51. The Reverend Dr. J. J. Tompkins was responsible for the general development of the work in its early stages. Indeed, it might be regarded as a result of the particularist character of maritime activity. With the failure of a movement making for the federation of universities in Nova Scotia, Dr. Tompkins, the Vice-President of St. Francis Xavier University, and others of the staff took charge of parishes in the province. At Canso, Dr. Tompkins stimulated an interest in coöperation, in part by making the wheat pool of the Canadian Prairie Provinces a matter of public information in Nova Scotia. He was instrumental in bringing about a demand for the Royal Commission of 1928, which made numerous recommendations. The success of the coöperative centers at Dover, at Harbor Bouché, and at Little Bras d'Or and in other places owes much to the assistance of the clergy. The strong support given by the Carnegie Foundation has been an important factor in the adult-education movement. Better roads and the automobile have enormously facilitated the work of organization. See *What Fishermen Are Doing,* Reports made to the Rural and Industrial Conference at Antigonish on August 19, 1937.

trend toward centralization which accompanied modern industrialism in the fishing industry. It has gained through a concentration on luxury products such as lobster and salmon, and because of the demands of the United States market for pickled fish for processing into boneless fish and other products, particularly since the recovery from the depression. More recently the movement has extended to Newfoundland and also to New Brunswick, where the federal government has given its support to the organization of the fisheries along coöperative lines under Father Ciasson at Shippigan.

The federal government has likewise attempted to increase the internal market for fresh fish by arranging for a system of subsidies for the betterment of transportation, the education of producers and consumers, and research work to aid in the solution of technical problems.[52] An escape from the problems of competition in the export markets has been sought in the extension of the domestic market for fresh fish. "The biggest problem is to get more and more good-quality fish on the tables of the consumers in this country."[53] The necessity of more efficient retail outlets has been constantly urged as a means of meeting competition from other food products. Intensive advertising campaigns have been carried on in Ontario and Quebec. Fish-inspection legislation was enacted in 1914 and has been improved both by amendments and in its enforcement, especially since inspection was made compulsory in 1933. The number of trawlers has been reduced as a means of controlling production. The provincial government of Nova Scotia has taken an increasingly active part in the fisheries, in spite of federal control. Federal-provincial loans have been arranged for hook-and-line fishermen, and totaled $157,090 in 1936–37. In the same year a subsidy was paid by the province of $1 a quintal on dry cod and 86⅔ cents on scalefish. It was claimed that this relieved the pressure on the pickled-fish market in Gloucester and increased prices for that product,[54] and that it also relieved the pressure on the fresh-fish markets of the interior, and did as much for the lobster and mackerel markets. In 1938 loans totaled $450,000, of which two thirds was provided by the federal government. Three trawlers were licensed to the Maritime-National Fish Company on condition that it purchased through the United Maritime Fishermen the catch of the inshore fishermen of southeastern Nova Scotia from

52. See the *Proceedings of the Nova Scotia Fisheries Conference,* pp. 74, 86, and *A Summary of the Report on the Marketing of Canadian Fish and Fish Products* (Ottawa, 1932); also R. F. Grant, *op. cit.*

53. *Proceedings of the Nova Scotia Fisheries Conference,* pp. 96–97.

54. MacKenzie and Zwicker, *op. cit.,* pp. 38, 57; also *Proceedings of the Nova Scotia Fisheries Conference,* pp. 55–57; also *Reports of the Nova Scotia Economic Council* (Halifax, 1938), XI, 11–38.

October to April at Halifax prices less transportation cost. In 1939 a federal act provided for the establishment of a salt-fish board to assist exports of salt fish to the extent of 25 per cent of the value.

The increasing importance of the lobster and the fresh- and frozen-fish industries, following the development of refrigeration facilities and the introduction of gasoline engines, was a factor contributing to the decline of the dried-fish industry. Other factors were the demands for labor which were inherent in the construction of railways in western Canada, the increasing industrialization of eastern Canada, including the growth of coal, iron, and steel industries in Nova Scotia, and the increasing importance of the cash system which accompanied improved transportation. The war period brought demands for transport, and—followed by the sharp decline in the prices of dried fish in 1921, and the profits made by rumrunning during the prohibition period in the United States—a further decline in the Lunenburg bank fishery. Fluctuations in prices characteristic of the North American continent involved wide swings in the demand for tropical products such as sugar and coffee; and the effects of the disparity made themselves felt in revolutions, in low standards of living in tropical regions, and also in low standards of living for producers of dried fish, who were forced to sell to those tropical regions and to purchase supplies from the temperate.

The task of maintaining a balance in the fishing industry has been a delicate one. Restriction of production by limiting the number of trawlers may increase the demand for labor, especially in the outlying ports; but the difficulties of rapid transportation may likewise decrease the consumption of fresh fish in the interior. In the United States, Canadian fish has greater difficulty in competing with the more highly mechanized industry and, in Canada, with products in the interior. The migration of labor to the United States has been checked by the quota regulation. An increase in consumption in the interior by extensive advertising, subsidies, and general improvements will in part offset a tariff detrimental to Nova Scotia though favorable to the interior; but it is doubtful economy to attempt to encourage the consumption of a product which is not as fresh as efficient mechanization can make it. Alternative methods of increasing consumption include the movement of population to the fishing centers, something of the sort which—though differing in kind and importance—may be seen in what takes place in the so-called "tourist trade." It would also seem important that attempts be made to determine and standardize grades of fresh fish. The restriction of consumption in the interior by the failure to provide the best possible product increases the production of pickled and dried fish for the difficult world markets. Labor involved in these industries is compelled on the one hand to meet competition from the mechanized industries of some countries, and from the lower standards of living in others. This is particularly the case in countries to which Canada has limited direct-shipping facilities, to which she is not able to offer an important market for goods imported in return under preferential treaties, and in which unfavorable tariffs combine with a low standard of living to restrict buying power. In spite of the importance of the cured-fish branch of the industry to the price structure and to the stability of outlying villages, the most effective economy will be in the direction of increased mechanization, in more rapid transportation, and in greater efficiency in the marketing organization of fresh fish on the North American continent. But a major revolution in the fishing industry, involving a shift from dependence on low-standard-of-living countries to dependence on high-standard-of-living areas in North America, can be accomplished only with tremendous effort.

FRANCE

To the flotilla of barks engaged in a common fishery there corresponds the big Breton village. And, since men are many, and the fishing ground is limited, people must stand together to defend it from "foreigners." Now the "foreigners" in this case are the people who live in the adjacent village; and villages are in swarms all along the coast. . . . Every little agglomeration of humankind depends upon the size, the wealth in fish, and the distance of the fishing ground from the village. . . . Another thing: The greater the extension of the fishery, the greater the concentration of the fisherfolk.

HÉRUBEL, *Pêches maritimes* (PARIS)

NEWFOUNDLAND, having succeeded in restricting the French bank fishery by passing the Bait Act, proceeded to press for the withdrawal of French fishermen from the so-called "French Shore." The restriction of the French hastened the introduction of trawlers and concentration on a small number of ports in France. The schooner was to be relegated to a position of minor importance, with disastrous effects for St. Pierre.

The Bait Act became effective in January, 1887, and the arrangement reached between France and Great Britain in 1885 was rejected by Newfoundland in March, 1887. It was claimed that the Bait Act blocked the French fishery and reduced competition in foreign markets to such an extent that the price of fish increased from 12 shillings to 15. Since the French fishery flourished, however,[55] it was probably the fail-

55. Ferdinand Louis-Legasse, *Evolution économique des Iles Saint-Pierre et Mi-*

ure of the Norwegian fishery in 1880 and its subsequent difficulties which brought about a rise in price. The number of schooners at St. Pierre increased from 186 in 1889 to 210 in 1897, and the number of small boats totaled 420 in 1900. The population increased from 5,929 in 1887 to 6,482 in 1902. Fishermen from St. Pierre had the advantage of an intimate knowledge of, and proximity to, the fishery. They were able to make four voyages a year to the Banks. Small capitalists with schooners valued at 12,000 to 15,000 francs dried their fish in St. Pierre or sent it to be dried at Bordeaux, and then carried it to the West Indies. Exports declined up to 1891 but increased thereafter.[56] Three-masted vessels which came from France to the fishery[57] increased in size and efficiency and heavy boats were replaced by light American dories.[58] The ships from France had the advantage of supplies of salt which cost from 60 to 75 per cent less than it cost at St. Pierre, and of provisions and supplies cheaper by from 25 to 45 per cent, partly as a result of a protective tariff introduced at St. Pierre in 1892. They also had the advantage of cheap labor. In 1898, 177 vessels sailed from France, of which Fécamp sent 53, Granville 30, St. Malo 58, Cancale 15, and Binic 15.

The French Shore was subjected to steady encroachments as a result of the increase in the population of Newfoundland. In 1880 Newfoundland began to develop a lobster fishery; and, in 1882, 1883, and 1889, factories were built on the French Shore at, respectively, St. Barbe, Port Saunders, and Meagher's Cove. A protest was lodged by France in September, 1886; and in the following year a French warship destroyed property first at Port Saunders, and in 1889 at Meagher's Cove. In 1887 French fishermen from Binic and St. Brieuc complained of encroachments by the English and of their use of cod traps. In the fol-

quelon (Paris, 1935), chap. iii; see also, for a valuable account of the French fisheries, George Roché, *Les Grandes pêches maritimes modernes de la France* (Paris, 1894); Edgar Aubert de la Rüe, "Le Territoire de Saint-Pierre et Miquelon," *Journal de la Société des Américanistes,* XXIX (2), 1937, 239–272.

56. French Exports of Cod from St. Pierre and Miquelon (*in hundredweights*)

1886	909,953	1892	434,858	1898	583,139
1887	756,144	1893	522,056	1899	628,011
1888	559,529	1894	486,586	1900	682,779
1889	531,457	1895	593,008	1901	562,230
1890	505,595	1896	734,124	1902	594,935
1891	411,887	1897	678,292	1903	419,748

57. After 1888 only a small number of bankers went to St. George's Bay because of the handicap of distance and time. Vessels turned to the use of salt herring and shellfish taken on the Banks with improved devices.

58. Huard, *op. cit.,* p. 188.

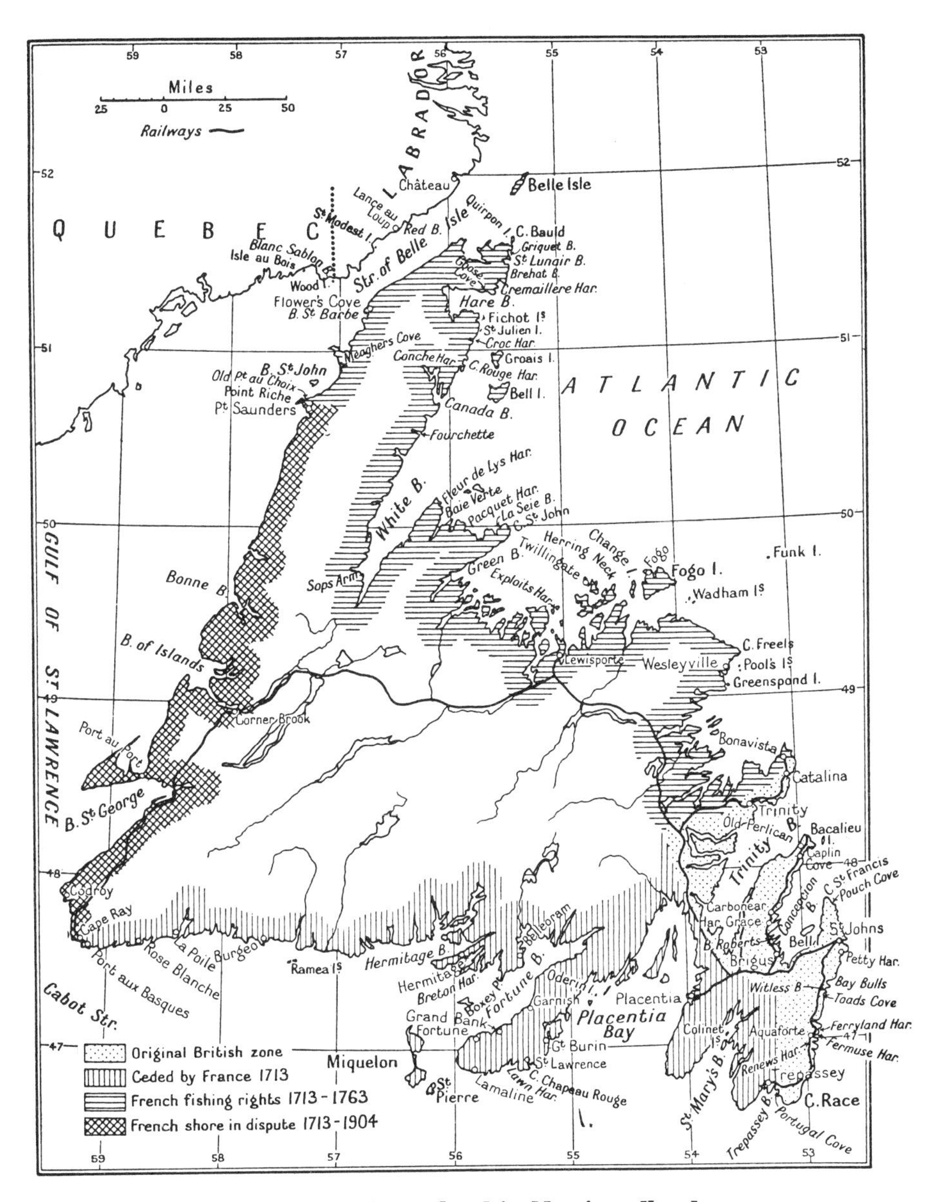

France and England in Newfoundland

lowing year, Newfoundland protested against the interference of the French and against the construction of French lobster factories; and in 1889 the use of cod traps on the French Shore was outlawed. When France and England agreed upon a *modus vivendi* to become effective on July 1, 1889, Newfoundland in her turn protested against it and claimed that the French were entitled only to the right "of fishing upon the coast and of drying the fish upon the land."[59] As the *modus vivendi* was renewed from year to year, Newfoundland renewed her protests annually, and refused to arbitrate. Under the *modus vivendi* each lobster packer was given a specified strip of coast under the control of British[60] and French commodores; but scarcity of space and a falling off in the lobster catch brought on difficulties, particularly since illegal packers were at work. British lobster factories steadily increased until they numbered 59 in 1897. The French fishery dwindled from 14 stations with 15 vessels and 649 men in 1894 to 6 stations with 6 vessels and 326 men in 1904,[61] together with 97 small-boat fishermen from St. Pierre. Lobster factories owned in France shipped 1,980 cases of lobster, and factories owned in St. Pierre 1,030. These factories had begun to purchase lobsters from Newfoundland fishermen and rigorous measures had been taken in 1902 and 1903 to check this trade. A treaty was finally signed on April 8, 1904, which terminated the right of the French to land and dry fish, in return for a payment of 1,375,000 francs.[62] French

59. The Newfoundland legislature objected to the agreement on the grounds that it had not been consulted. A public mass meeting was held on March 26 which supported the legislature and resolved "that it is absolutely necessary to the prosperity of the inhabitants of this colony that the last vestige of French rights shall be removed." Delegates were sent to Canada and to Great Britain to state the case against the restriction that had been put upon Newfoundland's expansion on the French Shore. *French Treaty Rights in Newfoundland, the Case for the Colony Stated by the People's Delegates* (London, 1890). See also *Documents diplomatiques; affaires de Terre-Neuve* (Paris, 1891); Charles de la Roncière, *La Question de Terre-Neuve;* and *Journals of the Assembly, Newfoundland,* 1890, Appendix, pp. 409 ff.; also *idem,* 1891, pp. 555 ff.; *idem,* 1892, pp. 522 ff.; *The Newfoundland French Treaties Act 1891* (London, 1891); *Correspondence with the Newfoundland Delegates Respecting the Proposed Imperial Legislation for Carrying out the Treaties with France* (London, 1891); *Further Correspondence Respecting the Newfoundland Fisheries, 1890–91* (London, 1891); *Further Correspondence Respecting the Newfoundland Fisheries* (London, 1891); *Further Correspondence Respecting the Newfoundland Fisheries, 1891–92* (London, 1892); *Further Correspondence Respecting the Newfoundland Fisheries* (London, 1893).

60. In 1834, on the lapse of the Newfoundland Fisheries Act of 1824 (5 Geo. IV, c. 51), naval officers carried out orders and instructions, but the Supreme Court negatived the decisions of these officers, and Sections XII and XIII of 5 Geo. IV, c. 51, were revived.

61. The more important were Red Island, Long Point, Tweed Island, Port aux Choix, Rouge, St. John's Island. See Louis-Legasse, *op. cit.,* chap. iv.

62. See *Journals of the Assembly, Newfoundland, 1904,* pp. 226 ff.; also P. T. McGrath, *Newfoundland in 1911* (London, 1911), chaps. xix, xx; and Lord Birkenhead,

owners of property were compensated by the British government, and arrangements were made for concurrent rights as provided for in the North Sea Conventions of 1881 and 1887.

After the turn of the century bad fishing seasons, heavy losses in wrecked schooners, and the introduction of trawlers led to the disappearance of the St. Pierre fleet.[63] The number of schooners declined from 151 in 1904 to 40 in 1909, and to one in 1915. The number of small boats with motor equipment totaled 288 in 1921. The number of ships from France engaged in the bank fishery had dropped from 226 in 1904 to 214 in 1909, had increased to 227 in 1914, had declined anew and to the low point of 53 in 1917, then had gradually increased to 129 in 1924, and declined once more to 44 in 1934. All ports except St. Malo, St. Servan, and Paimpol ceased to send ships during the war period, and these sent only 4. In 1925 Fécamp sent 16, Granville 9, Cancale 4, St. Brieuc one, Binic one, and St. Malo and St. Servan together sent 86. The decline of the St. Pierre fishery and the French schooner fishery was paralleled by the rise in the number of trawlers. Limitations in the supply of bait gave them a decided advantage in contrast with the use of setlines. The French catch with a bounty of 10 francs a quintal increased from 57 million pounds in 1918 to 344 million in 1925. French trawlers appeared on the Banks in 1904.[64] Following an exceptional fishery in 1908, they had migrated from Iceland, and by 1909 they numbered 32. Sixteen were from Boulogne, 4 from Fécamp, 2 from Le Havre, and 8 from Arcachon. In the years following 1914, they were used for war purposes; but after the war they increased again to 38 in 1920; and then, following a decline, to 47 in 1928. In 1934 there were 37. In 1927 Boulogne's trawlers had disappeared, but Fécamp sent 19, Bordeaux 8, and St. Malo 7. The size of trawlers increased from a maximum of 200 tons in 1909 to 2,000 tons in 1926. In the case of St. Malo, the average catch of a schooner was 5,000 quintals and of a trawler 18,000. The advantages of Fécamp as a port increased, and labor and skill were attracted to it. Improved regulations to protect labor, especially in the legislation of 1907,[65] and the conse-

The Story of Newfoundland (London, 1920), chap. x. For a full account see Emile Hervé, *Le French-Shore et l'arrangement du 8 Avril, 1904* (Rennes, 1905).

63. See Louis-Legasse, *op. cit.,* chap. v; also M. Bronkhorst, *La Pêche à la morue* (Paris, 1927), pp. 30 ff.

64. M. G. Massenet, *Technique et pratique des grandes pêches maritimes* (Paris, 1913), chap. vii.

65. Bronkhorst, *op. cit.,* pp. 45 ff., 152 ff.; J. Kerzoncuf, *La Pêche maritime* (Paris, 1917), pp. 89 ff.; G. de Raulin, *L'Industrie de la pêche* (Paris, 1925), pp. 98–99, chaps. vi, viii; also M. A. Hérubel, *Pêches maritimes* (Paris, [1911]), pp. 205, 222, 228–232, 240–241. On the cruelty, dangers, and long hours of the fishery see *Sur les Grand Bancs* (Paris, 1905); Pierre Loti, *Pêcheur d'Islande;* Léon de Seilhac, *Marins*

quent increase in the cost of vessel construction and operation, with the continuation of the bounties and duties on fish, went to increase the use of the trawler. Hardships have been imposed even on the trawlers by recent social legislation by the French government and the fall of the franc. The increase in trawlers was accompanied by an increased concentration of drying at Bordeaux.[66]

The rise of industrialism has caused far-reaching changes in the organization of the French fishery.[67] The small villages[68] of Brittany which dominated the dried-fish industry have fallen away, and in their place there is a centralization in ports that are large and specialized. The rise in the prices of food and supplies which accompanied a post-war policy of self-sufficiency has hampered the industry. The effects of the mechanical revolution on St. Pierre have been disastrous. The Canadian Hazen Bill of 1913 (3–4 Geo. V, c. 14) restricted the rights of fishing vessels in Canadian ports, and compelled trawlers to carry larger quantities of supplies from the home ports. In 1917, with government support, a refrigerating plant was established at St. Pierre at a cost of more than 15,000,000 francs, but the difficulty of getting the larger trawlers into the harbor rendered it of little importance. With the passing of the American Volstead Act, effective on January 1, 1920, the energies of the colony were turned to the illicit trade in wines and spirits. The repeal of the Prohibition Amendment and the depression were sources of serious problems, and on August 23, 1933, the situation culminated in a riot. Laboring under enormous deficits, the government has become increasingly centralized in the Conseil Supérieur des Colonies. The difficulties of the trawlers from France, the fall of the franc,[69] and the preference granted by the Dominican Republic on French fish in 1936 are streaks of light in the dark sky which hangs over St. Pierre.[70]

pêcheurs (Paris, 1899); and on legislation see Pierre Romet, *Etude sur la situation économique et sociale des Marins-Pêcheurs* (Paris, 1901); Henri Cuny, *Essai sur la condition des marins-pêcheurs* (Paris, 1904); Jules Prigent, *Loi du 17 Avril 1907 sur la sécurité de la navigation et la réglementation du travail à bord des navires* (Paris, 1910); E. Coué, *De la Sécurité et du travail dans la marine de commerce* (Brest, 1912); and J. M. Grossetête, *La Grande pêche de Terre-Neuve et d'Islande* (Rennes, 1921); Isaac Tual, *L'Engagement des marins pour la grande pêche* (Paris, 1907).

66. Massenet, *op. cit.,* chap. v.

67. Kerzoncuf, *op. cit.,* pp. 2–3.

68. The fishery in France, in contrast with that of Great Britain, was scattered among "a great number of little ports"; it was not centered "in a few great ports." M. A. Hérubel, *Pêches maritimes,* pp. 185–187. "In our Brittany tiny fishing ports are in great numbers. But, on the other hand, there is not a single great commercial port." For the difference between fishing and commercial ports, see *idem,* p. 177.

69. See P. F. Vineberg, *The French Franc and the Gold Standard* (Montreal, 1938).

70. Louis-Legasse, *op. cit.,* chaps. vi, vii.

CHAPTER XIV

CAPITALISM IN NEWFOUNDLAND, 1886–1936

To have abandoned the principle of democracy without accomplishing economic rehabilitation is surely the unforgivable sin.

THOMAS LODGE, *Dictatorship in Newfoundland*

THE abandonment by France of the Newfoundland shore, the decline of St. Pierre, and the migration of schooners from Nova Scotia in the prewar period had been favorable to Newfoundland. After the war the increasing efficiency of the French trawler fishery together with the growth of the Iceland trawler fishery caused serious competition, particularly in the markets for soft-cured cod.

Industrialism spread from Nova Scotia to Newfoundland. The vigorous, independent outlook of Nova Scotia evident in her demand for better terms in the Canadian federation appeared again in the repeal election of 1886, and in the effective demands upon the federal government for the construction of a railway from Point Tupper to Sydney, a line that was completed in 1890, and for a car ferry connecting with the mainland. This was completed in 1893. With the completion of the line to Sydney, Mr. (later Sir) Robert Reid turned his attention to the task of linking St. John's with Sydney by a combination of steamship service and transinsular railway. In 1881 he signed a contract with the Newfoundland Railway Company, and completed a line to Placentia in 1888. The firm of Reid and Middleton secured a contract in 1890 for the construction of a line north from Placentia Junction, as one part of the transinsular project. The line had been practically finished to Exploits River when, in 1893, a new contract was drawn up, and a new route was chosen, one that went to the Humber River valley and thence south to Port aux Basques. The completion of this railway in 1896 was followed successively by the extension of steamship lines along the coast northward to the Labrador and southward to Port aux Basques, and by the construction of railway branches from St. John's to Trepassey; from Brigus Junction to Harbor Grace, Carbonear, and Bay de Verde; from Whitbourne to Hearts Content; from Shoal Harbor to Bonavista; and from Notre Dame Junction to Lewisporte.[1] The numerous bays of the coast were linked to St. John's by the railways during the

1. See *The Royal Commission on the Natural Resources, Trade and Legislation of Certain Portions of His Majesty's Dominions: Newfoundland* (London, 1915), pp. 10–28, 34–37.

winter and by the steamships during the summer. Telegraph lines were built, and later communication was made more complete by radio transmission. Industrial development and an increased population followed improved transportation and communication;[2] and industries began effectively to compete with the fishery for labor.[3]

With her growing settlements, Newfoundland not only compelled the withdrawal of France but pressed for a greater control over the fishery, as against Canada and the United States. Following the abrogation of the Treaty of Washington, a duty was imposed by the United States on Newfoundland fish; and Newfoundland, like Canada, adopted the *modus vivendi* which required United States vessels to pay $1.50 per registered ton for annual licenses. Licenses granted in either Newfoundland or Canada were accepted in both countries and the revenue was divided. Canadians protested against possible interference, under the Newfoundland Bait Act, with the bank fishery and the Labrador herring fishery; but on April 20, 1887, this reassuring message was sent by Newfoundland: "Your fishermen are on same footing as our own, under bait bill, and no practical impediment in way of either." It was later claimed that American and Canadian vessels took advantage of the new arrangements to sell fish to St. Pierre; and Canadians were charged with selling herring from Cape Breton, the Magdalen Islands, and even from Newfoundland. A proclamation was issued by Newfoundland on April 3, 1890, giving notice that "all foreign, and British vessels not belonging to this colony, which require bait from our coasts for the prosecution of the cod fishery" could purchase one barrel of bait per ton register, but only on paying a license fee of one dollar per registered ton, which license would be good for only three weeks. A new license was required under similar terms at the end of that time. It was explained that "the government had no alternative but to put all outside vessels on the same footing, thus securing to the colony the advantages of a trade that others were engaged in at our expense." Lunenburg protested that her fleet of 80 sail, with vessels averaging from 80 to 100 tons, which visited

2. Population of Newfoundland and Labrador: 1884, 197,335; 1891, 202,040; 1901, 220,984; 1911, 242,619; 1921, 263,035; and 1933, 289,516.

3. The expansion of the iron and steel industry in Nova Scotia resulted in the development of the iron mines at Wabana on Bell Island in Conception Bay. W. J. A. Donald, *The Canadian Iron and Steel Industry* (Boston, 1915). A paper mill was constructed at Grand Falls in 1909, with a port at Botwood, by the Anglo-Newfoundland Development Company, controlled by Lord Northcliffe; and a pulp mill was built at Bishop's Falls by Albert E. Reed and Company of London. See P. T. McGrath, *Newfoundland in 1911* (London, 1911); D. C. Seitz, *Newfoundland* (London, 1927); also J. R. Smallwood, *The New Newfoundland* (New York, 1931), *passim;* and *The Book of Newfoundland,* ed. J. R. Smallwood (St. John's, 1937).

Newfoundland from three to five times a year, would be subject to a tax of $40,000. Canadians pointed out that Newfoundland vessels were admitted to Canadian inshore fisheries and to port privileges, whereas their vessels were required to pay light, harbor, and pilotage dues, and that they maintained five lights and four fog signals on the Newfoundland coast free. On June 20, 1890, Newfoundland introduced regulations permitting vessels requiring capelin or squid to obtain a license at the rate of one dollar per barrel up to forty barrels.

In March, 1890, the Newfoundland government began to consider an attempt to negotiate a reciprocity treaty with the United States independently of Canada; and in September, 1890, it secured the consent of Great Britain for what was to be called the Blaine-Bond Convention.[4] The Honorable A. W. Harvey asserted

> that in all previous negotiations, more particularly those of 1854, 1871 and 1888, Newfoundland was not represented. That while it was true that she was given the option of becoming a party to such arrangements as had been effected, yet it was equally true that her interests had been sacrificed in each case. That she had watched with interest the negotiations made in 1888 between Canada and the United States, and attributed their failure, not to diverse trade interests, so much as to other questions in dispute between the two countries. He considered that the failure of 1888 was due almost entirely to the irritated state of public feeling in the United States with reference to such questions as the Canal Tolls and Behring Sea difficulty. In view of this it was thought desirable by the government of Newfoundland to enter into negotiations on her own account. With this in view she made application and eventually received the consent of the imperial government to enter upon such negotiations.

Sir John A. Macdonald protested in October; and on December 9, 1890, Messrs. J. S. D. Thompson and C. H. Tupper united in a statement which said that if the Blaine-Bond Convention were adopted "the singular case would be presented of one colony in the empire admitting foreign vessels to privileges in her ports and excluding the vessels of the neighboring colonies as well as of the mother country from the like privileges." The Newfoundland Assembly made answer that they were "aware of the interference of Canada in relation to this matter and they cannot fail to appreciate the same as a menace to the independence of the colony. They emphatically protest against the interests of this col-

4. *Papers in Reference to Various Questions Affecting Newfoundland and Canada* (Ottawa, 1893); also *S.P.* (*Sessional Papers*) Nos. 23, 1892, and 20, 1893. For an account of earlier difficulties over the alleged abuse by Newfoundland fishermen of privileges granted to them in the case of their herring imports to Canada see *S.P.* No. 54, 1884.

ony being made subservient to those of the Dominion of Canada."[5] And on March 20, 1891, it was provided that licenses should be granted only to Newfoundland and American fishing vessels. On March 30 Sir William Whiteway wired to C. H. Tupper: "Greatly regret your government's recent action in matter United States convention evidencing hostility not cordiality—to Newfoundland." Canada protested against the tariff changes of Newfoundland which allowed of the fact that "small fish usually sold by fishermen from the Banks in exchange for bait and supplies . . . [were] made free when sold by United States fishermen." But in 1885 Canada had passed a statute conferring the needed authority to impose duties upon fish imported from Newfoundland. And late in 1891 Newfoundland complained that "at a period of the year when exports of flour, etcetera, from Canadian ports are invariably suspended until the ensuing spring, and knowing that such exports had ceased for a while and therefore retaliation by Newfoundland could not immediately affect any interest in Canada, the Canadian government placed a heavy duty on fish products entering ports of the Dominion from this Colony." Newfoundland imposed a duty of $1.05 a barrel on Canadian flour, while American flour was admitted at 30 cents. Sir William Whiteway wrote to Sir Robert Herbert on May 4, 1891:

The unfriendly and unjustifiable action of the Dominion government in urging and so far with success, upon Her Majesty's government the withholding assent to the Newfoundland and United States convention has aroused a bitter spirit of resentment on the part of a large majority of the people of the colony. The Dominion government having failed to procure an arrangement with the United States is most unfairly in our opinion exercising its influence to debar Newfoundland from obtaining an advantage because Canada cannot induce the United States to concede the same privileges to her. The subordination of the interests of Newfoundland to those of Canada is working great injury to the former.

Great Britain became annoyed at the controversy and mildly complained of Canadian pertinacity. Lord Knutsford wrote to Lord Stanley on February 11, 1892, to say that

while, however, Her Majesty's government have, in view of the negotiations about to be commenced at Washington, informed the Newfoundland government that the conclusion of the convention must be again deferred, they feel that in justice to that colony they cannot postpone the ratification indefi-

5. *Journals of the Assembly, Newfoundland, 1891,* pp. 490 ff.; also *idem,* 1892, pp. 413 ff. See *Correspondence Relating to a Proposed Convention to Regulate Questions of Commerce and Fishing between the United States and Newfoundland* (London, 1891).

nitely, and should your ministers not succeed in obtaining a satisfactory arrangement with the United States, the attitude of Her Majesty's government in regard to the signature of the convention will have to be reconsidered.

In the meantime, in view of the deplorable results accruing both to the Dominion and Newfoundland from the relations at present subsisting, I would venture to urge strongly upon your ministers to consider, whether by personal communication with the government of Newfoundland and a mutual agreement not to further discuss past controversies, some amicable arrangement cannot be made.

Of this there was sufficient promise in a dispatch to Lord Stanley, sent by Sir Terence O'Brien on May 27. It read: "Notice inserted in Gazette to-day that extra duties levied under section 13 Revenue Act 1891, will not be collected on and after this date. Dominion government having removed duties on fish and fish products exported from Newfoundland into Dominion of Canada, notice has been given by telegraph to officials to grant bait licenses to Dominion fishermen upon same terms as to Newfoundland fishermen giving similar bonds."

In November of the same year a conference between Newfoundland and Canadian delegates was held at Halifax.[6] Sir John Thompson stated the Canadian position. If the arrangements "were refused on the ground of Canada's fidelity to the interests of the empire, Canada could not be blamed for asking that the protection of Her Majesty's government should still be extended to her people against a convention which would injure their interests." For Newfoundland, Mr. Harvey referred to Sir John Thompson's statement "that in the traditions regarding the treatment of the fisheries in British North America, they had always been considered the property of the empire and not the property of the provinces to which they were adjacent";[7] and he replied that "this usage had first been violated by Canada when in 1885 she adopted a statute which gave authority to levy duty upon fish imported from Newfoundland. . . . It hardly became Canada to attribute to Newfoundland a violation of the traditional usage."

Canada's enforcement of the Convention of 1818 made her own reciprocity negotiations impossible and her insistent opposition to the Blaine-Bond Convention bitterly disappointed Newfoundland.

As Nova Scotia, through such representatives as Thompson and Tupper, attempted to press an aggressive policy upon Canada as a means of securing favorable arrangements from the United States by

6. *Journals of the Assembly, Newfoundland, 1893,* pp. 279–350; also D. W. Prowse, *A History of Newfoundland* (London, 1896), pp. 532–533.

7. J. C. Hopkins, *Life and Work of the Right Honorable Sir John Thompson* (Brantford, 1895), chap. xv.

blocking attempts on the part of Newfoundland to secure such arrangements, so Newfoundland had reached the stage in her development where her independent and aggressive policy resented Canadian aggressiveness and continued to follow independent lines. Again, as the independent development of New England hastened the independent development of Nova Scotia, so did Nova Scotia's hasten that of Newfoundland. The geographical differences of New England, Nova Scotia, and Newfoundland were accentuated by their varying stages of development, and the significance of these divergencies was concentrated in the commercial centers of Boston, Halifax, and St. John's and reflected in the policies of their respective governments.

The bad fishery of 1893 and the depression of the 'nineties had serious results for a community dependent on a single export and heavily burdened with railroad debts and charges. The death of Hall, a partner in the firm of Messrs. Prowse, Hall and Morris, the London agents of the firms exporting fish to European markets, was followed in 1894 by the collapse of both the Union and the Commercial banks. The financial difficulties of the Newfoundland government led to negotiations with Canada looking toward federation. But the difference in the character of Newfoundland's development, as shown especially in her lack of local government and direct taxation, the increasing influence of the continent on Canadian policy, and the growing independence of Newfoundland—coupled with her bitterness toward Canada—proved to be insuperable difficulties.[8] The herculean efforts of Sir Robert Bond finally succeeded in arranging a loan in Great Britain, and in 1898 a new contract was made for the construction of the railway by the government. But this contract was also canceled and remodeled as a result of the attacks by Sir Robert Bond in 1901.

The increasing independence of Newfoundland was strengthened by the severe trials of the 'nineties, and was evident in the failure of federation agreements, in the new contract with the railway, in Newfoundland's hostility to France, and in the Hay-Bond treaty of 1902. This practically repeated the Blaine-Bond proposals of 1891 in seeking to arrange for the free admission of Newfoundland fish into the United States in return for permitting American vessels to purchase bait and supplies in Newfoundland ports. In 1904 the projected treaty was defeated by the American Senate. Because of this defeat, in 1905 Newfoundland brought to an end the *modus vivendi* agreement which per-

8. *Newfoundland Royal Commission, 1933 Report* (London, 1934), pp. 25–27, and Appendix E; and *Journals of the Assembly, Newfoundland,* 1894–95, pp. 369–434; also Lord Birkenhead, *The Story of Newfoundland* (London, 1920), chaps. viii, ix. For the proposed terms see Prowse, *op. cit.,* pp. 552–555.

mitted American vessels to purchase bait and supplies and to ship crews under license. The Foreign Vessels Fishing Act prohibited both the sale of bait and supplies and the shipping of Newfoundland crews by foreign vessels. To secure markets for her products Newfoundland attempted to use against New England the same tactics that Nova Scotia had used following the abrogation of the Reciprocity and Washington treaties. Canadian vessels, chiefly from Lunenburg, continued to obtain bait with licenses issued under the Bait Act of 1888, and frozen herring was admitted to United States ports in American bottoms free of duty. Large numbers of vessels were engaged in the autumn fishery on the south and west coasts of Newfoundland, and Americans evaded hostile legislation by employing Newfoundland fishermen at Sydney and outside the three-mile limit. Rather than purchase fish, they proceeded to take their own catches. Legislation enacted in 1906 then forbade Newfoundland fishermen to take employment on foreign vessels.[9] But protests from the United States compelled Great Britain to disallow the act and led to the adoption of a *modus vivendi* on October 6, 1906. This permitted the use of purse seines and the engaging of Newfoundland fishermen outside the three-mile limit. Newfoundland refused to become a consenting party, holding to Great Britain's statement in 1857 that "the consent of the community of Newfoundland would be regarded by Her Majesty's government as the essential preliminary to any modification of their territorial or maritime rights."[10] At the Imperial Conference in 1907 Sir Robert Bond suggested that the problem should be submitted to the Hague Tribunal, and on September 7, 1910, an arbitration commission gave its award, the findings of which were, in 1912, incorporated in a treaty.[11] The insistence of Newfoundland on her right to regulate the fisheries within her own territories had been sustained. "The right of Great Britain to make regulations without the consent of the United States as to the exercise of the liberty to take fish referred to in Article 1 of the treaty of October 20, 1818, in the form of municipal laws, ordinances, or rules of Great Britain, Canada, or Newfoundland is inherent to the sovereignty of Great Britain." Among other points settled it was

9. See *Correspondence Respecting the Newfoundland Fisheries* (London, 1906); *Further Correspondence Relating to the Newfoundland Fishery Question* (London, 1907); also *Canada and Its Provinces,* VIII, 681–723; and *The Cambridge History of the British Empire,* Vol. VI, chap. xxviii; Thomas Hodgins, *Fishery Concessions to the United States in Canada and Newfoundland* (Toronto, 1907).

10. See *Journals of the Assembly, Newfoundland, 1908,* pp. 218 ff.; also the Right Honorable Sir Robert Bond, *Speech on Modus Vivendi* (St. John's, 1907).

11. *North Atlantic Coast Fisheries,* I, 65–113; also C. E. Cayley, "The North Atlantic Fisheries in the United States: Canadian Relations," doctor's thesis, University of Chicago, 1931; McGrath, *op. cit.,* chap. xxi; and P. E. Corbett, *The Settlement of Canadian-American Disputes* (New Haven, 1937), pp. 34–49.

laid down that "in case of bays the three marine miles are to be measured from a straight line drawn across the body of water at the place where it ceases to have the configuration and characteristics of a bay. At all other places the three miles are to be measured along the sinuosities of the coast." The United States was conceded the right to fish in bays on the southern and western coasts of Newfoundland and in the waters of the Magdalen Islands.

Newfoundland's final attempt to assert her position in the adjustment of boundaries came with the appeal before the Privy Council for the settlement of the Labrador boundary dispute. Canada placed a very narrow interpretation on the meaning of the word "coast," whereas Newfoundland insisted that it included the territory drained by rivers flowing to the Atlantic. Here, in 1927, Newfoundland's contention was sustained by the Privy Council.[12] St. John's continued to be the metropolitan center of the economic activity of the coastal region and of the territory drained by the rivers of that region.

The depression of the 'eighties and 'nineties was marked not only by Newfoundland's assertiveness in excluding Canada, the United States, and France from her fishery and trade but also by her efforts to increase her own production. In 1887, when the Bait Act was passed, legislation was also enacted[13] granting a bounty on shipbuilding to aid in the development of the bank and Labrador fisheries.[14] In 1888 a Fisheries Commission was established, and in 1893 a Department of Fisheries. Large-scale operations involved a catch of poorer cod than those taken by hook and line and also greater difficulties in curing. From 1888 to 1898 cod traps were prohibited on the northeast and the south coasts as a result of complaints of fishery exhaustion and of the decline in prices because of bad curing. Bultows or trawls were prohibited in parts of Placentia Bay. A hatchery for cod spawn was started at Dildo in 1890. Bait freezers were installed at Burin and Presque in the early 'nineties, and, following an appropriation for cold-storage facilities for bait, at convenient points in 1902. By 1904 there were likewise bait freezers at Petty Harbor, Cape Broyle, Bay Bulls, Ramea, and on the Labrador. In 1880 and later the herring fishery had suffered from com-

12. See *Privy Council, in the Matter of the Boundary between the Dominion of Canada and the Colony of Newfoundland in the Labrador Peninsula;* also W. G. Gosling, *Labrador* (London, 1910), chap. xvii.

13. See 50 Vic., c. 24, and 53 Vic., c. 25. Improved ships were a result of legislation in 1891 requiring all vessels to be passed upon by a surveyor appointed by Lloyds. C. Fane, "Newfoundland," *Journal of the Society of Arts,* April 7, 1893.

14. In 1876, 4 vessels were fitted out; in 1877, 7; in 1878, 10; and in 1879, 28. In 1885 schooners from the Bonavista and Fogo districts were engaged on the banks near the Funk Islands.

petition by Cape Breton and, again at the beginning of the new century, when the herring deserted the Labrador. In 1898 the government established a minimum price for herring and rules and regulations for this fishery were drawn up, extended, and enforced. But difficulties were numerous. The salmon fishery was compelled to meet competition from British Columbia, although improved transportation brought numbers of tourists interested in fly-fishing. Conservation measures were introduced in the salmon fishery and went into effect in 1900 on the Gander and other rivers.

The Labrador fishery was subject to wide fluctuations as shown in the declines in 1886 and 1888.[15] The number of schooners in the fishery —"floaters," as they were called[16]—decreased from 825 in 1894 to 470 in 1898. The number of traps, however, increased from 2,588 in 1891 to 4,182 in 1901.

After the turn of the century, because of a rise in prices, the withdrawal of the French from Newfoundland, the decline both of the French fishery and of the American dry fishery, the shift from schooners to steamships in New England and Nova Scotia, and the migration of the former to Newfoundland, "floaters" increased to a high point of 1,432 in 1908. In 1905 prices were quoted as the highest since 1813. "The decided decline of the French bank fisheries has had a salutary effect on the price of Labrador fish in all the European markets where the product of the French fisheries known as 'lavé' comes into direct competition with our Labrador catch."[17] Spain, Italy, and Greece were able to absorb larger quantities of the Labrador product and were added to the basic markets of Portugal and Brazil. In 1905 the efforts of the Newfoundland and British governments secured a reduction in the Grecian duty on cod of from 6*s.* 4*d.* a quintal to 2*s.* 6*d.*, in return for the exemption of Grecian currants from duty in Newfoundland. In 1907 bad curing weather in the Labrador resulted in complaints from

15. For a description of the Labrador fishery see G. F. Durgin, *Letters from Labrador* (Concord, 1908), *passim;* E. C. Robinson, *In an Unknown Land* (London, 1909); and C. W. Townsend, *Along the Labrador Coast* (Boston, 1907).

16. Schooners with crews which proceeded to various points depending on the state of the fishery.

17. EXPORTS FROM FRANCE

(*in hundredweights*)

	1899	1904
Algeria	21,936	24,328
Spain	181,128	73,929
Portugal	4,390	4,119
Italy	104,444	50,109
The Levant	28,543	17,309
Foreign American States	1,119	3,510

Mediterranean countries and in that year, as a result of the improvement in the French fishery, large quantities of French fish were exported to Oporto. Difficulties in grading were increased by the growing use of traps and a rush to get fish to market. The number of traps totaled 6,530 in 1911. "The use of traps is now universal on the Labrador and the fish taken generally small and owing to the shortness of season cannot be made into hard dry salt fish. It does not keep well and is all rushed off to market together, with the result that the markets are always glutted and the returns small. . . . The fishery has now become a trapping voyage only."[18] Setlines or trawl lines had been used to a slight extent by bank schooners on the Labrador, but the problem of bait supply was insuperable in a fishery dominated by traps. It had become impossible to secure suitable boats, gear, and bait for the prosecution of the trawl or the hook-and-line fishery, and the decline of these methods brought further deterioration in the cure. "On an average of years, the sale of codfish shipped from the coast has been unremunerative to exporters. To this fact must be principally ascribed the marked decline in the outfit for this fishery."[19]

The subsidence of the wave of obsolescent schooners from Nova Scotia was hastened by the inauguration of steam service on the Labrador and to the markets. In the sealing steamships of large firms in St. John's the crews were taken to their respective stations, those of Job Brothers going to Blanc Sablon, those of Baine, Johnston and Company to Battle Harbor, those of F. C. Jerrett to Smokey, and Captain William Bartlett's to Turnavik. At the end of the season on the Straits of Belle Isle they proceeded to the green fishery in northern Labrador. "Stationers," or crews which had proceeded to the Labrador to carry on the fishery from a given point, and those who manned "floaters" complained that crews were brought in by steamships to the territory north of Battle Harbor and monopolized the fishing berths.[20] Legislation was enacted in 1910 designed to limit the Labrador fishery to schooners, but it had little effect. In the district from Battle Harbor to Cape Harrigan "stationers" dried fish on the coast and sent it directly to market. At Domino Run and Indian Tickle "many fishing schooners were anchored, and several big square riggers flying the flags of Norway, England and countries of the Mediterranean. These were here to

18. (1914.) Gosling, *op. cit.*, p. 424.

19. *Fifth Annual Report of the Newfoundland Board of Trade* (St. John's, 1914), pp. 11–12.

20. See the Report of the Select Committee on the Bill to Prohibit the Prosecution of the Labrador Fishery in Steam Vessels, *Journals of the Assembly, Newfoundland,* 1905.

buy fish." The extension of a government steam service to northern Labrador enabled large numbers to become "stationers,"[21] especially with the introduction of powerboats. In 1914 it was estimated that 4,000 boats had engines chiefly imported from Canada.

I know of no invention that has appeared in connection with our fisheries that has minimized labor for our men more than the motor boat.[22]

The Newfoundland sailing fleet has been sadly depleted and as the losses are not likely to be replaced, and as the supply of sailing vessels from the neighboring provinces is too intermittent, shippers are more and more turning to steamer freights. This also has the advantage of more regularity in arriving and a shorter duration of the sea voyage.[23]

As a result of the increasing importance of the steamship, Conception Bay firms which supplied the schooners began to decline.[24] St. John's

21. In 1930 the steamers carried planters, crews, and supplies from Conception Bay down to stations on the Labrador coast. The crews hired generally on a share basis. From northern bays, such as Notre Dame, schooners carried down their salt and supplies, thus saving the freight, and caught fish along the northern part of the Labrador coast. Fish were exported directly from the coast in chartered steamers, particularly by planters, crews, and stationers, which, it was alleged, involved dumping on the European market; or, particularly in the case of "floaters," the catch was taken to their home ports or to St. John's and sold as "Labrador," or cured in Newfoundland (12 hogsheads of salt to 100 quintals) and sold as "Labrador shore." The powerboat enabled families to come down on the steamships on payment of a nominal fare and carry on the fishery in their own interest. This trend continued as a result of dissatisfaction with the share or planter system, in which one merchant employed several crews. With the race down in the spring, berths for traps were the objective, and were obtained in part by drawing lots. The cash system enabled the fishermen partially to avoid being dependent on the planter or middleman who received supplies from the large merchants. The sale of fish to the supplier or planter, and in turn to the merchant, necessitated classification for various markets; and the absence of an efficient grading system weakened the selling position of the product. The lack of grades was also a handicap to outright sales and the building up of an exchange or speculative market. The banks carried the product at least from March to November at rates of 6½ to 8 per cent. Fish could be kept for periods longer than a year, but deterioration, storage, and carrying charges were factors contributing to rapid sales.

22. Abraham Kean, *Old and Young Ahead* (London, 1935), pp. 119–120.

23. "Many of the vessels were not built in this country but bought from Nova Scotia at about one eighth of their original cost. . . . Because of the low price of fish it would never have paid to have had new schooners built." Smaller boats of 15 to 20 tons "have practically become obsolete except perhaps at Placentia, St. Mary's and Fortune Bays. . . . The 20-quintal boats that were in the northern bays are almost forgotten, motor-boats having taken their place. . . . Coastal boats carried upwards of three thousand fishermen from Conception Bay, Trinity Bay, Bonavista Bay, on to the so-called French Shore, including Gray Islands, Fishott Islands, Quirpon and Cook's Harbour, farther up the straits and Labrador." *Idem,* pp. 150–155.

24. "Evidence of a considerable decline . . . is reflected in the depopulation of some of the Conception Bay towns." *Fifth Annual Report of the Newfoundland Board of Trade.* For an invaluable account of the decline of these firms and increas-

gained at their expense as an export center for various markets. An increase in exports to Brazil[25] because of the prosperous state of the cotton and sugar market was accompanied by increasing shipments from New York. For the West Indian trade Halifax occupied an important position because of better steamship facilities. In 1913 "the increase in fish shipped in packages continues, and a greater quantity is yearly being shipped by steamers, either direct or by way of New York or Liverpool." Transshipments meant loss of time and damage, and were superseded by arrangements in which shippers chartered vessels to proceed directly from St. John's to Spain, Italy, and Sicily, thus reducing the time to between eleven and fifteen days. The Newfoundland Shipping Company operated a direct line of steamers in 1910, and the

ing dependence on St. John's, see Nicholas Smith, *Fifty-two Years at the Labrador Fishery* (London, 1936). Schooners took crews and supplies for Conception Bay firms to the Labrador fishery, but as steamers became of more importance to Labrador and Labrador markets, St. John's firms supplied the planters. See the Report of the Commission on the Fishery, *Journals of the Assembly, Newfoundland, 1915.* The decline of the outports was a result of improved transportation and the demands of large-scale capital equipment. In 1897 the cannon and harpoon were introduced in the whaling industry, and factories were established on the south coast by Bowring, Job Brothers, MacDougall, and Harvey. The area was quickly depleted and the plants abandoned. Later, with the support from Norwegian capital, a plant was established at Gready's Harbor on the Labrador. Philip Tocque, *Newfoundland* (Toronto, 1878), p. 302; J. G. Millais, *Newfoundland and Its Untrodden Ways* (London, 1907), p. 163 and *passim;* also *The Royal Commission on the Natural Resources, Trade and Legislation of Certain Portions of His Majesty's Dominions: Newfoundland* (London, 1915), p. 42. In the seal fishery the first iron steamer, the *Adventurer,* was introduced in 1905 and was rapidly followed by others. Wireless telegraphy and airplanes increased the returns of the industry in the postwar period. The number of steamers declined and concentrated on St. John's. The refining plants in the outports disappeared. Large-scale capital equipment and concentration were followed by the organization of strikes and political agitation leading to measures improving labor conditions. See *Chafe's Sealing Book* (3d ed., St. John's, 1923); Kean, *op. cit.;* and W. H. Greene, *The Wooden Walls among the Ice Floes* (London, 1933). The lobster fishery declined steadily from a peak in 1905. See *Reports of the Newfoundland Fishery Commission,* II, No. 1, pp. 42 ff. The herring fishery declined on the south coast, but the development of the "Scotch cure" brought about an expansion of markets. In 1914 about one half was sold as pickled and one half, produced by the winter fishery on the west coast, as bulk and frozen herring.

25. "These two articles, jerked beef and codfish, are the two preserved foods used extensively in Brazil. They get the jerked beef from the Argentine, and it is by the price of this jerked beef that the price of codfish is set. . . . It would not matter what religion they were; they would be obliged to have this form of preserved food." Fish had the advantage in Lent but jerked beef was otherwise a serious competitor. *Royal Commission on the Natural Resources, Trade, etc.,* p. 45. "We have new markets and better steam shipping facilities and that helps it out. Then the railway going through Brazil and other countries that we ship to enables them to distribute the fish there better. Furthermore all articles of food have gone up in price and there has been a great deal in that way to help price." *Idem,* p. 42. "When salt codfish passes a certain price the consumer will turn to other suitable foods obtainable at less price than codfish."

Red Cross and Furness lines also combined to do the same in the case of Italy and Spain.[26] "Spain and Italy have been paying higher prices for Labrador fish than England would or could pay, the difference being from 2s. a quintal upwards in favour of marketing abroad." "The shipments by these direct steamers have enlarged the market in several ways, first of all by getting the new season's fish to market some fortnight or three weeks earlier than it was possible to do by way of Liverpool, thus giving a lengthened period of consumption."[27] On the other hand, indirect shipments meant a steady increase of exports in casks and the development of standardized packages which made it possible to send fish in much better condition and on a through bill of lading. Purchasers were able to buy in smaller consignments but with direct shipments difficulties returned. "Unfortunately, the good effect which the exporting merchants may have derived from having a recognized standard article which they could sell for cash has been minimized by the practice which has again become general of sending large consignments to market for sale. The practice has caused the markets to be in a steadily overstocked condition and has largely put a stop to outright sales to foreign merchants." Many testified as to the increasing quanti-

26. *Royal Commission on the Natural Resources, Trade, etc.*, p. 37.

27. *Annual Report of the Newfoundland Board of Trade,* 1911. "The fact that Newfoundland exports an unusually large proportion of her products, and imports an unusually large proportion of her requirements results in a considerable foreign trade per head of population; and as a result the means of communication with the outside world are of the utmost importance to the welfare of the colony. . . . The major portion of Newfoundland produce is exported to countries not embraced within the British Empire. The imports, on the other hand, are drawn principally from the United States, Great Britain, and Canada, which countries are her customers to a comparatively limited extent. It constantly happens, therefore, that tonnage carrying exports has to return in ballast while steamers conveying imports would get but little outward cargo were it not for through freight in transit for foreign ports. . . . A large amount of fish is still exported in small sailing vessels, but the tendency is towards the steady elimination of these in favour of steamer shipment wherever this is possible." *Royal Commission on the Natural Resources, Trade, etc.,* p. 16. "The providing of these steamers, giving ample carriage capacity for much of the imports and exports of the colony, has resulted in complete transformation in our methods of shipping fish to European markets, where formerly a large fleet of British sailing vessels known locally as 'English' schooners was employed for this purpose, and where as many as 120 in the old days visited St. John's every year; but now they are practically unknown, a few plying to Labrador in the summer months to take cargoes of salt-cured fish direct to Europe. But steamers have even invaded that trade, the great bulk of fish from here being now shipped in casks or other packages via the Allan or Furness steamers to Liverpool or Glasgow, and thence transferred to steamers which sailed to Mediterranean ports, through freights being given and greater speed and certainty in the transport of the commodities to destination being insured." *Idem,* pp. 22–23. "The colony exports about 85,000 quintals of codfish to the West Indies by steamer *via* Halifax or New York, and imports from Barbados about 10,000 puncheons of molasses by sailing vessels returning from Brazil." *Idem,* p. 17.

ties of fish sold tal qual.[28] "The purchase of fish tal qual is absolutely demoralizing to fishermen." "That the method of buying fish tal qual has much to do with the decline in the good cure of fish cannot be questioned, since it does not differentiate between the fish which is of good quality, the result of great care and labor, and that which is indifferently attended to."[29] The Board of Trade report for 1913 stated that "it may truthfully be said that there is no cull of fish to be standardized, all fish being bought tal qual." "The utmost cleanliness should . . . be observed in the operations of splitting, washing from the knife, and dry-salting as opposed to the pernicious practice of pickling, which in many localities is but too common."

The increase in trap fishing on the Labrador and the expansion of the fishery on the coasts of Newfoundland which followed the withdrawal of the French,[30] the effectiveness of the Bait Bill, the introduction of steamships, and the completion of the railway were accompanied by an increase in the production of soft-cured fish. The consequent low prices of soft-cure grades coincided with the decline of the dried fishery in Canada and the United States. Nova Scotia bought large quantities, it was claimed, without culling, and thus offered a direct incentive to lowering the standards of cure. In 1910, American firms were engaged in purchasing green fish—now known as "salt bulk," i.e., fish wet-salted in piles—in Newfoundland, particularly with the encouragement given by the exemption from duties of fish imported in American bottoms. In 1913, with lower tariffs, exports of this cure increased rapidly, and brought 3¼ cents a pound, equal to $7.50 a quintal tal qual "dry." "Not only is the business good in itself but it . . . reduces the quantity of bank fish to be marketed in Europe."[31]

Improved transportation facilities by railway and steamship brought about changes in the internal market. The retail trade of St. John's tended to become less, as it was carried on by small merchants in the outports, but wholesale trade increased. Canadian banks which dominated Newfoundland banking after the crash of 1894 established branches in the outports and the cash system was gradually extended. "The people of the colony are becoming yearly more independent of credit from the merchant."[32] These trends strengthened the basic de-

28. Taking them as they come.
29. *Third Annual Report of the Newfoundland Board of Trade,* p. 17.
30. For a valuable description of the south-coast fishery see Millais, *op. cit.* Bankers operated until June 15 with trawls and herring bait. Capelin came in from June 10 to August 1, and traps and seines were employed near the shore. From August 1 to October 1 hand lines and trawls were used.
31. *Fifth Annual Report of the Newfoundland Board of Trade,* p. 10.
32. Kean, *op. cit.,* pp. 202–205.

centralization of the economic and political structure of Newfoundland. The absence of local government except in St. John's meant that ecclesiastical organizations became active in education, and that medical and social services were extended by private agencies. Sir Wilfred Grenfell built up his hospital organization in northern Newfoundland and in the Canadian and Newfoundland Labrador.[33] In 1896 he started a co-operative organization at Red Bay, which was followed by others at St. Anthony's, St. Modest, and Flower Cove.

The trend toward decentralization became evident in the success of Mr. W. F. (later Sir William) Coaker. He organized a telegraph operators' union in 1895 and directed his energies toward building up a center that would be independent of control of St. John's firms. Following the low prices and poor grades of fish in the "black fall" of 1908 he organized, in November, the first local of the Fishermen's Protective Union at Herring Neck.[34] After a year of organization activities the first union convention was held in November, 1909. A union store to supply provisions and heavy goods was proposed for St. John's, and supplies such as flour, salt, and molasses were to be imported directly by the outports. "Wonderful savings are possible by buying and selling in bulk," Coaker told his supporters; "a store at headquarters will enable the smaller branches to derive the benefits as well as the larger, and will supply what cannot be imported in large quantities by the districts."[35] His followers advocated political action and voiced their opposition to French trawlers, to the introduction of gasoline engines, to the use of "rinds"—fir-tree bark—in drying fish, and to forest exploitation within three miles of the shore. They asked for regular price information[36] and the appointment of a trade agent in South America, and were in favor of cold-storage facilities for both bait and fresh fish. In 1910 they began to publish a paper, and a union office was opened in St. John's.[37] In the following year the Fishermen's Union Trading Company was organized with a capital of $100,000 and four union cash stores were started. In 1912 ten cash stores were established in the out-

33. See *A Labrador Doctor* (New York, 1919); F. L. Waldo, *With Grenfell on the Labrador* (New York, 1920); Anne Grenfell and Katie Spalding, *Le Petit Nord* (New York, 1920); H. P. Greeley and F. E. Greeley, *Work and Play in the Grenfell Mission* (New York, 1920).

34. See J. R. Smallwood, *Coaker of Newfoundland* (London, 1927), and *Twenty Years of the Fishermen's Protective Union of Newfoundland*, ed. the Honorable Sir W. F. Coaker (St. John's, 1930).

35. *Twenty Years of the Fishermen's Protective Union*, p. 3. The information on the Fishermen's Protective Union has been obtained from this volume.

36. See Kean, *op. cit.*, pp. 152–153, on the problem of securing accurate information in Newfoundland.

37. The offices were moved to Port Union in February, 1918.

ports, and in 1913 twelve, or a total of twenty-six permanent and seven temporary branches; and one more store was started in St. John's. In 1912 the company handled 8,000 hogsheads of salt, 8,000 barrels of flour, and 6 shiploads of coal. In 1913 the totals were 15,000 barrels of flour, 400 barrels of beef and pork, 150,000 pounds of butter, 40,000 pounds of tea, and 20,000 pounds of tobacco. In all, business had increased from $250,000 to $400,000. In 1913 Crosbie's waterside premises in St. John's were acquired, a clothing factory was started—but closed in 1914—and the steamship *Kintail*, later the *Can't Lose*, was purchased. Attempts were made to increase the price of fish and oil.[38]

The effect of the opening up of new regions could be seen in new trading organizations and in new political alignments. In 1905 the West Coast had complained of the restrictions imposed on American trade. An election had taken place in May, 1909; and, in spite of Sir Robert Bond's defense of Newfoundland in international disputes, he had been defeated by Sir Edward Morris. In the next election, in 1913, Mr. W. F. Coaker's group elected eleven members and, in the Sealing and Logging Bills, he succeeded in introducing legislation which improved labor conditions. The war solved the problem of regional pressures temporarily when a Union government was formed in 1916.

The early years brought about further disruption in the Newfoundland fishery. Its decline as a result of the war, the withdrawal of men from the fishery for war service, and an increased demand for foodstuffs resulted in a rise in prices.[39] The effect of an embargo on the market in Greece in November and December, 1915, and difficulties in the same market in 1916 were offset by a decline in the Norwegian fishery and, in the Portuguese market, by a reduction of the duty on Newfoundland fish to 2 shillings, or to that on Norwegian. Prices of provisions and supplies also increased. All commodities save salt were withdrawn from the free list and drawbacks on kerosene oil and gasoline used in the fishery were canceled. The scarcity of tonnage sent shipping rates so high that they stimulated local shipbuilding.[40] During the war more than 100 vessels were launched and commissioned, including 20 of from 120 to

38. It was claimed that in 1912 an alleged combine of St. John's merchants was broken by means of sales made to R. H. Silver, and Smith and Company of Nova Scotia.

39. For the three years 1915, 1916, and 1917, prices in quintals for North Shore cod were $7.50, $8.50, and $10.50. For Labrador fish they were $6.30, $6.70, and $9.00. See *Annual Report of the Department of Marine and Fisheries . . . 1917* (St. John's, 1918).

40. The new ships were chiefly fore-and-aft rigged, and it became difficult to get certified mates and masters, as the earlier regulations applied to square-rigged ships

500 tons. The use of wooden sailing vessels increased rates of insurance and led to the formation of local marine-insurance companies and to government assistance. Coastwise shipping advanced rapidly in price and ocean tonnage was commandeered for war purposes, with the result that heavy burdens were imposed alike on the main railway line and its branches. A ministry of shipping and a board of food control were created.

In 1917 high prices and the peak year of production in the history of the fishery in Newfoundland brought about unprecedented prosperity. New equipment was introduced and, following the failure of the policy of bonusing bait freezers, save for the single unit at Fogo, a cold-storage plant for bait was opened by Harvey and Company at Rose Blanche. Demands for foodstuffs led to an increase in the production of frozen fish. The Newfoundland Atlantic Fisheries completed a cold-storage plant at St. John's in 1918.[41] "The past season has been the first in which there have been any successful results from the cold storage of our fishes."

The increasing importance of steamships went with changes in marketing structure. Large organizations in St. John's introduced more effective methods for external markets. In 1909 George Hawes had capitalized the possibilities of improving marketing by setting up an organization in Spain. Two years later fifteen firms, including Job Brothers and Company, Harvey and Company, Bowring Brothers, The Monroe Export Company, George M. Barr, The Tors Cove Trading Company, that is, A. and W. Goodridge, and The Grand Bank Fisheries, formerly The Harris Export Company, formed a group; and they appointed him as their agent in Spain, to handle their fish on consignment and thus avoid losses through claims on outright sales. Newfoundland's exports to Spain rose from 110,000 hundredweight in 1905 to 270,000 in 1908, but declined to 180,000 in 1911. Exporters to Italy appointed H. Le Messurier to work along lines similar to those of George Hawes in Spain. Northern Italy was an important market for Labrador fish, and exports from Newfoundland to Italy had increased from 110,000 hundredweight in 1905 to a peak of 380,000 in 1909. They had declined to 140,000 in 1911, increased to 220,000 the following year, and declined to 150,000 in 1915. The war meant a parallel introduction of boards for food control in European countries. As a result of high prices due to a cornering of the market in 1917, a government marketing organization, the Consorzio per L'Importazione e

41. The interests of capital were largely associated with Sir Robert Reid, and actual control, till then exercised from Hull, was taken over in 1920. *Tenth Annual Report of the Newfoundland Board of Trade* (St. John's, 1918), pp. 9–10.

la Distribuzione dei Merluzzi e Stoccofissi,[42] was formed in September, 1918, for the purpose of buying and distributing fish in Italy. The Newfoundland government was, in 1919, strengthened in its dealings with the Consorzio by a voluntary agreement among exporters as to prices and conditions of shipment; but the problem of grading and inspecting Labrador fish proved insuperable. High prices during the war were attributed to control organizations, and attempts were made to revive them. The Coaker group increased in influence under a Union government but resigned as a protest against the government marketing legislation proposed for the fishing industry. In May, 1919, after the elections of that year, Coaker became Minister of Marine and Fisheries and took an active part in developing more effective devices for the control of exports. Under the Imports and Exports Restriction Act of 1918 a proclamation was issued on November 20, 1919, which required all exporters of cod to secure licenses from the Minister of Marine and Fisheries before shipment. The Supreme Court declared the proclamation illegal as issued under that act; and on January 20, 1920, a second proclamation to the same effect was issued under the War Measures Act of 1914. Under these proclamations, and with a view to preventing sales in foreign markets by irresponsible and ill-informed parties, circulars were sent out worded in part as follows:

> The intention is to attempt to regulate our selling prices in accordance with circumstances, not necessarily interfering with present methods and channels of making sales, except in certain cases where it is apparent that drastic action is essential. The regulation of such prices will be made with a view to the increase of consumption by lowering prices when this is deemed desirable, the restricting of consumption by raising prices when it is clear that this can be done advantageously . . . and the protection of buyers abroad who purchase cargoes or parcels of fish, on a cash against documents basis, from fraudulent or careless shippers.[43]

Hawes was to be appointed adviser in the case of the Mediterranean markets and as sole agent for Italy. A second circular letter, of December 2, 1919, was specific regarding individual markets: "No permits for export," it read, "will be granted except upon the following conditions as to prices, terms of sale and arrangements regarding consignments."[44] Minimum prices and outright sales were specified for Lisbon and Figueria, with all consignments to Messrs. Lind and Couto at Oporto. Minimum prices in the case of two grades were specified for Spain. Ship-

42. See Riccardo Bachi, *L'Alimentazione e la politica annonaria in Italia* (Bari, 1926), pp. 488–491.

43. See *Eleventh Annual Report of the Newfoundland Board of Trade* (1919), pp. 39–40.

44. *Idem,* p. 42.

ments to Greece, the United Kingdom, the United States, Brazil, and the West Indies were to be made only after evidence had been received that prices were being adhered to. A third circular reassured the "so-called small exporter" that the regulations would not interfere with his interests, and referred to minor modifications of the second circular.

Opposition to the autocratic character of the above proclamations and to the subsequent regulations led to the passing successively of acts "to Regulate the Exportation of Salt Codfish," "to Provide for the Standardization of Codfish," and also of an act "for the Better Obtaining of Information Respecting the Cod Fishery." A codfish-exportation board was set up, and annually, in September, it was required to discuss with exporters the conditions of sale. On September 2 and 3, 1919, the first meetings were held. As a result of discussion, prices were drawn up for various markets and for various grades, and conditions were laid down for the differing markets.[45] Trade commissioners were appointed in Portugal, Italy, Greece, the West Indies, the United States, Liverpool, and Malaga. A contract was signed with the Consorzio in Italy, in January, 1920, and the Newfoundland government purchased Labrador fish to keep up prices.

But this system of export-price controls was to last for only a year. Serious disturbances in Portugal, Spain, Italy, Greece, and Rumania and the sharp increases in exchange rates between the currencies of those countries and sterling brought about a collapse in January, 1921, and the recall of the trade commissioners.[46] The Italian lira fell to 97 to the pound and the Greek drachma to 90. Brazil exchange also declined. Markets were in part lost to Halifax during the controls period. Readjustments of transportation facilities made themselves felt in the heavy losses suffered by sailing vessels, the increasing importance of steamships,[47] and the reduction in railway earnings. An export tax of 20 cents a quintal had been put on Labrador fish if shipped in British bottoms and of 40 cents if in foreign. Heavy import duties also added to the burdens of Newfoundland. In 1921 the Consorzio was abolished. Duties were imposed by Spain and Portugal on Norwegian fish, and an agreement was reached with the Portuguese board.

The mechanization of the industry after the war narrowed markets

45. *Rules and Regulations Made under the Codfish Exportation Act;* also *Rules and Regulations Relating to the Standardization of Codfish, and Including Proceedings of the Convention of Licensed Codfish Exporters* (St. John's, 1920).

46. Merchants were by no means unanimous in their support of the policy of export controls. Protests were made in St. John's and from places such as Burin; and it was charged that the plan collapsed because of sales made by two exporters, A. E. Hickman and W. A. Munn.

47. In 1922 insurance rates to Portugal on shipments by sailing vessels averaged from 7½ to 8½ per cent, and by steamer were usually under one half of 1 per cent.

for Newfoundland fish. Iceland[48] had the advantage of a nine months' season and of being able to place her product on the market three months earlier. The use of motor vessels and trawlers, the standardization of the cure, improved methods, the expansion of the coöperative movement after about 1880, and the displacement of the truck system by loans contributed to a marked increase in production and a marked improvement in quality.[49] Accessibility to the European markets made easier the sale of fresh and green fish to England and other countries. Control by an export union improved the position of Iceland fish in foreign markets and lowered costs of production for the purchase of supplies. Trawlers from Grimsby and Hull proceeded to Iceland, and the consumption of Iceland fish increased in England. As a result of better preparation, such cod sold at much higher prices in Liverpool than the Labrador fish. Exeter and Bristol began to purchase salt-bulk cargoes from Newfoundland to be cured as Labrador and sold in English and Mediterranean markets.

Norway possessed an advantage in having made large profits during the war, which were diverted with government support to the extension of the fishery. With excellent steamship connections and a strong bargaining position due to her ability to purchase from other countries, she could offer stiff competition.[50] Fish were sold green to the merchants who were thus "enabled to treat considerable quantities of these several products (cod, roes, tongues, sounds) at one operation and at very little expense," whereas in Newfoundland "every fisherman is for himself; he has always caught, cured and marketed his own fish." Waste products "could only be obtained in small quantities from each fisherman, and the cost of collecting would make an extensive trade in them prohibitive."[51] A winter fishery was supplementary to agriculture, while in Newfoundland a summer fishery competed with it. A surplus population employed in the winter season "can give greater attention to the

48. Margaret Digby, "The Organization of the Icelandic Fisheries," *Yearbook of Agricultural Cooperation* (1935), pp. 182–193; *idem* (1937), pp. 412–414.

49. "All the evidence points to the fact that the standards of the Newfoundland cure are not getting worse but are remaining stationary, while more and better supplies are being produced elsewhere." *Reports of the Newfoundland Fishery Research Commission* (St. John's, 1933), II, No. 1, pp. 13 ff.

50. See Report of W. F. Penny on his trip to Norway. *Journals of the Assembly, Newfoundland,* 1921, pp. 418 ff.

51. *The Royal Commission on the Natural Resources, etc.: Newfoundland* (London, 1915), pp. 38, 48. Nova Scotia had similar difficulties, particularly in the pickled-fish industry. "You have thousands of different men in different localities with different ideas, putting up the same thing for the same market." *The Royal Commission on the Natural Resources etc.: The Maritime Provinces,* p. 131. For figures on the distribution of population about the coast of Newfoundland see E. B. Shaw, "Population Distribution in Newfoundland," *Economic Geography,* July, 1938, pp. 239–254.

by-products of the fishery,[52] whereas in Newfoundland unless the prices for the by-products are remunerative it pays our people far better to catch fish." The concentrated handling of fish in Norway facilitated the collection of more accurate statistics and, accordingly, control over production.

Portugal made further efforts to develop her own fishery. Trawlers were introduced and the production of her bank fleet increased from a low point of 7,000,000 pounds in 1919 to 25,000,000 in 1921 and to 39,000,000 in 1924. Portuguese fish were brought home in November and December and cured, and came into competition with Newfoundland "dry shore" and south-coast bank fish. The increase in the production of soft-cure grades[53] and the disorganization of markets after the war placed Newfoundland at a disadvantage in competition with other countries. In Oporto, damp fish could not be handled because of lack of facilities. And in Spain tal qual fish were dumped on a consignment market. Regions which by virtue of climate provided a monopoly market for shore fish were encroached upon by cod from other countries as a result of the introduction of cold-storage facilities.[54] Italian consumption con-

52. In the manufacture of cod-liver oil Norwegians improved the technique introduced into Newfoundland in 1848. By it oil was obtained by means of indirect heat through pans with hot-water bottoms. The Norwegians developed a nonfreezing process in 1885 through which the stearine was removed by pressing or filtering. In 1903 Newfoundland introduced a process for rendering oil by direct steam in a single pan. But in spite of improvements in refining, the wide fluctuations in medicinal-oil prices and in the yield, with limitations on production added, were heavy handicaps. Experimental work at Bay Bulls has been directed toward widening markets for medicinal and for low-grade oils. On cod-liver oil see also *Reports of the Newfoundland Fishery Research Commission,* I, No. 1, pp. 13 ff., and II, No. 1, pp. 32 ff.

53. "Our fishermen have been too long accustomed to carry on their voyages in large schooners with costly traps and skiffs and with no trawling gear. . . . It is very questionable whether the trap fishery has paid its way in recent years and it is undoubtedly true that trap fishing is far from being as remunerative as it was in years gone by." *Seventeenth Annual Report . . . of the Newfoundland Board of Trade* (1925), pp. 10–11.

54. The provinces of Granada and Malaga were markets for English and American cod. "Merchants prefer them for this reason: They keep, and if they must stand over in the warehouse the merchant's mind can be at ease. It is not the same with the French cod. Though of finer quality than the foreign fish, less salt goes into its curing, it is softer than the English cod, which grows harder after it has been stored for a certain time. If the temperature rises the French cod very quickly begins to be the worse for it. Around the bones reddish spots commence to come out. They are the first symptoms of decomposition, and it takes place because the fish are not dry enough to stand climates that are hot and humid. Cod is the standard food everywhere in these regions; and though the fisheries of England and North America hold the market now, France could share it if her fish were cured in a different way. However, Andalusia being a colder country, the French cod can find a market there; and, to add to that, a market which, though still limited, is growing steadily, is now open in the province of Granada." M. A. Hérubel, *Pêches maritimes* (Paris, [1911]), p. 331.

tinued low, because of the unfavorable exchange and the country's increasing trend to agriculture. In Greece, exchange declined in 1922 to 400 drachmas to the pound. Northern Brazil was reported to be turning to the consumption of jerked beef. Under the West Indian agreement, Canada had been given a preference from 1920 to 1924[55] in the Jamaica market. The Fordney Tariff of 1922 increased the duty in Puerto Rico to $1.25 a quintal. The high cost of outfits, low prices, fluctuation in exchanges, taxation by importing countries, and social unrest contributed to depressed conditions in Newfoundland. The prices of shore and Labrador fish dropped from $10 and $8 a quintal in 1921 to $6 and $4 in 1922.

"During the years 1921–1924 many failures occurred; and by the elimination of these weak concerns the commercial outlook was improved and business has shown a tendency to gravitate towards established houses which have the capital, credit and ability to conduct their affairs along approved lines."[56] To offset the effects of the depression a preferential tariff agreement was arranged between England and Spain which lowered duties on Newfoundland fish about 60 cents a quintal, or from 72 to 24 pesetas a hundred kilos. At Alicante, Hawes and Company established cold-storage facilities for handling consigned fish, and in addition purchased fish on the Labrador for direct exportation. The export tax for scientific research was lowered to 20 cents and 10 cents a quintal, respectively, on foreign and British bottoms in 1922 and dropped in 1923. The influence of the Fishermen's Protective Union was evident in 1923 in the removal of duties on flour, pork, beef, molasses, gasoline, and kerosene.[57] They were imposed again in 1924, with the ex-

55. Duties on Newfoundland herring were 2*s.* 8*d.* a hundred pounds, and on cod 4*s.* 8*d.* On Canadian herring and cod, duties were, respectively, 2 shillings and 3*s.* 6*d.*

56. *Seventeenth Annual Report . . . of the Newfoundland Board of Trade* (1925), p. 5.

57. It was estimated that duties on a Labrador outfit involved a burden of 12 per cent.

The Fishermen's Union Trading Company claimed in 1922 that it forced down the price of salt from $2.40 to $2.00 a hogshead, that it forced up the price of herring, that it purchased 75,000 quintals of cod, and supplied 50 Labrador schooners and 5,000 fishermen. In 1923 the company purchased premises at Harbor Breton on the south shore and bought fish on that coast. Permanent representatives were appointed in Spain, Italy, and the West Indies. In the following year it built premises at Musgravetown, purchased the *Beulah Mae* on the south shore, and bought fish at Burin and Point aux Gail. The shipyards at Port Union continued to build schooners, and by the end of 1925 had completed a total of 25, which by 1928 had risen to 30, with three motor vessels and a sealing plant. In 1925 a cold-storage plant was acquired at Patras in Greece and a steel motor vessel, the *Elly,* for the Brazilian trade. Prices of fish in the north had been raised to the level of St. John's prices and were $1 a quintal more than those paid in the south in 1925; but complaints were numerous that there had been wide variations in the prices paid to individual fishermen for the

ception of those on flour and gasoline, and those on beef, pork, and kerosene were again removed in 1925.[58] Duties imposed in the same year on lines, tobacco, butter, and other commodities were reduced or abolished by 1928, when representatives of the Union in the Assembly increased from 8 to 13. On schooners bounties were paid of $30 a ton up to 50 tons. In 1924 it could be said, "For the first time for several years the local price for a quintal of shore fish exceeded the cost of a barrel of flour, and a marked improvement in general conditions is the result."[59] In 1925, stable and more favorable exchanges in relation to sterling improved the markets in Portugal, Spain, and Brazil, and weakened Norway as a competitor. The surplus Norwegian catch was sold to Russia and Germany; and Italy absorbed fish which would otherwise have been forced on a falling market in Brazil.[60] But at the same time Aberdeen fish were sold in Brazil in competition with those of Newfoundland; and other exchange problems, together with the difficulties of shipping the softer cures across the equator, resulted in a sharp decline in consumption. The high prices of 1924 and the introduction of closed seasons for lobster, which were maintained from 1925 to 1928, turned labor toward the cod fishery and increased the number of fishermen. The demand for Newfoundland fish manifested itself plainly in a marked increase in the number of foreign fish merchants buying in the local market and doing direct exporting.

The recovery had been of brief duration. With the rise in prices, Iceland encroached on the market for Labrador cod in Greece and Italy, and on that of shore-dry cod in Spain. There was a continued extension of large-scale production, and Newfoundland felt the effects of competition in grades where large-scale methods were most effective.[61] France became a less serious competitor after the stabilization of the

season. "In Lunenburg all fishermen receive prices alike for their voyages." In 1937 the Union claimed to have brought about an increase in the returns from sealing by encouraging competition from New York.

58. 12 and 13 Geo. V, c. 17, 1922, and 15 Geo. V, c. 23, 1925.

59. *Sixteenth Annual Report . . . of the Newfoundland Board of Trade* (1924), p. 4.

60. From 1920 to 1924, inclusive, Pernambuco imports, in drums, fell year by year from 155,000 to 132,000, 89,000, 82,000, and finally to 63,000.

61. The *Hawes Report on the Salt-fish Trade for 1927–28* noted that, in the case of Newfoundland, markets are being steadily lost to Iceland, its chief competitor in the soft-cure trade. It is true that the shore-fish or hard-cure markets appear to be impregnable against all competition; this is due to the undoubtedly better eating qualities of Newfoundland fish. There is no reason why these qualities should not be of the same advantage to Newfoundland in its soft-cure trade, except that the consumers of this cure are influenced to a much greater extent by the appearance of the fish. It is no exaggeration to say that Newfoundland is behind the times in every branch of its salt-fish industry.

franc in 1927, but production in Iceland was growing as a result of an increase in trawlers. In metric tons it rose from 51,085 in 1925 to 65,596 in 1928, and to 70,573 in 1930. Labrador fish were forced out of the markets in northern Italy in 1930, while imports from Iceland and the Faroe Islands increased from 3,600 metric tons in 1927 to 7,000 in 1931. In Portugal, Norway tended to dominate the Lisbon market; and Iceland and the Faroe Islands increased exports to Oporto from 700 metric tons in 1927 to 8,750 in 1931. In the markets of southern Spain —Alicante, Cartagena, Malaga, Seville, and Valencia—Labrador-cure imports declined from 10,200 metric tons in 1927 to 4,400 in 1931.

Newfoundland was compelled to turn to markets in South America and the West Indies. In 1927 shipments were made by direct steamer to Brazil; and exports to the northern markets of Brazil increased from 125,000 hundredweight in 1925 to 345,000 in 1929. But Norwegian cod also began to compete with Newfoundland's "Madeira"[62] fish in this market. "Brazilian buyers have been 'bearing' fish owing to the poor coffee crop this year (1929), and also owing to the low rate of exchange now existing. These reasons may be justifiable, but it must also be considered that on account of our poor trap fishery from which we get our supplies for their market, stocks are lower today than for a very long time."[63] The percentage of fish exported from Newfoundland to the West Indies increased from 5.9 in 1907 to 11.6 in 1928, and to 25.8 in 1931.[64]

> Jamaica consumes a large quantity of fish and often gives higher returns than the Oporto market for the same grade. It is also helpful in avoiding glutting the Oporto market. A considerable amount of consumption was lost to Canada this season owing to shippers holding out for prices which possibly market conditions did not warrant. . . . The practice continues of consigning to Barbados excessive quantities of fish which that market cannot consume. Barbados is forced as a consequence to consign the surplus fish to other islands, and these consignments extend as far south as Trinidad and Demerara, thus making it difficult to sell outright to these places direct from Newfoundland.[65]

An increase in fish production by the application of more efficient methods in areas which had been formerly restricted in the industry in-

62. A grade formerly suited to the Madeiras but later sold to South America and the West Indies.

63. *Twenty-first Annual Report . . . of the Newfoundland Board of Trade* (1929), p. 16.

64. *Reports of the Newfoundland Fishery Research Commission,* III, No. 1, p. 13.

65. *Eighteenth Annual Report . . . of the Newfoundland Board of Trade* (1926), pp. 9–10.

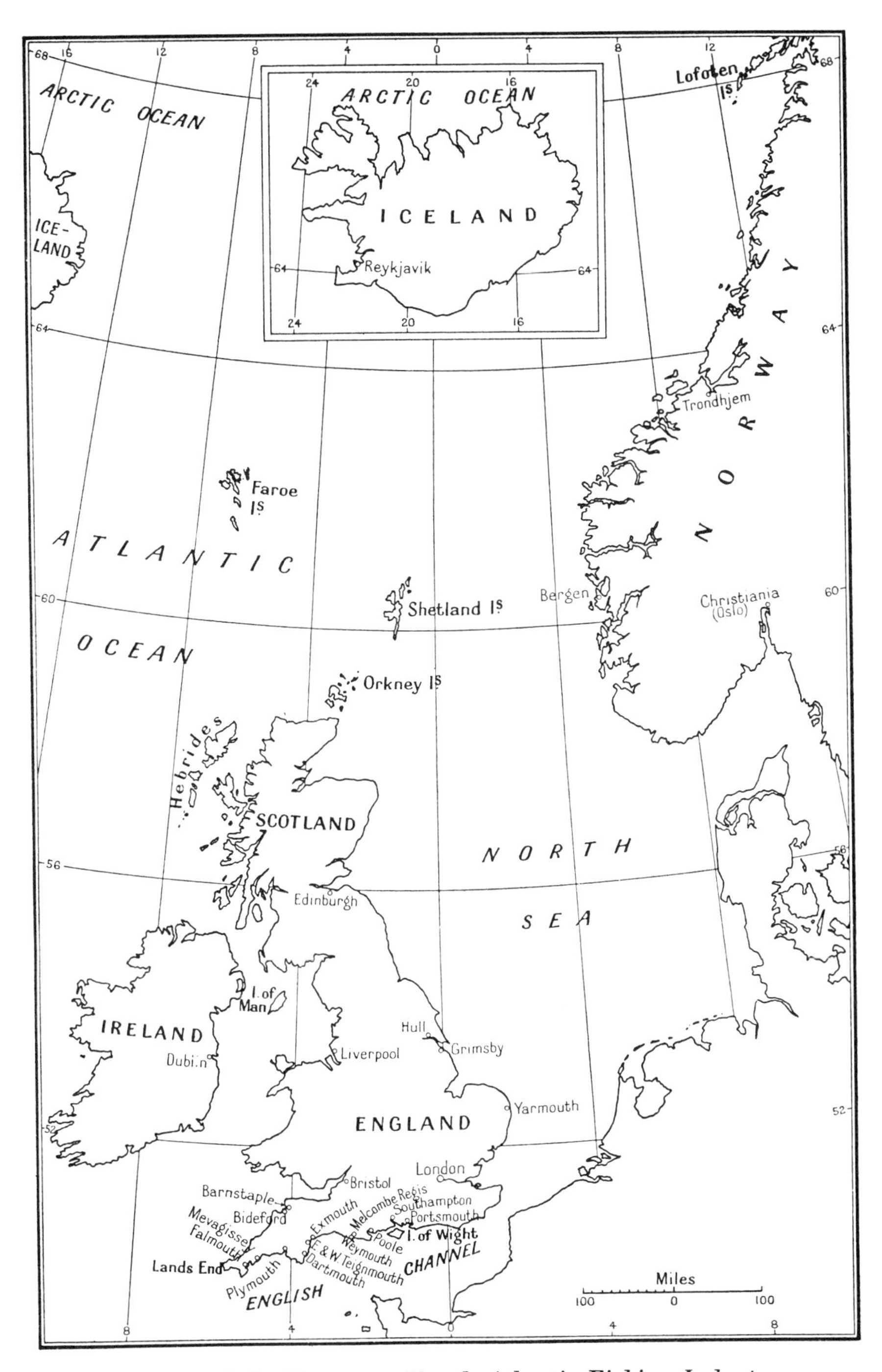

Centers of the European North Atlantic Fishing Industry

volved an increased production of grades which were turned out quickly and adapted to climates[66] less suited to the production of hard-cured fish. Newfoundland's large-scale production had been achieved by set-lines and traps in areas less suited to drying. The use of trawlers by other countries brought about severe competition for the Newfoundland soft-cured fish, which in turn had its effect on the hard-cured product[67] produced chiefly in the northeastern part of the island for the more fastidious markets of southern Italy, Spain, Portugal, and southern Brazil.

> Bank fish cannot be compared with shore fish[68] at any time. There is no fish in the country better than that produced at Fogo. Bonavista comes second. . . . I consider the whole fishery of Newfoundland is badly handled and the fish is cured worse than it was 40 years ago. People are getting away from the good standard of curing fish. The old-fashioned people worked on the fish themselves from sunrise to sunset. These people made good money because they made good fish. Take cod trap fish; that is never as good as fish caught on the line. . . . West Coast fish is not to be compared with northern fish.[69]

As in Canada, the pressure of the increased production of soft-cured fish not only weakened the market for dry-cured fish but hastened the trend toward pickled and fresh fish. Fresh fish was sent from the southwest coast to Halifax and Sydney in spite of high freight charges. Newfoundland salt-bulk fish was exported for processing into boneless; and other species of fish increased in importance. The Scotch cure for herring was introduced before the war, and in the postwar period quanti-

66. "By far the greater part of the difficulties in obtaining a product of high quality lies in the unfavourable drying conditions prevailing." The humidity was more pronounced in Newfoundland than in Iceland. *Reports of the Newfoundland Fishery Research Commission* (St. John's, 1933), II, No. 1, p. 19. Problems of drying were regarded as responsible for the large number of grades of fish and for difficulties in marketing. *Idem,* p. 29.

67. The *Report of the Commission of Enquiry* (St. John's, 1937) has stressed the importance of hard-cured fish in Newfoundland production, and the difficulties of alternative products. At the same time it recognized the problems involved in concentrating upon the hook-and-line fishery, and in improving the quality of the product. Light-salted fish had numerous advantages over heavy-salted.

68. Fish taken by small boats, brought in and given the light-salted cure.

69. The Honorable P. Templeman in the *Proceedings of the Convention of Licensed Codfish Exporters,* September 2, 3, 1920, p. 65. It was claimed that flakes were used to a less extent. Kean, *op. cit.,* p. 100. Soft-cure could be dried and shipped as hard-cure for less discriminating markets. South-coast fish were sent to Halifax and dried for the West Indies. This grade included soft-cure fish on the Labrador, the Straits, and the French Shore. It required from 16 to 20 hogsheads of 54 gallons each to make 100 quintals of fish; and Newfoundland shore-cure called for only 8 hogsheads. It is claimed that about 100 Labrador fish or 20 Newfoundland shore fish made one quintal. Drying the fish reduced the weight from about 2¼ to 1. The cod-liver oil was expected to pay for the salt, and, later, both the gasoline and the labor of salting and

ties were sent to New York for the Jewish trade, although it suffered by competition from Pacific-coast herring. The trade in frozen herring declined with the construction of large storage plants in the United States which handled fish taken on the Atlantic coast in the late summer and fall. The Hudson's Bay Company acquired an interest in the firm of Job Brothers in 1927, and began the export of fresh frozen salmon on a large scale.[70] Their ship, the *Blue Peter*, acted as a floating factory or mothership for handling salmon gathered by small boats operating along the coast. Harvey and Company, the Monroe Export Company, and Baird and Company also became more directly concerned with this fishery. In 1929 about 2,500,000 pounds of frozen and chilled salmon were produced and about 600,000 pounds of other frozen- and chilled-fish products. The cold-storage industry was also extended to the production of about 3,000,000 pounds of frozen and chilled blueberries.

As a result of the depression the value of the fisheries declined from more than $10,000,000 in 1929 to $3,000,000 in 1931. "We doubt," it could be said of the latter year, "if the average price for the whole catch of shore fish has exceeded $2.50 and that of the Labrador, $1.50."[71] In

curing. Cadiz salt, matured by a year's storage, was regarded as most satisfactory for Straits fish shore-cured, but was not sufficiently strong for Labrador, for which Santa Pola salt was required.

SALT USED FOR GRADES OF FISH

Grade	Fish per Hogshead of Salt (in quintals)
Labrador (heavy-salted)	6¼– 6¾
Straits of Belle Isle	7 – 8
Grand Bank (west coast)	6½– 7½
Inshore (light)	8 –10

Reports of the Newfoundland Fishery Research Commission, II, No. 1, p. 11. In the Labrador fishery "where the fish are 'thin' . . . the water is largely removed by salt action and partly by drying." *Idem,* p. 9. The salter "has to judge the amount [to] put on each fish according to its size. Such a degree can only be acquired by long practice; and apprenticeship to such an occupation ought to be of long duration involving clear instruction to the youthful pupil from the experienced operator. It is doubtful if this thorough apprenticeship prevails, and this state of affairs can be fixed upon as one very weak link in the chain of curing operations." *Idem,* p. 10.

70. Exports of pickled salmon declined from a high point of 1,679,700 pounds in 1922 and exports of frozen and fresh salmon increased to 4,500,000 in 1930. See *Reports of the Newfoundland Fishery Research Commission,* I, No. 2, pp. 11 ff.

71. *Twenty-third Annual Report . . . of the Newfoundland Board of Trade* (1931), p. 18.

LABRADOR PRICES
(*per quintal*)

	St. John's	Off the Coast
1929	$5.50	$4.50
1930	3.80–3.00	2.75
1931	2.50–1.80	1.50

See a description of conditions in Newfoundland in 1933, *idem,* pp. 21–24. *Newfoundland Royal Commission Report 1933* (London, 1934), pp. 78 ff.

1931 the catch of salmon declined 50 per cent below that of 1930. The repeal of the Eighteenth Amendment and the disastrous tidal wave on November 18, 1929, increased the difficulties of the south coast.[72]

The imposition of tariffs by importing countries was accompanied by exchange controls, for example in Greece and Brazil, and by Newfoundland's increasing dependence on the West Indian market to which over 265,000 quintals were exported in 1931, Puerto Rico alone taking more than 90,000. The low United States tariff on high-moisture-content fish facilitated exports of Newfoundland-Labrador fish to the latter market at the expense of Canadian fish.

> If we go back about twenty years we will find that we were shipping much larger quantities of fish to Spain, Italy, Brazil, etc., in other words to markets that required good fish and paid fair prices. We have now got down to the point where 25 per cent of our fish is of so poor a quality that it is only fit for the West Indies, and the prices we have to accept in these markets are so low that there is no profit to the exporter and [it] means a heavy loss to the fisherman, as the price of small West India for the last two years has been about $1.25 to the fisherman. The prices for which our fish has been sold in the West Indies have largely driven out the competition of any other fish, particularly Nova Scotia fish, and it would appear that the time will come when this country will be catching fish for the West India markets and Nova Scotians will be catching fish for their own and the American markets, in other words, selling their fish fresh or frozen. . . . The one market that has been a great boon to this country for the last few years is Porto Rico which used to be largely controlled by the Canadians, but our very low prices have almost completely driven them out.[73]

Barbados purchased fish chiefly from Newfoundland. Canada lost her preference in Jamaica in 1924 and Newfoundland steadily increased her share of the market. The downward trend of sterling exchange added itself to other difficulties in 1932 and early in 1933.

The depression was met by fresh efforts to develop a policy of export controls. The coffee crisis in Brazil led to the formation of a mutual exporters' association including Job Brothers, Bowring Brothers, the A. E. Hickman Company, A. H. Murray and Company, and Jas. Baird, which association appointed an agent, Captain G. Power, in Brazil. Shippers outside the association included Baine Johnston, the Fishermen's Union Trading Company, Crosbie and Company, T. Hallett and Company, and W. and J. Moore. Both groups worked together

72. See *MacDermott of Fortune Bay Told by Himself* (London, 1938), chaps. xx, xxv, and *passim.*

73. *Twenty-fourth Annual Report . . . of the Newfoundland Board of Trade* (1932), p. 14.

in 1932 and later years. Brazilian exchange weakened in 1933 with the weakening of the Canadian dollar and the strengthening of the pound; and in 1934 exchange regulations compelled importers to buy 40 per cent of foreign bills on the free market at 76 milreis to the pound sterling, and the remainder on the official market fixed by the government, or at 60 milreis to the pound. Quota regulations were further introduced giving 85 per cent of the official exchange to "preferred" countries, not including Great Britain and the empire. In addition, duties on dried cod were increased by 100 per cent as a protection for the jerked-beef industry. In February, 1935, regulations compelled buyers to secure all exchange on the official market, which increased the price from 80 milreis to 135 milreis a drum. In July, buyers were allowed to obtain exchange on free markets and small shippers were placed in a position to offer. To prevent a possible break, an agreement was drawn up among buyers in September which held the price at 28 shillings, and this was renewed in October. In November it was learned that one of the signatories to the agreement had broken the price by offering fish at the same price to retailers as to wholesalers and there was a temporary break. It was said that the association enjoyed lower freight rates to Brazil than those set for Canadian shipments through New York. In spite of the activities of the organization, exports to Brazil declined from 330,000 drums in 1929 to 168,000 in 1935. The possibility of Italian demands for jerked beef improving the market was offset by difficulties in Italy which restricted the demand for dried fish. In 1934 and 1935, sales from Great Britain[74] and Norway to southern Brazil, where very hard dried fish were in demand, also declined. Northern Brazil, with its demand for small light-salted fish, was dominated by Newfoundland in spite of the antagonism to the association. But new taxes in 1937 and 1938 narrowed the Brazilian market.

Attempts were made to set up boards for the marketing of fish in Europe, especially upon the establishment of the Commission Government in Newfoundland. Preliminary drafts, in 1931 and 1932, of legislation governing the export of salt codfish ended with the enactment of the Salt Codfish Act of July 7, 1933. A chairman was appointed and, under him, inspectors and cullers of fish in various districts. An advisory committee of the Salt Codfish Exporters Association was also appointed on

74. The agreement between Brazil and Great Britain in 1936 excluded Newfoundland from preferential treatment. The latter was apparently not regarded as a colony and had little to offer as a market for Brazilian products. The advantage of the fishery to Great Britain for her purchases of tropical products and sales of supplies for the fishery, which was conspicuous under the colonial system, is in contrast with the development of Newfoundland as a region unable to purchase large quantities of products from tropical regions.

August 1, and later a joint shipping committee with three representatives of the Hawes group, two of the Mutual Shippers Association, and two from shippers outside these groups. On August 24 regulations were enacted prohibiting the purchase of fish on a tal qual basis and defining certain standards. On September 12 further regulations[75] prohibited the export of fish to Oporto in any direct steamer after September 30 without permission from the Salt Codfish Exportation Board. These regulations were intended to prevent the arrival of large cargoes and a consequent glutting of the market, and to encourage smaller shipments, preferably in sailing vessels. In 1934 Portugal introduced further measures to increase the consumption of fish taken by her own fleet. Moreover, Oporto buyers began to purchase, for drying in Portugal, salt-bulk fish in larger quantities, chiefly from Iceland and Norway, but also from Newfoundland and St. Pierre. An act was passed requiring all importations to be made under license and all purchasers of foreign-caught fish to take a prorata proportion of national fish. Under a trade agreement Norway was granted a quota of 40 per cent of the foreign fish, and purchases were made from a committee of Norwegian exporters. Buyers in Oporto and Lisbon were organized into purchasing groups called *gremios* (guilds) under a government control board made up of two importers, two owners of domestic fishing vessels, and a chairman. The Oporto *gremio* offered 32 shillings for large and medium new bank fish and 25 shillings for small, with payment of 80 per cent against documents in May, 1935, but lowered this to 30 shillings and 24 shillings; to 28 shillings and 23 shillings at the end of June, and to 26 shillings and 21 shillings at a later date. As a result of the decline, a committee of Newfoundland exporters was formed about the end of July; and they supported the Salt Codfish Board in designating Portugal as a controlled market to which sales were prohibited except under permit, and on terms stated in it. In spite of this, organization sales were made at the end of August at 25 shillings and 20 shillings and Portugal made heavy claims for a reduction on the grounds of bad curing. Difficulties between Portugal and Norway, the latter being reluctant to accept low prices, were offset by the removal, in September, of a tax imposed on French fish in April, and the allocation to France of a quota of 10 per cent. In 1935 the Salt Codfish Board was abolished,[76] and a new organi-

75. See *Rules and Regulations Respecting the Fisheries of Newfoundland, 1933.* The Bait Act was suspended for one year.

76. For copies of the commission's legislation see *Report of the Commission of Enquiry* (St. John's, 1937), pp. 214–225. A Supreme Court decision permitting an exporter to ship to Puerto Rico in spite of a contract of Porto Rico Exporters, Ltd., with a larger importer in Puerto Rico has led to amendments to an act in 1936 further strengthening the control of the commission. *The Financial Post,* November 26, 1938.

zation was set up under the Department of Natural Resources of the Commission Government, which absorbed the former Department of Marine and Fisheries. The new Salt Codfish Board included three government representatives and six representatives of the trade appointed in 1935 by the government but thereafter elected by the Salt Codfish Exporters Association. This organization was included in the Newfoundland Fisheries Board in 1936. The Oporto *gremio* attempted to fix prices in 1936 at 25 shillings and 20 shillings but the board insisted on 30 shillings and 25 shillings, and after a period of considerable difficulty a compromise was reached at 27 shillings and 22 shillings.[77] An arrangement was made by which Hawes was appointed agent for exporters to Portugal, but members not in the Hawes group have made numerous protests against the terms of the agreement.

Almost prohibitive duties were introduced in Spain in 1933 and were followed by import quotas in 1934 which allocated to Newfoundland 198,230 quintals, based on the average for the three years' shipments from 1931 to 1933 inclusive. An attempt was made to reserve the quota for the higher-priced shore-cured fish and to limit shipments of Labrador cure. A duty of 17 shillings a hundredweight was imposed on foreign cod, and exchange restrictions added a further handicap, although it was reduced by an exchange-clearing arrangement between Spain and Great Britain. The outbreak of the Civil War in 1936 had disastrous effects on the Spanish market. In March, 1935, Italy received limited imports of dried cod. They totaled only 20 per cent of the amount imported in the same period in the previous year. At the end of March, through the intervention of the Dominions Office, Newfoundland was granted a quota of 70 per cent of the quantity imported in each quarter of the preceding year. Exchange restrictions which delayed payments for four and five months, limited imports, and the imposition of sanctions by the League of Nations closed the market. Although it was reopened in 1936, exchange limitations were unsatisfactory. In Greece, exchange restrictions have favored Newfoundland and France. European governments have exercised increasing control over imports and have attempted to increase national production not only by the encouragement of trawlers but also by imports of salt-bulk fish from Iceland, France, and elsewhere, such fish to be dried in the importing country. Newfoundland, with a small population and limited bargaining power, has been compelled to rely to an increasing extent on organiza-

77. Charges to be deducted would include freight, 75 cents, in Portuguese bottoms (schooners 55 cents to 65 cents), commission and discounts, 34 cents, loading 10 cents, sales tax 2 cents, and marine insurance 10 cents.

tion and on the assistance of Great Britain, particularly through the Dominions Office.

The Commission Government[78] has attempted to meet the problems by the improvement of selling organizations which export to the West Indies, South America, and Europe, and by various other forms of assistance. In 1934 bait was supplied by the government at St. John's as well as by private plants at Burin and Holyrood, and depots were increased in numbers.[79] The steamship *Malakoff* was used as a floating bait depot. In 1935 a trawler, the *Imperialist*, was operated by the Newfoundland Trawling Company, and extensive drying operations were carried on at Harbor Grace. With the support of these developments, production on the Banks has increased.[80] Fishing inspectors were employed to improve the cure of the fish. The tariff was revised to lower the burden on the fishery in 1934; and in 1936 rebates were granted on gasoline used in the fisheries. In the fall of 1934 arrangements were made to finance the construction of schooners, with the result that large numbers were engaged in the Labrador fishery in 1935.[81] The government has further supported a research program begun under the Empire Marketing Board. It includes a station at Bay Bulls[82] for carrying on studies in the movements of fish, as also studies of improved methods of producing cod-liver oil and other fishing products. Coöpera-

78. See *The Report of the Commission of Enquiry* (St. John's, 1937), pp. 226–233 and *passim;* also Henry King, "Report on the Seafisheries of Newfoundland," *Canadian Journal of Economics and Political Science,* May, 1938, pp. 219–222; and Henry King, *The Marketing of Fishing Products in Newfoundland* (St. John's, 1937).

79. Bait freezers were installed at Quirpon, Joe Batt's Arm, Bonavista, Bay de Verde, Bay Bulls, and St. Mary's.

80. See an account of a trip in a banker on a spring cruise in 1932. *Reports of the Newfoundland Fishery Research Commission,* II, No. 1. Of a fishing fleet at Fortune Bay it could be said: "These fifty schooners are practically the last of the once famous North Atlantic fleet of salt cod fishermen that used to range over the offshore banks from Hatteras to Labrador." *Idem,* Appendix C.

81. The outlay on schooners is inevitably high, as the efficiency of the schooner tends to increase with the size. In 1937, 323 vessels and 2,580 men went to the Labrador and, in 1938, 261 vessels and 2,243 men. In the latter year, in a total catch of 1,147,125 quintals, 506,091 were shore fish, 233,387 deep-sea, and 407,647 Labrador. Export of salt bulk, chiefly to Portugal, Italy, and Greece, increased from 65 thousand hundredweight in 1937 to 207 thousand in 1938. Exports of fresh codfish in 1938 reached 363 thousand pounds, codfish fillets 782 thousand, halibut 522 thousand, herring 4,185 thousand (a phenomenal increase), lobsters 1,385 thousand, salmon 4,227 thousand. The seal fishing declined to 97,345 skins. Two companies with nine ships participated.

82. See H. Thompson, *A Survey of the Fisheries of Newfoundland and Recommendations for a Scheme of Research,* December, 1930; *Reports of the Newfoundland Fishery Research Commission,* I, No. 1; also *The Report of the Imperial Economic Committee on the Marketing and Preparing for Market of Foodstuffs Produced within the Empire, Fifth Report, Fish* (London, 1927).

tive efforts have been given support.[83] Encouragement has been given to the production of frozen fish and to improvements in marketing other products. In 1934, when complaints were made of the disappearance of open-market operations in New York, a herring board was set up and began to function in 1935. The board attempted to maintain New York prices, but the fishery has not been successful. In 1937 a rebate of $1.00 per hogshead was given on salt and in 1938 one of $1.50. In September of the latter year it was decided to guarantee a fixed price of $3.00 for Labrador fish. This involved a subsidy of $450,000. In 1939 the salt rebate was discontinued, and a guaranteed price was established ranging from $2.85 for Labrador No. 2 to $5.50 for large merchantable. Arrangements have been worked out for the establishment of a processing and cold-storage plant by the General Sea Foods Company at Poile Bay on the American "treaty shore." Under responsible government the French were refused permission to establish plants on the "treaty shore." Under commission government Americans are being encouraged.

The fundamental difficulties of Newfoundland are inherent in her position as a producer of a commodity that is consumed in tropical countries with a large Catholic population and low purchasing power. She is at the same time subject to the effects of industrialism in the marked concentration of the tropics on the production of products such as sugar, coffee, and bananas. She is in competition with recently industrialized fishing regions and is affected by prices of supplies determined largely by the North American continent. Prices of cod rise more sharply and fall farther than prices of other products, and the result is that expansion is apt to be more pronounced and depression more acute. The products of industrialized agriculture imported from continental North America, such as flour, pork, and beef, increased in price earlier, more rapidly, and at a more sustained rate than cod. After the turn of the century and with the advent of improvements in transportation, Great Britain tended to be displaced by Canada and the United States.[84] Loans were floated in North America, the price structure was

83. The Cape St. Francis Cooperative Society secured a grant of $3,000 from the commission and built a plant at Pouch Cove in 1936. Another coöperative society was started at Ferryland. Margaret Digby, *Report on the Opportunities for Cooperative Organization in Newfoundland and Labrador* (London, 1934). The commission has appointed Mr. Gerald Richards to develop coöperative organization along the lines of the St. Francis Xavier movement.

84. The absence of steamship competition has been held responsible for high rates to Great Britain and the decline in British trade. Imports from Great Britain fell from 44 per cent of the total in 1888 to 27.5 per cent in 1912–13. *Royal Commission on the Natural Resources, Trade, etc.*, p. 29.

linked to the continent through Canadian banks, and transportation rates by land or sea were part of the American system. In the postwar and depression periods the economy has tilted back toward Great Britain. Newfoundland was squeezed between two civilizations. She produced for tropical countries with low standards of living, and had to compete with other foodstuffs and goods purchased from highly industrialized countries. "It does not seem probable that there is any other country of equal size and importance that has to import from abroad practically the great mass of the necessaries of life."[85] Consuming areas made little vocal demand for improvements in production, but accepted readily the improvements in technique of Newfoundland's competitors. Disturbance in production and in markets has meant that "the results of labor are likely to be comparatively unevenly divided." The merchant and the truck system[86] served as buffers between continental fluctuations and those of the tropical regions, with the result that standards of living could be forced down more sharply in Newfoundland[87] than in

85. McGregor's Report, *Journals of Assembly, Newfoundland,* 1907.

86. For a valuable discussion of the truck problem see *The Report of the Commission of Enquiry* (St. John's, 1937), pp. 128 ff.

87. Labor migrated to the centers of industrial development in Newfoundland. A second paper mill was built at Corner Brook in 1935, and was later acquired by the International Paper Company, and recently by the Bowater-Lloyd interests; and lead and zinc deposits were developed at Buchans by the American Mining and Smelting Company and the Anglo-Newfoundland Development Company. Migration became more rapid because of the character of the fishery and that of the labor as well. "The adaptability of the Newfoundlander who is in turn a fisherman, a lumberman, a miner, and who does some farming on a small scale and with a limited variety of products, has enabled him heretofore to speedily suit himself to every new industry that has been introduced into the colony." *Royal Commission on the Natural Resources, Trade, etc.,* p. 3. "The codfishery . . . is almost unique as an industry in that the class which owns the capital in it has managed . . . to throw the whole risk, or very nearly the whole risk . . . on to the shoulders of the working classes. Moreover it is a striking example of an industry in which the real capitalist has gone very far towards making a profit the just charge on the proceeds of the sale of a manufactured article, taking precedence even of a bare subsistence for the primary producer." Thomas Lodge, *Dictatorship in Newfoundland,* pp. 49–50. As a result hoarding "is the fisherman's instinctive method of throwing back on the capitalist class the real capitalist risk." *Idem,* p. 59. In 1915 it was estimated that a half million dollars was hoarded. *The Royal Commission on the Natural Resources, Trade, etc.,* p. 46. The problem of how to render assistance was a difficult one "in a system under which the major part of the revenue of a community is derived from a general tariff, that capital expenditure, if it takes the form of wage payments on objects not directly remunerative (as it largely did in Newfoundland) tends to swell unnaturally current revenue. . . . Capital expenditure on imported money, whether borrowed or given, inflates the current revenue and goes on inflating it owing to the delayed action of wage expenditure. . . . Its cessation brings about an inevitable corresponding deflation which operates disastrously and continuously in the reverse direction while the expenditure itself arouses a misleading feeling of optimism in the general public." Lodge, *op. cit.,* pp. 37–38.

Nova Scotia. Increased production for European markets compelled the more cheaply cured fish of Newfoundland to go to the markets of the West Indies and South America.[88] In these markets the decline in the price of sugar and the difficulties of coffee production which followed the war have had a continuously depressing influence on the price of cod.[89] Less favored producers of salt cod, such as Canada, turned to an increasing extent toward the production of fresh, frozen, and pickled fish, and favored areas of Newfoundland exported fresh fish to Canada.[90]

The tragedy of Newfoundland followed the late development of commercial activity and responsible government. The encirclement by capitalism using the railway, the steamship, and the trawler introduced rigidities in fixed charges and rates which the limited development of governmental institutions based on a commodity subject to wide fluctuations in returns was unable to withstand. Inability to retreat from the effects of the depression on world markets weakened the economy to the point where an Anglo-Saxon population accepted the verdict that responsible government should be dispensed with as a constitutional nicety. In its place commission government[91] has introduced control over commercialism in export boards.

88. Exports in 1936–37 totaled 956,829 quintals, of which Brazil received 245,267, Porto Rico, 196,821, Portugal, 153,193, and Jamaica, 107,699.

89. "The price paid for bananas and the rate paid for labor [in Jamaica] are controlling factors in the consumption of cod fish." *Twenty Years of the Fishermen's Protective Union,* p. 371.

90. "We can import Newfoundland cod fish," said a spokesman for North Sydney, "at thirty and forty cents less than we pay the local fishermen." Speaking for Lunenburg another witness said: "We have had to bring fishermen from Newfoundland since post-war days because we could not get men here." *Idem,* p. 112. Restrictions on Newfoundland labor in Canada reduced "the yearly emigration of the fishermen on the south coast of Newfoundland to the Lunenburg fisheries." *MacDermott of Fortune Bay,* p. 215.

91. See R. A. MacKay, "Newfoundland Reverts to the Status of a Colony," *American Political Science Review,* October, 1934, pp. 895–900; also Lodge, *op. cit.*

CHAPTER XV

CONCLUSION

GEOGRAPHICAL BACKGROUND

THE economic history of the regions adjacent to the submerged areas extending to the northeast of America's north Atlantic seaboard is in striking contrast to that of the continental regions. In the continent's northern area the St. Lawrence facilitated expansion westward and a concentration on fur, lumber, and wheat; in the submerged areas innumerable small, drowned river valleys in the form of bays and harbors facilitated expansion eastward and a concentration on fish. Drainage basins bring about centralization, and submerged drainage basins decentralization. Development westward followed the St. Lawrence and its tributaries; development eastward had its origin in the scattered harbors and bays formed by the drowned system of rivers and their tributaries. Again, staple products coming down the St. Lawrence system made for a centralization of exports, whereas fishing from the numerous ports of an extended coast line made for decentralization. Unity of structure in the economic organization of the St. Lawrence was in strong contrast with the lack of unity in the fishing regions. Penetration up the St. Lawrence brought a succession of exports, fur being displaced by lumber, and lumber by pulp and paper, wheat and minerals. On the Atlantic, expansion was along the seaboard with an increasing concentration on the fisheries. In the interior, economic history was marked by changes to new staple industries; on the Atlantic, changes were centered in a single industry.

The contrast between the economic organization of the continent and that of the fringes of the submerged areas has been evident in the economic history of regions tributary to the submerged areas in so far as they have reflected the influence of the continent and of the sea. The influence of North America became progressively weaker as the fishing regions extended to the northeast and were offset by the increasing influence of Europe. New England, attached to the continent, Nova Scotia, almost an island, and Newfoundland, an island, reflect the progressive effects of a geological disturbance in their relative isolation from the continent, in the extent of the resistant formations in the land area, the extent of the submerged area, and consequently in the more severe limitations upon agriculture and lumbering and the increasing importance of sea fishing. Recent industrialism, in the shape of mining and the pulp and paper industries, has struck most forcibly the region which had been most closely connected with the fisheries, and with disastrous results for its political and economic life. The greater importance of seasonal variations toward the northeast impressed itself in the migrations and the varying sizes of the cod and, consequently, in the greater variability of the industry. In the Gulf of Maine the fishery[1] developed as a winter industry; in Nova Scotia it was limited to the early spring and summer; and in Newfoundland, to a shorter season.

Climatic limitations restricted the activities of European nations. The New England fishery was prosecuted most effectively by settlers resident in the winter and with vessels capable of carrying on the industry at a distance from the shore. As the number of vessels increased, the summer fishery was extended to the banks off Nova Scotia. The large cod cured in cool weather in New England were suited to Bilbao and a Spanish market; but cod taken in the summer and cured in Nova Scotia were of poorer quality and, until a century ago, were largely sold as food for the slaves in the West Indies. Great Britain had no supplies of solar salt and consequently she was better adapted to exploit the summer fishery of Newfoundland. Fish were cured and sent to Portugal, southern Spain, and the Mediterranean. France, with abundant domestic supplies of salt and a market which, extending from Paris, with its demands for large bank fish in the "green" state, to La Rochelle, to the Bay of Biscay ports, and to Marseilles on the Mediterranean, with their demands for dry fish, carried on the fisheries under wide variations of sea and climate. Vessels from the Channel ports proceeded to the Banks to secure fish for the Lenten season, and to the Petit Nord and Gaspé for the cod required in the Mediterranean markets. Vessels from the Bay of Biscay ports went to Cape Breton and Nova Scotia to take fish to domestic, colonial, and foreign markets alike. The Channel ports, having a demand for Mediterranean products, added the dry fishery to their spring bank fishery. Specialization of production, as called for by the different markets, took differing forms throughout the history of the industry, but it was always important. The geographical background, represented by climate, the salt available, the technique of the industry, and the size and abundance of the fish, was a stabilizing factor of supply. In tropical countries, demands

1. Large cod taken in the Gulf of Maine were handled successfully in the cool winter season, but off the coast of Nova Scotia and on the Grand Banks in the summer the quantity of salt required varied with the season and the size of the fish. Smaller cod taken near shore in Nova Scotia, the Gaspé Peninsula, and Newfoundland, brought in daily and placed on flakes exposed to wind and sun, could be cured with relatively little salt. To the north, along the Labrador, salt and cool weather produced a soft cure.

for a higher protein content in the diet of the people, the difficulties of cold storage, the rigidity of customs and consumers' preferences, and the dominance of Catholicism were stabilizing factors of demand.

COMMERCIALISM AND THE FISHERY[2]

In the sixteenth century the fishing industry of the New World involved a marked extension of the economic frontiers of the nations of Europe. The widely scattered ports of France sent vessels north to Newfoundland and the Straits of Belle Isle and south to Cape Breton and Nova Scotia. By 1550 this fishery was prosecuted on the Banks, and by the end of the century at Gaspé in the Gulf of St. Lawrence. The story of the Portuguese fishery paralleled that of the French fishery in point of time; but the Portuguese fishery had been largely limited to southeastern Newfoundland. The Spanish fishery grew rapidly after the middle of the century, and Portugal was absorbed by Spain in 1581. Excluded from Iceland, the English fishery expanded in Newfoundland and became established with the defeat of Spain, the opening of Spanish markets, and access to supplies of specie and to cheap supplies of salt. Expansion called for an increase in ships and seamen and a greater production of food adapted to naval demands and long voyages in tropical regions. France and England were quick to take advantage of the decline of Spanish control and the rise in prices which accompanied the inflow of American treasures; and, in place of the Spanish, who made their fish "all wet and do drie it when they come home," the English and French dried it in the New World and carried it to Spain. The English fishery extended itself along the Avalon Peninsula and later into New England waters. The fishery of France expanded to the Gaspé coast and to areas suited to the production of dry fish. West Country commercial interests fought not only for open markets for fish but also against monopoly control of production and trade. The struggle continued in the seventeenth century and was successful both in breaking down an English monopoly of Spanish trade and in defeating attempts to establish a monopoly in Newfoundland. The fishing industry in Newfoundland provided a new frontier; and in its development, with the increase in ships, seamen, and trade, it broke the rigid chains of centralized control. The seasonal character of the industry in Newfoundland strengthened the position of West Country interests and enabled ships

2. For a general summary see H. A. Innis, "An Introduction to the Economic History of the Maritimes (including Newfoundland and New England)," *The Report of the Canadian Historical Association,* 1931, pp. 85–96, revised in R. F. Grant, *The Canadian Atlantic Fishery* (Toronto, 1934), pp. xii–xxi, and in the article, "The Fishing Industry," *Encyclopaedia of Canada.*

to participate in trade with Spain, and later to trade in the Mediterranean during the off-peak period of the winter season.

The extension of fishing to the New England coast and the development of the winter fishery facilitated the growth of settlements, shipping, and trade, and the emergence of a second vigorous commercial organization. Ships carried fish to the markets of northern Spain and brought back manufactured products from England. During the summer season they followed the cod to Nova Scotia, and, with the expansion of sugar production in the West Indies, sold the poorer grades to that region. The increase in the fishing industry and trade brought about conflict with France. The French fishery was confined to the summer season but expanded as a result of the Civil War in England. Placentia was established and, under Colbert, France became aggressive in New France and the French West Indies. But recovery in Newfoundland and expansion in New England restricted the French not only in production but also in markets. Carried on over wide areas and from scattered seaports, and concerned largely with the green fishery and domestic markets, the French fishery had no focus in settlements, shipping, or trade. Attempts at building up control by chartered companies and under government auspices meant conflict with the separatist character of the fishery and weakened its opposition to English expansion. In 1713 the Treaty of Utrecht saw the withdrawal of the French from Placentia, and from Nova Scotia to Louisburg in Cape Breton. New England commercialism began to conflict with that of the West Country, in Newfoundland, by giving help to its increasing trade and settlements; but in 1763 the main results became evident in the defeat of the French and the Treaty of Paris. Commercialism in New England provided an aggressive nucleus for the expansion of the British Empire and accentuated the difficulties of France in building an empire in North America. The geographical handicaps proved insuperable to commercialism as a basis of support for a French empire, and commercialism based on the fishery was diffused among a number of widely scattered ports in France.

With the elimination of the French from the North American continent, friction between the commercialism of New England and that of England itself became more serious and led to the American Revolution. The extension of the New England fishery to Nova Scotia and Cape Breton was accompanied by expansion of trade to Europe and to the West Indies. Sugar production in the less exhausted lands of the French West Indies was encouraged by trade from New England. New England commercialism came increasingly into conflict with British West Indian sugar interests and, in Newfoundland, with the commercial interests of the West Country, because the trade of New England supported more settlers and lessened the importance of the fishing ships. Hostilities with the French tempered and concealed the conflict between New England and England. But the defeat of the French intensified it and laid it bare, and it ended with the independence of the colonies. The influence of France declined; and its decline strengthened the divisive trends at work within the British Empire.

The fishing industry of New England flourished because it possessed abundant resources and supported a powerful commercial group engaged in the trade, first, of the Atlantic and eventually, after the American Revolution, of the Pacific. New England attempted by bounties and duties to extend her fishery and to save herself from exclusion from British waters and from the West Indian markets. The geographical isolation of Nova Scotia, the limitations of sailing vessels, and Nova Scotia's position as a bulwark against the French favored the growth of strong commercial interests concentrated at Halifax. After the Treaty of Versailles in 1783, commercial interests attempted to strengthen resistance to the United States and seized on the advantages offered by Great Britain's hostility to the Americans. Aggressive commercialism, based on the fishing industry and shipping and rooted in New England traditions, came into evidence in the efforts of Nova Scotia, through its influence on Imperial policy, to exclude imports of manufactured products from the United States by tariffs, to restrict the rights of New England fishing vessels, to prevent smuggling, to limit the production of fish in the United States, to discourage direct trade between the United States and the British West Indies, and to encourage its own direct trade. With the increasing importance of Nova Scotia within the empire, there came a relative decline in the influence of the British West Indies—a decline made plain and brought about by the successful competition of East Indian sugar, by measures for the abolition of slavery, and by other measures favorable to Nova Scotia and unfavorable to the West Indies. It coincided further with the westward expansion of the United States and the decreasing influence of New England in American policy, as was shown by the restrictions imposed on her fishery in the Convention of 1818.

Expanding markets in the protected slave states of the South and the withdrawal by Great Britain of restrictions on American trade in the British West Indies hastened the expansion of the mackerel fishery in the United States. This caused a marked increase of activity in the Gulf of St. Lawrence and encroachments on the inshore waters of the British American colonies, accompanied by smuggling. Anti-American aggressiveness on the part of Nova Scotia, which went with the extension of her commerce after the American Revolution, became more conspicuous in legislation such as the Hovering Act and in the steps taken for its enforcement.

The influence of Nova Scotia on British commercial policy declined with the trend toward free trade in Great Britain, and this hastened the increasing concentration of Nova Scotia upon the problem of barring New England fishing vessels from British waters. The disappearance of preferences in the West Indian market and growing competition from the United States resulted in increasing attempts to restrict American fishing by the enforcement of the Convention of 1818. Free trade in Great Britain and the removal of the shelter of commercial preferences intensified a feeling in Nova Scotia that she must seek protection through her own efforts. Having modified and contributed to the disappearance of the British colonial system, the colonists attempted to modify the commercial policy of the United States. The success of their tactics was registered in the Reciprocity Treaty of 1854, which admitted fish from Nova Scotia free of duty and permitted American vessels to fish in Canadian waters. The reluctance of Great Britain to help led Nova Scotia to turn to coöperation with New Brunswick and Canada, and, after the abrogation of the Reciprocity Treaty, led eventually to Confederation. The growth of nationalism in the colonies was the inevitable accompaniment of the British policy of free trade. The attempt to mold Imperial policy to the demands of Nova Scotian commercial interests and to exclude the United States from the fishery went with the winning by Nova Scotia of an increasing control over her own internal policy. The varied character of her resources, her proximity to Prince Edward Island, New Brunswick, and the Gulf of St. Lawrence, and the importance of wooden sailing vessels hastened the growth of various ports, the development of agriculture, the construction of roads, and the expansion of wooden shipbuilding. The centralization of the commercial group at Halifax was offset by the decentralizing tendencies of trade, fishing, and other industries. The inherited influence of New England, evident in Nova Scotia's commercial spirit, was also evident in the province's political development. New England's influence on the commercial center and in western Nova Scotia was accompanied by a growing Scottish and, to a limited extent, Acadian influence, particularly in eastern Nova Scotia and Cape Breton. Commercial interests from the Channel Islands organized the trade and fishery of Cape Breton and the Gulf of St. Lawrence and restricted expansion from Quebec and Halifax. The convergence of differing racial, religious, and economic groups, together with the Imperial influence that was likewise concentrated in Halifax as a naval center, contributed to the "intellectual awakening of Nova Scotia"[3] and to that balancing of interests which made her the cradle of responsible government[4] in the "Second Empire." Her position, sensitive alike to competition from New England and to the effects of Imperial commercial policy, gave Nova Scotia a great influence with the administration in Great Britain.

The end of reciprocity was followed by renewed and vigorous efforts on the part of Nova Scotia's political leaders, Howe[5] and Tupper, to secure for her fish free entry to the United States market. License fees were increased, and the support of Canada after Confederation was enlisted to enforce the restrictions of the Convention of 1818. The growing importance of the steamship increased the effectiveness of regulations over an industry dominated by sailing vessels. The assertiveness of Nova Scotia made itself felt in the terms of Confederation, when the fisheries were placed under the Dominion in order to strengthen the position of the industry in international negotiations. It was also felt in Nova Scotia's insistence on the construction of the Intercolonial Railway and in the vesting in the provinces of control over natural resources by the terms of the British North America Act. Later, it was likewise apparent in the compromise tariff, the successful demand for better terms in 1869,[6] the repeal election of 1886, the selection of finance ministers from the Maritime Provinces, and the increasing strength of those provinces in the federal system. The tendency of Great Britain to temporize in negotiations with the United States was resisted; and, in the Treaty of Washington and the subsequent arbitration proceedings, Nova Scotian interests succeeded in securing, through Canadian representatives, a recognition of their position.[7] But Nova Scotia's insistence, working through the Dominion, upon rights recognized under the Convention of 1818 defeated its own ends, particularly upon the decline in importance of the inshore mackerel fishery. To the withdrawal of New England interests from the salt-fish industry and the increasing importance of the fresh-fish trade, dependent as it was on more accessible fishing grounds, were added the expansion of the Nova Scotian fishery under the Washington Treaty and a rapid increase in shipbuild-

3. D. C. Harvey, *Dalhousie Review,* XIII, pp. 1–22.

4. W. R. Livingston, *Responsible Government in Nova Scotia,* University of Iowa Studies, 1930; and Chester Martin, *Empire and Commonwealth* (Oxford, 1929).

5. See Joseph Howe's Detroit address advocating reciprocity, of August 14, 1865. *The Speeches and Public Letters of Joseph Howe,* ed. J. A. Chisholm (Halifax, 1909), II, 438–455.

6. D. C. Harvey, "Incidents of the Repeal Agitation in Nova Scotia," *Canadian Historical Review,* March, 1934, pp. 48–56.

7. L. B. Shippee, *Canadian-American Relations 1849–1874* (New Haven, 1939).

ing. When, however, in 1885 the United States abrogated the Washington Treaty, acute difficulties arose a year later; for Canada insisted on the enforcement of the Convention of 1818. There followed an attempt to introduce a second Washington Treaty which failed in the United States Senate because of protests from New England. During the period of high license fees which had followed the ending of reciprocity in 1866, the American mackerel fishery in the Gulf of St. Lawrence had conferred advantages on Prince Edward Island in the shape of trade and transshipment profits, and had led her mercantile interests to refuse to coöperate in excluding Americans from the inshore fishery. But this conflict of interests disappeared with the Washington Treaty, the rise of the fishery in Prince Edward Island, and the decline of the American fishery in the Gulf; and in 1873 the Island entered Confederation.

In Newfoundland, settlements grew and the fishery expanded, first with the support of New England, and then with the difficulties between England, France, and the United States which terminated in 1815. The aggressive commercialism of the West Country based on the fishing industry had restricted the growth of settlements in Newfoundland, even as New England's aggressive commercialism had accelerated *her* expansion and her break from the empire, and as, in the case of Nova Scotia, her aggressive commercialism had fostered the growth of independence within the empire. Increasing settlements represented a trend that received aid from the Maritime Provinces and Canada, while the disappearance of the fishing ships and the rise of an aggressive lesser merchant class, especially in St. John's, led to the struggle for responsible government. The emergence of commercialism in Newfoundland was accompanied by trends similar to those of other fishing regions. The readjustment of the political structure to meet the demands of settled areas and commercialism was coupled with fiscal arrangements to provide capital for the construction of roads and the development of agriculture.

After the Napoleonic wars the revival of the French fishery, supported by bounties and duties, narrowed markets for cod from the Banks. The introduction of the bultow system meant demands for bait from the south shore of Newfoundland. Competition by the increasing quantities of French cod caught with bait supplied by Newfoundland fishermen, who were thus drawn away from the cod fishery, led to demands for protective legislation. As the commercial interests of Nova Scotia resisted the United States, so did the commercial interests of Newfoundland resist France, Canada, and the United States. The assertiveness of Newfoundland came out in her refusal to accept treaties between Great Britain and France in 1857 and 1885, in her collecting customs duties on the Labrador, and in her exclusion both of West Country firms and traders from the United States and Nova Scotia. Sir Charles Tupper's attempts to secure the free entry of Canadian fish into the United States market in the second Washington Treaty led, in 1887, to the suggestion of acceptable proposals for federation with Newfoundland; but failure there, and Newfoundland's frustrated attempts to secure the free entry of fish through the Blaine-Bond Convention, when Canada insisted on delay by Great Britain, aroused the Island's enmity once more. Resentment against Canadian interference in this case limited the possibilities of success in later Confederation negotiations. Nova Scotia's insistence upon independence had led to Confederation, and it intensified Newfoundland's insistence upon independence, which led to isolation. Continued pressure from the settlements on the French Shore and protests from Newfoundland finally brought about the agreement in 1904 under which the French withdrew. The same assertiveness was further evident in Newfoundland's protests against encroachments by the United States, in the termination of the *modus vivendi* agreement in 1905, in the negotiations which ended with the award by the North Atlantic Fisheries Arbitrations, in resistance to Canadian encroachments in Labrador, and the decision of the Privy Council in the Labrador boundary dispute.

THE RISE OF CAPITALISM

THE "pull" of the continent has been made manifest by the importance of the capital equipment essential to the development of the fresh- and frozen-fish industries. *Le sel a cédé le pas à la glace* (salt has yielded to ice), and the schooner to the steamship, the railway, and the trawler. Increasing diversification of the industry in the more southerly areas, with wider markets, the possibilities of continuous yearly operation, and the availability of larger fishing resources, as offered by lobster, haddock, halibut, herring, and other fish, have meant increasing mechanization. Pickled cod has become more important in view of the demands of the continental market for the processed product. Capital equipment and financial control have been extended to the north, from New England to Nova Scotia, and to Newfoundland and the Gulf of St. Lawrence; and displaced, obsolescent capital equipment has been swept away by it as by a frontal wave. Sailing vessels have betaken themselves from New England to Nova Scotia, and from Nova Scotia to Newfoundland. With the increasing demands of continental America and increased mechanization, capital equipment has moved northward, and labor has moved southward. Newfoundland labor has been available to man the Lunenburg fleet, and both Newfoundland and Nova Scotia have witnessed a pronounced migration to New England, British Columbia, and elsewhere.

Newfoundland, as the region most distant from the continent and most dependent on the cod, has been less influenced by the demands for fresh fish. The decline of the dry fishery in the United States and Canada has contributed to her increasing specialization on salt cod. On the other hand, the expansion of population along the coasts of Newfoundland and Labrador, the development of the interior following the construction of the railway, the introduction of steam navigation, and the use of motor engines contributed to the introduction of large-scale technique, such as fish traps, and to increased production in areas less suited to drying, with the result that larger quantities of soft-cure fish were produced. The spread of industrialism, especially after the war, brought about an increase in production through the use of trawlers by Iceland, as also by France and other European countries, and the use of motor engines in Norway and elsewhere. The increased competition of soft-cure fish pushed Newfoundland out of the European markets, but at the same time it pushed them into those of Brazil and the West Indies, and hastened the trend in Canada toward the production of fresh and frozen fish. The expansion of the North American market during the prosperity of the 'twenties intensified the decline in production of dry fish because of the competing demand for labor and equipment. Labor was attracted to industries in Canada and Newfoundland, and fishing vessels were easily adapted to the exigencies of prohibition.

Newfoundland has been placed in a vulnerable position.* The demand for salt cod in tropical and Catholic countries has been more directly exposed to the effect of fluctuations in economic activity incidental to regions producing tropical commodities. These tropical products, being luxuries, are subject to wide variations of demand from countries in the temperate zone. Such variations are due to many things—to cyclical business disturbances; to the influence of mechanization on tropical commodities as, for example, citrus fruits, bananas, sugar, and coffee; to the weakness of government machinery in countries whose peoples have low standards of living, as is made evident in bankruptcies, exchange rates, and revolutions—and these variations are also due to the possibilities of competition from fish produced as a by-product in mechanized countries. Demand for luxury products fluctuates sharply, as it does for dried cod, whereas fluctuations in the cost of provisions and supplies such as flour, salt pork, and salt beef from temperate continental areas have been less pronounced. Areas engaged in the production of fish by manual labor are squeezed severely during depressions.

*See Notes to Revised Edition, p. 509.

Wide fluctuations in income, in the catch and price of fish, and the limitations of governmental machinery, together with the absence of a speculative market, involved that extensive use of credit which manifested itself in the truck system. Dependence on the disequilibrium of international trade, in the case of exports, added to the internal burden put upon the truck system.[8] The commercial organization provided a crude insurance system which balanced declines in one district against profits in another, losses in one season against gains in other seasons, and the losses suffered by some fishermen against the gains made by others.[9] It has been charged that the cost of this insurance system has been unduly high. Numerous evils have been attributed to it and large profits have been gained; but periodically many firms have disappeared, and the losses have been devastating. No large-scale organization persisted in the fishery with the tenacity shown by the Hudson's Bay Company in the fur trade.[10] The fishing industry of the North Atlantic has been exogenous in its development. Settlements have extended laterally along the coast and have looked to the sea.[11] Newfoundland is a striking illustration of the effect of the impact of capitalism on a country's economy. Settlements fringing the coast were without navigation facilities during the winter season until the 'eighties and 'nineties brought them the railway. The divisiveness which characterized the fishing region was epitomized in Newfoundland. With a development of agriculture that was limited because of geographical conditions and the efficiency of commercial organization in taking advantage of cheaper supplies of agricultural products from the continent, specialization in the fishing industry was pronounced. "The nature of the fishing business is such that it ties them down pretty closely to that speculation alone. They do not branch out into outside speculation." Limitations upon internal de-

8. In 1903 a fisherman at Dumpling on the Labrador, on receiving 50 cents for ten fish, remarked, "In all my long life, fishing always, I never sold a fish for money before." G. F. Durgin, *Letters from Labrador* (Concord, 1908).

9. See W. G. Gosling, *Labrador* (London, 1910), pp. 468–469, for an account of the limitations of small organizations.

10. See *The Book of Newfoundland,* ed. J. R. Smallwood (St. John's, 1937), II, 227 ff.

11. "The people of the islands of Groix, Sein, Lofoden, and the Orkneys and Shetlands have concentrated on the fishery for three centuries—from 1500 to 1825. Newfoundland is still unexplored; life is restricted to the sea-coast. In our time a quarter of its people, or fifty thousand of them, are engaged in taking and salting cod and the remaining three quarters depend indirectly on the fishery. In Nova Scotia the land is not cultivated; and the foreshore of the Norwegian massive is peopled with fishermen. There, as in Newfoundland, social organization shows the effects of occupational grouping. The men are few and the fishing grounds are enormous; therefore there is no coalescence of social groups. There are no villages. Every fisherman lives in a house apart. Collectivity does not extend beyond the family." M. A. Hérubel, *Pêches maritimes* (Paris, [1911]), p. 18.

velopment, the relatively late development of government, the absence of those traditions of local government which are characteristic of New England and Nova Scotia, and dependence on the export of a staple product sold to tropical countries in the Atlantic basin necessarily resulted in the growing up of a type of government that was dependent for its revenues solely on the tariff and was without adequate machinery for handling problems that are entrusted to municipal institutions. St. John's emerged, in 1888, as the only municipal government on the island. The balancing of interests through divergent resources which provided a setting for the growth of responsible government was absent in Newfoundland. In Nova Scotia commercialism lent its support to it. In Newfoundland the same commercialism made it extremely difficult. In Newfoundland sectarian and racial divisions were intensified, whereas in Nova Scotia they provided the basis for tolerant compromise. Government emerged in the struggle with West Country merchants; and, following the decline of their influence, it was concerned with the struggle between the outports and St. John's and between the fishermen and the merchants, the latter a struggle intensified by, and intensifying, religious and racial differences.

Internal difficulties were obscured in the period of expansion accompanying the prolonged efforts to assert rights against France, Canada, and the United States. The success of those efforts and the extension of settlements to more remote areas, such as the west coast, meant increasing difficulties of control from St. John's in the face of competition from less remote areas such as Halifax and Montreal, particularly with the depression. The improvement of steamship connections and the construction of the railways strengthened control, but at the cost of a direct increase in debt and an indirect burden because of the disappearance of wooden sailing vessels. For railways built as a result of community pressure had extended branches to tap the heads of deep bays, and the value of the sailing vessel had suffered accordingly. The development of mining, of hydroelectric power, and of pulp and paper plants was not enough to offset the financial burden represented by the railways, particularly since these developments gave access to more direct steamship services, as at Botwood, Corner Brook, and Wabana. As in the other maritime regions, too, rates were kept down by water competition, and earnings were restricted by limited resources. But, with this, there was none of that extensive continental development to support the railway which had grown up in New England and, in part, in Nova Scotia, and no larger government unit to assume the burden of the debt as was also the case in Nova Scotia. The basic importance of water transportation restricted railway traffic, and the strong competitive position of steamship lines which forced down railway rates necessitated dependence on the tariff for revenue to meet railway deficits. A lower tariff with stable railway rates encouraged trade with England, and with Halifax on the south and west coasts, whereas a high tariff and stable railway rates made for trade with the continent. An absence of taxation on land, incidental to the absence of local governments in the outports, makes it hard for St. John's merchants to compete, burdened as they are by the need of meeting the costs of municipal government. The extension of trade in the outports and the development of industries subordinate to the fishery, i.e., packing and the like, have offset the trend toward centralization incidental to the railway and steamship lines and increased the burden of the tariff and of unemployment. The encouragement of industrialism as a source of traffic income and the lightening of the burden of debt incurred by the construction of transportation facilities attracted labor[12] from the fishing industry; and government, adapted to the fishing industry, was further weakened by its efforts to meet the problems involved in the extension of large-scale corporate activities from continental America and Europe, as illustrated, for example, by the character of the concessions it has made in the shape of its natural resources, such as its minerals and pulp and paper.

The dominance of commercial organization apparent in the character of government institutions involved serious handicaps in meeting problems of long-term debt incidental to investments in railways and capital equipment. Government machinery dependent on a tariff connected with the fishing industry was not adequate to meet the peculiar demands of other industries. The fishing industry provided government revenues primarily through the tariff, and the commercial organization became in a sense a collecting agency. In the absence of local government, expenditures were distributed by members of the Assembly, which meant the payment, over a large area, of an enormous number of small sums of money, and a problem of administration which placed heavy burdens on the central executive and called for reliance on such local members, with consequent charges of patronage on a wholesale scale. The commercial interests acted as purchasers of fish, distributors of goods, and collecting agencies.

As a result of the strain of the increasing burden of long-term debt during a serious depression a commission dominated by civil servants from Great Britain has been set up.[13] The weakness of the state, the strength of sectarian influences, a reliance on the support of Great

12. *The Royal Commission on the Natural Resources, Trade, etc.: Newfoundland* (London, 1915), p. 3.

13. "Newfoundland, Economic and Political," *Canadian Journal of Economics and*

Britain, due particularly to her support in the long struggle with France and in the struggle to maintain her position in European markets, and, finally, geographical proximity were factors which wrought for loyalty to Great Britain. It was a loyalty which expressed itself very plainly in the Island's acceptance of the report of the Newfoundland Royal Commission of 1933 that recommended the abolition of responsible government and the establishment of the present government by commission. Those regions most directly connected with the fishing industry have been most drastically affected by capitalism and the depression. Riots and the disappearance of responsible government characterized St. Pierre as they did St. John's. The regions most seriously affected by disturbed foreign markets, by changes in technique, and by dependence on protected continental areas for supplies possess the least possibility of securing protection. St. Pierre and the fishing villages of Brittany declined with the schooner and the emergence of industrialism under the guise of the trawler. An industry which flourished with commercialism and an international economy was crushed by the demands of capitalism and nationalism.

The impact was less striking in Nova Scotia because of her more divergent resources and also because of influences inherent in Confederation. The names of Thompson, Tupper, and Borden, all of whom became prime ministers, indicate the pervasive influence of Nova Scotia. Tupper's energies were directed toward Confederation, the active extension of a transcontinental railway, when he was the minister of railways and canals, and the development of the coal, iron, and steel industries in Nova Scotia. The preferential agreements, the establishment of the principle of bounties for iron and steel and its extension to other commodities, the tariff policy of the Fielding regime, and the extension of the National Transcontinental Railway to Halifax reflected the demands of Nova Scotia during the Laurier administration. The defeat of reciprocity in 1911 and election of a prime minister from Nova Scotia were followed by the appointment of a finance minister from Toronto.[14] The postwar period has been marked by continued and successful efforts to ensure the consideration of maritime demands within the federal structure. The Royal Commission on Maritime Claims, the Royal Commission of Economic Inquiry, and the White Commission all pointed to the effect of industrialism on the commercialism of Nova Scotia. They meant adjustments in Nova Scotia, whereas the Royal Commission on Newfoundland meant political suicide.

Political Science, February, 1937, pp. 58–85; J. L. Paton, "Newfoundland, Its Plight and Its Pill," *Contemporary Review,* January, 1934, pp. 56–63; also Thomas Lodge, "Oligarchy in Newfoundland," *The Fortnightly,* October, 1938, pp. 475–484.

14. See E. M. Macdonald, *Recollections: Political and Personal* (Toronto, 1938).

The tradition of Nova Scotia's assertiveness prior to Confederation contributed to the conspicuous advance in extraterritorial sovereignty after Confederation. Agreements with the West Indies, which in 1915 provided improved communications and preferential arrangements by which fish excluded from the United States could be sold more advantageously in the West Indies, the increasing control over foreign policy during the war,[15] the birth of the British Commonwealth of Nations, and such developments of status as were evident in the establishment of legations in Washington, Paris, Tokyo, and other centers can be traced in part directly to the influence of Nova Scotia. The Department of Marine and Fisheries has been foremost in pressing for the extension of national rights in Imperial and international conventions dealing with shipping and the fishing industry. The Halibut Treaty was the first to be signed without the assistance of the British legation. The North Atlantic Fisheries dispute was the first major case to be settled before the Hague Tribunal.[16] The holding of the North Atlantic Fisheries Conference, with its representatives from the United States, Canada, and France, is a further illustration of the role played by the fisheries in the development of international machinery.

The cultural stability of Nova Scotia which emerged from the conflict of divergent interests contributed significantly to Canadian cultural growth. The stabilizing effect of the School Act of 1864 in the federal political field could be seen in the struggle between Sir John Thompson,[17] a Roman Catholic, and D'Alton McCarthy, who was an Orangeman and the head of the Equal Rights Association, for the leadership of the Conservative party; and it was also evident in the resignation of Sir Clifford Sifton from, and the support given by Fielding to, the Liberal administration in the case of the Saskatchewan and Alberta school legislation of 1905. The opening of the West in Canada and the relative decline of the Maritimes were followed by a migration to western Canada and to the Central Provinces. As a result, the less formal influence of Nova Scotia[18] on education is conspicuous in its having been the birthplace of men who were later to be university presidents in Canada and elsewhere. Ecclesiastical policy has been tempered by the mem-

15. See Right Honorable Sir R. L. Borden, *Canada in the Commonwealth* (Oxford, 1929), chaps. x–xiv, and especially Robert Laird Borden, *His Memoirs* (Toronto, 1938).

16. Mr. J. S. Ewart, the Canadian legal advocate, became a foremost advocate of Dominion status.

17. Fred Landon, "D'Alton McCarthy and the Policies of the Later Eighties," *Canadian Historical Association Annual Report,* 1932, pp. 43–50.

18. See R. H. Whitbeck, "A Geographical Study of Nova Scotia," *Bulletin of the American Geographical Society,* 1914, pp. 413–419.

bership of those of Maritime birth in ecclesiastical hierarchies. Commercial and financial organizations have taken root and extended from Nova Scotia to all Canada, to the West Indies, and elsewhere. The post-war period has been marked in part by a return movement of "Maritimers" and immigration of Canadians from other parts of Canada. A renaissance has been apparent in the coöperative movement sponsored by St. Francis Xavier University, in the activities of Maritime universities, in the rejuvenation of the interest in cultural growth, in the development of museums, in the preservation of archives, and in a revival of pride in a notable past.

New England has played a similar and more striking role in the cultural life of the United States. She has been securely entrenched in American economic development, but even here attempts have been made to offset the influence of the continent proper by the formation of the New England Conference and Council in 1925.[19] The influence of New England was evident in the omission of the fishery from the Reciprocity Treaty of 1935[20] and in the slight reduction of the tariff on fish in the agreement of 1938.

Commercial organization based on the fishing industry contributed to the expansion and influential character of New England and Nova Scotia in the development of the continent; but they contributed to the isolation of Newfoundland. New England and Nova Scotia reinforced and strengthened the division of North America dictated by the St. Lawrence. The separatist character of the economic life of the Maritimes has been written into the federal constitutions[21] of the continent and has been fundamental in the position of Massachusetts and Nova Scotia. St. Pierre remains not only as a recognition of the position occupied by France but also—a thing made plain by the problems of prohibition—as a weak joint in continental projects that is capable of causing serious damage. The peculiarities of each region have been stressed by competition. Division has involved competition and competition has involved division.

19. N. S. B. Gras, "Regionalism and Nationalism," *Foreign Affairs,* VII, 454–467; *New England's Prospect, 1933* (New York, 1933); *Regional Planning, Part III, New England* (Washington, July, 1936).

20. H. C. Goldenberg, "The Canada–United States Trade Agreement 1935," *Canadian Journal of Economics and Political Science,* May, 1936, pp. 209–212.

21. See V. F. Barnes, *The Dominion of New England* (New Haven, 1923); A. E. Morse, *The Federalist Party in Massachusetts to the Year 1800* (Princeton, 1909); S. B. Harding, *The Contest over the Ratification of the Federal Constitution in the State of Massachusetts* (Cambridge, 1896); C. A. Beard, *An Economic Interpretation of the Constitution of the United States* (New York, 1929), pp. 95–96; also V. G. Setser, *The Commercial Reciprocity Policy of the United States 1774–1829* (Philadelphia, 1937), pp. 99–100.

PROBLEMS OF EMPIRE

The fishery was effective as a "nursery for seamen" through the support it gave to freedom, and to the expansion of trade and shipping rather than in its direct contribution to the navy. There is much evidence that fishermen were not good sailors. The fishery evaded control in Iceland, in Newfoundland, in New England, and in Nova Scotia. The West Country opposed the formation of settlements in Newfoundland to the point of hastening the rise of the fishing industry in New England. Nova Scotia, in turn, resisted the control of New England and accentuated the isolation of Newfoundland. West Country fishing ships met the demands of the navy in men and ships, and the defense of Newfoundland ultimately depended on the navy rather than the army. But effective protection by the navy in war periods was at the expense of the fishing ships and an aid to the settlements. Every separate region became the nucleus of resistance to other regions, and all had the support of the fishery, shipping, and trade. These various regions combined against the French or sought alliances from one and the other as circumstances directed, but they always strove for independence, an effective indication of naval strength. Ships were essential to the prosecution of the fishery, and the industry made for flexibility of organization, mobility of labor and technical equipment, and specialization as the basis of trade. The carrying trade was the direct and indirect support of naval power. The expansion of the French fishery in the late seventeenth century failed to bring naval dominance, in part because it was a result of the expansion of the English carrying trade, with the effective support it gave to the navy. West Country support of the naval strength of Great Britain was of less value than the support of New England in the defeat of the French. What could be accomplished by New England naval strength was made apparent in the struggle for independence.[22] Nova Scotia profited through privateering when New England was in difficulties.

Dried fish, as food for seamen, gave shipping the range of the tropics and played its part in the activities of Spain and Portugal in the New World. Specie was secured in return for the fish and, in its turn, it increased the liquidity and effectiveness of the commercial organization of the countries which received it.* England, thus supplied with specie, could trade more effectively with India. While mercantilism opposed exports of specie and stressed the necessity of importing it as a basis of expansion when trading in other commodities, the East demanded

22. See R. G. Albion, *Forests and Sea Power* (Cambridge, 1926).

specie. Neither England nor her colonies had gold or silver mines. "We have no treasure but by trade, for mines we have none." The demands of the East provided a sheet anchor for the commercial activity of the Atlantic basin. The demands of the Catholic peoples for cod and of metropolitan areas for luxury goods gave greater meaning and value to the anchorage. "It is certain that the vices and follies of mankind support and encourage trade and manufactures much more than the bare necessities of life." The effectiveness of specie in developing trade was evident in the production of commodities of high value and light bulk in the New World ranging from furs to tobacco and sugar. The former depended on the aborigines of North America, the latter on the aborigines of Africa. Slaves brought to the New World to produce tobacco and sugar meant a demand for the poorer grades of dried fish and the expansion of trade from New England.

Encouragement given to the production of tobacco and sugar by England was a means of draining specie from Spain. Encouragement was given to the importing of wine from Portugal rather than from France, in order to secure a return cargo which was high in bulk and low in value as compared with fish. The fishing industry, as carried on from Newfoundland, meant employment of labor, construction of ships, and imports of specie to England. As carried on from New England, it meant lending aid to the production of sugar and tobacco. Mercantilism gave its support to the Newfoundland fishery as a means of attracting specie from Spain, and its support to the fishery of New England as a means both of increasing the production of sugar in the British West Indies and decreasing imports from Spain.[23] Newfoundland fish were sold directly for specie, and New England fish either directly for sugar or indirectly for specie.

Exports of sugar from the British West Indies led to the emergence of vested interests which fostered legislation opposed to the trade of New England and the colonies and stimulated the growth of smuggling in other sugar-producing regions. Direct trading between the West Indies and England flourished at the expense of the auxiliary trading between the colonies and the West Indies. Monopolies obtained by consumers' goods such as sugar came into conflict with profits from producers' goods such as fish. Fish exported from Newfoundland to Spain was a "producer's good" for specie for England, and fish from New

23. For a description of mercantilism in relation to the colonies, see C. M. Andrews, *The Colonial Period of American History* (New Haven, 1938), Vol. IV, chap. x; E. F. Heckscher, *Mercantilism* (London, 1935), and Jacob Viner, *Studies in the Theory of International Trade* (New York, 1937), are less concerned with the colonial aspects.

England, a "producer's good" for slaves and sugar both. In Newfoundland, provisions and supplies from New England stimulated the increase of population and were opposed by West Country shipping interests. Its demands as an active commercial region for bills of exchange in Newfoundland corresponded to its demands for specie in the West Indies and to England's demands for specie in Portugal. The export of producers' goods to the West Indies and to Newfoundland exposed New England to the indirect effect of policies dictated by groups directly concerned with Great Britain. An expanding commercial system broke the bonds of a rigid political structure defended by vested interests. The genius of Adam Smith foresaw and hastened the trend of expanding trade beyond the bounds of empire.

The activity of commercialism, which played its part in the disappearance of the colonial system but had also been intensified by it, loosened the powerful forces of competition in the Atlantic basin, and made it inevitable that systems of Imperial control should defeat their own ends. Competition, enforced by commercialism based on the fishery, and encouraged by Parliament with an interest in the shipping and the fishing industries as a basis of naval support, was effective because of the mobility of labor, ships, and markets. Disunity in the fishing industry was in contrast with the concentration of interests in the production of other staples. The short-term credit typical of commercialism based on the fishery was in contrast with the long-term credit typical of capitalism in the fur trade, the timber trade, and the sugar industry. Capital control involved concentration on such sources of long-term funds as London, and commercial control concentration on such instruments as ships continuously searching for new sources of trade. After 1783 the second British Empire stressed the importance of staples characterized by long-term credit. The success of the American colonies dependent on commercialism paralleled the success of a new empire in Newfoundland, the West Indies, Nova Scotia, the territories of the Hudson's Bay Company, and Canada—an empire that was dependent on capitalism and a less aggressive commercialism. A more permanent basis of expansion had been reached for both the United States and the British Empire. The new empire was more firmly based on direct exports to Great Britain in return for finished products, and the monopolies of the old empire became impossible because of the importance of trade with the United States. The second empire was saved by the loss of the thirteen colonies, which made impossible an integration such as New England had provided for the old one. Monopolies were checked by the United States as a competitor.

The dominance of the fur trade on the St. Lawrence and of the fishing industry on the Atlantic in the French Empire made enormous demands on the energies of the mother country. The inherent antagonism between settlement and fur trade necessitated the abandonment of company organization in the St. Lawrence valley, and the assumption of direct responsibility by France in 1663. The supply of furs was a result of the rapidity of cultural change among the hunting Indians of North America, and the demand was a result of changes in taste in the metropolitan markets of Europe. The uncertain elasticity of supply was linked to an uncertain elasticity of demand. With fortifications, wars, and the cost of fostering the formation of settlements as a means of supporting the fur trade there had been periods of inflation and difficulty. The demand for fish was a result of a relatively inelastic demand for specie, and supply responded directly because of the mobility of labor and ships. New France was unable to supply provisions for the fishery, the fortified gateway to the St. Lawrence at Louisburg in Cape Breton, and for the French West Indies, and they were forced to depend on the English colonies. Without support from New France and with restrictions on smuggling to the English colonies, the French fishing industry in the New World was compelled to depend on ships which came out annually from France rather than on settlements. The trade in products between the French West Indies and New England, through Louisburg, sapped the strength of New England's fishery at Canso. Acadia had been a region which had served as a French support for Louisburg. And when, in the war between England and France, the French were expelled from it, their expulsion was marked by a cruelty which itself was characteristic of the fishery.

After the conquest of New France the demands of the fur trade on the St. Lawrence, reinforced by the aggressiveness of colonial and British merchants in extending it to the interior, checked the possibilities of lending support to Nova Scotia. With the removal of restrictions imposed by the French, English colonies which had been hampered in trade with the interior, and had extended their settlements to Nova Scotia and their trade to the regions of the Atlantic basin and beyond, rapidly expanded and reached new frontiers. Montreal, now under English control, reasserted its competitive position against New York. The antagonism of Acadian and French settlements in the Gulf of St. Lawrence to New England traders—an antagonism born of wars and the ruthless expulsion of the Acadians—made easier the occupation of the region by traders from the Channel Islands. Attempts on the part of Great Britain to restrict the expansion of the English colonies to the interior by encouraging trade from Montreal, and to limit their trade with the French West Indies and Newfoundland, contributed to the Revolution.

A sudden expansion in the demand for West Indian products, resultant upon the collapse of New France and the opening of the fur trade in the interior, conflicted violently with the policy of the British West Indies and Great Britain in refusing to increase production by the acquisition of the French West Indies. The Sugar Act, the Quebec Act, and Palliser's Act checked the colonies by land and by sea. Because of the success of the Revolution, the energies of the United States poured inward, and outward to the Pacific. Nova Scotia emerged to occupy the gap in empire trade vacated by New England and she attempted vigorously to elbow the United States out of the trade with Newfoundland and the British West Indies. But expansion of the settlements and the fishery of Newfoundland led to increased trade from the Maritimes. This enhanced the difficulties of the British West Indies and necessitated freedom of trade with the United States.

The shift of the fur trade from the St. Lawrence to Hudson Bay after 1821 and the development of the timber trade under preferences granted during the Napoleonic wars reversed the trend of a fur trade opposed to immigration and to agriculture. The expansion of the internal market in the United States was followed by the recovery and the expansion of the New England fishery after the Napoleonic wars, by encroachments on British fishing grounds, by reciprocity, Confederation, the Washington Treaty, and retaliatory tariffs on the part of both nations. As a result of competition from New England, Nova Scotia turned toward the St. Lawrence. Confederation was hastened by the prospects of increased trade in agricultural products as a means of competing with the United States in the West Indies, and by the prospect of markets for coal on the St. Lawrence. Where the French Empire had failed to link up the fur trade on the St. Lawrence, the fishing industry of the Maritimes, and the sugar plantations of the West Indies, Canada, with an agricultural base on the St. Lawrence, was successful.

This success was achieved because of the commercialism of New England, which compelled a realignment of the colonial system, and the commercialism of Nova Scotia, which directed the modification. The inherent contradictions in legislation which encouraged trade and shipping through the Navigation Acts and restricted it through monopolies favoring the sugar interests were fostered by the shipping interests and the West Indian interests in England. The rebuff to Parliament in the revolt of the colonies was followed by legislation abolishing the slave trade, rotten boroughs, and the preferences and duties protecting vested interests.

Slavery in the British West Indies declined in the face of competition from more efficient slavery in the French West Indies,[24] and cheaper production of sugar in other parts of the empire and of the world.*The disappearance of slavery in the British Empire and other countries weakened the market for poorer grades of dried fish. The wooden sailing vessel declined with slavery and with modern naval equipment which rendered obsolete the argument that the fishery was a nursery for seamen.

The commercial interests of Nova Scotia pressed in the direction of free trade, but they were opposed by the demands of the Canadian vested interests in timber and flour. The repeal of the Corn Laws in 1846, of the Navigation Acts in 1849, and of preferences on colonial timber came with the winning of responsible government by the colonies, and control over their customs and their natural resources. The success of industrialism in England evident in the advent of freedom of trade compelled the colonies to rely on their own devices, which included tariffs, railways and canals, debts, and Confederation. Tariffs became a source of the revenue needed for the construction of transportation systems and a source of protection against the competition of an increasingly efficient industrial area under free trade.

The difficulties of the French and British empires[25] were made plain by the maladjustment of political structures to economic expansion in the New World. Chartered companies disappeared before and after the time of Adam Smith when faced with competition from the flexible organization called forth by the commercialism of the West Country and of New England. They were followed by the navigation system, condoned by Adam Smith but in itself sufficiently rigid to promote the growth of vested interests in staple products, and to lead to the clash with commercialism which broke the first empire and molded the second. France provided no basis for a flexible and diversified economic structure. Commercialism was less effective in wresting specie from Spain and Portugal, and the failure of company organization was followed not by commercialism as in New England but by government control as in New France. New France depended on the French Empire but New England revolted from the British Empire. The place of the Crown in New France involved the direct link between trade and public finance and the difficulties which accompanied the inflations after 1700 and 1750. The inability of the French Empire to withstand the inroads of English commercialism led to its breakdown. The collapse of the external empire contributed to the internal breakdown in the French Revolution. France

24. Adam Smith, *Wealth of Nations* (New York, 1937), pp. 553–554.

25. See H. E. Egerton, *A Short History of British Colonial Policy* (London, 1908); and *The Commonwealth of Nations*, Part I, edited by Lionel Curtis (London, 1917).

failed to link diverse regions, and England was unable to repress the aggressive integration of diverse regions.

The activity of commercialism based on the fishing industry and the relative activities of shipbuilding and trade fostered by the navigation system had significant implications for constitutional development.[26] The right of free fishing in the New World was assumed to be under the control of the Crown in 1621; but the industry was affected by parliamentary legislation in the Navigation Acts and, after the Revolution of 1688, in the Act of 1699 relating to Newfoundland, and also in subsequent legislation. While such legislation coincided with the demands of West Country commercial interests, it clashed with the interests of the colonies under the Crown. With the addition of New France to the empire the New World was divided into the territory of the Hudson's Bay Company under the Crown, the old Province of Quebec and Newfoundland under King and Parliament, and the diverse constitutions of the colonies under which protests were made against the increased powers of Parliament. It proved impossible for Great Britain to combine these elements. Monopolies under the Crown were suited to continental regions such as India and the region tributary to Hudson Bay, but were inadequate to Maritime areas. It proved difficult for both France and England to combine the commercial type of organization of the Maritime areas with the capital type of organization in the company. New France and the West Indies under companies and the Crown were difficult to combine with the fishing industry dominated by private enterprise in the ports of France. The problem of empires was one of constitutional as well as of economic organization.

The growth and decline of commercial nuclei around the regions concentrating on the fishing industry and the intense character of their activity implied numerous adjustments. The competition between small units, together with flexibility of organization, weakened government control. The commercial activity of the French ports was subject to control because of the scattered character of the industry, the restricted size of the developments, and the dominance of the French Crown. Government support was enlisted in the interests of a free fishery by such devices as bounties and regulations ostensibly designed to strengthen the navy. In the West Country, regulations were opposed to settlements in Newfoundland; in New England and Nova Scotia, they were opposed to restrictions on trade. After the Seven Years' War and the American Revolution bounties were granted in both areas. Political weapons became increasingly effective with the growth of the commer-

26. See W. P. M. Kennedy, *Essays in Constitutional Law* (London, 1934), pp. 3–23.

cial centers. France was driven in turn from Nova Scotia, Cape Breton, the Gulf of St. Lawrence, and the French Shore; New England was driven from Nova Scotia, the Gulf of St. Lawrence, and Newfoundland; and Nova Scotia was driven from Newfoundland. Nova Scotia ceded her rights to the fishery to Canada as a means of resisting the United States more effectively, and Newfoundland insisted on her control over the fishery as a means of resisting France, the United States, and Canada. The long and endless disputes over fishing rights arose from the conflict between rising and falling commercial nuclei. The character of the disputes varied with the intensity of commercialism, and they ranged through a gamut which comprised the attacks on the formation of settlements in Newfoundland, those on Louisburg by New England, and those on the French Shore by Newfoundland; and they even included the Labrador boundary dispute and the enforcement of customs regulations. Encroachment and resistance depended on the relative strength of commercialism in the West Country, New England, Nova Scotia, France, and Newfoundland. Frontiers were difficult to establish in the fishing industry and on the seas. Unity of effort under Confederation secured the Treaty of Washington and the Halifax Award; but eventual success, in the shape of the completion of the Intercolonial Railway, the deepening of the St. Lawrence waterway, and the extension of railroads to the Pacific, together with the disappearance of the sailing vessel, attracted labor from the fishing industry. Nova Scotia turned to the interior in Canada and the United States, and retreated from world markets when she found herself in competition with the capacity of large-scale fish production in other important countries. The results of the retreat were evident in the revolution[27] from an economy facing the sea with a large number of ports to an economy dependent on a central port and railways to the interior. Her recovery has involved the construction of railways to the outports and the restoration of ports abandoned with the sailing vessels when first they encountered that new power inherent in the railroad.

The disappearance of an active commercial region as a result of the impact of machine industry has been a major calamity to the fishing re-

27. The trawler problem has risen in part in Nova Scotia, in contrast with New England, because of Nova Scotia's position as a peninsula. The approach to the continent is narrowed, as in a funnel, and concentrates upon Halifax. Moreover, the success of the trawler in the fresh-fish industry in the United States while lessening the difficulties of Nova Scotia by involving a retreat from the dried-fish market has increased them by strengthening resistance to a lowering of tariffs on fresh fish from Canada. On the other hand, government support of the dried-fish industry in Newfoundland has been met by government support in Canada. These considerations may have the serious effect of lessening the interest in the problems of the fresh-fish market.

gions of France, New England, Nova Scotia, and Newfoundland. The readjustment of an economy dependent upon the sea to one dependent on the land has involved many difficulties in New England. It has also called out an extensive policy including government support for the Intercolonial Railway, the maritime freight-rates act, the West Indies agreement, coal subventions, extensive subsidies and aggressively active measures for Nova Scotia's fisheries, and the collapse of responsible government in Newfoundland. The transition from dependence on a maritime economy to dependence on a continental economy has been slow, painful, and disastrous. The tremendous initiative which characterized commercialism based on the fishing industry could be measured in the collapse of West Country company control over trade and the fishery, in the history of Newfoundland and New England, the defeat of France and the breakdown of the colonial system, the disappearance of the Navigation Acts, and even the rise of responsible government and the establishment of Confederation. This is an initiative which cannot be suddenly replaced. The effects of the tragedy of the replacement of commercialism by capitalism call for a long period of expensive readjustment and restoration, and this cannot take place without policies which foster the revival of initiative under responsible governments.

NOTES TO THE REVISED EDITION

(References are marked with asterisks in the text. Where more than one note has been added to a page, the line on which each occurs is indicated.)

p. 50. Dry fishery essential to trade. Green fishery characteristic of domestic trade. Foregn country could sell green fish with difficulty as it would involve further process of drying. Drying in new world, rather than old, part of trade between nations.

p. 51. Up to 1536 fishmongers [were] divided in [to] two companies—salt-fishmongers and stock-fishmongers. Nuns kept dried cod in storehouse, stock "because it nourisheth no more than a dried stock" (Erasmus). Large quantity of dry habarden (salted cod) and small quantity of "Iceland fish." Eileen Power, *English Nunneries* (Cambridge, 1922), p. 139.

p. 103. Cheap provisions from New England to Newfoundland probably stimulated the byeboat fishery.

p. 104. Newfoundland not controlled by Crown but by parliament especially after 1688 and definitely linked to Great Britain and thought [of] as part [of Great Britain] under Navigation Acts but other colonies controlled by charters from Crown reluctant to accept parliament, and first to be influenced by parliament, and this [non-acceptance of parliament was] insisted on to overcome difficulties of [arising from] Navigation Acts by colonies. Problem of land where Crown dominant contrasted with problem of sea where parliament dominant. Expansion by sea weakened Crown and strengthened parliament. British institutions dominated by sea.

p. 119. C. P. Gould, "Trade between the Windward Islands and the Continental Colonies of the French Empire 1683-1763," *Mississippi Valley Historical Review,* March, 1939, pp. 473-490.

p. 122. [The] Dutch captured [the] fishing fleet off St. Jean de Luz and Ciboure, on its return from Newfoundland in 1672.

p. 132. [The purpose of these attempts being] to encourage trade with Portugal.

p. 135. [The commodity in question being] cod.

p. 137. See John U. Nef, *The Rise of the British Coal Industry* (London, 1932), on [the] enormous importance of the coal trade to shipping, 1550-1700.

p. 199. I.e. after 1774, Quebec Act restoring [having restored] Labrador to Canada.

p. 213. Sir John Pringle to Hume, July 8, 1775, London: Americans forced "at all times (even in time of war) to come with their cargo of wine taken up in Spain or Portugal to the Isle of Wight, or other English ports, unload it and put it again on board, before they could carry it home. The porters at such places could only gain while the Provincials were unnecessarily the sufferers." Birkbeck Hill, *Letters of David Hume to William Strahan* (Oxford, 1888), p. 292.

p. 220. J. D. Phillips, "The Routine Trade of Salem, 1783-1789," *The American Mercury,* October, 1941, pp. 345-351.

p. 228. "The lumber and provisions of the United States are more necessary to our West India Islands than the rum and sugar of the latter are to the former. Any interruption or restraint of commerce would hurt our loyal much more than our revolted subjects. Canada and Nova Scotia cannot justly be refused at least the same freedom of commerce which we give to the United States." Adam Smith to William Eden, December 15, 1783, John Rae, *Life of Adam Smith* (London, 1895), p. 385.

p. 230, l. 7. See History and Description of the Southern and Western townships of Nova Scotia in 1795 by Rev. James Monro, Appendix C, Report of Board of Trustees of Public Archives of Nova Scotia, 1946.

p. 230, addition to footnote 9. A whaling company was started at Shelburne, in 1784, with 9 firms, one Halifax, one London, remainder Shelburne, capital £8350. A third of capital lost by 1789. Shelburne exports 1788, 13,141 qtls. cod, 4,192 casks pickled fish, 61 casks smoked salmon, 149 bbls. fish oil, 14,798 gals. sperm oil. In 1791 [the company was] driven out of West Indies by Newfoundland fish and New England lumber [and] turned to carrying trade between Nova Scotia and Newfoundland.

p. 283. [Le Bouthillier] failed 1924.

p. 356. Fruing closed 1912.

p. 400, addition to footnote 69. See Lambert de Boilieu, *Recollections of Labrador Life* (London, 1861), for a description of the fishery, sealing, and trapping, at an Establishment at Lewis Bay and of a trip across to Sandwich Bay.

p. 426 First plant in Gaspé to ship fresh cod fillets on Canadian market 1928 secured more favorable rates in competition with Lunenburg and Lockeport, 1932. Consignments of fresh fish led to dumping. In 1932 provincial government began to build refrigeration plants along coast [of] Magdalen Islands [and] on south shore. Sharp increase in shippers of fresh fish on Canadian market. Greater skill in getting more fillets from fish and in speed of cutting fillets. Sharp decline in salt fish especially in Gaspé.

p. 430. Louis Bérubé, *Coup d'Oeil sur les pêcheries de Québec* (Ste Anne de la Pocatière, 1941), pp. 17-18, 27.

p. 493. Problem of Newfoundland pushed out of Europe after war, to West Indies and South America or regions with poorly organized societies unable to resist the effects of depression and marginal to more powerful areas. Toughness of cultural background fundamental in withstanding effects of severe strain. Marginal poorly organized regions suffer acutely with depression. Loss of responsible government in Newfoundland. Marginal staple-producing countries dependent on prices of raw material. Luxury goods high in price, subject to wide fluctuations [and] fixed income charges to make whole very unstable in New France, Brazil, West Indies. As transportation improves bulkier lower value goods increase in importance and, because of character of necessities, [are] less subject to severe fluctuations.

p. 500. See H. A. Innis, "Liquidity Preference as a Factor in Industrial Development," in *Political Economy in the Modern State* (Toronto, 1946), pp. 168-200.

p. 505. German bounty paid [on] sugar after 1870 ruined West Indies and Cuba [and] led to revolt and Spanish American War. Brooks Adams, *America's Economic Supremacy* (New York, 1900), pp. 62-65.

INDEX